# ELECTROMAGNETIC AND ELECTROMECHANICAL MACHINES

# Harper & Row
# Power & Machinery Series

**Mablekos:** *Electrical Machine Theory for Power Engineers*

**Matsch/Morgan:** *Electromagnetic and Electromechanical Machines,* Third
   Edition

**Shultz/Smith:** *Introduction to Electric Power Engineering*

# ELECTROMAGNETIC AND ELECTROMECHANICAL MACHINES

**Third Edition**

**Leander W. Matsch, Late**

**J. Derald Morgan**
New Mexico State University

**HARPER & ROW, PUBLISHERS, New York**
Cambridge, Philadelphia, San Francisco,
London, Mexico City, São Paulo, Singapore, Sydney

*1817*

Sponsoring Editor: Peter Richardson
Project Editor: Ellen MacElree
Cover Design: Lawrence R. Didona Art & Design
Text Art: Vantage Art, Inc.
Production: Delia Tedoff
Compositor: Tapsco, Inc.
Printer and Binder: The Maple Press Company

**Electromagnetic and Electromechanical Machines,** Third Edition

Copyright © 1986 by Harper & Row, Publishers, Inc.

**Library of Congress Cataloging in Publication Data**

Matsch, Leander W.
  Electromagnetic and electromechanical machines.

  Previous editions by Leander W. Matsch.
  Includes bibliographies and index.
  1. Electric machinery.   I. Morgan J. Derald.
II. Title.
TK2182.M37  1986      621.31′042      85-16348
ISBN 0-06-044271-9

85  86  87  88  9  8  7  6  5  4  3  2  1

*Dedicated to:*

*The memory of Leander W. Matsch, his wife and family;*

*The family of J. Derald Morgan: June, Laura, Kimberly, Rebecca, John, Jr., and his mother and father, Avis and John;*

*Our students, past, present, and future.*

# Contents

# 5   The Induction Motor                                        246

# 6   Direct-Current Machines                                    317

# 7    System Applications of Synchronous Machines          430

# 8    Special Machines                                     474

# Preface

The philosophy of this third edition is similar to that of the first—major emphasis is placed on physical rather than mathematical concepts in the analysis of conventional electrical machinery such as electromagnets, reactors, transformers, rotating electromagnetic machines, and machine system performance.

The book is intended for electrical engineering students in their junior or senior years as well as for professional engineers. Because its scope is greater than can be fully covered in any one three-semester-hour course, the instructor has latitude in the selection of material to be presented.

The SI system of units is used exclusively throughout the text, since that system has gained acceptance in commercial practice as well as in technical publications. Relying wholly upon this system in textbooks has the advantage of confronting students with only one system of units instead of confusing them with as many as three systems as, for example, has been common in dealing with magnetic circuits.

Nearly all the subject matter of the second edition has been retained. The third edition is a text that contains a beginning chapter on circuits to provide a fundamental basis of nomenclature for the text. It provides a power circuits introduction or review as appropriate for a given curriculum. It may be covered in detail or in part, left as an exercise to the student, or skipped entirely. This edition:

- Focuses only on the electromagnetic properties of devices that are essentially electromechanical.
- Introduces AC machines before DC machines after an introduction to electromagnetic energy conversion and transformer theory.

- Provides material on system theory. Universities with only one course in power may use this text in their one course to introduce both machines and power systems.
- Provides material for those who, after introducing basic machine theory, like to stress the many applications of energy conversion of a conventional and unconventional nature.

## ACKNOWLEDGMENTS

The authors acknowledge the role our former editor, Mr. Carl McNair, played in bringing us together to revise this important work. We express our gratitude to our wives and families for their patience with us in our respective contributions to the material and to our many colleagues who have provided encouragement and valuable suggestions. During the production of this third edition Leander W. Matsch passed away, and therefore this edition is dedicated as a memorial to his significant contributions to the profession.

I wish to thank my secretary at the University of Missouri—Rolla, Betty Cleveland, for her patient help with the manuscript. I also thank my friends and colleagues who listened to my ideas and provided suggestions. They are Dr. David Cunningham, Professor George McPherson, Dr. Stanley Marshall, Dr. William Tranter, and Dr. Roger Ziemer, each of whom is a successful author, teacher, and researcher in his own right. Others who often helped me were Professor Morris, Dr. Ronald Fanin, Mr. Ed Hart, and my Dean at the University of Missouri—Rolla, Dr. Robert L. Davis. A special note of thanks is due Ms. Linda Laub, who produced the solutions in this text, and to Linda Boswell and Janice Spurgeon, who typed the solutions manual.

The following people reviewed the outline and draft manuscript and provided valuable suggestions and comments: Peter Sauer, University of Illinois-Urbana; Sarma S. Mulukutla, Northeastern University; Stewart Stanton, Montana State University; Wayne Knabach, South Dakota State University; Robert Harrington, George Washington University; F. W. Schott, UCLA; Harry M. Hesse, Rensselaer Polytechnic Institute; Kerwin C. Stotz, Virginia Military Institute; A. Chandrasekaran, Rochester Institute of Technology; Edwin Cohen, New Jersey Institute of Technology; Wayne Gilchrest, Clemson University; and Lee L. Nichols, Virginia Military Institute.

My thanks go to the staff at New Mexico State University, who stepped in after my move from the University of Missouri—Rolla to help me complete the final details of editing. To my secretary Charlotte Beene and my student assistant Kristine Ford at NMSU I am most appreciative.

I am grateful to the many people at Harper & Row who worked on this project. I would especially like to thank Ellen MacElree, the project editor, who provided excellent editorial assistance and help in the production of this book. I would also like to thank Pete Richardson, the sponsoring editor, for his assistance and patience.

J. Derald Morgan

# Chapter 1

# Basic Concepts of Power Circuits

The engineer who is required to work with power circuits must be intimately familiar with ac circuit analysis. In particular he or she must understand and be able to apply the concepts of steady-state and transient analysis for single-phase and three-phase ac circuits. In this chapter the fundamental ideas associated with ac circuits are reviewed, the notation to be used in this text is introduced, and the concepts of three-phase circuits, complex power, and per-unit quantities are established.

## 1-1 PHASOR DIAGRAMS

Sinusoidal waveforms of the same frequency can be represented by phasors, which may be added and subtracted like coplanar vectors. These waveforms may be functions of time, as in the case of ac voltages and currents of constant amplitude and constant frequency, as well as space waves of magnetomotive force (mmf) and flux density of constant-amplitude frequency and wavelength.

The instantaneous value of the voltage $e = \sqrt{2}E \sin \omega t$ may be considered as the imaginary component of a coplanar vector, called a phasor, which rotates counterclockwise at the constant angular velocity $\omega$ as shown in Fig. 1-1, where $E$ is the effective value of the sinusoidal voltage.

Consider the rotating phasors of constant amplitudes $A$ and $B$ in Fig. 1-2(a). Their vertical projections, when plotted against $\omega t$ along the horizontal axis, will trace out the sinusoids in Fig. 1-2(b), which are defined by

$$a = A \sin \theta_a = A \sin (\omega t + \alpha) \tag{1-1}$$

$$b = B \sin \theta_b = B \sin (\omega t + \beta) \tag{1-2}$$

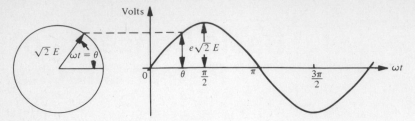

**Figure 1-1**   Generation of a sinusoid.

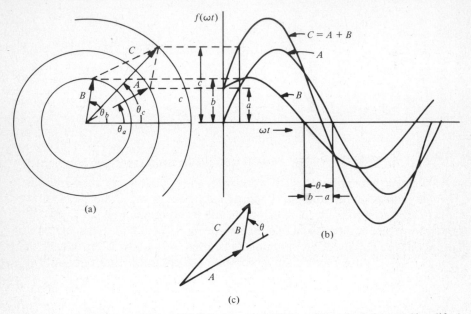

**Figure 1-2**   (a) Phasor addition. (b) Step-by-step addition of two sinusoids. (c) Simplified completion of parallelogram.

Phasor **A** lags phasor **B** by $\theta = \theta_b - \theta_a = \beta - \alpha$. A step-by-step process may be used to add the two sinusoids $a$ and $b$ to obtain the third sinusoid $c$, as indicated by

$$c = a + b \doteq A \sin \theta_a + B \sin \theta_b = C \sin \theta_c \qquad (1\text{-}3)$$

This laborious process, illustrated in Fig. 1-2(b), is sidestepped by completing the rotating parallelogram in Fig. 1-2(a) or by the simpler construction in Fig. 1-2(c), both of which perform phasor addition. In the case of ac voltages and currents, the magnitudes of the phasors usually represent the root mean square (rms) values rather than the amplitudes. It is this value that is normally displayed by voltmeters and ammeters.

Sinusoidal distributions of mmf and flux density in the air gap of electric machines are also treated as phasors. Then $\theta_a$ and $\theta_b$ in Eqs. 1-1 and 1-2 represent space angles rather than time angles. However, the magnitude of a space phasor is made equal to the amplitude of the wave.

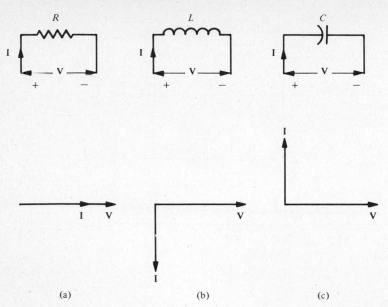

**Figure 1-3** Phasor diagrams showing phase relationships between voltage and current. (a) Noninductive resistance. (b) Pure inductance. (c) Pure capacitance.

The phasor diagrams in Fig. 1-3 show the phase displacement between the voltage and current for (a) a noninductive resistance where the voltage and current are in phase, (b) a purely inductive reactance where the current lags the voltage by 90°, and (c) a totally capacitive reactance where the current leads the voltage by 90°.

Figure 1-4(a) shows a series-parallel circuit and the corresponding phasor diagram, in which the voltages and the currents are added in accordance with the convention of Fig. 1-2(c). The phasor sums are expressed by

$$\mathbf{V}_{ac} = \mathbf{V}_{ab} + \mathbf{V}_{bc}$$

and
$$\mathbf{I}_1 = \mathbf{I}_2 + \mathbf{I}_3$$

The current phasor $\mathbf{I}_3$ is shown twice in Fig. 1-4(b), although the upper $\mathbf{I}_3$ phasor is not required for performing the addition and is normally omitted from phasor diagrams. It is shown here for reasons of clarification.

## 1-2 AC CIRCUIT RELATIONSHIPS

For purposes of analysis, the voltage at the bus† of a power system or piece of power equipment can be for practical purposes assumed to be a sinusoidal of constant frequency. Most of the theory developed in this text is dependent on the phasor representation of sinusoidal voltages and currents. Boldface capital letters with appropriate subscripts are used to represent phasors, with lighter

† A circuit node or connection point is referred to in power-system terminology as a bus.

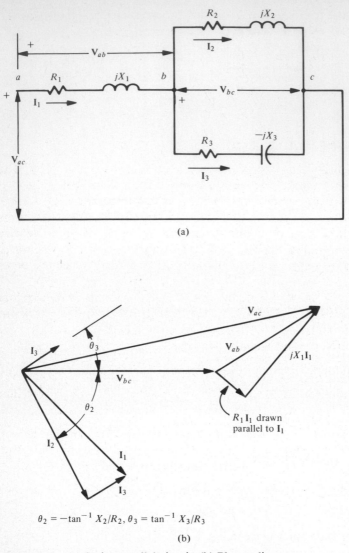

**Figure 1-4**    (a) Series-parallel circuit. (b) Phasor diagram.

capital letters indicating the rms magnitude of the quantity. Where a generated voltage is specified, the letter $E$ is used to indicate the electromotive force (emf). The letter $V$ is used to indicate a potential difference. Table 2-1 provides the reader with a list of symbols, units, and nomenclature used throughout this text.

If a voltage and current are expressed as a function of time as

$$v(t) = 14.14(\omega t + 30°) \tag{1-4}$$

$$i(t) = 70.7\omega t \tag{1-5}$$

the maximum values are $V_{\max} = 14.14$ V and $I_{\max} = 70.7$ A. It is common practice to define the magnitude of sinusoidal waveforms by the rms value, which is equal to the maximum value divided by $\sqrt{2}$. Therefore, for the $v(t)$ and $i(t)$ given in Eqs. 1-1 and 1-2 the magnitudes are

$$V = \frac{14.14}{\sqrt{2}} = 10 \text{ V rms}$$

$$I = \frac{70.7}{\sqrt{2}} = 50 \text{ A rms}$$

## 1-2.1 Single-Subscript Notation

In Fig. 1-5 the emf is $E_g$ and the voltage between nodes $a$ and $o$ is $V_T$. The current in the series circuit is the load current $I_L$. The voltage across the load impedance $Z_L$ is between nodes $b$ and $n$ and is $V_L$. Polarity markings (+) for voltages and an arrow for positive direction of current flow are essential for phasor representation. In an ac circuit the terminal marked (+) is positive with respect to the other terminal marked (−) for one half-cycle of the voltage wave-form and is negative with respect to the terminal marked (−) for the other half-cycle.

In Fig. 1-5 during one half-cycle, the voltage $V_T$ is positive when the terminal marked (+) is at a higher potential than the terminal marked (−). In the next half-cycle the terminal marked (+) is negative and $V_T$ is negative.

The arrow used to depict current has the same function as the polarity markings, which is to show the assumed positive direction of current. The actual current flow reverses direction each half-cycle. The arrow is used to point in the assumed positive direction of current. When the current flows in a direction opposite the arrow, the current is negative.

The phasor current in Fig. 1-5 is

$$I_L = \frac{V_I - V_L}{Z_T} = \frac{V_L}{Z_L} = \frac{E_g}{Z_g + Z_T + Z_L} \tag{1-6}$$

and the phasor voltage at the generator terminal is

$$V_T = V_L + I_L(Z_T) \tag{1-7}$$

or $\qquad\qquad V_T = E_g - I_L Z_g \tag{1-8}$

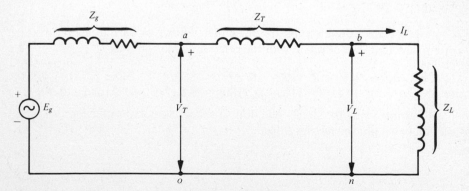

**Figure 1-5** Circuit using single-subscript notation.

Notice that in Fig. 1-5 certain nodes were assigned letters. The voltages associated with these nodes may be designated by use of a single-letter subscript where the voltage associated with that node is assumed to be positive with respect to a reference node. It is common practice to use lowercase letters to define instantaneous values. In Fig. 1-5, where node $a$ is positive with respect to reference, nodes $a$ and $b$ are at a higher potential than nodes $o$ and $n$, and the instantaneous, amplitude, and phasor voltages are written as

$$v_a = v_T = v_a - v_o \qquad v_b = v_L = v_b - v_n$$
$$V_a = V_T \qquad\qquad V_b = V_L$$
$$\mathbf{V}_a = \mathbf{V}_T \qquad\qquad \mathbf{V}_b = \mathbf{V}_L$$

Figure 1-5 depicts an ac circuit. The generated voltage is shown as a circle. It is common practice to show an assumed positive voltage for an emf and an assumed direction of current to establish a reference from which to develop a solution.

### 1-2.2 Double-Subscript Notation

Three-phase circuit relationships are clarified and simplified by the adoption of a system of double subscripts to replace the polarity marks and arrows. The basic convention is simple.

To denote the direction of current flow, the order of the subscripts denotes the positive direction of flow. For example, in Fig. 1-5 the arrow pointing from $a$ to $b$ defines the direction of positive current for $\mathbf{I}_L$. When double-subscript notation is employed, $\mathbf{I}_L$ is positive when the current flows from $a$ to $b$ and is expressed as $\mathbf{I}_{ab}$. The current $\mathbf{I}_{ba} = -\mathbf{I}_{ab}$.

Where voltages are expressed in double-subscript notation the letters describe the two points between which a voltage exists. It is common to state that the first letter in the subscript denotes the voltage of a node with respect to the voltage at the node depicted by the second subscript. The voltage $\mathbf{V}_{ab}$ in Fig. 1-5 is the voltage of node $a$ with respect to node $b$. It denotes that $\mathbf{V}_{ab}$ is positive during the half-cycle when $a$ is at a higher potential than $b$. It further states that when $a$ is at a higher potential than $b$, $\mathbf{V}_{ab}$ is the voltage drop across $Z_T$. The phasor voltage and current relationship is given by

$$\mathbf{V}_{ab} = \mathbf{I}_{ab}Z_T$$

Reversing a double subscript for either voltage or current gives a voltage or current 180° out of phase with the original value. That is,

$$\mathbf{V}_{ab} = \mathbf{V}_{ba}\underline{/180°} = -\mathbf{V}_{ba}$$

or

$$\mathbf{I}_{ab} = \mathbf{I}_{ba}\underline{/180°} = -\mathbf{I}_{ba}$$

Summarizing the single- and double-subscript notation for Fig. 1-5 yields the following phasor relationships:

$$\mathbf{V}_T = \mathbf{V}_a = \mathbf{V}_{ao}$$
$$\mathbf{V}_L = \mathbf{V}_b = \mathbf{V}_{bo}$$
$$\mathbf{I}_L = \mathbf{I}_{ab}$$

Kirchhoff's voltage law written using double subscripts for the circuit of Fig. 1-5 yields

$$\mathbf{V}_{oa} + \mathbf{V}_{ab} + \mathbf{V}_{bn} = 0 \tag{1-9}$$

where the subscripts are in the order of tracing the closed path around the circuit. Replacing $\mathbf{V}_{oa}$ by $-\mathbf{V}_{ao}$ and noting that $\mathbf{V}_{ab} = \mathbf{I}_{ab}Z_T$,

$$-\mathbf{V}_{ao} + \mathbf{I}_{ab}Z_T + \mathbf{V}_{bn} = 0 \tag{1-10}$$

so that

$$\mathbf{I}_{ab} = \frac{\mathbf{V}_{ao} - \mathbf{V}_{bn}}{Z_T} \tag{1-11}$$

## 1-3 THREE-PHASE CIRCUITS

An ideal three-phase source generates three sinusoidal voltages of equal magnitudes displaced from each other by an angle of 120° in time phase. The elementary three-phase generator shown in Fig. 1-6(a) has three identical stator coils, of one or more turns, displaced from each other by a space angle of 120°. Only two sides of each coil and terminals $aa'$, $bb'$, and $cc'$ are shown. The rotor is a two-pole magnet driven counterclockwise at a constant speed. In this simple generator, each of the three stator coils constitutes one phase. Figure 1-6(b) and (c) shows the waveforms and the phasors of the three voltages.

Three-phase sources and loads may be connected in delta ($\Delta$) or in wye (Y). A delta connection is effected for the armature of the generator in Fig. 1-

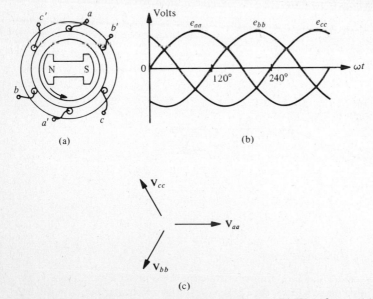

Figure 1-6 (a) Elementary three-phase generator. (b) Waveforms. (c) Phasor representation of generated voltages.

6(a) by connecting terminals $a'$ to $b$, $b'$ to $c$, and $c'$ to $a$, the three junctions forming the three generator terminals. A wye connection can be made by connecting together either all three primed terminals or all three unprimed terminals to form the neutral of the wye. The free terminals of the three phases are then the generator terminals. The neutral of such a generator is sometimes used as a fourth terminal. The wye connection is commonly used for generators.

## 1-3.1 Delta-Connected Impedances

Figure 1-7(a) shows three impedances, $Z_{ab}$, $Z_{bc}$, and $Z_{ca}$, connected in delta and supplied from a three-phase source. The voltages impressed on the impedances are the line-to-line voltages, equal in magnitude and 120° apart. Note that in three-phase circuits the line-to-line voltages are those between any pair of phase voltages. If the three-phase system is balanced, the voltages between phases will

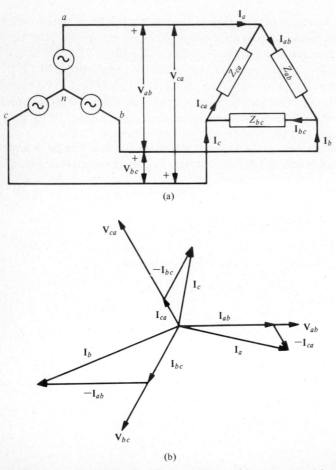

(a)

(b)

**Figure 1-7** (a) Schematic diagram of three delta-connected impedances supplied from a wye-connected three-phase source. (b) Phasor diagram of voltages and currents in a noninductive resistance load.

be equal for any pair. The symbol used to represent the line-to-line voltage is $V_{LL}$. If the phasor $\mathbf{V}_{ab}$ is assumed to lie on the real axis,

$$\mathbf{V}_{ab} = V_{LL}\underline{/0°}$$

$$\mathbf{V}_{bc} = V_{LL}\underline{/-120°}$$

$$\mathbf{V}_{ca} = V_{LL}\underline{/-240°} \tag{1-12}$$

If there are no mutual impedances between $Z_{ab}$, $Z_{bc}$, and $Z_{ca}$, the currents in these three impedances are independent of each other and are expressed by

$$\mathbf{I}_{ab} = \frac{V_{LL}}{Z_{ab}}\underline{/0°}$$

$$\mathbf{I}_{bc} = \frac{V_{LL}}{Z_{bc}}\underline{/-120°}$$

$$\mathbf{I}_{ca} = \frac{V_{LL}}{Z_{ca}}\underline{/-240°} \tag{1-13}$$

where $Z_{ab}$, $Z_{bc}$, and $Z_{ca}$ are expressed in complex form.

According to Kirchhoff's law, the line currents must be

$$\mathbf{I}_a = \mathbf{I}_{ab} - \mathbf{I}_{ca}$$

$$\mathbf{I}_b = \mathbf{I}_{bc} - \mathbf{I}_{ab}$$

$$\mathbf{I}_c = \mathbf{I}_{ca} - \mathbf{I}_{bc} \tag{1-14}$$

The sum of the line currents $\mathbf{I}_a$, $\mathbf{I}_b$, and $\mathbf{I}_c$ is equal to zero, which is true for any three-wire system. A phasor diagram is shown in Fig. 1-7(b) for three unequal noninductive resistances connected in delta.

## 1-3.2 Balanced Delta-Connected Load

When the three impedances are equal (i.e., $Z_{ab} = Z_{bc} = Z_{ca} = Z$), the load is said to be balanced. Then from Eqs. 1-12 and 1-13 it follows that

$$\mathbf{I}_{ab} = \frac{V_{LL}}{Z}\underline{/0°}$$

$$\mathbf{I}_{bc} = \frac{V_{LL}}{Z}\underline{/-120°}$$

$$\mathbf{I}_{ca} = \frac{V_{LL}}{Z}\underline{/-240°} \tag{1-15}$$

and from Eqs. 1-14 and 1-15 we get

$$\mathbf{I}_a = \frac{V_{LL}\underline{/0°} - V_{LL}\underline{/-240°}}{Z} = \frac{(1.5 - j0.866)V_{LL}}{Z}$$

$$= \frac{\sqrt{3}V_{LL}\underline{/-30°}}{Z} = \sqrt{3}\mathbf{I}_{ab}\underline{/-30°} \tag{1-16}$$

similarly

$$I_b = \sqrt{3}I_{bc}\underline{/-30°}$$

$$I_c = \sqrt{3}I_{ca}\underline{/-30°}$$

The relationships in Eq. 1-16 are illustrated for a balanced noninductive resistive load by the phasor diagram in Fig. 1-7(b). Equations 1-16 show that, for balanced voltage and balanced load, the magnitude of the line current is $\sqrt{3}$ times that of the current in the delta with a phase lag of 30°.

## 1-3.3 Wye-Connected Impedances

Figure 1-8(a) shows three wye-connected impedances $Z_a$, $Z_b$, and $Z_c$ supplied from a wye-connected source with a switch in the neutral conductor.

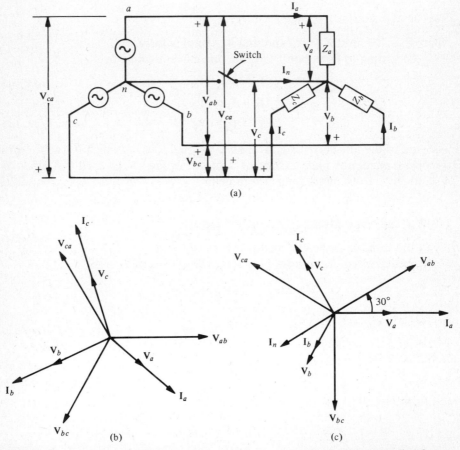

**Figure 1-8** (a) Schematic diagram of three wye-connected impedances supplied from a wye-connected source. (b) Phasor diagram with neutral open. (c) Phasor diagram with neutral closed.

## 1-3.4 Three-Wire, Wye-Arrangement (Neutral Connection Open)

The neutral current $I_n$ is zero when the switch is open. The impedances are then supplied from a three-wire wye-connected source which might just as well be delta-connected. Either supply connection will deliver line-to-line voltages $V_{ab}$, $V_{bc}$, $V_{ac}$. Then from Kirchhoff's laws,

$$Z_a I_a - Z_b I_b = V_{ab} = V_{LL} \underline{/0°}$$
$$Z_b I_b - Z_c I_c = V_{bc} = V_{LL} \underline{/-120°}$$
$$Z_c I_c - Z_a I_a = V_{ca} = V_{LL} \underline{/-240°} \qquad (1\text{-}17)$$

Also from Kirchhoff's voltage law,

$$V_{ab} + V_{bc} + V_{ca} = 0 \qquad (1\text{-}18)$$

The voltages across the three impedances are

$$V_a = Z_a I_a$$
$$V_b = Z_b I_b$$
$$V_c = Z_c I_c \qquad (1\text{-}19)$$

Making use of Eqs. 1-17 and 1-18, after some algebraic manipulation it is found that

$$I_a = \frac{(Z_c - Z_b \underline{/-240°}) V_{LL} \underline{/0°}}{D}$$

$$I_b = \frac{(Z_a - Z_c \underline{/-240°}) V_{LL} \underline{/-120°}}{D}$$

$$I_c = \frac{(Z_b - Z_a \underline{/-240°}) V_{LL} \underline{/-240°}}{D} \qquad (1\text{-}20)$$

where $\qquad D = Z_a Z_b + Z_b Z_c + Z_c Z_a$

## 1-3.5 Balanced Wye-Connected Load

For balanced load $Z_a = Z_b = Z_c = Z$, and on that basis, Eq. 1-19 is simplified to the following expressions:

$$I_a = \frac{V_{LL}}{\sqrt{3}Z} \underline{/-30°}$$

$$I_b = \frac{V_{LL}}{\sqrt{3}Z} \underline{/-150°}$$

$$I_c = \frac{V_{LL}}{\sqrt{3}Z} \underline{/-270°} \qquad (1\text{-}21)$$

Then from Eqs. 1-20 and 1-21, the line-to-neutral voltages are expressed by

$$\mathbf{V}_{an} = \frac{V_{LL}}{\sqrt{3}} \underline{/-30°}$$

$$\mathbf{V}_{bn} = \frac{V_{LL}}{\sqrt{3}} \underline{/-150°}$$

$$\mathbf{V}_{cn} = \frac{V_{LL}}{\sqrt{3}} \underline{/-270°} \qquad\qquad (1\text{-}22)$$

where          $\mathbf{V}_a = \mathbf{V}_{an}$, $\mathbf{V}_b = \mathbf{V}_{bn}$,  and  $\mathbf{V}_c = \mathbf{V}_{cn}$

Equation 1-22 shows that under balanced conditions the magnitude of the line-to-line voltage is equal to that of the line-to-neutral voltage times $\sqrt{3}$ with a phase lead of 30°. The line-to-line voltages are expressed in terms of the line-to-neutral voltages as phasors by

$$\mathbf{V}_{ab} = \sqrt{3}\,\mathbf{V}_{an}\underline{/30°}$$

$$\mathbf{V}_{bc} = \sqrt{3}\,\mathbf{V}_{bn}\underline{/30°}$$

$$\mathbf{V}_{ca} = \sqrt{3}\,\mathbf{V}_{cn}\underline{/30°} \qquad\qquad (1\text{-}23)$$

### 1-3.6  Four-Wire Arrangement (Load Neutral Connected to Source Neutral)

When the neutrals of the load and of the source are connected by closing the switch in Fig. 1-8(a), the line-to-neutral voltages at the load are balanced (i.e., equal to the balanced line-to-neutral voltages of the source). This is true whether the three wye-connected impedances are equal or not, and Eq. 1-22 applies. Neutral current expressed by

$$\mathbf{I}_n = -(\mathbf{I}_a + \mathbf{I}_b + \mathbf{I}_c)$$

results when the impedances are unbalanced but is zero for balanced conditions. It therefore makes no difference whether the neutrals of the load and source are interconnected or not for balanced loads.

The phasor diagram in Fig. 1-8(b) represents current and voltage relationships for three unequal noninductive resistances connected in wye with the load neutral isolated from the source neutral, and Fig. 1-8(c) shows a phasor diagram for the same load resistance but with the two neutrals connected.

### 1-3.7  Phase Sequence

It is standard practice in the United States to designate the phases *abc* such that under balanced conditions the voltage and current in *a* phase leads the voltage and current in *b* phase by 120° and in *c* phase by 240°. This is known as *abc phase sequence*. If the rotation of the generator is reversed or any two of the three leads from the armature (not including the neutral) to the generator terminals are reversed, the reversed phase sequence becomes *acb*.

## 1-4 COMPLEX POWER

An alternating current of constant frequency expressed by

$$i = \sqrt{2}\omega I \sin \omega t \tag{1-24}$$

induces an emf when it flows in a circuit that has a constant inductance of $L$ henries which is

$$e = \frac{d(Li)}{dt} = 2\omega LI \cos \omega t = \sqrt{2}X_L I \cos \omega t \tag{1-25}$$

where $I$ is the rms value of the current and $X_L = \omega L$, the inductive reactance in ohms. The rms value of the induced voltage is

$$E = X_L I \tag{1-26}$$

and when Eq. 1-26 is substituted in Eq. 1-25, we have

$$e = \sqrt{2}E \cos \omega t \tag{1-27}$$

The current and voltage expressed by Eqs. 1-24 and 1-27 are sinusoids of the same frequency with the current wave lagging the voltage wave by 90°.

Since the real power in ac circuits, under steady-state conditions at constant frequency, is the average of the instantaneous power taken over one or more half-cycles, the real power associated with the above current and voltage waves is zero:

$$P = \frac{1}{\pi} \int_0^\pi ei \, d(\omega t) = 0$$

Real power is also expressed by

$$P = EI \cos \theta \tag{1-28}$$

where $\theta$ is the angle between the current $I$ and the voltage $E$. In a purely inductive circuit $\theta = 90°$ and the real power is zero.

However, an inductive circuit is said to consume reactive power, which is expressed by

$$Q = EI \sin \theta \tag{1-29}$$

where $\theta$ is the angle by which the current $I$ lags the voltage $E$.

Magnetic circuits excited by alternating current consume reactive power. Reactive power may therefore be regarded as required for the production of magnetic flux. On the other hand, the current in capacitive circuits leads the voltage, causing $\theta$ to be negative, from which it follows that capacitive circuits generate reactive power. Capacitors are frequently used in industrial power systems to furnish reactive power, an arrangement that is also known as power-factor correction.

The complex power in a circuit is expressed by

$$\mathbf{S} = P + jQ \tag{1-30}$$

where $P$ and $Q$ are the real power and reactive power. If **E** and **I** are the voltage and current expressed as complex quantities, then

$$\mathbf{S} = \mathbf{EI}^* \tag{1-31}$$

where **I*** is the conjugate of the current phasor **I**. The real part of a complex quantity equals the real part of its conjugate, and its imaginary part is equal in magnitude but of opposite sign to that of its conjugate. In Fig. 1-9 the voltage phasor is expressed by

$$\mathbf{E} = E \cos \alpha + jE \sin \alpha = E\epsilon^{j\alpha} \tag{1-32}$$

and the current phasor by

$$\mathbf{I} = I \cos \beta + jI \sin \beta = I\epsilon^{j\beta} \tag{1-33}$$

Then the conjugate current phasor is

$$\mathbf{I}^* = I \cos \beta - jI \sin \beta = I\epsilon^{-j\beta} \tag{1-34}$$

and the product of Eqs. 1-32 and 1-34 is the complex power

$$\mathbf{S} = EI\epsilon^{j(\alpha-\beta)} = EI\epsilon^{j\theta}$$
$$= EI \cos \theta + jEI \sin \theta \tag{1-35}$$

Hence, the real power in watts is the real part of Eq. 1-35.

$$P = \mathrm{Re}\,(\mathbf{EI}^*) = EI \cos \theta \tag{1-36}$$

and the reactive power in VARs is the imaginary part of Eq. 1-30.

$$Q = \mathrm{Im}\,(\mathbf{EI}^*) = EI \sin \theta \tag{1-37}$$

The power factor is expressed by

$$\mathrm{PF} = \frac{P}{|S|} = \cos \theta$$

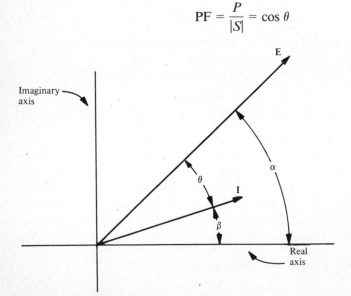

**Figure 1-9**   Voltage and current phasors.

If $S_1, S_2, S_3, \ldots, S_n$ expresses the complex power of $n$ parallel components of a system, the total complex power of that system is

$$S_T = S_1 + S_2 + S_3 + \cdots + S_n$$
$$= P_1 + P_2 + P_3 + \cdots + P_n + j(Q_1 + Q_2 + Q_3 + \cdots + Q_n)$$
$$= P_T + jQ_T \tag{1-38}$$

## 1-4.1 Power Triangle

It has just been shown that for several components in parallel in a system the total $P$ will be the sum of the average powers of the individual components and the total $Q$ is likewise the sum of the average reactive power of the individual components. The triangle drawn in Fig. 1-10 is for an $S_T = P_T + jQ_T$ where the total power of the system is inductive. If the system were capacitive, the $Q_T$ would be the negative of that shown in Fig. 1-10.

Figure 1-11 is given to demonstrate the resultant power triangle when an inductive load with power triangle $a$ of $S_1$, $P_1$, and $Q_1$ and a lagging power factor is combined with a capacitive load with a leading power factor and a power

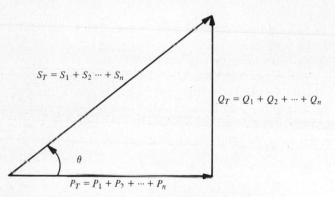

Figure 1-10  Power triangle for an inductive system.

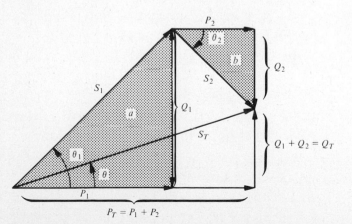

**Figure 1-11**  Power triangle of a parallel inductive and capacitive load.

triangle $b$ comprised of $\mathbf{S}_2$, $P_2$, and $Q_2$. The resultant system $\mathbf{S}_T$, $P_T$, and $Q_T$ remains inductive when these two loads are placed in the system in parallel as shown in the figure.

**EXAMPLE 1-1**

In the figure for this example, two induction motor loads are connected in parallel at the end of a three-phase line with balanced impedance. The voltage at the input terminals of the motors is a balanced 440 V.

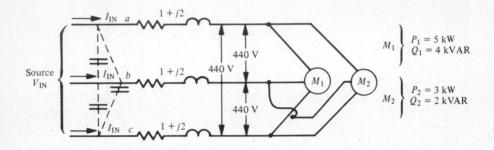

    (a) Find the $|V_{\text{IN}}|$ and $|I_{\text{IN}}|$ and the input power factor. Sketch the power triangle for the system.

    (b) In order to correct the power factor, consideration is being given to the connection of a three-phase delta-connected capacitor bank (shown by dotted lines). Find the values of capacitance per phase of the delta necessary to bring the input power factor to a 0.9 lagging. Also, compare the values of input current before and after power-factor correction. Assume $V_{IN}$ is constant at the value found in part a of this problem.

*Solution*

(a)
$$P_T = P_1 + P_2 = 5 + 3 = 8 \text{ kW}$$
$$Q_T = Q_1 + Q_2 = 4 + 2 = 6 \text{ kVAR}$$
$$\mathbf{S} = P_T + jQ_T = 8 + j6 \text{ kVA}$$
$$|S| = \sqrt{P_T^2 + Q_T^2} = \sqrt{8^2 + 6^2} = 10 \text{ kVA}$$
$$|S| = \sqrt{3}|V_L|\,|I_{\text{IN}}| = 10 \text{ kVA}$$
$$|I_{\text{IN}}| = \frac{10 \text{ kVA}}{\sqrt{3}(440)} = 13.13 \text{ A}$$
$$P_{\text{LINE}} = 3|I_{\text{IN}}|^2 R = 3(13.13)^2(1) = 518 \text{ W} = 0.518 \text{ kW}$$
$$Q_{\text{LINE}} = 3|I_{\text{IN}}|^2 X = 3(13.13)^2(2) = 1036 \text{ VAR} = 1.036 \text{ kVAR}$$
$$P_{\text{IN}} = P_{\text{LINE}} + P_L = 0.518 + 8 = 8.518 \text{ kW}$$
$$Q_{\text{IN}} = Q_{\text{LINE}} + Q_L = 1.036 + 6 = 7.036 \text{ kVAR}$$
$$\mathbf{S}_{\text{in}} = 8.518 \text{ kW} + j7.036 \text{ kVA}$$
$$|S_{\text{IN}}| = \sqrt{(8.518)^2 + (7.036)^2} = 11.05 \text{ kVA}$$

or
$$|S_{IN}| = \sqrt{3}|V_{in}|\,|I_{IN}| = 11.05 \text{ kVA}$$

$$|V_{IN}| = \frac{11.05 \text{ kVA}}{\sqrt{3}(13.13)} = 486 \text{ V}$$

$$\theta_{IN} = \tan^{-1}\frac{Q_{IN}}{P_{IN}} = 39.6°$$

$$PF_{IN} = \cos 39.6° = 0.771 \text{ lagging}$$

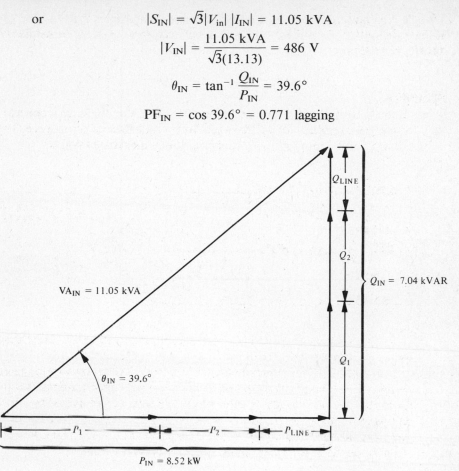

(b)  To correct the power factor to 0.9 lagging, the new $\theta_{in}$ must be such that $\theta_{in} = \cos^{-1} 0.9 = 25.8°$. This is a reduction from 39.6 to 25.8°. This is accomplished by adding negative (capacitive) VARs. (Note that the real power does not change during this process if one assumes that the capacitor is lossless.)

$$Q_{IN} = P_{IN} \tan \theta_{IN} = 8.52 \tan 25.8 = 4.12 \text{ kVAR}$$

$$Q_{added} = Q_{IN} \text{ (with capacitance)} - Q_{IN} \text{ (without capacitance)}$$

$$Q_{added} = 4.12 - 7.04 = -2.92 \text{ kVAR (negative sign indicates capacitive)}$$

$$Q_{added/phase} = \frac{Q_{added}}{3} = \frac{2.92}{3} = 0.973 \text{ kVAR}$$

$$Q_{added/phase} = \frac{|V_{IN}|^2}{X_C}$$

$$X_C = \frac{|V_{IN}|^2}{Q_{added/phase}} = \frac{(486)^2}{973} = 243 \text{ }\Omega$$

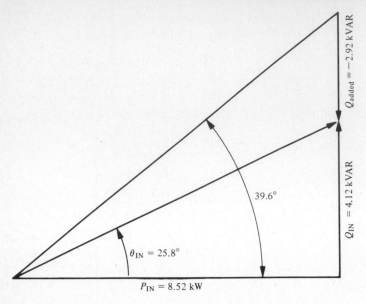

$$C = \frac{1}{2\pi f X_C} = \frac{1}{2\pi(60)(243)} = 10.9 \ \mu\text{F}$$

The magnitude of the input current with the capacitors added is

$$|I'_{\text{in}}| = \frac{P_{\text{in}}}{\sqrt{3} V_{\text{in}} \times \text{PF}} = \frac{8518}{\sqrt{3}(486)(0.9)} = 11.25 \ \text{A}$$

The reduction in current is

$$\Delta I = 13.13 - 11.25 = 1.88 \ \text{A}$$

nearly a 15 percent reduction. In large power situations with high load (current), wiring costs can be reduced with proper maintenance of a power factor of 0.85 or greater.

## 1-4.2 Power Flow

The relation between $PQ$ and bus voltage $V$ or generated voltage $E$ is important when the flow of power is considered. The fundamental issue involves the question of direction of power flow, that is, whether power is being absorbed by a system component or delivered to the system for a specified voltage and current. Consider the circuits of Fig. 1-12 to be those of an ac ideal voltage source of constant magnitude and frequency with zero impedance with polarity markings shown and as described earlier in this chapter.

It would be expected for Fig. 1-12(a) to be a generator, since current is out of the positive terminal. To fully understand whether Fig. 1-12(a) is indeed a generator, the phasor relationship of **E** and **I** must be evaluated. If the real part of **EI\*** is positive, the circuit is a generator. However, if the real part of **EI\*** is

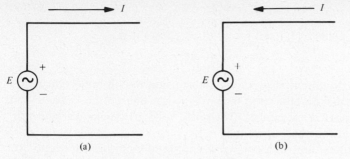

**Figure 1-12**   AC circuit with polarity markings.

negative, the circuit of Fig. 1-12(a) is functioning as a load on the system and is absorbing power as a motor.

Similarly for Fig. 1-12(b) it would be expected that this condition is one of power absorption. This is true if for the conditions shown the real part of $\mathbf{EI}^*$ is positive. If the real part is negative, the circuit is generating power.

Consideration of the sign of the imaginary part of $\mathbf{EI}^*(Q)$ for each case gives information on the absorption or supply of reactive power. For Fig. 1-12(a) when $Q$ is positive, the load is inductive and the generator supplies inductive reactive power. If $Q$ is negative, the generator is receiving reactive power and the load is capacitive. Table 1-1 illustrates the various conditions.

### 1-4.3 Power in Balanced Three-Phase Circuits

If $V_\phi$ and $I_\phi$ are the voltage and current per phase in a balanced three-phase circuit, the complex power per phase is

**TABLE 1-1.   MOTOR AND GENERATOR ACTION SUMMARY**

| Assumed circuit polarity | | $\mathbf{EI}^*$ calculation |
|---|---|---|
| 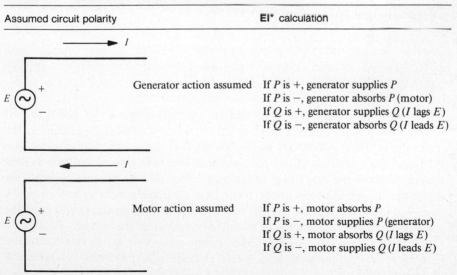 | Generator action assumed | If $P$ is +, generator supplies $P$<br>If $P$ is −, generator absorbs $P$ (motor)<br>If $Q$ is +, generator supplies $Q$ ($I$ lags $E$)<br>If $Q$ is −, generator absorbs $Q$ ($I$ leads $E$) |
| | Motor action assumed | If $P$ is +, motor absorbs $P$<br>If $P$ is −, motor supplies $P$ (generator)<br>If $Q$ is +, motor absorbs $Q$ ($I$ lags $E$)<br>If $Q$ is −, motor supplies $Q$ ($I$ leads $E$) |

$$S_\phi = V_\phi I_\phi^* \tag{1-39}$$

where $I_\phi^*$ is the conjugate of $I_\phi$, and for all three phases it is

$$S = 3S_\phi = 3V_\phi I_\phi^* \tag{1-40}$$

Three-phase equipment is generally rated using line-to-line voltage and line current. In the delta connection $V_\phi = V_{LL}$, the line-to-line voltage, and $I_\phi = I_L/\sqrt{3}$, where $I_L$ is the line current. The phase voltage in the wye connection cis $V_\phi = V_{LL}/\sqrt{3}$ and the phase current $I_\phi = I_L$.

Then the complex three-phase power is expressed for either connection by assuming a voltage reference angle of zero and

$$S = \sqrt{3}V_{LL}I_L(\cos\theta + j\sin\theta) \tag{1-41}$$

where $$I = I(\cos\theta - j\sin\theta) \tag{1-42}$$

The real power and reactive power are

$$P = \sqrt{3}V_{LL}I_L\cos\theta \tag{1-43}$$

and $$Q = \sqrt{3}V_{LL}I_L\sin\theta \tag{1-44}$$

where $\theta$ is positive when the phase current lags the phase voltage.

If $S_1, S_2, \ldots, S_n$ are loads supplied from the same source, the total power supplied by that source is

$$S_T = P_1 + P_2 + \cdots + P_n + j(Q_1 + Q_2 + \cdots + Q_n) \tag{1-45}$$

This equation is valid whether all the loads are wye-connected or delta-connected, also if some are wye-connected and others delta-connected.

## 1-5 PER-UNIT QUANTITIES

Power systems use components that have many different ratings. Transmission and distribution voltages cover a range from a few thousand volts to a million volts. Since many solutions of power-system problems require only a single-phase representation of the three-phase system to achieve a useful solution, it is convenient to use a normalizing method called *per-unit* (pu). The reader should keep in mind that all voltages and currents in a power system are related by the turns ratio of the transformers in the system. Voltages, currents, voltamperes, and impedances are related in such a manner that selection of any two as base values establishes the other two. In addition, since all voltages and currents are related directly or inversely as the turns ratio of transformers in any part of a power system, all voltages, currents, voltamperes, and impedances will have the same per-unit values regardless of where they appear in the system.

Beginning with the single-phase representation of a power system and selecting a base voltampere and a base voltage, the following relationships are derived:

$$\text{Base current, A} = \frac{\text{base VA}_\phi}{\text{base } V_{LN}} = \frac{\text{base kVA}_\phi}{\text{base kV}_{LN}} = \frac{\text{base MVA}_\phi}{\text{base MV}_{LN}} \qquad (1\text{-}46)$$

$$\text{Base impedance, } \Omega = \frac{\text{base } V_{LN}}{\text{base } I} = \frac{\text{base kV}_{LN}}{\text{base } I \times 1000} \qquad (1\text{-}47)$$

$$\text{Base impedance, } \Omega = \frac{(\text{base } V_{LN})^2}{\text{base VA}_\phi} = \frac{(\text{base kV}_{LN})^2}{\text{base MVA}_\phi}$$

$$= \frac{(\text{base kV}_{LN})^2 \times 1000}{\text{base kVA}_\phi} \qquad (1\text{-}48)$$

$$\text{Base power, kW}_\phi = \text{base kVA}_\phi \qquad (1\text{-}49)$$

$$\text{Per-unit impedance (of a circuit element)} = \frac{\text{actual impedance, } \Omega}{\text{base impedance, } \Omega} \qquad (1\text{-}50)$$

In the equations above the subscripts $LN$ and $\phi$ denote line to neutral and per phase. Since equipment and system data is most often provided as three-phase quantities, the three-phase relationships are developed in the following equations, where the subscript $LL$ denotes line-to-line values and $3\phi$ denotes three-phase values:

$$\text{Base current} = \frac{\text{base kVA}_{3\phi}}{\sqrt{3} \times \text{base voltage, kV}_{LL}} \text{ A} \qquad (1\text{-}51)$$

From Eq. 1-48

$$\text{Base impedance} = \frac{(\text{base voltage, kV}_{LL}/\sqrt{3})^2 \times 1000}{\text{base kVA}_{3\phi}/3} \Omega \qquad (1\text{-}52)$$

$$\text{Base impedance} = \frac{(\text{base voltage, kV}_{LL})^2 \times 1000}{\text{base kVA}_{3\phi}} \Omega \qquad (1\text{-}53)$$

$$\text{Base impedance} = \frac{(\text{base voltage, kV}_{LL})^2}{\text{base MVA}_{3\phi}} \Omega \qquad (1\text{-}54)$$

**EXAMPLE 1-2**

A 220- to 110-V single-phase transformer rated at 3 kVA has a high-voltage winding impedance of 0.24 $\Omega$. Find the per-unit impedance of the transformer winding.

*Solution.*  Select 220 V and 3 kVA as the base quantities. From Eq. 1-46:

$$\text{Base current} = \frac{3000 \text{ VA}}{220 \text{ V}} = 13.64 \text{ A}$$

Substituting the base current in Eq. 1-47,

$$\text{Base impedance} = \frac{220 \text{ V}}{13.64 \text{ A}} = 16.13 \text{ } \Omega$$

and dividing the actual impedance by the base impedance per Eq. 1-50 gives the per-unit impedance of the transformer winding

$$\text{Per-unit impedance} = \frac{0.24}{16.13} = 0.015 \text{ pu}$$

$$\text{Percent } Z = 1.5 \text{ percent}$$

Note that winding impedance changes as the turns ratio squared; therefore, the low-voltage winding impedance is $0.24/2^2 = 0.06$ pu.

Assume low voltage of 110 V as base. Therefore, the base current

$$\text{Base current} = \frac{3000 \text{ VA}}{110 \text{ V}} = 27.27 \text{ A}$$

$$\text{Base impedance} = \frac{110}{27.27} = 4.03 \text{ } \Omega$$

$$\text{Per-unit impedance} = \frac{0.06}{4.03} = 0.015 \text{ pu}$$

## STUDY QUESTIONS

1. Define what is meant by a phasor in an ac circuit. Explain the difference between a phasor and a vector.

2. List reasons why three-phase circuits are used for the transmission of large quantities of power as opposed to single-phase circuits.

3. If three-phase circuits are more advantageous for the transmission of power, then possibly a six-phase circuit would be even more advantageous. Describe the advantages and disadvantages of transmitting power using a higher number of phases than three.

4. If six-phase power were to be used, how would six-phase power be generated?

5. In a three-phase circuit transformers are connected delta and wye. List the advantages and disadvantages of connecting transformers in either delta or wye.

6. If the primary winding of a transformer is connected delta and the secondary is connected wye, what is the phase difference in the voltage $V_{ab}$ from the primary to the secondary side?

7. In connecting power circuits using delta-wye transformations, an engineer must take precautions to assure that an equal number of delta-wye transformations occur in each segment of the transmission line that is to be tied together. Can you explain why?

8. Assume for practical purposes that loads connected to a power system are balanced. Describe a condition under which loads could be reasonably assumed to be balanced and another condition under which loads could reasonably be assumed to be unbalanced.

9. If a three-phase-circuit, four-wire arrangement is serving an unbalanced load, a neutral current will be present. Estimate or show how to calculate the voltage that will be impressed on the neutral during unbalanced conditions.

10. In this chapter there was a discussion of complex power. What is there in the power

system that causes the power expression to be complex with a largely inductive component?

11. Are there any parts of the power system contributing a capacitive reactance to offset the inductive reactance of the power system? If so, what are they and in what relative magnitudes do they contribute to the overall reactance of the system?

12. List as many reasons as you can for the use of power-factor correction in a power system.

13. If power-factor correction is used, list the economic factors that must be included in deciding whether or not to use power-factor correction.

14. What is the usefulness of the per-unit system in power-system analysis?

15. Does the per-unit system have as great a value today, with the widespread use of computing capacity, as it had in years past?

# PROBLEMS

**1-1.** If $v = 141.4 \cos (\omega t + 60°)$ V and $i = 14.14 \sin (\omega t + 120°)$ A, find for both $v$ and $i$:
  (a) The maximum value.
  (b) The rms value.
  (c) The phasor expression using voltage as the reference.
  (d) The phasor expression using current as the reference.
  (e) Is the circuit in which the voltage and current exist capacitive or inductive?

**1-2.** Two voltages are given as $v_1 = 100 \sin (\omega t + 60°)$ and $v_2 = 50 \sin (\omega t + 30°)$. Express $v_1$ and $v_2$ as voltage phasors. Find the resultant phasor $V = V_1 + V_2$. If $\omega t$ for $v_1$ is not equal to $\omega t$ for $v_2$, can $v_1$ and $v_2$ be added as phasors?

**1-3.** Express each of these currents as a phasor:
  (a) $800 \cos (\omega t + 130°)$ A.
  (b) $480 \cos \omega t - 640 \sin \omega t$ A.
  (c) $550 \sin (\omega t + 40) + 420 \sin (\omega t + 110°)$ A.

**1-4.** A voltage $v = 141.4 \cos 377t$ V is applied to a resistive circuit with $R = 25\ \Omega$.
  (a) Write the expression for $i$ as a function of time.
  (b) What is the frequency of the applied voltage and the current through the resistor?
  (c) Write the expression for power as a function of time.
  (d) What is the frequency of the power?
  (e) Calculate the average power. Show that it is equal to $EI \cos \theta$.

**1-5.** For this problem use the circuit of Fig. 1-4 and an applied voltage $V_{ac} = 70.7 \sin (\omega t + 30°)$. If the frequency of the applied voltage is 60 Hz and $R_1 = 10\ \Omega$, $R_2 = 5\ \Omega$, $R_3 = 1\ \Omega$, and $L_1 = 5$ mH, $L_2 = 10$ mH, and $C_3 = 20\ \mu$F, find:
  (a) $I_1$, $I_2$, and $I_3$.
  (b) $V_{ab}$ and $V_{bc}$.
  (c) Draw the voltage and current phasor diagram using $I_1$ as the reference.
  (d) Calculate the average input power to the circuit.

**1-6.** For the circuit of Fig. 1-5 where the internal impedance of the generator is $Z_g = 2 + j6$ and $Z_T = 1 + j2$ is the transmission-line impedance which supplies the load impedance $Z_L = 2 + j5$, draw a phasor diagram using $E_g$ as the reference with a value of $500\ \underline{/0°}$ V showing:

     (a) $I_L$, $V_T$, and $V_L$.

     (b) What is the value of the phasor voltage $V_{ab}$?

     (c) On the phasor diagram show the components of the voltage $V_{ab}$ which are $I_L R_T$ and $I_L X_T$.

     (d) Calculate the total power generated.

     (e) Calculate the power absorbed in the load. Is it different from the power generated? Why?

**1-7.** Using the circuit of Fig. 1-5 and the circuit parameters of Prob. 1-6, and $E_g = 500$ $\underline{/0°}$, find the change in values of voltages $V_T$ and $V_L$, current $I_L$, and power if a capacitor of 10 $\mu$F is placed in parallel with $Z_L$ between nodes $b$ and $n$. Assume a supply frequency of 60 Hz.

**1-8.** Determine the line current drawn from a three-phase 440-V line by a three-phase 10-hp motor operating at rated load. The motor has an efficiency of 91.6 percent and a power factor of 85 percent lagging. Also calculate the $P$ and $Q$ drawn by the motor.

**1-9.** If the three lines connecting the motor in Prob. 1-8 have impedances of $0.5 + j1.0$ $\Omega$, find the line-to-line voltage at the bus that supplies the motor. Assume that the motor bus voltage is 440 V.

**1-10.** For the system in Fig. 1-7, assume that the load impedances are 12 $\Omega$ per phase and that the supply lines have an impedance of $2 + j5$ $\Omega$ each. If the supply voltage is 480 V, find the line current and the line voltage at the terminals of the load resistors.

**1-11.** A three-phase delta-connected motor bank draws 1000 kVA at 0.65 power factor lagging from a 480-V source. Determine the kVA rating of capacitors to make the combined motor-capacitor bank power factor 0.85 lagging. Calculate the line current before and after power-factor correction.

**1-12.** In Prob. 1-11 power-factor correction was used to correct a low power factor. The line current changed as a result. If the supply-line impedance is $1 + j2$, calculate the change in line losses when power-factor correction is used. If the motor operates continuously over the year and power costs $0.04 per kWh, what is the dollar savings per year because line losses are reduced?

**1-13.** In Fig. 1-8 a three-phase source of 4160 V supplies a wye-connected load of $Z_a = Z_b = Z_c = 5 + j10$. Calculate:

     (a) $V_a$, $V_b$, $V_c$.

     (b) $I_a$, $I_b$, $I_c$.

     (c) Draw the phasor diagram using $V_a$ as the reference.

**1-14.** For Prob. 1-13 with the switch in Fig. 1-8 closed prove that $I_n = 0$.

**1-15.** Show that for a balanced three-phase wye-connected system as shown in Fig. 1-8 the neutral current is always zero whether or not the switch is open or closed (open-switch three-wire system, closed-four wire system).

**1-16.** For the system of Fig. 1-8, with the neutral switch closed $Z_a = 1 + j2$, $Z_b = 2 + j1$, $Z_c = 1 + j1$, and a balanced supply voltage of 480 V, calculate $I_a$, $I_b$, $I_c$, and $I_n$.

**1-17.** A single-phase transformer is rated 220/440 V, 5 kVA. The leakage reactance measured on the low-voltage side is 0.06 $\Omega$. Find the per-unit leakage reactance based on the transformer rating. Find the leakage reactance viewed from the high-voltage side of the transformer.

**1-18.** A transmission line has a characteristic impedance of 320 $\Omega$. The line has a nominal rating of 360 kV and 1400 MVA. What is the per-unit line impedance based on the nominal line rating?

**1-19.** A 30-MVA, 13.8-kV, three-phase generator has a synchronous reactance of 7.5 $\Omega$. Calculate the per-unit synchronous reactance on the generator base. Recalculate the per-unit synchronous reactance on a 100-MVA base.

**1-20.** A voltage source $E_{an} = 10 \: \underline{/30°}$ kV supplies a current from the source of $I_{na} = -100 \: \underline{/240}$ A. Find $P$ and $Q$ and identify the source as delivering or receiving $P$ and $Q$. Is the source a generator or motor?

**1-21.** Two ideal voltage sources are designated as machines 1 and 2 connected through a line with an impedance $Z = 0 + j1 \: \Omega$. If $E_1 = 100 \: \underline{/0}$ V and $E_2 = 100 \: \underline{/30°}$ V, determine:
   (a) Which machine is generating and which machine is consuming power.
   (b) Which machine is receiving and which is supplying reactive power.
   (c) The $P$ and $Q$ absorbed by the line impedance.

**1-22.** A three-phase supply has terminals labeled $a$, $b$, and $c$. Between any pair the voltage measures 208 V. A resistor of 100 $\Omega$ and a capacitive reactance of 100 $\Omega$ at the supply frequency are connected in series from $a$ to $b$ with the resistor connected to terminal $a$. The connection point between the capacitor and the resistor is called $n$. What will the voltage be between $c$ and $n$ for an $abc$ phase sequence? What will it be for an $acb$ phase-sequence supply? Show the result graphically by phasor analysis.

## BIBLIOGRAPHY

Gross, Charles A. *Power System Analysis.* New York: John Wiley & Sons, Inc., 1979.

Matsch, Leander W. *Electromagnetic & Electromechanical Machines,* 2d ed. New York: Dun-Donnelley Publishing Co., 1977.

Schultz, Richard D., and Richard A. Smith. *Introduction to Electric Power Engineering.* New York. Harper & Row, Publishers, Inc., 1985.

Stevenson, W. D., Jr. *Elements of Power System Analysis,* 4th ed. New York: McGraw-Hill Book Company, 1982.

# Chapter 2

# Energy Conversion

Industrial processes generally require the conversion of energy from a readily available form to another form more suitable for utilization. Devices that effect such energy conversion are sometimes called *transducers,* of which there exists an enormous variety—far too many to fall within the scope of this textbook. The treatments in this book are largely confined to the more common types of transducers, such as electromagnets, transformers, electric motors, and electric generators.

In the transformer two or more stationary electric circuits are coupled by means of a magnetic field so that electrical energy may be converted from voltage and current values in one circuit to more suitable values in the other coupled circuits. Electric motors and generators in commercial use are almost without exception electromagnetic devices that effect electromechanical energy conversion by using the magnetic field as a coupling medium between a stationary member and a moving member. The same principle is true for electromagnets. Although there are devices that use the electric field as a coupling medium for electromechanical energy conversion, their use is confined to a limited number of special applications that are not treated in this text.

The greater ease with which energy may be stored in magnetic fields accounts largely for the wide use of electromagnetic devices for electromechanical energy conversion. To gain some measure of the energy capacity or the value of energy density that might be attained in a dielectric, consider mica with a typical value of relative dielectric constant of 7.5 and a dielectric strength up to about $2 \times 10^8$ volts per meter. On that basis the maximum energy density† in the electric field would be

† For an elementary discussion of electric fields in simple configurations, see L. W. Matsch, *Capacitors, Magnetic Circuits and Transformers* (Englewood Cliffs, N.J.: Prentice-Hall, Inc., 1964), Chap. 2.

$$\frac{ED}{2} = \frac{k_r \epsilon_0 E^2}{2} = 7.5 \times 8.854 \times 10^{-12} \times \frac{(2 \times 10^8)^2}{2}$$

$$= 1.327 \times 10^6 \text{ joules per cubic meter}$$

where $E$ = electric field intensity, volts per meter

$D$ = electric flux density, coulombs per square meter

$k_r$ = relative dielectric constant

$\epsilon_0$ = $8.854 \times 10^{-12}$, dielectric constant of free space, farads per meter

This amount of energy density is equal to $1.33 \times 10^6$ joules per cubic meter, a value that is unrealistically high because the dielectric material would not be capable of supporting an electric field intensity as high as $2 \times 10^8$ V/m for an appreciable time under practical conditions. However, this same value of energy density can be readily stored magnetically in air at a magnetic flux density of 1.82 webers per square meter (teslas). Nevertheless, the capacitor affords a simple means for illustrating the elementary principles which relate force to the energy stored in a field (an electric field in a capacitor), as shown in the following section.

At this point it is well to mention that, unless otherwise specified, derivations in this textbook are based on the International System (SI) of units (Table 2-1).

Although the United States and other countries are at present in the process of converting to the SI system of units, two older systems, the centimeter-gram-second (cgs) and mixed English system, are still in use in the United States. For that reason and because much of the technical literature of the past makes use of one or the other of these older systems, conversions of their magnetic units into those of the SI system are presented in Table 2-2. Other conversion factors and constants are listed in Appendix B.

## 2-1 FORCE IN A CAPACITOR

Figure 2-1 shows a capacitor with two plates that are free to move with respect to each other. The charges on the plates resulting from the applied voltage $v$ produce a *force of attraction* between the plates. If the left-hand plate is fixed, the force $f$ tends to move the right-hand plate to the left, and if the plate moves under this force, electrical energy is converted to mechanical energy. If there are no losses in the dielectric and the resistance of the plates and the connecting leads are zero, the differential energy input from the electrical source in the differential time $dt$ is

$$\boxed{dW_e = dW_\psi + dW_{\text{mech}}} \tag{2-1}$$

where $dW_\psi$ is the differential energy stored in the electric field and $dW_{\text{mech}}$ the differential mechanical *output* of the capacitor. The differential charge $dq$ is supplied by the electric source, and the electrical input is therefore

$$dW_e = v\, dq = v\, d(Cv)$$
$$= v^2\, dC + vC\, dv \tag{2-2}$$

**TABLE 2-1.** INTERNATIONAL SYSTEM OF UNITS (SI SYSTEM)

| Quantity | Unit | Abbreviation | Symbol |
|---|---|---|---|
| Basis of the system | | | |
| Length | meter | m | $l$ |
| Mass | kilogram | kg | $M$ |
| Time | second | s | $t$ |
| Electric current | ampere | A | $I$ |
| Thermodynamic temperature | degree kelvin | K | $T$ |
| Luminous intensity | candela | cd | |
| Derived units | | | |
| Area | square meter | m² | $A$ |
| Volume | cubic meter | m³ | $V$ |
| Frequency | hertz | Hz | $f$ |
| Density | kilogram per cubic meter | kg/m³ | |
| Power | watt | W | $p$ |
| Electric charge | coulomb | C | $q$ |
| Current | ampere | A | $i, I$ |
| Voltage | | | |
|    Potential difference | volt | V | $v, V$ |
|    Electromotive force | volt | emf | $e, E$ |
| Electric field strength | volt per meter | V/m | $E$ |
| Electric resistance | ohm | Ω | $r$ |
| Capacitance | farad | F | $c$ |
| Magnetic flux | weber | Wb | $\phi$ |
| Inductance | henry | H | $L$ |
| Magnetic flux density | tesla (Wb/m²) | T | $B$ |
| Magnetic field strength | ampere per meter | A/m | $H$ |
| Magnetomotive force | ampere | A | $F$ |
| Energy | joule | J | $W$ |
| Force | newton | N | $F$ |

**TABLE 2-2.** MAGNETIC-UNIT CONVERSION

| Multiply | By | To obtain |
|---|---|---|
| Tesla (Wb/m²) ($B$) | $6.45 \times 10^4$ | Maxwells or lines/square inch ($B$ in.) |
| Ampere turns/m ($H$) | $1.257 \times 10^{-2}$ | Oersteds |
| Ampere turns/m ($H$) | $2.54 \times 10^{-2}$ | Ampere turns/inch ($H$ in.) |
| Ampere turns/($F$) | $0.4\pi$ | Gilberts |

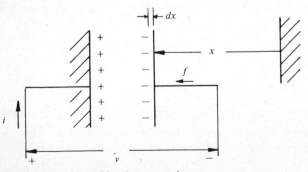

**Figure 2-1** Movable-plate capacitor.

where $C$ is the instantaneous value of the variable capacitance. The energy stored in the electric field is

$$W_\psi = \frac{v^2 C}{2} \qquad (2\text{-}3)$$

$W_\psi$ is the stored energy for the particular value of $C$ at a given instant regardless of the value of $C$ prior to that instant or any subsequent value that it might attain.

Keeping in mind that $C$ is a variable, we find that the differential energy stored in the dielectric, according to Eq. 2-3, is

$$dW_\psi = vC \, dv + \frac{v^2}{2} \, dC \qquad (2\text{-}4)$$

Then from Eqs. 2-1, 2-2, and 2-3,

$$v^2 \, dC + vC \, dv = vC \, dv + \frac{v^2}{2} \, dC + dW_{\text{mech}}$$

from which

$$dW_{\text{mech}} = \frac{v^2}{2} \, dC \qquad (2\text{-}5)$$

The force $f$ displaces the right-hand plate through the differential distance $dx$ and we have

$$dW_{\text{mech}} = f \, dx \qquad (2\text{-}6)$$

From Eqs. 2-5 and 2-6 the force is found to be

$$\boxed{f = \frac{v^2}{2} \frac{dC}{dx}} \qquad (2\text{-}7)$$

Also,

$$q = vC \qquad (2\text{-}8)$$

and when Eq. 2-8 is substituted in Eq. 2-7, the result is

$$\boxed{f = \frac{q^2}{2C^2} \frac{dC}{dx}} \qquad (2\text{-}9)$$

If the voltage is held constant, $dv = 0$ in Eq. 2-4 and the differential mechanical energy output is equal to the differential stored energy. This 50–50 division of the energy input holds only for constant voltage and only for dielectrics in which the relative dielectric constant $k_r$ is constant (i.e., independent of $E$).

The energy relationships associated with constant charge are interesting in that, for one thing, they demonstrate the reversibility of the energy stored in a capacitor. The differential mechanical energy is expressed in terms of charge on the basis of Eq. 2-9 by

$$dW_{mech} = \frac{q^2}{2C^2} \, dC \tag{2-10}$$

and since $v = q/C$, the differential stored energy from Eq. 2-3 by

$$dW_\psi = \frac{q \, dq}{C} - \frac{q^2}{2C^2} \, dC \tag{2-11}$$

If $q$ is held constant as, for example, by disconnecting the capacitor from the electrical source,

$$dW_\psi = - \frac{q^2}{2C^2} \, dC \tag{2-12}$$

Equation 2-12 shows the stored energy to be reduced by an amount equal to the gain in the mechanical energy output. This is to be expected because the electrical energy input is zero, since $dq$ is zero and the mechanical energy must therefore be abstracted from that stored in the electric field of the capacitor. The electrical energy stored in the capacitor is thus converted into mechanical energy.

The process is reversible. If a force is applied to increase the separation of the plates while the charge remains constant, the voltage increases and the energy stored in the capacitor increases accordingly. If, on the other hand, the capacitor is connected to a source of constant voltage while the plates are separated, the charge in the capacitor must decrease; $dq$ is negative and energy is fed back into the electrical supply. One-half of the energy returned to the constant-voltage supply is furnished by the mechanical source and the remaining half is given up by the capacitor. Now we have mechanical energy converted to electrical energy.

Equations 2-7 and 2-9 show that the forces developed by the electric field are in directions such as to increase the capacitance. Hence, in a capacitor, motor action is associated with an increase in capacitance, whereas generator action requires a decrease in capacitance.

If rotational rather than translational forces are developed,

$$dW_{mech} = T \, d\theta \tag{2-13}$$

From Eqs. 2-5 and 2-13 the torque is expressed by

$$T = \frac{v^2}{2} \frac{dC}{d\theta} = \frac{q^2}{2C^2} \frac{dC}{d\theta} \tag{2-14}$$

where $d\theta$ is the differential angular displacement in radians.

The capacitor microphone (known also as the condenser microphone or electrostatic microphone) depends for its operation on changes in capacitance due to the motion of a diaphragm responding to forces produced by sound waves. This, however, is a case in which the amounts of power involved are quite small.

## 2-2 THE TOROID

Figure 2-2 shows a *toroid* in the form of a hollow cylinder with a uniformly distributed winding of $N$ turns carrying a practically constant current of $i$ A. If the number of turns $N$ is large, the magnetic flux lines produced by the current are concentric circles confined to the toroid. This is evident from the direction of the flux lines through the plane of a rectangular loop carrying current as illustrated in Fig. 2-2(b). When Ampère's circuital law is applied to the circular path of radius $x$ and thickness $dx$ in Fig. 2-2(a), the magnetomotive force (mmf) is

$$\mathscr{F} = \oint \mathbf{H} \cdot \mathbf{dl} = Ni \qquad \text{ampere turns} \qquad (2\text{-}15)$$

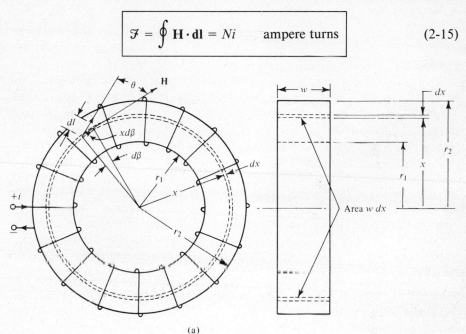

(a)

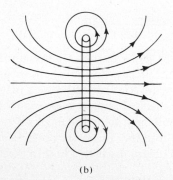

(b)

**Figure 2-2**  (a) Toroid with a current-carrying winding. (b) Magnetic flux lines of a rectangular current loop.

where $\mathbf{H} \cdot \mathbf{dl} = H \cos \theta \, dl$ in Fig. 2-2(a) and

$$\cos \theta \, dl = x \, d\beta \tag{2-16}$$

The quantity $H$ is constant everywhere in the circular path if the material in the toroid has constant permeability, as is the case for free space and most nonferrous materials. Then, on the basis of Eqs. 2-15 and 2-16,

$$\mathcal{F} = H \oint x \, d\beta = Hx \int_{\beta=0}^{\beta=2\pi} d\beta = 2\pi Hx = Ni \tag{2-17}$$

and the magnetic field intensity is

$$H = \frac{Ni}{2\pi x} \qquad \text{ampere turns per meter} \tag{2-18}$$

The flux density in the circular path is

$$B = \mu_0 \frac{Ni}{2\pi x} \qquad \text{teslas} \tag{2-19}$$

where $\mu_0 = 4\pi \times 10^{-7}$ H/m, the permeability of free space. The magnetic flux† through the differential area $da = w \, dx$ in Fig. 2-2(a) is

$$d\phi = \mathbf{B} \cdot \mathbf{da} = B \cos \theta w \, dx \tag{2-20}$$

and since the direction of $\mathbf{B}$ is tangential, $\theta$ must equal zero. Accordingly, we get

$$\phi = \int_{x=r_1}^{x=r_2} d\phi = \frac{\mu_0 Niw}{2\pi} \int_{r_1}^{r_2} \frac{dx}{x} = \frac{\mu_0 Niw}{2\pi} \ln \frac{r_2}{r_1} \tag{2-21}$$

Ferrous materials and certain alloys of metals are known as *magnetic materials*. Their magnetic permeability is usually much greater than that of free space and is commonly expressed as the product $\mu_r \mu_0$ where $\mu_r$ is called the *relative permeability* and $\mu_0$ is the *permeability of free space*. Hence, the flux in a toroid that has a uniform relative permeability of $\mu_r$ is

$$\phi = \frac{\mu_r \mu_0 Niw}{2\pi} \ln \frac{r_2}{r_1} \tag{2-22}$$

The relative permeability of a given magnetic material varies with the magnetic field intensity, a feature that places a limitation on its usefulness for analytical derivation. It is generally more convenient to use a magnetization curve, as shown in Fig. 2-6 for calculations involving ferrous materials.

In many applications it is sufficient to divide the mmf by the mean length of flux path to obtain $H$ and to multiply the corresponding value of $B$ by the area normal to the mean path to obtain the flux. Thus, in the case of the toroid, this approximation yields

† For an elementary discussion of the basic laws applying to slowly varying magnetic fields, see L. W. Matsch, *Capacitors, Magnetic Circuits and Transformers* (Englewood Cliffs, N.J.: Prentice-Hall, Inc., 1964), pp. 89–101.

$$H = \frac{Ni}{\pi(r_2 + r_1)}$$

and since the cross-sectional area $A = (r_2 - r_1)w$,

$$\phi = \frac{\mu_r\mu_0 Ni}{\pi} w \frac{r_2 - r_1}{r_2 + r_1} \qquad (2\text{-}23)$$

## 2-3 SERIES AND PARALLEL MAGNETIC CIRCUITS

Generally the steady or slowly varying flux in magnetic circuits of homogeneous material with a uniform cross-sectional area can be approximately expressed by

$$\phi = Ni \frac{\mu_r\mu_0 A}{l} = \mathscr{F} \frac{\mu_r\mu_0 A}{l} \qquad (2\text{-}24)$$

where $A$ is the cross-sectional area and $l$ the mean length of the flux path. The quantity $\mu_r\mu_0 A/l$ is called the *permeance* and its reciprocal is the *reluctance*. Equation 2-24 can be then abbreviated to

$$\boxed{\phi = \mathscr{F}\mathscr{P} = \frac{\mathscr{F}}{\mathscr{R}}} \qquad (2\text{-}25)$$

where $\mathscr{P}$ and $\mathscr{R}$ are the permeance and the reluctance, respectively.

The counterpart of Eq. 2-24 for the electric circuit is

$$I = V \frac{A}{\rho l}$$

where $V$ is the voltage drop over the length $l$ of a straight conductor that has a uniform cross-sectional area $A$ and a resistivity $\rho$. Equation 2-25 is comparable to $I = V/R$ for the electric circuit.

If $\mathscr{R}_1, \mathscr{R}_2, \ldots, \mathscr{R}_n$ are the reluctances of $n$ components, the total reluctance when these are in series is

$$\mathscr{R}_T = \mathscr{R}_1 + \mathscr{R}_2 + \cdots + \mathscr{R}_n \qquad (2\text{-}26)$$

and when they are in parallel, the total permeance is

$$\mathscr{P}_T = \mathscr{P}_1 + \mathscr{P}_2 + \cdots + \mathscr{P}_n \qquad (2\text{-}27)$$

The use of Eqs. 2-26 and 2-27 is generally restricted to computations pertaining to unsaturated magnetic structures that contain air gaps. In such structures the reluctance of unsaturated iron is usually small enough in relation to that of the air gaps to be negligible. However, when the reluctance of iron components is appreciable, it is generally more convenient to compute the mmfs for each component for a given flux on the basis of the magnetization curve for that component.

Although some of the relationships in magnetic circuits are similar to those in electric circuits, it is generally more difficult to suppress leakage fluxes in magnetic circuits than it is to suppress leakage currents in electric circuits. The leakage flux in magnetic circuits that contain air gaps may be very pronounced, particularly when the leakage path is in parallel with an air gap.

**EXAMPLE 2-1**

An electromagnet with a cylindrical plunger is shown in Fig. 2-3. Neglecting the reluctance of the iron, leakage flux, and fringing of flux at the air gap, determine the flux in the magnet when the 800-turn coil carries a current of 3.5 A.

*Solution.* There are two air gaps in series, the 0.25-cm gap within the coil and a concentric gap of 0.025 cm radial length between the throat and the plunger. The magnetic reluctance of the 0.25-cm air gap is

$$\mathcal{R}_1 = \frac{l_1}{\mu_0 A_1} = \frac{0.25 \times 10^{-2}}{4\pi \times 10^{-7}\pi(3.0 \times 10^{-2}/2)^2}$$

$$= 28.1 \times 10^5$$

and that of the 0.025-cm air gap between the plunger and the throat of the magnet is

$$\mathcal{R}_2 = \frac{l_2}{\mu_0 A_2} = \frac{0.025 \times 10^{-2}}{4\pi \times 10^{-7}\pi(3.0 + 0.025)(1) \times 10^{-4}}$$

$$= 2.1 \times 10^5$$

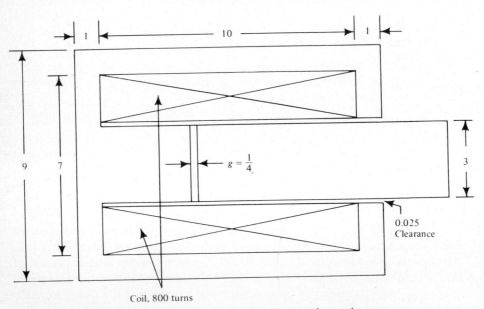

**Figure 2-3**   Plunger-type electromagnet. Dimensions are in centimeters.

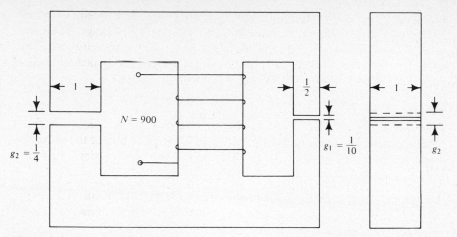

**Figure 2-4**   Electromagnet with two air gaps in parallel. Dimensions are in centimeters.

The total reluctance

$$\mathscr{R}_T = \mathscr{R}_1 + \mathscr{R}_2 = (28.1 + 2.1) \times 10^5 = 3.02 \times 10^5 \qquad \frac{1}{\text{henries}}$$

and the flux is

$$\phi = \frac{\mathscr{F}}{\mathscr{R}_T} = \frac{Ni}{\mathscr{R}_T} = \frac{800 \times 3.5}{3.02 \times 10^6}$$

$$= 9.3 \times 10^{-4} \text{ Wb}$$

The value of flux is somewhat high because the reluctance of the iron is neglected. This is offset at least in part by neglecting the effect of fringing† at the 0.25-cm air gap. (See Sec. 2-5.)

**EXAMPLE 2-2**

Figure 2-4 shows an electromagnet with two air gaps in parallel. Neglect leakage‡ and the reluctance of the iron but correct for fringing by adding the length of the air gap to each of the other two dimensions and determine the flux in each air gap and the total flux when the current in the 900-turn winding is 0.2 A.

*Solution.*   Since there are two air gaps in parallel, their combined permeance is the sum of their permeances. The permeance of the 0.10-cm air gap $g_1$ is

___

† Fringing—the phenomenon of flux spreading to a larger cross section when it leaves a high-permeability material entering a relatively low-permeability material such as air in an air gap (see Sec. 2-5).

‡ Leakage—the portion of flux that does not link the two magnetic circuits being coupled is called leakage flux or simply leakage (see Sec. 2-5).

$$\mathcal{P}_1 = \frac{\mu_0 A_1}{l_1} = \frac{4\pi \times 10^{-7}(0.5 + 0.1)(1.0 + 0.1) \times 10^{-2}}{0.1}$$

$$= 8.31 \times 10^{-8} \text{ H}$$

and the flux in the 0.1-cm air gap is

$$\phi_1 = \mathcal{F}\mathcal{P}_1 = Ni\mathcal{P}_1 = 900 \times 0.2 \times 8.31 \times 10^{-8}$$
$$= 1.50 \times 10^{-5} \text{ Wb}$$

The permeance of the 0.25-cm air gap $g_2$ is

$$\mathcal{P}_2 = \frac{\mu_0 A_2}{l_2} = \frac{4\pi \times 10^{-7}(1.0 + 0.25)(1.0 + 0.25) \times 10^{-2}}{0.25}$$

$$= 7.85 \times 10^{-8} \text{ H}$$

and the flux in the 0.25-cm air gap is

$$\phi_2 = \mathcal{F}\mathcal{P}_2 = 900 \times 0.2 \times 7.85 \times 10^{-8} = 1.41 \times 10^{-5} \text{ Wb}$$

The total flux, which is that in the middle leg, is

$$\phi_T = \phi_1 + \phi_2 = (1.50 + 1.41) \times 10^{-5} = 2.91 \times 10^{-5} \text{ Wb}$$

The total flux can also be found from the total permeance and the mmf as follows:

$$\mathcal{P}_T = \mathcal{P}_1 + \mathcal{P}_2 = (8.31 + 7.85) \times 10^{-8} = 16.16 \times 10^{-8}$$
$$\phi_T = 900 \times 0.2 \times 16.16 \times 10^{-2} = 2.91 \times 10^{-5} \text{ Wb}$$

The leakage flux in the magnetic circuit of Fig. 2-4 is appreciable, particularly that through the air path in parallel with the 0.25-cm air gap $g_2$. The flux that links the exciting winding is therefore appreciably greater than the calculated value if the iron is unsaturated.

## 2-4 MAGNETIC MATERIALS

Magnetic materials include certain forms of iron and its alloys in combination with cobalt, nickel, aluminum, and tungsten. These are known as *ferromagnetic materials* and are easy to magnetize since they have a high value of relative permeability $\mu_r$.

Materials that have a relative magnetic permeability $\mu_r$ not appreciably greater than unity are considered nonmagnetic. In the atoms of *nonmagnetic materials*, the magnetic effect of electron angular momentum or electron spin in one direction is completely offset by equal electron angular momentum in the opposite direction. However, in the case of ferromagnetic materials, compensation of electron angular momentum is not complete and tiny, completely magnetized regions called *domains* exist in the crystals of such materials. The application of low values of magnetic field intensities causes the domains to undergo a boundary displacement.† An increase in the magnetic field intensity

† R. M. Bozorth, *Ferromagnetism* (Princeton, N.J.: D. Van Nostrand Company, Inc., 1951).

produces a sudden orientation of the domains toward the direction of the applied field. Further increases result in a slower orientation of the domains, and the material is said to become *saturated*. Figure 2-5 shows these three regions.

Above saturation the *intrinsic flux* density remains constant. The intrinsic flux density $B_{int}$ is $B$ minus $\mu_0 H$ and is a measure of the flux density contributed by the magnetic material itself. For example, in Fig. 2-6, $B \approx 2.08$ for $H = 10^5$, for which $B_{int} = B - \mu_0 H \approx 2.08 - 4\pi \times 10^{-7} \times 10^5 \approx 1.95$; and for $B = 2.01$, $H = 5 \times 10^4$ and $B_{int} \approx 2.01 - 4\pi \times 10^{-7} \times 5 \times 10^4 \approx 1.95$. The difference between $B$ and $B_{int}$ is negligible in the operating range of conventional iron-cored equipment.

The magnetic permeability of ferromagnetic materials varies with magnetic field intensity, starting at a relatively low value and increasing to a maximum, then falling off with increasing saturation. A typical magnetization curve is shown in Fig. 2-6. The permeability is also generally different for increasing flux density than for decreasing flux density at the same value of magnetic field intensity, as exhibited by the hysteresis loop in Fig. 2-7, when magnetization is carried through a complete cycle.

Magnetic materials form an indispensable part of numerous electromagnetic devices, such as electric generators, power transformers, telephone receivers, relays, loudspeakers, and magnetic recorders.

Magnetic cores that operate at low frequencies (in the audio-frequency range) are usually comprised of silicon steel laminations. However, ferrite cores molded from a mixture of metallic oxide powders are used in many high-frequency operations.

## 2-5  IRON AND AIR

Magnetic circuits are constructed in a variety of shapes. The magnetic cores that carry ac fluxes and those in dc devices requiring rapid rates of response are

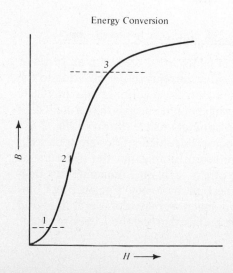

Energy Conversion

**Figure 2-5**  Magnetization curve showing three regions of domain behavior.

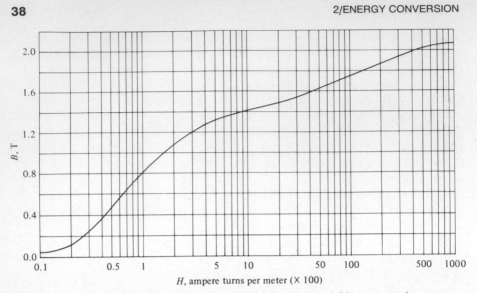

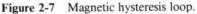

**Figure 2-6**   DC magnetization curve for M-19 fully processed 29-gauge steel.

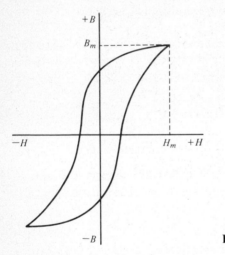

**Figure 2-7**   Magnetic hysteresis loop.

assembled from laminations the surfaces of which are coated with an oxide or with an insulating varnish to prevent excessive currents in the iron due to the time-varying flux. Smaller devices that operate on direct current frequently have laminated cores because the laminated construction is cheaper than a solid core of the same general configuration. Typical laminations are shown in Fig. 2-8.

Magnetic calculations for the more complex configurations are somewhat involved because the flux density is not uniform throughout the structure, as some parts may be highly saturated while others carry only moderate flux densities. Figure 2-9 shows small magnetic cores built by Arnold Engineering Company. A single computer or data-processing machine may use several thousand of the smallest of these cores. Such cores are also used in high-frequency magnetic amplifiers where high gain is needed.

**Figure 2-8**    Shapes of steel laminations. (Courtesy of U.S. Steel Corporation.)

## 2-5.1 Magnetic Leakage and Fringing

Magnetic circuits in which an unsaturated ferromagnetic core or iron core is excited by only one winding generally have negligible leakage flux if the iron is not interrupted by an air gap. The toroid in Fig. 2-2 is an example of a magnetic circuit with small leakage. However, depending upon the location of the winding, high values of leakage flux may be present in magnetic circuits in which the iron contains an air gap.

Figure 2-10(a) shows a steel core with an air gap of length $g$. If there were no leakage flux, the flux in all parts of the iron and in the air gap would have the same value. The dashed line in Fig. 2-10(a) indicates approximately the mean path of the flux when there is no leakage. Placing the winding on that leg of the core which contains the air gap, as shown in Fig. 2-10(b), results in low leakage if the iron is unsaturated. If the winding is placed on a leg of the core that does not contain the air gap, as in Fig. 2-10(c), the leakage will be appreciable.

**Figure 2-9**    Magnetic cores. (Courtesy of the Arnold Engineering Company.)

Only a small value of mmf is required to maintain the flux from $a$ to $a'$ in a clockwise direction in Fig. 2-10(b), and if the core had infinite permeability, this mmf would be zero, since the entire mmf is then consumed by the air gap. However, the mmf required to maintain the flux along the path from $b$ to $b'$ must be high to overcome the reluctance of the air gap even if the iron had infinite permeability. This mmf produces leakage flux in the air spaces that parallel the air gap, as, for example, the window in the core of Fig. 2-10(c). Although the electromagnets in Fig. 2-10(b) and (c) are identical except for the location of the exciting winding, the former has much lower magnetic leakage than the latter.

Because it is not practical to have the winding cover the air gap of all magnetic structures, such structures may have appreciable leakage flux. Sizable leakage fluxes may exist in magnetic circuits with and without air gaps when the iron is saturated. Precise magnetic calculations of such structures must take leakage into account. Such calculations are seldom straightforward and are therefore not within the scope of this textbook.†

The mmf required by an air gap is much greater than that for an equivalent length of iron and therefore produces appreciable flux through the air near the sides of the air gap. This effect is known as *fringing* and increases the effective

† Methods of taking leakage flux into account are shown in H. C. Roters, *Electromagnetic Devices* (New York: John Wiley & Sons, Inc., 1941).

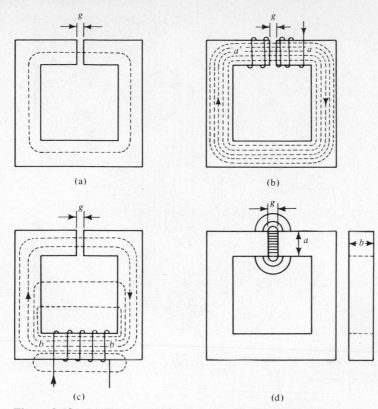

**Figure 2-10** (a) Ferromagnetic core with air gap. (b) Exciting winding surrounding leg with air gap. (c) Exciting winding on leg not containing air gap. (d) Fringing at air gap. [Fringing is not indicated at air gap in (a), (b), or (c) ]

area of the air gap. Fringing is taken into account for short air gaps empirically by adding the length of the air gap to each of its other two dimensions. Thus, the corrected area of the air gap in Fig. 2-10(d) is

$$A = (a + g)(b + g)$$

**EXAMPLE 2-3**

The magnetic circuit in Fig. 2-11 is comprised of a laminated core of steel with an air gap $g = 0.25$ cm and has an exciting winding of 350 turns. The magnetization curve for the iron is shown in Fig. 2-6. the presence of nonmagnetic material between laminations is taken into account by a stacking factor of 0.93 for this core. Neglect leakage but correct for fringing and calculate the current in the exciting winding to produce a flux of 5.0 $\times 10^{-4}$ Wb in the core.

*Solution.* Since magnetic leakage is neglected, there is no parallel flux path and the flux is confined to a path through the iron and air gap in series.

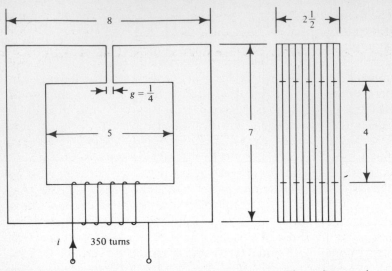

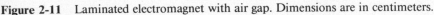

**Figure 2-11**   Laminated electromagnet with air gap. Dimensions are in centimeters.

Net area of core $= (1.5 \times 2.5 \times 0.93) \times 10^{-4} = 3.49 \times 10^{-4}$ m$^2$

Mean length of flux path in iron $= 2(6.5 + 5.5) \times 10^{-2} = 0.24$ m

Flux density in iron, $B_{iron} = \dfrac{\phi}{A_{iron}} = (5.0 \times 10^{-4})/(3.49 \times 10^{-4})$

$$= 1.43 \text{ T}$$

H for iron from Fig. 2-6 $= 1000$ ampere turns/m

Mmf for iron $\mathcal{F}_{iron} = H_{iron}l_{iron} = 1000 \times 0.24 = 240$ ampere turns

$A_{air}$ (corrected area of air gap) $= (a + g)(b + g)$

$$= (1.5 + 0.25)(2.5 + 0.25) \times 10^{-4}$$
$$= 4.81 \times 10^{-4} \text{ m}^2$$

Flux density in the air gap, $B_{air} = \dfrac{\phi}{A_{air}} = \dfrac{5.0 \times 10^{-4}}{4.81 \times 10^{-4}} = 1.04$ T

It should be noted that the stacking factor applies only to laminated iron and not to the air gap:

$$H_{air} = \frac{B_{air}}{\mu_0} = \frac{1.04}{4\pi \times 10^{-7}} = 8.27 \times 10^5 \text{ ampere turns/m}$$

Mmf for air gap $\mathcal{F}_{air} = H_{air}g = 8.27 \times 10^5 \times 0.25 \times 10^{-2}$
$$= 2070 \text{ ampere turns}$$

Total mmf for the iron and air gap in series:

$$\mathcal{F}_t = \mathcal{F}_{iron} + \mathcal{F}_{air} = 240 + 2070 = 2310 \text{ ampere turns}$$

and the current is

$$i = \frac{\mathcal{F}_t}{N} = \frac{2310}{350} = 6.6 \text{ A}$$

It is interesting to note that the air gap in Example 2-3 requires about 8.6 times the mmf required by the iron, although the length of flux path through the iron is 96 times as great as the air-gap length. This shows that air gaps are undesirable in magnetic circuits where low mmfs are required for given values of flux. Air gaps are eliminated from punchings in the form of hollow disks, as, for example, in the toroidal core of Fig. 2-2. This construction is impractical, except for small cores, because of the prohibitive cost of large dies. The use of E and I laminations in electromagnets as shown in Fig. 2-12 makes for an economical construction. The exciting winding is usually placed on the center leg of such three-legged cores. The flux passes through the center leg and divides equally between the outer legs as shown. The effect of air gaps at the joints can be minimized by alternating the positions of the E and I pieces in successive layers. Overlapping L-shaped pieces are used in some cores and overlapping I-shaped pieces in others. In today's automated fabrication the I piece is pressed against the E piece with great compression on the order of tens of tons, and the edges are welded with an electric-arc welder often operated by a robot. This method also reduces the air-gap effect.

## 2-5.2 Graphical Analysis

The calculation of mmf for such simple structures as in Fig. 2-11 is quite simple for a given value of flux. However, it is not quite as straightforward to determine the flux when the mmf is given because of the nonlinear characteristic of the iron. Graphical methods based on the following are used for such calculations.

The total mmf is

$$Ni_t = Ni_{iron} + Ni_{air}$$

from which

$$Ni_{iron} = Ni_t - Ni_{air}$$
$$= Ni_t - \mathcal{R}_{air}\phi \tag{2-28}$$

where $\mathcal{R}_{air}$ is the reluctance of the air gap. When both sides of Eq. 2-28 are plotted as functions of the flux $\phi$, the result is as shown in Fig. 2-13. The intersection of the two plots determines the value of the flux.

**EXAMPLE 2-4**

A laminated core similar in shape to that in Fig. 2-11 contains a 0.12-cm air gap. The characteristic of the iron is shown in Fig. 2-6. The core is stacked to a thickness of $b = 3.0$ cm, the width $a = 2.5$ cm, and the mean

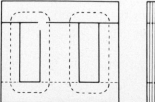

**Figure 2-12**   Core comprised of E and I laminations with overlapping butt joints.

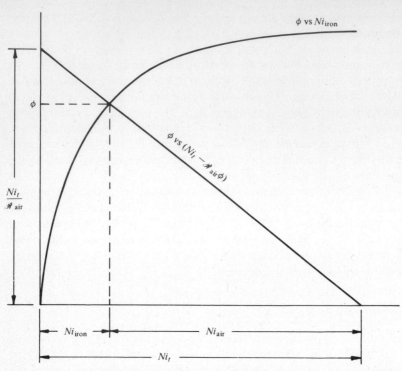

**Figure 2-13**   Graphical method for determining flux in iron core containing air gap.

length of flux path is 30 cm. The winding has 500 turns and carries a current of 3.45 A. Assume a stacking factor of 0.93.

*Solution.*   The flux-versus-mmf characteristic is plotted in Fig. 2-14 based on the magnetization curve in Fig. 2-6, making use of the following: $Ni_{iron} = 30.0 H_{iron}$ and $\phi = 3.0 \times 2.5 \times 0.93 B_{iron}$. The area of the air gap corrected for fringing is

$$A_{air} = (2.50 + 0.12)(3.0 + 0.12) \times 10^{-4} = 8.17 \times 10^{-4} \text{ m}^2$$
$$l_{air} = 0.12 \times 10^{-2} \text{ m}$$

and the reluctance of the air gap is therefore

$$\mathcal{R}_{air} = \frac{l_{air}}{\mu_0 A_{air}} = \frac{0.12 \times 10^{-2}}{4\pi \times 10^{-7} \times 8.17 \times 10^{-4}}$$
$$= 1.17 \times 10^6$$

The intercept on the flux axis is found by dividing the total mmf by $\mathcal{R}_{air}$, thus:

$$\frac{Ni_t}{\mathcal{R}_{air}} = \frac{500 \times 3.45}{1.17 \times 10^6} = 14.7 \times 10^{-4} \text{ Wb}$$

$$\mathcal{F} = Ni_t = 500 \times 3.45 = 1725 \text{ ampere turns}$$

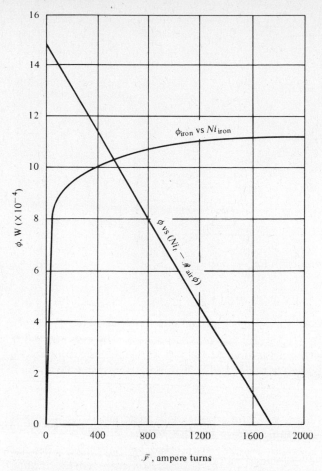

**Figure 2-14** Graphical construction.

The intercept on the mmf axis is the total mmf of 1725 ampere turns. The flux in the magnetic circuit is determined by the intersection of the two characteristics in Fig. 2-14 at a value of $10.3 \times 10^{-4}$ Wb.

The mmf for the iron is 500 ampere turns and for the air gap, 1225 ampere turns. If the reluctance of the iron were neglected, the flux in the magnetic circuit would have a value of $14.7 \times 10^{-4}$ Wb instead of $10.3 \times 10^{-4}$ Wb and would be in error by 43 percent. This rather large error is due to saturation of the iron.

In a given structure the flux may be obtained for other values of mmf by shifting the air-gap line parallel to itself so that it intersects the given values of mmf on the abscissa as shown in Fig. 2-15. To calculate the flux for a given value of mmf and different values of air-gap length, it is necessary only to adjust the slope of the negative air-gap line to accord with the reluctances of the air gap. These procedures apply only to fairly simple structures; complex structures require more involved calculations.

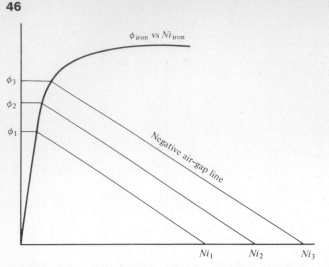

**Figure 2-15**    Graphical construction for various mmfs.

## 2-5.3  Core Losses

Time-varying fluxes produce losses in ferromagnetic materials, known as *core losses,* which consist of hysteresis and eddy-current losses. The cores of transformers and reactors and the armatures of dc and ac machines are therefore subject to core losses. No such losses occur in iron cores that carry flux which does not vary with time.

**Hysteresis Loop**   The nonlinear properties of magnetic materials as characterized by the hysteresis loop were mentioned in Sec. 2-4. Numerous devices, such as self-excited dc generators, magnetic amplifiers, and peaking transformers, depend for their operation upon the nonlinear characteristics of their magnetic circuits. In other cases nonlinearity is undesirable because it may distort the waveforms of voltages and currents in ac circuits. *Hysteresis loss,* which is proportional to the area of the hysteresis loop and the frequency of the flux in hertz, is also a disadvantage in most cases.

If the core in the toroid in Fig. 2-2(a) consists of a magnetic material that is initially completely unmagnetized and is then subjected to a current in the winding, applied in the indicated direction, it will carry flux in the clockwise direction. Let the direction of the current and of the flux both be considered positive. In increasing current through a range of values such that the magnetic field intensity or magnetizing force reaches a value of $+H_{max}$ as shown in Fig. 2-16(a), the flux density in the core reaches a maximum value along the curve *oab.* If the current is decreased to zero, the flux density will have the value *of.* The application of a current in the reverse direction so as to produce the magnetizing force $-H_{max}$ reverses the direction of the flux density, as indicated by *e'b'* in Fig. 2-16(a). Now if the current is again reversed and adjusted to produce the initial value $+H_{max}$ for the magnetizing force, the flux density will be *ec,* which is lower than the initial value *eb.* Cycling the material (i.e., applying alternate equal values of $+H_{max}$ and $-H_{max}$) a number of times produces equal

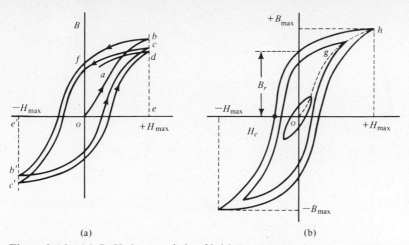

**Figure 2-16**  (a) *B–H* characteristic of initially unmagnetized iron. (b) Hysteresis loops.

values of $+B_{max}$ and $-B_{max}$ and results in a symmetrical hysteresis loop. If the core is worked through a succession of decreasing loops, the positive tips of these loops will be on the *normal magnetization curve* indicated by *ogh* in Fig. 2-16(b).

The flux density $B_r$ at $H = 0$ is the *residual flux density* and the magnetizing force $H_c$ at $B = 0$ is the *coercive force*. For example, in the case of M-19 steel, which has the dc magnetization curve shown in Fig. 2-6, a value for $H_{max}$ of 160 ampere turns/m results in $B_{max} = 1.0$ T, $B_r = 0.64$ T, and $H_c = -29$ ampere turns/m, as determined from the hysteresis loop supplied by the manufacturer but not shown in this text.

**Hysteresis Loss**  If the magnetizing force applied to a magnetic material is carried through a complete cycle from $+H_{max}$ to $-H_{max}$ and back to $+H_{max}$, the *B–H* characteristic is described by a hysteresis loop as shown in Fig. 2-17(a). The area of the loop represents the energy loss in a unit cube of the core material during one cycle.

When $H$ is increased from zero at point 1 in Fig. 2-17(b) to $H_{max}$ at point 2, the energy absorbed by the unit cube is

$$W_1 = \int_{-B_r}^{B_{max}} H \, dB \qquad (2\text{-}29)$$

as represented by the area 1-2-4. A subsequent reduction in $H$ from $H_{max}$ to 0 is accompanied by a reduction in the stored energy as represented by the shaded portion 2-3-4 in Fig. 2-17(c).

$$W_2 = \int_{B_{max}}^{B_r} H \, dB \qquad (2\text{-}30)$$

The energy input is negative, since $H$ is positive but $dB$ is negative. The energy absorbed by the core during the half-cycle from $-B_r$ to $B_r$ is represented by the area 1-2-3 of one-half the hysteresis loop. The same amount of energy is absorbed

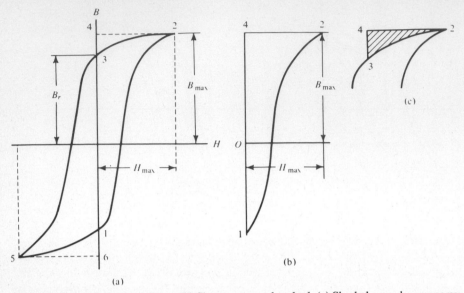

**Figure 2-17** (a) Hysteresis loop. (b) Shows energy absorbed. (c) Shaded area shows energy given up by steel.

during the next half-cycle. The hysteresis loss per cycle in a core of volume Vol that has a uniform flux density $B$ throughout its volume is therefore

$$W_h = \text{Vol} \oint H\, dB \qquad (2\text{-}31)$$

where the line integral represents the area of the loop.

A cyclic variation of the flux at $f$ hertz results in $f$ hysteresis loops per second and the power is

$$P_h = fW_h = \text{Vol}\, f \times \text{area of loop} \qquad (2\text{-}32)$$

The hysteresis loss is expressed empirically using a relationship from Charles P. Steinmetz that $\oint H\, dB = \eta B^n$ so that

$$P_h = \eta\, \text{Vol}\, fB_{\text{max}}^n \qquad (2\text{-}33)$$

The values $\eta$ and $n$ are determined by the nature of the core material. The exponent may vary between 1.5 and 2.5 for different materials and is actually a function of $B_{\text{max}}$ in a given core. However, it is important to note that the hysteresis loss varies directly as the frequency for a given $B_{\text{max}}$.

**Reentrant Loop**  If after the positive value of $H$ has reached a value $H_a$, it is decreased to a value $H_b$ as shown in Fig. 2-18 and then increased to $H_{\text{max}}$, the reentrant loop $a$–$b$ is formed. The area of the reentrant loops represents an additional hysteresis loss over and above that represented by the main loop.

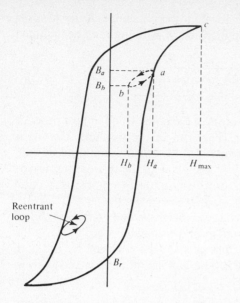

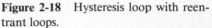

**Figure 2-18** Hysteresis loop with reentrant loops.

**Rotational Hysteresis Loss** The flux in transformers, reactors, and ac electromagnets oscillates along a path that is practically fixed while the rotors of rotary electromagnetic devices such as motors and generators are subjected to fluxes that change their direction by virtue of rotation. The rotational hysteresis loss is greater at low magnetization than the corresponding oscillating hysteresis loss, while at high flux densities the rotating hysteresis loss actually decreases, becoming quite low at very high flux densities.

**Maximum Flux Density under Sinusoidal Excitation** Consider a core such as the one in Fig. 2-10(b), with or without an air gap, linked by a winding of $N$ turns. Assume the flux in the core to be defined by

$$\phi = \phi_m \sin \omega t = AB_m \sin \omega t$$

where $A$ is the cross-sectional area of the core and $B_m$ is the maximum instantaneous flux density assumed to be uniform throughout the core. The induced voltage in the winding is

$$e = \frac{d\lambda}{dt} = \frac{N\,d\phi}{dt} = N\omega AB_m \cos \omega t$$

and the maximum instantaneous value is

$$E_m = N\omega AB_m$$

for which the rms value is

$$E = \frac{E_m}{\sqrt{2}} = \frac{N\omega AB_m}{\sqrt{2}} = 4.44 fNAB_m \qquad (2\text{-}34)$$

Generally in iron-core reactors and transformers the resistance of the winding is small enough so that the voltage $V$ applied to the winding may be assumed to be equal to the induced voltage:

$$V = E = 4.44fNAB_m = 4.44fN\phi_m \qquad (2\text{-}35)$$

**Eddy-Current Loss**   The ac flux induces emfs in the core, which in turn produce eddy currents that circulate in the iron. The iron in the magnetic circuits is laminated to prevent excessive eddy currents. Molded ferrites are used in solid form for some high-frequency applications. The plane of the laminations is parallel to the flux, thus confining the eddy currents to paths of small cross section and correspondingly high resistance.

A piece of lamination having a thickness $\tau$, height $h$, and width $w$ is shown in Fig. 2-19. The thickness $\tau$ is shown disproportionately large. If $B_m$ is the maximum flux density, the maximum flux in the area enclosed by the path 1-2-3-4 is very nearly $2hxB_m$, since $\tau \ll h$. The voltage induced in this path is, from Eq. 2-35,

$$E = 4.44fB_m2hx \qquad (2\text{-}36)$$

since the path corresponds to only one turn (i.e., $N = 1$ in Eq. 2-35).

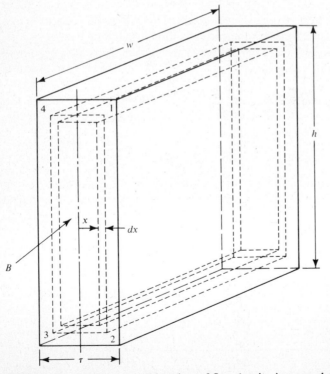

**Figure 2-19**   Lamination. Direction of flux density is normal to the eddy-current path 1-2-3-4-1.

The resistance of the path 1-2-3-4 is

$$R = \frac{\rho l_{\text{path}}}{A_{\text{path}}} = \frac{2h\rho}{w\,dx} \tag{2-37}$$

and the differential current in this path is

$$dI = \frac{E}{R} = \frac{4.44 f B_m wx\,dx}{\rho} \tag{2-38}$$

and the differential power loss is, from Eqs. 2-36 and 2-38,

$$dP_e = E\,dI = \frac{(4.44 f B_m)^2 2whx^2\,dx}{\rho}$$

The eddy-current loss in the entire lamination is

$$P_e = \int_0^{\tau/2} dP_e = \frac{2(4.44 f B_m)^2 wh}{\rho} \int_0^{\tau/2} x^2\,dx$$

$$= \frac{(4.44 f B_m)^2 wh\tau^3}{12\rho} \tag{2-39}$$

In Eq. 2-39 the quantity $4.44 = 2\pi \div \sqrt{2}$ and $wh\tau$ = the volume Vol of the lamination, so

$$\boxed{P_e = \frac{\text{Vol } \pi^2 f^2 \tau^2 B_m^2}{6\rho}} \tag{2-40}$$

The eddy-current loss can also be expressed in terms of the applied voltage and the number of turns in the exciting winding when the resistance of the winding is negligible. The expression is found, upon substitution of Eq. 2-35 in 2-40, to be

$$\boxed{P_e = \text{Vol }\frac{V^2\tau^2}{12\rho N^2 A^2}} \tag{2-41}$$

For a given winding and core Eq. 2-41 can be abbreviated to

$$\boxed{P_e = k_e V^2} \tag{2-42}$$

Equations 2-40, 2-41, and 2-42 are valid only for such values of frequency below which the flux-density distribution is not affected by the eddy currents themselves. When the frequency is increased beyond that value, the flux does not penetrate the laminations completely and thinner laminations should be used. A common thickness of lamination for continuous operation on 60 Hz is 0.036 cm.

Tests on silicon steel show the eddy-current losses to be about 50 percent greater than the values computed from Eqs. 2-40 and 2-41. This difference is due to the large grain size for silicon steel, as in general the eddy-current loss increases with grain size for a given material. However, Eqs. 2-40 and 2-41 are guides as to the relative effects of the various factors that influence eddy-current losses. It should be noted that for a given $B_{max}$ the eddy-current losses are proportional to the square of the frequency as well as to the square of lamination thickness. On the other hand, the hysteresis loss, for a given $B_{max}$, is directly proportional to the frequency and is independent of lamination thickness as long as flux penetration in the laminations is complete.

**High-Frequency Magnetic Materials** Eddy-current losses can be held to practical limits in the audio range from 100 to 10,000 Hz by using a lamination thickness of the order of 25 mm. However, for operation at radio frequencies, thicknesses of the order of 2.5 mm are used. Efficient cores are also manufactured by compressing powdered magnetic material bound together by an adhesive substance that insulates the powdered particles from each other. Other core constructions make use of a mixture of metallic oxides known as *ferrites,* which are molded into the desired size and shape for operation at high frequencies. Their physical properties and manufacturing operation are similar to those of ceramic materials. Such cores have dc resistivities more than $10^6$ times greater than metals, which makes them particularly suitable for high-frequency applications. While the relative permeabilities of these materials are as high as 5000, their use is somewhat limited because they saturate at less than one-half the flux densities at which annealed sheet steel saturates.

## 2-6 FLUX LINKAGE AND EQUIVALENT FLUX

Since magnetic flux lines close upon themselves, they can link electric circuits—as, for example, the winding of the toroid in Fig. 2-2(a) or perhaps a loop of wire as in Fig. 2-2(b). If each turn of an $N$-turn winding is linked by the same value of flux $\phi$, the flux linkage is

$$\lambda = N\phi \qquad \text{weber turns} \qquad (2\text{-}43)$$

The different turns in many windings do not all link the same flux. Some of the windings in electric motors and generators are distributed among slots so that different amounts of flux link the various turns. However, if the fluxes $\phi_1$, $\phi_2$, $\phi_3$, $\ldots$, $\phi_n$ link $N_1$, $N_2$, $N_3$, $\ldots$, $N_n$ turns, respectively, of a winding in which $N_1$, $N_2$, $N_3$, $\ldots$, $N_n$ may each be one or more turns, then from Eq. 2-43 we have the flux linkages

$$\lambda_1 = N_1\phi_1, \lambda_2 = N_2\phi_2, \lambda_3 = N_3\phi_3, \ldots, \lambda_n = N_n\phi_n$$

and the total flux linkage will be

$$\lambda = \lambda_1 + \lambda_2 + \lambda_3 + \cdots + \lambda_n$$

or

$$\lambda = \sum_{i=1}^{i=n} N_i\phi_i \qquad (2\text{-}44)$$

The *equivalent flux* is that value of flux required to link all the turns to produce a given flux linkage:

$$\phi = \frac{\lambda}{N_1 + N_2 + N_3 + \cdots + N_n}$$

## 2-6.1 Energy Stored in Magnetic Circuits

A change in the magnetic flux linking an electric circuit induces an emf in that circuit expressed by Lenz's law as

$$e = \frac{d\lambda}{dt} \qquad (2\text{-}45)$$

The polarity of the induced voltage is such that it would produce a current, if the circuit were closed, in a direction as to oppose any *change* in the flux. For example, an increasing current in the toroid of Fig. 2-2(a) produces an increasing flux and the induced emf if acting by itself would produce a current flowing in a direction opposite that of the increasing current. Thus, for the direction of the current shown in Fig. 2-2(a), an increase in flux would make the upper terminal positive as marked.

To pass a current through a winding that links a magnetic flux, it is necessary to apply the voltage

$$v = Ri + \frac{d\lambda}{dt} \qquad (2\text{-}46)$$

where $R$ is the resistance of the winding and $\lambda$ is the flux linkage of the winding. Currents in other circuits coupled magnetically with this winding, or even permanent magnets may contribute to $\lambda$.

If the winding has $N$ turns and is linked by the equivalent flux $\phi$, Eq. 2-46 may be written as

$$v = Ri + N\frac{d\phi}{dt} \qquad (2\text{-}47)$$

The electric power input to the circuit is

$$p = vi = Ri^2 + Ni\frac{d\phi}{dt}$$

and the differential energy input during the differential time $dt$ is

$$dW = p\,dt = Ri^2\,dt + Ni\,d\phi \qquad (2\text{-}48)$$

where $Ri^2\,dt$ is the irreversible energy converted into heat and $Ni\,d\phi$ the reversible energy stored in the field.

If the winding links a magnetic circuit of length $l$ and of uniform cross-sectional area $A$ and if the flux density is uniform, we have

$$\phi = AB \qquad (2\text{-}49)$$

Further, if the flux is due entirely to the current $i$ in the winding, then

$$H = \frac{Ni}{l} \tag{2-50}$$

The energy stored in the magnetic field is, from Eqs. 2-48, 2-49, and 2-50,

$$dW_\phi = Ni\, d\phi = HlA\, dB$$

But $lA$ is the volume, Vol, of the magnetic circuit, so

$$dW_\phi = \text{Vol}\, H\, dB \tag{2-51}$$

A change in flux density from a value of $B_1$ to $B_2$ requires an energy input to the field occupying a constant volume of

$$W_\phi = \text{Vol} \int_{B_1}^{B_2} H\, dB \tag{2-52}$$

If the relative permeability $\mu_r$ is constant in the range from $B_1$ to $B_2$, then

$$W_\phi = \frac{\text{Vol}}{\mu_r\mu_0} \int_{B_1}^{B_2} B\, dB = \frac{\text{Vol}}{\mu_r\mu_0} \frac{B_2^2 - B_1^2}{2} \tag{2-53}$$

An increase in the flux density from zero to $B$ stores an amount of energy expressed by

$$\boxed{W_\phi = \frac{\text{Vol}\, B^2}{\mu_r\mu_0}\frac{}{2}} \tag{2-54}$$

which may also be written as

$$W_\phi = \text{Vol}\, \mu_r\mu_0 \frac{H^2}{2} \tag{2-55}$$

or

$$\boxed{W_\phi = \text{Vol}\, \frac{BH}{2}} \tag{2-56}$$

For nonlinear magnetic circuits the stored energy is expressed by

$$\boxed{W_\phi = \text{Vol} \int_0^B H\, dB} \tag{2-57}$$

Equation 2-54 shows that the greater the value of $\mu_r$, the less energy is stored in the field for a given value of $B$. The energy stored in an air gap may be several times that stored in a much greater volume of iron. The value of $\mu_r$ in Example 2-3 is

$$\frac{B_\text{iron}}{\mu_0 H_\text{iron}} = \frac{1.43}{4\pi \times 10^{-7} \times 1000} = 1130$$

The maximum value of $\mu_r$ for certain transformer steels may exceed 9000, which is attained at a value of $B$ (about 0.5 T) that is too low for power transformers, since either a core or a winding of excessive size would be required.

## 2-6.2 Self-Inductance

An electric circuit in which the current links magnetic flux is said to have *inductance*. If the medium in the flux path has a linear magnetic characteristic, the inductance is defined as *flux linkage per ampere.* The parameter associated with the flux linkage produced by the current in the circuit itself is called the *self-inductance,* and that associated with the flux produced by the current in another circuit is known as the *mutual inductance.* The coefficients of self-inductance and mutual inductance simplify the expressions for the relationships between the current and voltage in magnetic circuits by relating the flux linkage to the current rather than to the flux as in Eqs. 2-46 and 2-47.

If the flux linkage $\lambda$ in a linear magnetic circuit as expressed in Eq. 2-46 is due only to its own current $i$, then we have

$$v = Ri + \frac{d}{dt}(Li) \qquad (2\text{-}58)$$

where $L$ is the self-inductance in henries. Comparison of Eq. 2-58 and Eq. 2-46 shows the emf of self-induction to be

$$e = \frac{d}{dt}(Li) = \frac{d\lambda}{dt} \qquad (2\text{-}59)$$

for $L$ independent of $i$ and $t$ and consequently,

$$\boxed{L = \frac{\lambda}{i}} \qquad (2\text{-}60)$$

The energy stored in the magnetic field can be expressed in terms of the self-inductance and the current as shown below. The differential energy input to the magnetic field during the differential time $dt$ is, on the basis of Eq. 2-59,

$$dW_\phi = ei\,dt = Li\,di = i\,d\lambda$$

and, if the initial values of $i$ and $\lambda$ are both zero, the energy input to the field during the time interval $t$ is

$$W_\phi = \int_0^t ei\,dt = L\int_0^i i\,di = \frac{Li^2}{2} \qquad (2\text{-}61)$$

In a linear magnetic circuit, then,

$$\boxed{W_\phi = \frac{Li^2}{2} = \frac{\lambda i}{2}} \qquad (2\text{-}62)$$

If the electric circuit has $N$ turns linking the equivalent flux, $\phi = \lambda/N$, Eq. 2-60 can be rewritten as

$$L = \frac{N\phi}{i} \tag{2-63}$$

and when Eq. 2-25 is substituted,

$$L = N^2\mathcal{P} = \frac{N^2}{\mathcal{R}} \tag{2-64}$$

$\mathcal{P}$ is the magnetic permeance and $\mathcal{R}$ the magnetic reluctance of the flux path. Thus, a flux path of uniform area $A$ and length $l$, in which the flux density is uniform and normal to $A$ throughout the length $l$, would have a permeance *as used in Eq. 2-64* of $\mathcal{P} = \mu_r \mu_0 A/l$, where $\mu_0 = 4\pi \times 10^{-7}$.

### 2-6.3 Mutual Inductance

The mutual inductance between two circuits may be defined as the flux linkage produced in one circuit by a current of 1 A in the other circuit. Figure 2-20(a) shows a schematic diagram of two inductively coupled circuits. A current $i_1$ in circuit 1 produces magnetic flux, part of which links turns of circuit 2.

Let $\lambda_{11}$ = flux linkage of circuit 1 produced by its own current $i_1$

$\lambda_{21}$ = flux linkage of circuit 2 produced by the current $i_1$ in circuit 1

$\lambda_{12}$ = flux linkage of circuit 1 produced by the current $i_2$ in circuit 2

$\lambda_{22}$ = flux linkage of circuit 2 produced by its own current $i_2$

Then, since inductance is defined as the flux linkage per ampere, we have

$L_{11} = \dfrac{\lambda_{11}}{i_1}$, self-inductance of circuit 1

$L_{22} = \dfrac{\lambda_{22}}{i_2}$, self-inductance of circuit 2

$L_{12} = \dfrac{\lambda_{12}}{i_2}$, mutual inductance, based on the flux linkage with circuit 1 for unit current in circuit 2

$L_{21} = \dfrac{\lambda_{21}}{i_1}$, mutual inductance, based on the flux linkage with circuit 2 for unit current in circuit 1

The equivalent magnetic circuit in Fig. 2-20(b) represents the coupled circuits shown in Fig. 2-20(a) and where $N_1$ and $N_2$ are the numbers of turns in circuits 1 and 2, respectively. The shaded portions represent an ideal magnetic material (i.e., one that has infinite permeability and infinite resistivity). The three air gaps have reluctances of $\mathcal{R}_1$, $\mathcal{R}_2$, and $\mathcal{R}_l$ as shown.

If a current of $i_1$ flows in circuit 1 while the current in circuit 2 is zero, we have the equivalent fluxes

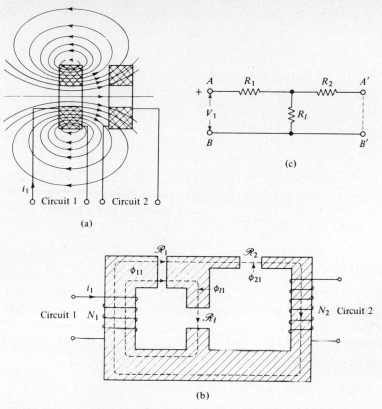

(a)

(c)

(b)

**Figure 2-20**  (a) Two circuits with mutual inductance. (b) Equivalent magnetic circuit. (c) Analogous electric circuit.

$$\phi_{11} \equiv \frac{\lambda_{11}}{N_1} \qquad \phi_{21} \equiv \frac{\lambda_{21}}{N_2} \tag{2-65}$$

and the equivalent leakage flux of circuit 1 does not link any turns in circuit 2:

$$\phi_{l_1} = \phi_{11} - \phi_{21} \tag{2-66}$$

Analysis of the magnetic circuit in Fig. 2-20(b) may be facilitated by comparing it with the electric circuit of Fig. 2-20(c), which is the equivalent T circuit for any three-terminal arrangement of pure resistances. Then, on the basis of the analogy between voltage and mmf, current and flux, and resistance and reluctance, we may write

$$\phi_{21} = \frac{\mathscr{R}_l}{\mathscr{R}_1 \mathscr{R}_2 + \mathscr{R}_1 \mathscr{R}_l + \mathscr{R}_2 \mathscr{R}_l} N_1 i_1 \tag{2-67}$$

and

$$\lambda_{21} = \frac{\mathscr{R}_l N_1 N_2}{\mathscr{R}_1 \mathscr{R}_2 + \mathscr{R}_1 \mathscr{R}_l + \mathscr{R}_2 \mathscr{R}_l} i_1$$

and

$$L_{21} = \frac{\lambda_{21}}{i_1} = \frac{\mathscr{R}_l N_1 N_2}{\mathscr{R}_1 \mathscr{R}_2 + \mathscr{R}_1 \mathscr{R}_l + \mathscr{R}_2 \mathscr{R}_l} \tag{2-68}$$

from an interchange of subscripts 1 and 2 in Eq. 2-68 we find that

$$\boxed{L_{12} = L_{21}}$$

showing the mutual inductance between two electric circuits coupled by a homogeneous medium of constant permeability $\mu_r$ to be reciprocal.

When there are only two inductively coupled circuits, the letter $M$ is frequently used to represent mutual inductance:

$$M = L_{12} = L_{21}$$

The self-inductances of the two circuits can be expressed in terms of the reluctances by

$$L_{11} = \frac{N_1 \phi_{11}}{i_1} = N_1^2 P_1 = \frac{N_1^2 (\mathcal{R}_2 + \mathcal{R}_l)}{\mathcal{R}_1 \mathcal{R}_2 + \mathcal{R}_1 \mathcal{R}_l + \mathcal{R}_2 \mathcal{R}_l} \tag{2-69}$$

and

$$L_{22} = \frac{N_2 \phi_{22}}{i_2} = N_2^2 P_2 = \frac{N_2^2 (\mathcal{R}_1 + \mathcal{R}_l)}{\mathcal{R}_1 \mathcal{R}_2 + \mathcal{R}_1 \mathcal{R}_l + \mathcal{R}_2 \mathcal{R}_l} \tag{2-70}$$

**Coefficient of Coupling**   If it were possible to arrange coupled circuits so that there were no leakage flux (i.e., $\mathcal{R}_l \rightarrow \infty$), the coupling would be perfect. However, some leakage flux is present in practical circuits. Let

$$k_1 = \frac{\phi_{21}}{\phi_{11}} = \frac{\mathcal{R}_l}{\mathcal{R}_2 + \mathcal{R}_l} \tag{2-71}$$

$$k_2 = \frac{\phi_{12}}{\phi_{22}} = \frac{\mathcal{R}_l}{\mathcal{R}_1 + \mathcal{R}_l} \tag{2-72}$$

Then the coefficient of coupling is defined by

$$k \equiv \sqrt{k_1 k_2} = \frac{\mathcal{R}_l}{\sqrt{(\mathcal{R}_1 + \mathcal{R}_l)(\mathcal{R}_2 + \mathcal{R}_l)}} \tag{2-73}$$

Since the reluctances must all be positive, it is apparent from Eq. 2-73 that the coefficient of coupling cannot exceed unity. However, values as high as 0.998 are not unusual in iron-core transformers, whereas $k$ in air-core transformers is generally less than 0.5. Comparison of Eqs. 2-68, 2-69, 2-70, and 2-73 shows that

$$\boxed{M = L_{12} = L_{21} = k\sqrt{L_{11}L_{22}}} \tag{2-74}$$

**EXAMPLE 2-5**

The rotary electromagnet† in Fig. 2-21 has windings of $N_1 = 400$ turns and $N_2 = 200$ turns on the stator and rotor, respectively. The resistance of

---

† Based on L. W. Matsch and W. E. Ott, "An Educational Generalized Electromagnetic Energy Converter." *IEEE Trans. Education* E-13 (November, 1970): 205–210.

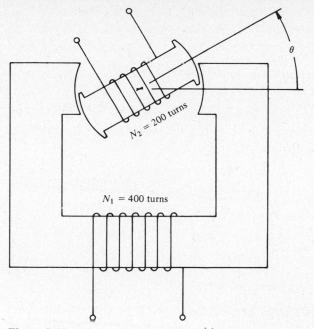

**Figure 2-21** Rotary electromagnet with two inductively coupled windings.

the windings is negligible. Tests were made at a frequency of 60 Hz with the rotor stationary and with its magnetic axis displaced from that of the stator at an angle of 25°. The test data follows.

Excitation applied to the stator (400-turn) winding, rotor open-circuited:

| Stator | | Rotor |
|---|---|---|
| Volts | Amperes | Volts |
| 80.0 | 0.93 | 26.0 |

Excitation applied to the rotor (200-turn) winding, stator open-circuited:

| Rotor | | Stator |
|---|---|---|
| Volts | Amperes | Volts |
| 40.0 | 2.50 | 70.0 |

Calculate (a) the self-inductance $L_{ss}$ of the stator winding, (b) the self-inductance $L_{rr}$ of the rotor winding, (c) the mutual inductance $M$ between the stator and rotor windings, and (d) the coefficients $k_1$ and $k_2$ for the stator and rotor windings and the coefficient of coupling $k$.

*Solution*

a. The self-inductance of the stator may be calculated from Eq. 2-60 as follows:

$$L_{ss} = \frac{\lambda_{ss}}{i_s}$$

The maximum value of $\lambda_{ss}$ can be determined from Eq. 2-35 in which the maximum instantaneous flux is $\phi_m = AB_m$ and the maximum instantaneous flux linkage $N\phi_m$, so that

$$\lambda_{ss_m} = \frac{E}{4.44f} = \frac{80}{4.44 \times 60} = 0.300 \text{ weber turn}$$

The maximum instantaneous current is

$$\sqrt{2}I_s = \sqrt{2} \times 0.93 = 1.32 \text{ A}$$

Hence,

$$L_{ss} = 0.300 \div 1.32 = 0.227 \text{ H}$$

b. The self-inductance of the rotor is

$$L_{rr} = \frac{40}{\sqrt{2} \times 2.5 \times 4.44 \times 60} = 0.0425 \text{ H}$$

c. The mutual inductance on the basis of current in the stator winding with the rotor open is

$$M = \frac{26.0}{\sqrt{2} \times 0.93 \times 4.44 \times 60} = 0.0742 \text{ H}$$

and on the basis of rotor current with the stator open

$$M = \frac{70.0}{\sqrt{2} \times 2.50 \times 4.44 \times 60} = 0.0742 \text{ H}$$

The two values of $M$ check.

d. From Eq. 2-71

$$k_1 = \frac{\phi_{21}}{\phi_{11}} \qquad \text{from Eq. 2-65}$$

$$= \frac{N_1}{N_2} \frac{\lambda_{21}}{\lambda_{11}}$$

$$= \frac{N_1}{N_2} \frac{E_2}{E_1}$$

$$= \frac{400}{200} \times \frac{26}{80} = 0.65$$

$$k_2 = \frac{N_2}{N_1} \frac{\lambda_{12}}{\lambda_{22}}$$

$$= \frac{200}{400} \times \frac{70}{40} = 0.875$$

and $\qquad k = \sqrt{k_1 k_2} = \sqrt{0.65 \times 0.875} = 0.754$

As a check,

$$M = k\sqrt{L_{ss}L_{rr}}$$
$$= 0.754\sqrt{0.227 \times 0.0425} = 0.0741 \text{ H}$$

## 2-7 MAGNETIC FORCE

The magnetic flux that crosses an air gap in a *magnetic material* produces a force of attraction between the faces of the air gap. The core in Fig. 2-22(a) contains an air gap of variable length $g$ as determined by the position of the pivoted member. A current of $i$ A flowing in the exciting winding of $N$ turns produces flux in the path approximately as shown in Fig. 2-22(a). Figure 2-22(b) shows an equivalent magnetic circuit in which the total equivalent flux $\phi$ is shown to consist of the two components $\phi_g$, the air-gap flux, and $\phi_l$ the leakage flux. So

$$\phi = \phi_g + \phi_l \qquad (2\text{-}75)$$

If the pivoted member undergoes a displacement, in response to the force of attraction $f$, such that the air gap is shortened by the differential length $dg$, the work done by the force is the mechanical differential *output* of the electromagnet

$$dW_{\text{mech}} = f\,dg \qquad (2\text{-}76)$$

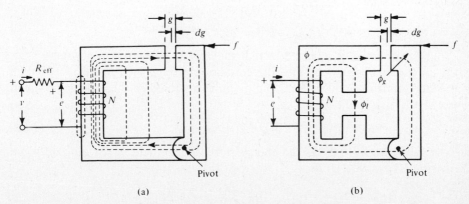

(a)                                    (b)

**Figure 2-22** Electromagnet. (a) Approximate flux path. (b) Simplified flux path.

while the electrical differential energy *input* is

$$dW_e = R_{eff}i^2 \, dt + i \, d\lambda \tag{2-77}$$

where $R_{eff}$ is the effective resistance that includes the ohmic resistance of the winding and takes into account core losses resulting from time variations in the flux.

The electrical differential input can be divided into components as follows:

$dW_e$ = all energy entering electromagnet

$dW_{mech}$ = mechanical energy leaving electromagnet

$dW_\phi$ = gain in reversible energy (i.e., gain in energy stored in the magnetic field)

$R_{eff}i^2 \, dt$ = gain in irreversible energy (producing temperature rise) plus energy leaving electromagnet in the form of heat

and Eq. 2-77 may therefore be rewritten as

$$dW_e = R_{eff}i^2 \, dt + dW_\phi + dW_{mech} \tag{2-78}$$

The quantity of $i \, d\lambda$ in Eq. 2-77 is called the *electromagnetic differential energy*. Then from Eq. 2-78 we get

$$\boxed{dW_{em} = i \, d\lambda = dW_\phi + dW_{mech}} \tag{2-79}$$

The reversible stored differential energy divides between the air-gap field and the leakage field:

$$dW_\phi = dW_{\phi g} + dW_{\phi l}$$

Also,

$$\lambda = N\phi = N\phi_g + N\phi_l$$

and Eq. 2-79 can be rewritten as

$$dW_{em} = Ni \, d\phi_g + Ni \, d\phi_l$$
$$= dW_{\phi g} + dW_{\phi l} + dW_{mech} \tag{2-80}$$

If the current is adjusted so that the air-gap flux remains constant while the air gap is shortened by the differential distance $dg$, then $Ni \, d\phi_g = 0$ and all the electromagnetic differential energy input is stored in the leakage field and we have $dW_{em} = Ni \, d\phi_l = dW_{\phi l}$, so

$$dW_{\phi g} + dW_{mech} = 0 \tag{2-81}$$

Equation 2-54 shows the energy stored in the field of the air gap, when fringing is neglected, to be

$$W_{\phi g} = \text{Vol}_g \frac{B_g^2}{2\mu_0} = A_g g \frac{B_g^2}{2\mu_0}$$

Then when $\phi_g$ is constant, $B_g$ must be constant and

$$dW_{\phi g} = \frac{B_g^2}{2\mu_0} d\text{Vol}_g = -\frac{A_g B_g^2}{2\mu_0} dg \qquad (2\text{-}82)$$

The force is found from Eqs. 2-76, 2-81, and 2-82 to be

$$\boxed{f = \frac{A_g B_g^2}{2\mu_0} \qquad \text{newtons}} \qquad (2\text{-}83)$$

Since the electrical input makes no contribution to the energy in the air gap, because of the constant air-gap flux, the mechanical energy must be extracted from that stored in the air-gap field. This is borne out by Eq. 2-81 and follows from the fact that the air gap gives up energy by virtue of its decreased volume. Consequently, no magnetic force is exerted on a nonmagnetic material (i.e., one that has a $\mu_r$ of unity). The force on the faces of an air gap or pole pieces is only that due to the flux that would exist over and above the flux that would exist if the pole pieces were replaced by air. The actual force is therefore proportional to the difference between the effect of iron and air, so

$$f = \frac{B_g^2 A_g}{2\mu_0} \left(1 - \frac{1}{\mu_r}\right) \qquad (2\text{-}84)$$

The term $1/\mu_r$ need be considered only in cases of high saturation or materials that are slightly magnetic. It is sometimes convenient to express the force in terms of the air-gap flux. Since $\phi_g = B_g A_g$, Eq. 2-83 can be rewritten as

$$\boxed{f = \frac{\phi_g^2}{2\mu_0 A_g}} \qquad (2\text{-}85)$$

### 2-7.1 Force and Torque in Singly Excited Magnetic Circuits

The special case of the magnetic forces on the parallel faces of a short air gap was discussed in Sec. 2-5. However, it is a general principle that the magnetic field produces forces that tend to change the configuration of the flux path such as to increase the self-inductance of the winding when there is only one excited winding. A variation in the self-inductance contributes to the emf of self-inductance, and Eq. 2-59 must be modified as follows:

$$\boxed{e = \frac{d\lambda}{dt} = L\frac{di}{dt} + i\frac{dL}{dt}} \qquad (2\text{-}86)$$

and the electrical power input to such a circuit is therefore

$$p_e = vi = Ri^2 + Li\frac{di}{dt} + i^2\frac{dL}{dt} \tag{2-87}$$

with a corresponding differential energy input of

$$dW_e = p_e\,dt = Ri^2\,dt + Li\,di + i^2\,dL \tag{2-88}$$

where $Ri^2\,dt$ is the differential energy dissipated in the winding in the form of heat. In general, part of the remaining energy is stored in the field. Equation 2-62 shows the stored energy to be

$$W_\phi = \frac{Li^2}{2}$$

a relationship that is valid whether the inductance is constant or time-variant *as long as the permeability is constant*. The power required to change the stored energy is

$$p_\phi = \frac{dW_\phi}{dt} = Li\frac{di}{dt} + \frac{i^2}{2}\frac{dL}{dt}$$

and the differential increase in the stored energy is

$$dW_\phi = Li\,di + \frac{i^2}{2}\,dL \tag{2-89}$$

If the change in the inductance is such as to produce differential displacement $dx$ with a developed force $f$, then the differential mechanical energy output is $f\,dx$. The differential electrical energy input must therefore be

$$dW_e = Ri^2\,dt + dW_\phi + f\,dx \tag{2-90}$$

which follows from the law of conservation of energy. Comparison of Eqs. 2-88, 2-89, and 2-90 shows that

$$f\,dx = \frac{i^2}{2}\,dL \tag{2-91}$$

and the developed force is, accordingly,

$$\boxed{f = \frac{i^2}{2}\frac{dL}{dx}} \tag{2-92}$$

Since the force $f$ produced by the magnet is in the direction of the displacement, electrical energy is converted into mechanical energy, and in the case of a singly excited magnetic circuit an increase in the inductance with displacement implies motor action. A decrease in inductance implies generator action as the *externally applied* force is in the direction of displacement.

The torque that results from a change in the configuration of the flux path, such as to produce rotational displacement, is expressed by

$$T = \frac{i^2}{2} \frac{dL}{d\theta} \qquad (2\text{-}93)$$

where $\theta$ is expressed in radians.

Force and torque developed in inductive circuits can also be expressed in terms of the variation of magnetic permeance with displacement by making use of Eq. 2-64 thus:

$$\frac{dL}{dx} = N^2 \frac{d\mathcal{P}}{dx}$$

and

$$f = \frac{(Ni)^2}{2} \frac{d\mathcal{P}}{dx} = \frac{\mathcal{F}^2}{2} \frac{d\mathcal{P}}{dx} \qquad (2\text{-}94)$$

Also,

$$T = \frac{\mathcal{F}^2}{2} \frac{d\mathcal{P}}{d\theta} \qquad (2\text{-}95)$$

It is sometimes convenient to express force and torque in terms of the variation in reluctance with displacement. Since $\mathcal{R} = 1/\mathcal{P}$,

$$f = \frac{i^2}{2} \frac{dL}{dx} = -\frac{(Ni)^2}{2\mathcal{R}^2} \frac{d\mathcal{R}}{dx} = -\frac{\phi^2}{2} \frac{d\mathcal{R}}{dx} \qquad (2\text{-}96)$$

and

$$T = \frac{i^2}{2} \frac{dL}{d\theta} = -\frac{\phi^2}{2} \frac{d\mathcal{R}}{d\theta} \qquad (2\text{-}97)$$

Equations 2-94 and 2-95 are convenient if the mmf is known and Eqs. 2-96 and 2-97 if the flux is known.

**EXAMPLE 2-6**

Each of the two stator poles of the rotary electromagnet in Fig. 2-23 carries a winding of 2000 turns. The magnetic structure is of iron that has negligible reluctance and has the following dimensions:

$R_1$ = 2.00-cm radius of rounded rotor portion
$W$ = 2.50-cm width of rotor-and-stator pole face perpendicular to the page
$g$ = 0.20-cm length of a single air gap
$\theta$ = angle between stator pole tip and adjacent rotor pole tip

a. Neglect leakage and fringing and calculate the self-inductance in terms of $\theta$ assuming a linear relationship between the area of the air gap and $\theta$.

b. Calculate the maximum torque for a current of 0.80 A.

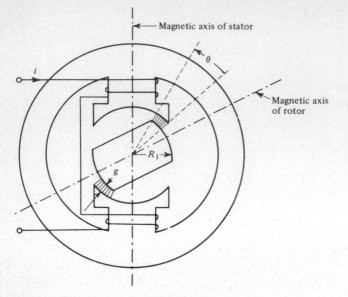

Magnetic axis of stator

$\theta$

Magnetic axis of rotor

$R_1$

$g$

$i$

**Figure 2-23**   Rotary electromagnet.

    c. Express the permeance in terms of $\theta$.
    d. Express the reluctance in terms of $\theta$.
    e. Express the energy stored in the magnetic field in terms of $\theta$ when the current is 0.80 A.

*Solution*

    a.   $L = \dfrac{\mu_0 N^2 A}{l}$     (from Eq. 2-64)

$$A = W\left(R_1 + \frac{g}{2}\right)\theta, \qquad \text{area of air gap, m}^2$$

$l = 2g$, length of double air gap, m

$\mu_0 = 4\pi \times 10^{-7}$

$N = 2000 \times 2 = 4000$ turns

$$L = \frac{4\pi \times 10^{-7}(4000)^2 2.50(2.00 + 0.10) \times 10^{-2}\,\theta}{0.40}$$

    b.   $= 2.64\theta$ H

    c.   $T = \dfrac{i^2}{2}\dfrac{dL}{d\theta} = \dfrac{(0.8)^2}{2} \times 2.64 = 0.844$ N-m/rad

$$\mathcal{P} = \frac{\mu A}{l} = \frac{4\pi \times 10^{-7} \times 2.5 \times 2.10 \times 10^{-2}\,\theta}{0.4}$$

$$= 1.65 \times 10^{-7}\,\theta \text{ weber per ampere turn}$$

d. $\mathcal{R} = \dfrac{1}{\mathcal{P}} = \dfrac{6.05 \times 10^6}{\theta}$ ampere turns per weber

e. $W_\psi = \dfrac{Li^2}{2} = 0.845\theta$ J

## 2-7.2 Force and Torque in Multiply Excited Magnetic Circuits

The electromagnetic energy (i.e., the amount of electrical energy in excess of that dissipated in the resistance of the circuit) supplied to an electric circuit carrying a current $i$ is, from Eq. 2-79,

$$dW_{em} = i\, d\lambda \tag{2-98}$$

The flux linkage in Eq. 2-98 can be due to the current $i$ alone, or it may be produced by currents in other circuits or by its own current in combination with currents in other circuits. When there are two coupled circuits, the differential electromagnetic energy input is

$$\begin{aligned} dW_{em} &= dW_{em1} + dW_{em2} \\ &= i_1\, d\lambda_1 + i_2\, d\lambda_2 \end{aligned} \tag{2-99}$$

where $\quad \lambda_1 = L_{11}i_1 + Mi_2$
$\qquad\quad \lambda_2 = L_{22}i_2 + Mi_1$

and further

$$dW_{em} = i_1\, d(L_{11}i_1 + Mi_2) + i_2\, d(Mi_1 + L_{22}i_2) \tag{2-100}$$

If the medium occupied by the magnetic field has uniform and constant permeability and if there is no change in the configuration of the magnetic circuit, no mechanical energy is involved and all the electromagnetic energy is stored in the field. On that basis, all inductances are constant and we have, from Eq. 2-100,

$$\begin{aligned} dW_\phi = dW_{em} &= L_{11}i_1\, di_1 + M(i_1\, di_2 + i_2\, di_1) + L_{22}i_2\, di_2 \\ &= L_{11}i_1\, di_1 + M\, d(i_1 i_2) + L_{22}i_2\, di_2 \end{aligned} \tag{2-101}$$

and the energy stored in the field is

$$W_\phi = L_{11} \int_0^{i_1} i_1\, di_1 + M \int_{0,0}^{i_1 i_2} d(i_1 i_2) + L_{22} \int_0^{i_2} i_2\, di_2$$

$$\boxed{W_\phi = \tfrac{1}{2}(L_{11}i_1^2 + 2Mi_1 i_2 + L_{22}i_2^2)} \tag{2-102}$$

Actually, Eq. 2-102 is valid whether the inductances are constant or variable as long as the magnetic field is confined to a medium of uniform and constant permeability.

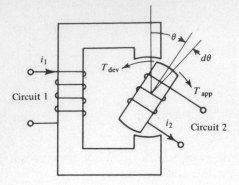

**Figure 2-24**   Rotary electromagnet with variable self- and mutual inductance.

When there are $n$ coupled circuits,[†] the energy stored in the field is expressed by

$$W_\phi = \sum_{i=1}^{n} \sum_{k=1}^{n} \frac{1}{2} L_{jk} i_j i_k \qquad (2\text{-}103)$$

Consider the rotary electromagnet in Fig. 2-24. The stator winding designated as circuit 1 is coupled with the rotor winding designated as circuit 2 for all values of $\theta$ except odd multiples of $\pi/2$. The currents $i_1$ and $i_2$ when flowing in the indicated directions produce magnetic force that produces torque in the counterclockwise direction for $0 < \theta < \pi/2$, and, if the rotor is free to move, electromagnetic energy is converted into mechanical energy. When the flux linkages $\lambda_1$ and $\lambda_2$ in Eq. 2-99 are expressed in terms of the inductances, which are all functions of $\theta$, the differential electromagnetic energy input is expressed by

$$dW_{em} = L_{11} i_1 di_1 + M(i_1 \, di_2 + i_2 \, di_1) + L_{22} i_2 \, di_2$$
$$+ \, i_1^2 \, dL_{11} + 2 i_1 i_2 \, dM + i_2^2 \, dL_{22} \qquad (2\text{-}104)$$

Part of this input is stored in the magnetic field and the remainder is converted into mechanical energy:

$$dW_{em} = dW_\phi + dW_{\text{mech}} \qquad (2\text{-}105)$$

The increase in the stored energy is found by differentiating Eq. 2-102 which results in

$$dW_\phi = L_{11} i_1 \, di_1 + M \, d(i_1 i_2) + L_{22} i_2 \, di_2$$
$$+ \, \tfrac{1}{2} i_1^2 \, dL_{11} + i_1 i_2 \, dM + \tfrac{1}{2} i_2^2 \, dL_{22} \qquad (2\text{-}106)$$

When Eq. 2-106 is subtracted from Eq. 2-104 the difference expresses the differential mechanical output in accordance with Eq. 2-105 as follows:

$$dW_{\text{mech}} = \tfrac{1}{2} i_1^2 \, dL_{11} + i_1 i_2 \, dM + \tfrac{1}{2} i_2^2 \, dL_{22} \qquad (2\text{-}107)$$

and the electromagnetic torque or developed torque by

[†] R. M. Fano, L. J. Chu, and R. B. Adler, *Electromagnetic Fields, Energy, and Forces* (Cambridge, Mass.: MIT Press, 1966), p. 306.

$$T_{em} = \frac{dW_{\text{mech}}}{d\theta} = \frac{1}{2} i_1^2 \frac{dL_{11}}{d\theta} + i_1 i_2 \frac{dM}{d\theta} + \frac{1}{2} i_2^2 \frac{dL_{22}}{d\theta} \qquad (2\text{-}108)$$

By a similar process the electromagnetic force for linear displacement is found to be

$$f_{em} = \frac{1}{2} i_1^2 \frac{dL_{11}}{dx} + i_1 i_2 \frac{dM}{dx} + \frac{1}{2} i_2^2 \frac{dL_{22}}{dx} \qquad (2\text{-}109)$$

In the case of $n$-coupled circuits, the torque on the basis of Eq. 2-103 is found to be

$$T_{em} = \sum_{j=1}^{n} \sum_{k=1}^{n} \frac{1}{2} i_j i_k \frac{dL_{jk}}{d\theta} \qquad (2\text{-}110)$$

Comparison of Eqs. 2-104 and 2-106 shows that, if the currents are held constant ($di = 0$), one-half of the electromagnetic input is stored in the magnetic field, the remainder being converted into mechanical energy. This equal division is true only for magnetic circuits that are linear and does not apply when there is appreciable saturation.

However, it is possible to obtain mechanical energy output even when there is no electromagnetic input (i.e., if the flux linkages are held constant, $d\lambda_1 = d\lambda_2 = 0$). Under this condition the mechanical energy is abstracted from the energy stored in the magnetic field.

### 2-7.3 Force and Energy in Nonlinear Magnetic Circuits

As shown in Sec. 2-5, the presence of magnetic materials causes magnetic circuits to be nonlinear. This effect is accentuated when the material is carried from the unsaturated region into the saturated region even if the magnetic circuit contains sizable air gaps. As a result, the value of inductance depends on whether it is defined on the basis of (a) flux linkage per ampere or $L_a = \lambda/i$, (b) the energy stored in the field or $L_\phi = 2W_\phi/i^2$, or (c) the slope of the tangent to the magnetization curve or $L_d = d\lambda/di$. Each of these definitions leads to the same value of inductance in magnetic circuits with air cores.

In the magnetization curve of Fig. 2-25 the current $oa$ produces the flux linkage $oe$. Three different values of inductance,[†] depending on the definition of inductance, are obtained for this value of current:

a. $L_a = \dfrac{\lambda}{i} = \dfrac{ab}{oa}$ $\qquad\qquad\qquad\qquad$ (2-111)

b. $L_\phi = \dfrac{2W_\phi}{i^2} = \dfrac{ac}{oa}$ $\qquad\qquad\qquad\qquad$ (2-112)

---

[†] L. T. Rader and E. C. Litscher, "Some Aspects of Inductance When Iron Is Present," *Trans. AIEE* 63 (1944): 133–139.

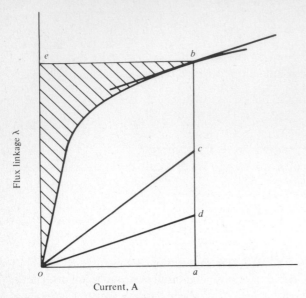

Figure 2-25    Magnetization curve.

c. $L_d = \dfrac{d\lambda}{di} = \dfrac{ad}{oa}$                                                    (2-113)

The shaded area *obe* represents the energy $W_\phi$ stored in the field. Line *od* is parallel to the tangent at *b*. From Eqs. 2-111, 2-112, and 2-113, and from Fig. 2-25 we find that $L_a > L_\phi > L_d$, an inconsistency which limits the usefulness of the concept of inductance when applied to nonlinear magnetic circuits. Nevertheless, useful approximations can be made by a choice of one of the three relationships depending upon the particular application. For example, Eq. 2-111 serves as a basis for approximating the saturated synchronous reactance of synchronous machines, while Eq. 2-112 is useful for evaluating the duty of contacts that interrupt inductive circuits and Eq. 2-113 may be used to determine the effectiveness of chokes and transformers in which small amounts of ac flux are superposed on relatively large amounts of dc flux. However, it usually is better to use the magnetization curve than to use inductance in the solution of nonlinear magnetic circuits.

**Energy Relations in Nonlinear Magnetic Circuits**    The energy stored in the magnetic field is expressed by $W_\phi = \int i \, d\lambda$ and is represented by the shaded area *obe above* the magnetization curve in Fig. 2-25. It is sometimes more convenient in the case of nonlinear systems to make use of a quantity known as the coenergy. In the case of magnetic circuits the coenergy is expressed by $W'_\phi = \int \lambda \, di$ and is represented by the area *oba under* the magnetization curve in Fig. 2-25.

Consider the plunger-type electromagnet illustrated in Fig. 2-26(a) and the magnetization curves *ob* and *od* in Fig. 2-26(b) for plunger displacements of $x$ and $x + \Delta x$. An increase of $\Delta x$ in the displacement corresponds to an equal

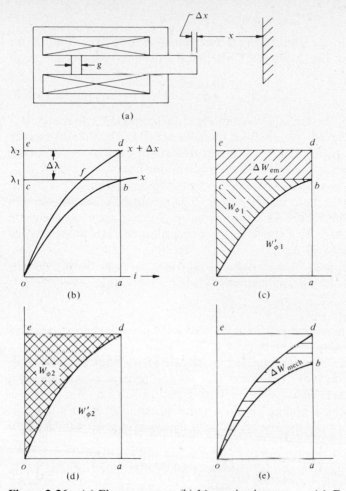

(a)

(b)

(c)

(d)

(e)

**Figure 2-26** (a) Electromagnet. (b) Magnetization curves. (c) Graphical representation of initially stored energy plus electromagnetic energy input, current constant. (d) Graphical representation of finally stored energy and coenergy. (e) Graphical construction showing electromechanical energy.

decrease in the length of the air gap $g$ which is accompanied by an output of mechanical energy. The electromagnetic energy input associated with the plunger displacement is expressed by

$$\Delta W_{em} = \Delta W_\phi + \Delta W_{mech} \qquad (2\text{-}114)$$

where $\Delta W_\phi$ is the increase in the energy stored in the magnetic field and $\Delta W_{mech}$ is the mechanical energy output. Further, if $W_{\phi 1}$ and $W_{\phi 2}$ are the values of stored energy before and after the displacement, then

$$\Delta W_\phi = W_{\phi 2} - W_{\phi 1} \qquad (2\text{-}115)$$

The mechanical energy output can therefore be expressed on the basis of Eqs. 2-114 and 2-115 by

$$\Delta W_{\text{mech}} = \Delta W_{em} + W_{\phi 1} - W_{\phi 2} \tag{2-116}$$

If the current in the winding of the electromagnet is held constant at the value *oa*, as shown in Fig. 2-26(b), (c), (d), and (e), the electromagnetic energy input is

$$\Delta W_{em} = i \, \Delta\lambda \tag{2-117}$$

and is represented by the rectangle *bced* in Fig. 2-26(b) and (c). The energy $W_{\phi 1}$ initially stored in the field is represented by the area *obc* in Fig. 2-26(b) and (c), and the sum $\Delta W_{em} + W_{\phi 1}$ in Eq. 2-116 is represented by the area *obde* in the same figures. The area *ode* in Fig. 2-26(d) represents the final stored energy $W_{\phi 2}$, and when this area is subtracted from the area *obde* in accordance with Eq. 2-116, the difference is the area *obd* in Fig. 2-26(e), which represents the mechanical energy output for the constant current *oa*. We may therefore conclude that, when the current is held constant, the mechanical energy output is equal to the increase in coenergy $\Delta W_{\phi}'$.

If the flux linkage is held constant at the value $\lambda_1 = oc$, during the displacement $\Delta x$, the electromagnetic input $\Delta W_{em}$ is zero and Eq. 2-116 then yields

$$\Delta W_{\text{mech}} = W_{\phi 1} - W_{\phi 2} \tag{2-118}$$

which means that the mechanical energy is abstracted from the energy stored in the magnetic field.

When the same process is applied to the case of constant flux linkage as to that of constant current, it is found that the mechanical energy output is represented by the area *obfo* in Fig. 2-26(b). Hence, when the flux linkage is constant, the mechanical energy output is equal to the decrease in the stored energy $\Delta W_{\phi}$. In both cases the average value of the electromagnetic force on the plunger is

$$f_{\text{av}} = \frac{\Delta W_{\text{mech}}}{\Delta x} \tag{2-119}$$

and for constant current

$$f_{\text{av}} = \frac{\Delta W_{\phi}'}{\Delta x}\bigg|_{i=\text{const.}} \tag{2-120}$$

while for constant flux linkage,

$$f_{\text{av}} = -\frac{\Delta W_{\phi}}{\Delta x}\bigg|_{\lambda=\text{const.}} \tag{2-121}$$

In the case of a rotary displacement $\Delta\theta$ the average torque is

$$T_{\text{av}} = \frac{\Delta W_{\phi}'}{\Delta\theta}\bigg|_{i=\text{const.}} \tag{2-122}$$

and $\qquad\qquad T_{\text{av}} = -\dfrac{\Delta W_{\phi}}{\Delta\theta}\bigg|_{\lambda=\text{const.}} \tag{2-123}$

If $\Delta x$ is made to approach zero, the area *fbd*, which represents the difference between the mechanical energy output for constant current and that for constant flux linkage, also approaches zero, and the electromagnetic force is then expressed by

$$f = \frac{\partial W'_\phi}{\partial x}(i, x) = -\frac{\partial W_\phi}{\partial x}(\lambda, x) \qquad (2\text{-}124)$$

Similarly, the torque is expressed by

$$T = \frac{\partial W'_\phi}{\partial \theta}(i, \theta) = -\frac{\partial W_\phi}{\partial \theta}(\lambda, \theta) \qquad (2\text{-}125)$$

where $W'_\phi$ is expressed explicitly as a function of the current $i$ and the displacement $x$ or $\theta$, while $W_\phi$ is expressed explicitly as a function of the flux linkage $\lambda$ and the displacement $x$ or $\theta$.

In arrangements where the excitation is furnished by more than one winding, say $n$ windings, we have

$$f = \frac{\partial W'_\phi}{\partial x}(i_1, i_2, \ldots, i_n, x) \qquad (2\text{-}126)$$

## 2-8 PERMANENT MAGNETS

Materials suitable for permanent magnets are said to be *magnetically hard.* These are difficult to magnetize but have high residual flux density and high coercive force as compared with the soft magnetic materials that are easy to magnetize and that generally have low residual flux density. A few of the many applications of permanent magnets are: magnetic clutches and couplings, loudspeakers, generators, television focusing units, magnetrons, measuring instruments, information storage in computers, and video recording.

Magnets in the form of a straight bar are usually magnetized by placing them between the poles of a powerful electromagnet. U-shaped or circular magnets may be magnetized individually or in groups by placing a conductor through their centers and applying a current surge of several thousand amperes to the conductor while the air gap of each magnet is bridged with a soft iron bar. The flux in the magnet is reduced somewhat when the bridging piece is removed. Upon closing again, the flux is increased but not to its previous value. Subsequent opening and closing of the air gap produces no appreciable change in the flux.

Present-day permanent-magnet materials are carbon-free alloys† of which the most generally used is the nickel–aluminum–iron group known as *Alnico.*

† For an extensive treatment, see R. J. Parker and R. J. Studders, *Permanent Magnets and Their Applications* (New York: John Wiley & Sons, Inc., 1962).

Older materials that are still used to a small extent are carbon steel, chrome, tungsten magnet, and cobalt magnet steels. Common shapes of Alnico magnets are rods, bars, and U shapes, although there are numerous other shapes as well. However, because these materials are difficult to machine, their shapes are generally made simple, and soft iron components in the more complex shapes are added to the magnetic circuit. Figure 2-27 shows the magnetic circuit of an indicating instrument that makes use of an Alnico magnet with soft iron parts that contain the air gap. A U-shaped magnet is shown in Fig. 2-28. Since the magnet generally produces flux through an air gap, the leakage flux across the air path between the vertical sections is large and the area carrying the greater amount of flux is built up as shown by the dashed lines in Fig. 2-28.

### 2-8.1 Operating Characteristics of Permanent Magnets

Permanent magnets operate on the demagnetization curve of the hysteresis loop. The greater the area ($OB_RPH_C$ in Fig. 2-30) under the demagnetization curve, the more effective is the material of the permanent magnet. A U-shaped magnet with a soft iron bar known as a *keeper* is shown in Fig. 2-29(a) and (b). The

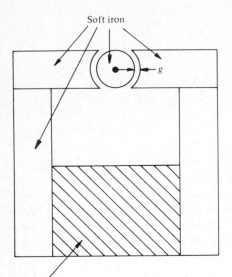

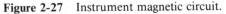

**Figure 2-27**   Instrument magnetic circuit.

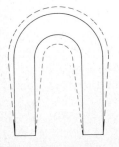

**Figure 2-28** U-shaped magnet. Solid lines indicate size if there were no leakage. Dashed lines show size and shape to maintain constant flux density when there is leakage flux.

exciting winding, which is usually removed after the magnet has been magnetized, has $N$ turns. The length of the permanent magnet (the U-shaped portion only) is $l$.

Suppose that a current is applied to the winding as indicated in Fig. 2-29(a) such as to produce the magnetizing force $H_{max}$ in Fig. 2-30, which shows the demagnetization curve of the material. The flux density is $B_{max}$ if leakage is neglected. Reducing the current to zero reduces the flux density to $B_R$, called the *retentivity*. To reduce the flux density to zero, it is necessary to apply a current in the reverse direction as shown in Fig. 2-29(b) and of such a value as to produce the coerciveness $H_C$ in Fig. 2-30. The region of interest in permanent magnets is represented by the demagnetization curve or the portion $B_R P H_C$.

If, after the application of $H_{max}$, a current is applied in the reverse direction [Fig. 2-29(b)] so as to produce the magnetizing force $H$ in Fig. 2-30, the flux density will have the value $B$. The same value of reduced flux density $B$ can be achieved with zero current in the winding by introducing an air gap $g$ of the proper length as shown in Fig. 2-29(c), assuming the leakage to be negligible.

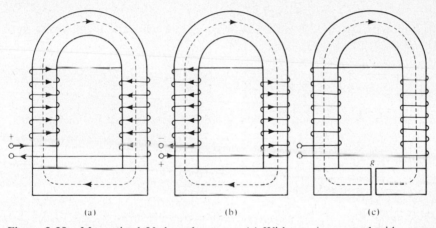

(a)                          (b)                          (c)

**Figure 2-29** Magnetized U-shaped magnet. (a) Without air gap and with current in magnetizing direction. (b) Without air gap and with current in demagnetizing direction. (c) With air gap and without current.

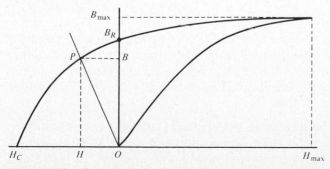

**Figure 2-30** Characteristic of a permanent-magnet material.

The mmf for the air gap is

$$\mathcal{F}_g = H_g g \qquad (2\text{-}127)$$

and the mmf for the permanent magnet is

$$\mathcal{F}_m = H l_m \qquad (2\text{-}128)$$

The total mmf around the closed path must be the sum of these two mmfs when the reluctance of the soft iron pieces on both sides of the air gap is neglected:

$$\mathcal{F}_t = \mathcal{F}_m + \mathcal{F}_g \qquad (2\text{-}129)$$

Since the current in the exciting winding is zero, we have, from Eq. 2-15,

$$\mathcal{F}_t = \oint \mathbf{H} \cdot \mathbf{ds} = 0$$

Therefore,

$$\mathcal{F}_m + \mathcal{F}_g = 0$$

and

$$\mathcal{F}_m = -\mathcal{F}_g \qquad (2\text{-}130)$$

Substitution of Eqs. 2-127 and 2-128 in Eq. 2-130 yields the magnetizing force of the magnet:

$$\boxed{H = -\frac{H_g g}{l_m}} \qquad (2\text{-}131)$$

which can also be expressed in terms of the flux density $B$ of the magnet by

$$\boxed{H = -\frac{B A_m g}{\mu_0 A_g l_m}} \qquad (2\text{-}132)$$

The straight line $OP$ in Fig. 2-30 represents Eq. 2-132 graphically. The intersection of $OP$ with the demagnetization curve determines the values of $B$ and $H$ for the permanent magnet. Since magnetic leakage is assumed to be negligible, the flux must be the same in all parts of the magnetic circuit and

$$\phi = B A_m = B_g A_g \qquad (2\text{-}133)$$

**EXAMPLE 2-7**

Figure 2-31(a) shows the demagnetization curve for the Alnico magnet of Fig. 2-31(b). The length of each of the two air gaps in Fig. 2-31(b) is 0.45 cm. The mean length of the Alnico magnet is 13 cm. Determine the flux in the air gaps; neglect leakage and the reluctance of the soft iron but allow for fringing at the air gaps.

*Solution.* The flux density in the permanent magnet is determined by means of the graphical construction illustrated in Fig. 2-31(a), and Eq. 2-132, where

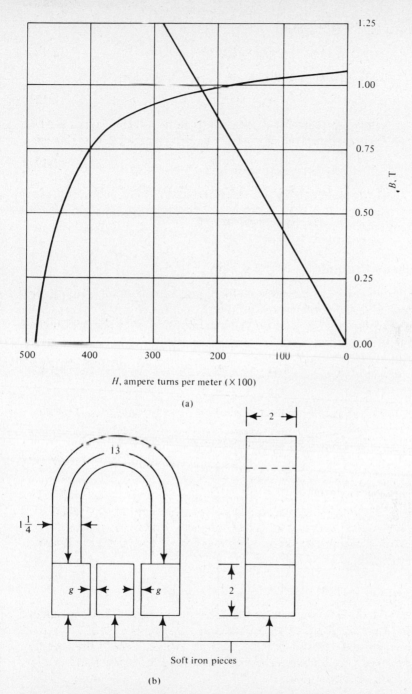

$H$, ampere turns per meter ($\times 100$)

(a)

Soft iron pieces

(b)

**Figure 2-31** (a) Demagnetization curve for Alnico V. (b) Alnico magnet with soft iron pole pieces and air gaps. Dimensions in centimeters.

$$1_m = 13 \times 10^{-2} \text{ m}, A_m = 1.25 \times 2.0 \times 10^{-4} = 2.5 \times 10^{-4} \text{ m}^2$$

$$2_g = 2 \times 0.45 \times 10^{-2} = 0.90 \times 10^{-2} \text{ m}$$

$$A_g = (2.00 + 0.45)(2.0 + 0.45) \times 10^{-4} = 6.0 \times 10^{-4} \text{ m}^2$$

From Eq. 2-132,

$$H = -\frac{BA_m(2g)}{\mu_0 A_g l_m} = -\frac{B \times 2.5 \times 10^{-4} \times 0.90 \times 10^{-2}}{4\pi \times 10^{-7} \times 6.0 \times 10^{-4} \times 13 \times 10^{-2}}$$

$$= -22{,}800B \text{ ampere turns/m}$$

This line plotted in Fig. 2-31(a) intersects the demagnetization curve at approximately $B = 0.97$ T and $H = 22{,}000$ ampere turns/m. The flux is therefore $\phi = BA_m = 0.97 \times 2.5 \times 10^{-4} = 2.42 \times 10^{-4}$ Wb.

## 2-8.2 Energy Product

The energy product for the magnetic material operating at point $P$ in Fig. 2-30 is $BH$. Figure 2-32 shows the energy product curve for Alnico V based on the demagnetization curve in Fig. 2-31(a). The maximum energy product is obtained at about $B = 0.85$ T and is equal to about $3.1 \times 10^4$.

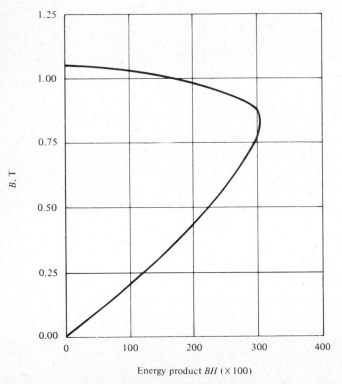

Energy product $BH$ ($\times 100$)

**Figure 2-32**   Energy-product curve for Alnico V of Fig. 2.31(a).

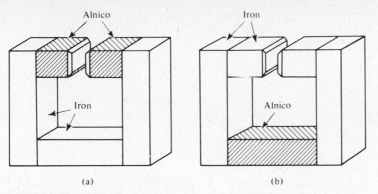

**Figure 2-33** Location of magnet material. (a) Good design, low leakage. (b) Poor design, high leakage.

If the effects of leakage are neglected, the size of the magnet is expressed in terms of the size of air gap by

$$\text{Vol}_m = \text{Vol}_g \frac{B_g H_g}{BH}$$

The volume of the magnet is minimum for a given air-gap volume when the energy product $BH$ is a maximum.

The effect of leakage flux on the volume of a permanent magnet may be taken into account by a correction factor $f_1$, which ranges from a value of 2 to about 20. Another factor, $f_2$, which takes into account the fringing in nonuniform air gaps, lies in a range from about 1.1 to 1.5. The volume of the magnet, when these factors are taken into account, is expressed by

$$\boxed{\text{Vol}_m = \text{Vol}_g f_1 f_2 \frac{B_g H_g}{BH}} \qquad (2\text{-}134)$$

Evaluation of the constants $f_1$ and $f_2$ requires judgment based on experience, as well as an understanding of the manner in which the magnet is to be used.†

Leakage may be minimized by locating the air gap in the permanent magnet material as shown in Fig. 2-33(a) rather than in the soft iron portion as shown in Fig. 2-33(b). This arrangement is comparable with that of placing the exciting winding in an electromagnet so as to cover the air gap as shown in Fig. 2-10(b).

The flux in a permanent magnet may undergo variations due to changes in the length of an air gap and due to external magnetizing forces. Figure 2-34 shows a device that depends for its operation on variation in the reluctance of an air gap. Current in the winding produces opposite poles at the ends of the armature, causing the armature to oscillate at the frequency of the current.

† For more complete treatment of design considerations, see R. J. Parker and R. J. Studders, *Permanent Magnets and Their Applications* (New York: John Wiley & Sons, Inc., 1962). Chaps. 4, 5; *Indiana Permanent Magnet Manual Number 6* (Valparaiso, Ind.: Indiana Steel Products Company); and E. M. Underhill, *Permanent Magnet Handbook* (Pittsburgh, Pa.: Crucible Steel Company of America, 1952), Sec. 2.

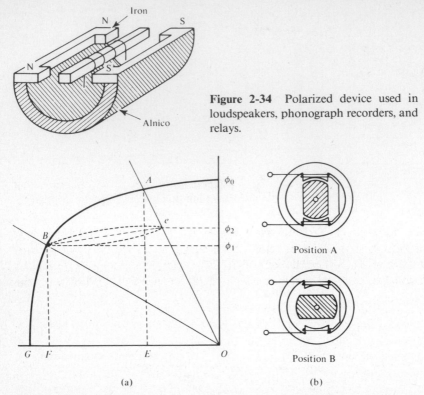

**Figure 2-34** Polarized device used in loudspeakers, phonograph recorders, and relays.

(a)

(b)

**Figure 2-35** (a) Flux variations of a permanent magnet due to changes in air gap. (b) Permanent-magnet generator.

The magnetic circuit and demagnetization curve of a permanent-magnet generator are shown in Fig. 2-35. The reluctance of the air gap is a minimum when the rotor is in position A and a maximum when it is in position B in Fig. 2-35(b). The flux $\phi_0$ in Fig. 2-35(a) is due to $B_R$ and would exist if there were no air gap. The air-gap line $OA$ corresponds to position A and the air-gap line $OB$ corresponds to position B with a flux $\phi_1 = BF$. The locus of operation is the minor hysteresis loop $Be$ between points $B$ and $e$ with flux oscillating between the values of $\phi_1$ and $\phi_2$ twice for each revolution of the rotor. Experience shows the slope of this line to be about equal to that of the tangent to the magnetization curve through $\phi_0$.

Figure 2-36 illustrates the effect of a demagnetizing force on the flux of a permanent magnet. Again $\phi_0$ is the flux due to $B_R$, and the introduction of the air gap produces a decrease in the flux from $\phi_0$ to $\phi_1$ as determined by the intersection of the air-gap line $OA$ with the demagnetization curve. The application of a demagnetizing force that has an mmf $OC$, in effect, causes the air-gap line to shift to the left parallel to itself and intersecting the demagnetization curve at point $B$. The result is a further reduction in the flux to $\phi_2$. Removal of the demagnetizing force does not restore the flux to its previous value, $\phi_1$, but

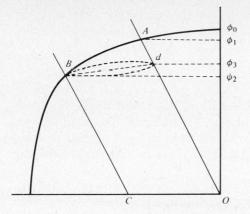
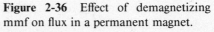

**Figure 2-36** Effect of demagnetizing mmf on flux in a permanent magnet.

to a lower value, $\phi_3$. Point $B$ follows along the lower portion of the minor hysteresis loop to point $d$. Repeated application and removal of the same demagnetizing force result in the relatively small flux variations of $\phi_3 - \phi_2$, the locus of operation being the minor hysteresis loop between points $B$ and $d$ on the

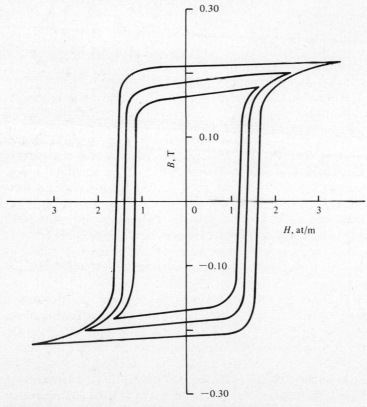

**Figure 2-37** Typical hysteresis loops for a ferrite core.

demagnetization curve. The axis $Bd$ of the minor hysteresis loop has about the same slope as the tangent to the demagnetization curve at $\phi_0$. Permanent magnets are stabilized by making use of this effect.

### 2-8.3 Square-Loop Ferrites

*Square-loop ferrites* are ferrites that, at low frequencies, exhibit hysteresis loops with nearly square corners, as shown in Fig. 2-37. Because of their square hysteresis loops, relatively low values of coercive force $H_C$, and high resistivities, these ferrite materials are suitable for high-speed memory storage and switching elements.

Figure 2-38 shows an elementary array of nine memory cores in the form of small rings, as small as $\frac{2}{3}$ cm o.d., for use in computers. Each core is linked by a *column* and a *row* magnetizing wire and a third wire, known as an *output wire*. If the residual flux in all cores is initially $-B_r$, then equal magnetizing currents $I_m/2$ applied simultaneously to both magnetizing wires through one

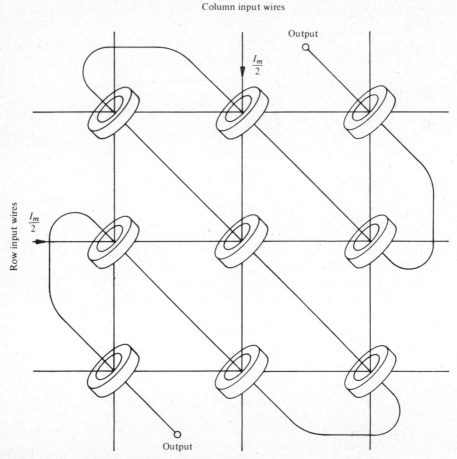

**Figure 2-38**   Simple array of memory cores.

core cause the flux in that core to reverse from $-B_r$ to $+B_r$ if $I_m$ is of such a value as to produce a magnetizing force of $H_m$. The other cores which are linked by only one current-carrying wire are relatively unaffected. Such an arrangement meets the requirement of a double-memory system for a binary code using the digits 1 and 0 only.

When all the wires are scanned in sequence by the application of negative current pulses $(-I_m)$, flux reversal occurs only in those cores that were previously magnetized in the positive direction, giving rise to a voltage pulse in the output winding.†

## STUDY QUESTIONS

1. What is a transducer? List as many practical types of transducers used as you can.
2. Why are most electromechanical devices magnetic-field devices?
3. List as many electromechanical devices as you can that use electric fields for energy storage and conversion.
4. What is meant by leakage flux?
5. What is meant by fringing of flux?
6. Where and under what conditions does fringing occur?
7. Where and under what conditions does leakage flux occur?
8. What is meant by the term reluctance in magnetic circuits?
9. Define the terms saturated and unsaturated iron.
10. What is permeance?
11. Make a list of ferromagnetic materials. Make a list of nonmagnetic materials.
12. What is meant by the term intrinsic flux density?
13. What is hysteresis?
14. Why are laminations used in electromechanical devices?
15. Why are air gaps undesirable in magnetic circuits?
16. List the core losses found in a ferromagnetic core structure. How are these losses minimized in good design?
17. What are eddy currents?
18. What effect does frequency have on magnetic material losses?
19. What does the term flux linkage mean?
20. What is magnetostriction?
21. How does magnetostriction affect the design of a magnetic device?
22. How is energy stored in a magnetic circuit?
23. Define the term (a) inductance, (b) self-inductance, (c) mutual inductance.
24. What is meant by a linear magnetic circuit? When does a linear magnetic circuit exist? When would a magnetic circuit not be linear?

† Various applications of ferrite cores are treated in C. J. Quartly, *Square-Loop Ferrite Circuitry* (Englewood Cliffs, N.J.: Prentice-Hall, Inc., 1962).

25. What is the coefficient of coupling?

26. What is the electromagnetic differential energy?

27. What is the term permeability, the term permeativity?

28. What is the law of conservation of energy?

29. What is meant by the term torque?

30. Define the relationships between stored energy, developed torque, and the angle of rotation.

31. Define what is meant by the term coenergy.

32. Describe how mechanical energy is abstracted from the energy stored in a magnetic field.

33. List as many materials as possible that are used as permanent magnets.

34. What is meant by the term permanent-magnet materials? How do they differ from ferromagnetic materials?

35. Define the term retentivity.

36. What is the importance of the energy product?

37. What is a demagnetization curve?

38. What is a square-loop ferrite?

39. List applications of square-loop ferrites.

40. What is meant by residual flux density?

41. What is meant by the term coercive force?

42. What is a magnetization curve? How is it used in design?

43. How can the hysteresis loop for a given material be measured?

44. What is rotational hysteresis loss?

45. What is a ferrite? What special properties does it have? How are ferrites made?

## PROBLEMS

**2-1.** For Prob. 2-1 assume you have two parallel plates in air. Assume that each plate has an area of 1 m$^2$ and a separation of 0.10 m. The plates are charged to a potential difference of 100 kV and then disconnected. (Assume the insulation is good enough that no leakage occurs and the charge is constant.) The plates are then moved to a separation of 0.15 m. Neglect fringing effects and calculate:
  **(a)** The voltage between the plates for the separation of 0.15 m.
  **(b)** The energy required to move the plates from 0.10 to 0.15 m.
  **(c)** The force between the plates initially and at the separation of 0.15 m.

**2-2.** For the parallel plates of Prob. 2-1 assume the voltage is held constant at 100 kV while the plate separation is decreased from 0.20 to 0.10 m. Neglect fringing at the plate edges and calculate:
  **(a)** The mechanical energy required to move the plates. Is this energy input to or an output from the plates?
  **(b)** The increase or decrease in the stored energy.
  **(c)** The electrical energy input or output.

**(d)** The average current if the time to change the spacing is 0.01 s. Show the direction of flow of current (in or out of the positive plate).

**2-3.** The figure for Prob. 2-3 shows a parallel-plate capacitor partially filled with a solid dielectric with a relative dielectric constant of $k_r$. Assume the solid dielectric is of the same thickness as the plate separation. Derive a general expression for the capacitance of the illustrated configuration.

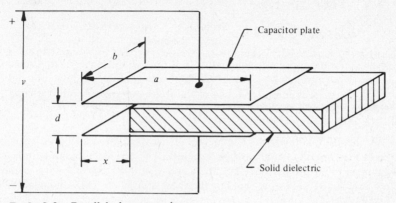

**Prob. 2-3**   Parallel-plate capacitor.

**2-4.** For the illustrated capacitor shown in Prob. 2-3 neglect fringing and derive an expression for the force on the solid dielectric slab in terms of the applied voltage and given dimensions if the applied voltage is:
**(a)** Constant at $V$ V.
**(b)** Sinusoidal excitation of $v = \sqrt{2}\, V \sin \omega t$ (express the force as a function of time and as an average value).

**2-5.** Through the use of Ampère's circuital law given in Eq. 2-15, find the $H$ field about a current $I$ in a filament conductor of infinite length. (Note that for a single filament $N = 1$.)

**2-6.** Using Ampère's circuital law, find the $H$ field inside and outside of an infinite length of conductor of finite cross section with radius $r_1$ carrying a current $I$ A uniformly distributed over its cross section.

**2-7.** The illustration for Prob. 2-7 shows two concentric nonmagnetic conductors separated by a dielectric that has a constant relative magnetic permeability of $\mu_r$. The currents $I_i$ and $I_o$ are uniformly distributed in the inner and outer conductors, respectively.
**(a)** Assume both currents to flow into the page (i.e., away from the observer), and evaluate the line integral of $H$ taken around a circular path concentric with the conductors, in terms of $I_i$, $I_o$, and the radius $r$ of the path of integration for (1) $r \leq r_1$, (2) $r_1 < r < r_2$, (3) $r_2 < r < r_3$, and (4) $r \geq r_3$.
**(b)** Repeat part (a) if the direction of the current in the outer conductor is out of the page (toward the observer) while that of the inner conductor is into the page as before.
**(c)** Repeat part (b) if $I_o = I_i$.
**(d)** How would the values of the line integrals of $H$ in each of the foregoing parts be affected if the relative magnetic permeabilities of the inner and outer conductors were $\mu_{ri}$ and $\mu_{ro}$?

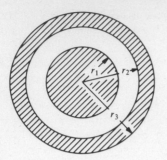

**Prob. 2-7**   Cross-sectional view of concentric conductors.

2-8. Express, for the concentric conductors in Prob. 2-7, the magnetic flux density $B$ in terms of the currents $I = I_i = -I_o$ in the inner and outer conductors and of the magnetic permeabilities $\mu_{ri}$, $\mu_{ro}$, $\mu_r$ of the inner and outer conductors and of the dielectric between them for (a) $r \le r_1$, (b) $r_1 < r < r_2$, (c) $r_2 < r < r_3$, and (d) $r \ge r_3$.

2-9. Two identical nonmagnetic conductors of rectangular cross section $(d)(w)$ m are embedded in a slot within an iron structure. The permeability of the iron may be assumed to be infinite. The currents $I_u$ and $I_l$ in the upper and lower conductors are distributed uniformly through the cross sections of the conductors and are assumed to flow into the page (i.e., away from the observer).

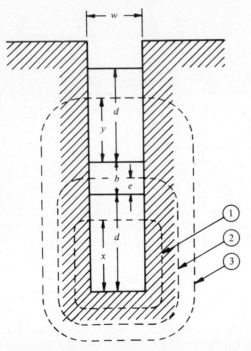

**Prob. 2-9**   Two conductors of rectangular cross section $d$ by $w$ embedded in a slot.

(a) Evaluate the line integral of $H$ around paths 1, 2, and 3 indicated by the dashed lines.

(b) Determine the magnetic flux that crosses the slot between conductors in terms of the currents and appropriate dimensions if the length of the embedded conductors is $L$, on the basis that the direction of $H$ within the slot is normal to the walls of the slot and that $\mu_r = 1$ for the material between the two conductors.

**(c)** Repeat part (b) if the direction of $I_l$ is reversed while that of $I_u$ remains unchanged.

**(d)** Evaluate the flux that passes across each conductor, in terms of the current and appropriate dimensions, for the condition of part (c).

**2-10.** The three-legged core shown in the figure for this problem has a winding of 100 turns on each outer leg and one of 50 turns on the center leg. Currents of 25 A, 10 A, and 15 A are applied to the three windings as indicated.

**(a)** Evaluate the line integral of $H$ around each of the paths 1, 2, and 3.

**(b)** Repeat part (a) but assume that the direction of the current in the center winding is reversed while the direction of the current in the outer windings is unchanged. The values of the currents are the same as in part (a).

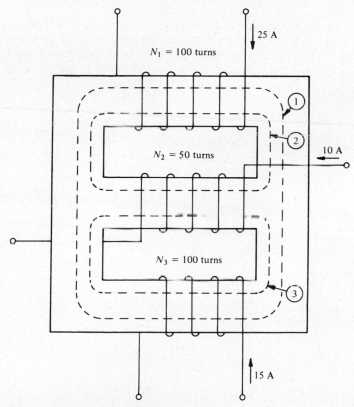

**Prob. 2-10**   Magnetic circuit.

**2-11.** The figure for this problem shows a two-pole magnetic structure that has an air gap of uniform radial length $g$ between two concentric iron cylinders. A winding with $N$ turns, uniformly distributed around the inner cylinder, carries a current $I$ A. The cross section of the axial winding conductors may be considered small enough so that the winding and its current may be regarded as a current sheet.

**(a)** Show that the line integral of $H$ around the path indicated by the dashed lines is

$$\oint H \cdot dl = \frac{2Ni}{\pi} \theta$$

considering $H$ positive when directed radially inward.

**(b)** Express the mmf across the air gap as a function of $\theta$ if the reluctance of the iron is negligible.

**(c)** Show that the waveform of $H$ versus $\theta$ when plotted for the developed view is triangular with an amplitude of $Ni/2g$. Assume the air-gap length $g$ to be small enough in relation to the diameter of the air gap that $H$ is uniform along $g$ at any angle $\theta$.

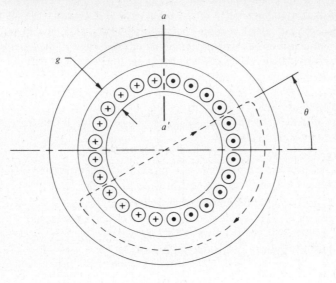

(a)

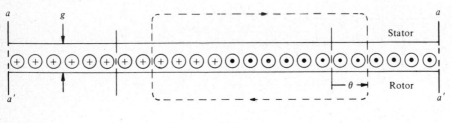

(b)

**Prob. 2-11** Two-pole magnetic structure with air gap of uniform length $g$. (a) Cross-sectional view. (b) Developed view.

**2-12.** Figure 2-11 shows an electromagnet with a small air gap. The material $BH$ curve is given in Fig. 2-6.

**(a)** What current must flow in the electromagnet winding to establish a magnetic flux of $4 \times 10^{-4}$ Wb?

**(b)** Neglecting leakage flux but accounting for fringing, what would be the flux if the gap length is doubled (increased to 0.5 cm) if the current in the winding is constant? *Hint:* Assume total mmf must be in the air gap as a first approximation and then compute $B_{air}$, $B_{iron}$, $F_{iron}$.

**(c)** What will be the flux for the conditions of part (b) if the air gap is reduced to 0.20 cm?

**(d)** Is the mmf in the iron a large percentage of the total?

**2-13.** The core loss in an iron-core reactor is 600 W, of which 400 W is hysteresis loss when the applied voltage is 220 V and the frequency is 60 Hz. Neglecting winding resistance:

**(a)** What is the core loss if the frequency is 10,000 Hz (audio range)?

**(b)** What is the core loss if the frequency is 60 Hz and the voltage is 440 V?

**(c)** What can one conclude from parts (a) and (b)?

**2-14.** The curve in Fig. 2-6 of the text is for M-19 29-gauge steel. The curve is plotted as a semilog plot and, as such, shows no range over which the material is linear. Replot Fig. 2-6 and

**(a)** Determine the region of linearity.

**(b)** Determine the number of ampere turns where one could consider the material to be saturated.

**(c)** What is the relative differential permeability in the linear range and in the region of saturation?

**2-15.** An air gap has a cross-sectional area of 0.1 m² and a length of 1 mm. What are the reluctance and permeance of this gap? What equivalent effect does fringing have on the reluctance and permeance of the gap?

**2-16.** In the figure for this problem a three-legged core is shown with an air gap in each leg of 0.25 cm. A winding of 300 turns draws a steady-state current of 10 A and is wound around an outside leg. Neglect leakage and the iron path reluctance. Assume a stacking factor of 0.95 for the core laminations. Correct for air-gap fringing and calculate the flux in each leg of the core.

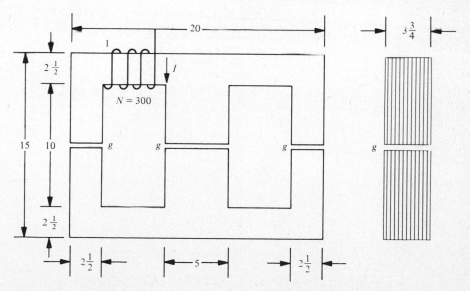

**Prob. 2-16**   Laminated three-legged core. Dimensions are in centimeters.

**2-17.** For Prob. 2-16 suppose that the 300-turn coil is moved to the center leg. With other conditions remaining the same, draw a schematic of the core. What is the flux in each leg?

**2-18.** For the core of Prob. 2-16 suppose that two coils of 200 turns each are placed on the outside legs of the core and the 300-turn coil is removed.

    **(a)** Draw a schematic of this new core.

    **(b)** Show the current into the coils such that the flux adds in the center leg.

    **(c)** What current is required in each coil to establish an air gap flux in the center leg of $7.54 \times 10^{-4}$ Wb?

    **(d)** Keep the current constant in each of the outer leg coils but reverse the polarity of one. What is the flux in each leg of the core? Neglect leakage flux but correct for fringing.

**2-19.** For the electromagnet with a cylindrical plunger of Example 2-1 illustrated in Fig. 2-3, calculate the energy stored in each of the air gaps of the electromagnet.

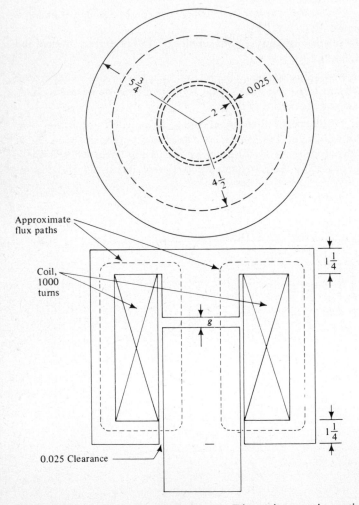

**Prob. 2-20**    Plunger-type electromagnet. Dimensions are in centimeters.

**2-20.** The figure shown is for a plunger-type electromagnet similar to that of Example 2-1. Neglect the effects of fringing and leakage flux. If the mmf in the iron is neglected, calculate the reluctance of the flux path including the two air gaps which includes the clearances of the throat and the gap at the end of the plunger.

**2-21.** The iron in the electromagnet of Prob. 2-20 has a magnetization curve through the data points given in the following table:

| F (ampere turns) | 25 | 50 | 100 | 200 | 300 | 500 | 750 | 1000 | 1500 | 2000 | 2500 |
|---|---|---|---|---|---|---|---|---|---|---|---|
| $\phi$ (Wb $\times 10^{-5}$) | | 50 | 85 | 110 | 128 | 140 | 155 | 163 | 168 | 175 | 180 | 185 |

  **(a)** Plot the iron magnetization curve.
  **(b)** Use graphical construction methods and calculate the flux for gap lengths of 0.25, 0.50, and 1.0 cm for a coil current of 5 A in the 1000-turn coil. (Note that the air-gap line in each case must include the reluctance of the air gap and the throat clearance as series air gaps.)
  **(c)** Calculate the force on the plunger at a current of 5 A when the gap is 0.25 cm.

**2-22.** The following data defines the upper half of the hysteresis loop for a sample of transformer steel:

| $B(T)$ | 0 | 0.155 | 0.310 | 0.465 | 0.620 |
|---|---|---|---|---|---|
| $H$ (ampere turns/m) | 33.5 | 39.4 | 46.1 | 57.1 | 74.0 |
| $B(T)$ | 0.698 | 0.775 | 0.852 | 0.93 | 1.0 |
| $H$ (ampere turns/m) | 85.0 | 98.5 | 117 | 144 | 175 |
| $B(T)$ | 0.93 | 0.852 | 0.775 | 0.698 | 0.620 |
| $H$ (ampere turns/m) | 101 | 57.8 | 31.4 | 14.6 | 1.97 |
| $B(T)$ | 0.415 | 0.310 | 0.155 | 0 | |
| $H$ (ampere turns/m) | $-13.8$ | $-23.6$ | $-28.7$ | $-33.5$ | |

Plot the complete hysteresis loop and determine:
  **(a)** The residual flux density.
  **(b)** The coercive force.
  **(c)** The energy product, $B = 0.31T$, on the decreasing portion of the loop.

**2-23.** The following no-load test data were obtained by exciting the 440-V winding of a 60-Hz transformer:

| Volts | Frequency, Hz | Watts |
|---|---|---|
| 440 | 60 | 103.5 |
| 220 | 30 | 45.2 |

Neglect the resistance of the winding:
  **(a)** Calculate (1) the 60-Hz hysteresis loss at rated voltage, and (2) the 60-Hz eddy-current loss at rated voltage.
  **(b)** If the volume and the kind of iron in the core remained unchanged but the thickness of the laminations were doubled, what would be the value of the core loss at (1) 440 V and 60 Hz, and (2) 220 V and 30 Hz?

**2-24.** A toroid such as that illustrated in Fig. 2-2 has a winding of two layers with 100 turns each wound on a core of nonmagnetic material. The dimensions of the core are $r_1 = 12.5$ cm, $r_2 = 15.0$ cm, and $w = 10.0$ cm. Assume the magnetic flux to be confined entirely to the core.

    **(a)** If the two layers of the winding are connected in series-aiding (with their polarities such that their mmfs add) and the current is 1 A, calculate (1) the magnetic flux in the core, (2) the flux linkage, (3) the self-inductance of the winding, and (4) the energy stored in the magnetic field.

    **(b)** Repeat part (a) if the two layers are connected in parallel-aiding and if the total current is 2 A.

**2-25.** A long, straight cylindrical conductor of radius $R$ meters and length $l$ meters has a relative permeability $\mu_r = 1$ and it carries a current of $I$ amperes distributed uniformly over its cross section. Determine in terms of $I$, $R$, and $l$:

    **(a)** The total magnetic flux within the conductor.

    **(b)** The magnetic energy density $BH/2$ within the conductor as a function of the distance $x$ from the center.

    **(c)** The energy stored in the magnetic field within the conductor.

    **(d)** The component of self-inductance due to the magnetic field within the conductor.

**2-26.** In Figure 2-11 an electromagnet with a single coil is shown. Neglect the reluctance of the iron and calculate the self-inductance.

## BIBLIOGRAPHY

Allegheny Ludlum Steel Corporation. 1961. *Electrical Materials Handbook.* Pittsburgh, Pa., 1961. The characteristics of various magnet materials including testing methods are treated.

Bozorth, R. M. *Ferromagnetism.* Princeton, N.J.: Van Nostrand Company, Inc., 1951.

Crosno, C. Donald. *Fundamentals of Electromechanical Conversion.* New York: Harcourt, Brace and World, Inc., 1968.

Crucible Steel Company of America. *Permanent Magnet Handbook.* Pittsburgh, Pa., 1957.

Dwight, H. B. *Electrical Elements of Power Transmission Lines.* New York: The Macmillan Company, 1954.

Fano, R. M., L. J. Chu, and R. B. Adler. *Electromagnetic Fields. Energy and Forces.* New York: John Wiley & Sons, Inc., 1960.

Fitzgerald, A. E., C. Kingsley, and A. Kusko: *Electric Machinery,* 3d ed. New York: McGraw-Hill Book Company, 1971.

General Electric Company. *Permanent Magnet Manual.* Magnetic Materials Business Section. Edmore, Mich.

Hayt, William. *Engineering Electromagnetics,* 2d ed. New York: McGraw-Hill Book Company, 1967.

Karapetoff, V. *The Magnetic Circuit.* New York: McGraw-Hill Book Company, 1911.

Matsch, L. W. *Capacitors, Magnetic Circuits and Transformers.* Englewood Cliffs, N.J.: Prentice-Hall, Inc., 1964.

Matsch, L. W. *Electromagnetic & Electromechanical Machines,* 2d ed. New York: Dun-Donnelley Publishing Co., 1977.

McPherson, George. *An Introduction to Electrical Machines and Transformers.* New York: John Wiley & Sons, Inc., 1981.

Meyerhoff, A. J., et al. *Digital Applications of Magnetic Devices.* New York: John Wiley & Sons, Inc., 1960.

MIT Staff. *Magnetic Circuits and Transformers.* New York: John Wiley & Sons, Inc., 1943.

Parker, R. J., and R. J. Studders. *Permanent Magnets and Their Applications.* John Wiley & Sons, Inc., 1962.

Polydoroff, W. J. *High-Frequency Materials.* New York: John Wiley & Sons, Inc., 1960.

Quartly, C. J. *Square-Loop Ferrite Circuitry.* Englewood Cliffs, N.J.: Prentice-Hall, Inc., 1962.

Rosa, E. B., and F. W. Grover. "Formulas and Tables for the Calculation of Mutual and Self Inductance," *NBS Science Paper 169.* Washington, D.C.: National Bureau of Standards, 1916.

Roters, H. C. *Electromagnetic Devices.* New York: John Wiley & Sons, Inc., 1941.

*Selection of Armco Electric Steels for Magnetic Cores.* Armco, Inc., Middletown, Ohio, 1984.

Shultz, Richard D., and Richard A. Smith. *Introduction to Electric Power Engineering.* New York: Harper & Row, Publishers, Inc., 1985.

Slemon, G. R. *Magnetoelectric Devices.* New York: John Wiley & Sons, Inc., 1966.

Spooner, Thomas. *Properties and Testing of Magnetic Materials.* New York: McGraw-Hill Book Company, 1927.

Timbie, W. H., and V. Bush. *Principles of Electrical Engineering.* New York: John Wiley & Sons, Inc., 1940.

U.S. Steel Corporation. *Non-oriented Electrical Steel Sheets* (ADUSS-2626). Pittsburgh, Pa.

Winch, R. P. *Electricity and Magnetism.* Englewood Cliffs, N.J.: Prentice-Hall, Inc., 1955.

Woodruff, L. F. *Principles of Electric Power Transmission.* New York: John Wiley & Sons, Inc., 1938.

# Chapter 3

# The Transformer

The transformer is an indispensable link in present-day commercial electric power systems and a vital component in many low-power applications, as in the case of some electronic circuits. Although the transformer is a static device, as its basic construction requires no moving parts, some of the principles underlying its operation are useful in analyzing the performance of electric motors and generators.

## 3-1 THE TWO-WINDING TRANSFORMER

The transformer may be defined as a device in which two or more stationary electric circuits are coupled magnetically, the windings being linked by a common time-varying magnetic flux. One of these windings, known as the *primary,* receives power at a given voltage and frequency from the source; and the other winding, known as the *secondary,* delivers power, usually at a different voltage but at the same frequency, to the load.

Transformers that operate in the audio-frequency range have laminated iron cores and come in a variety of core-and-winding configurations. Figure 3-1 shows common arrangements of cores and coils in transformers; the cores in Fig. 3-1(a) and (b) consist of stacks of steel laminations comprised of flat punchings, while those in Fig. 3-1(c) and (d) are wound of a long continuous strip of sheet steel in the direction in which the steel was rolled during manufacture. The wound cores are magnetized in the direction of rolling, thus making for lower core loss and lower exciting current than when magnetized across the direction of rolling. In large transformers, similar steel materials are also used,

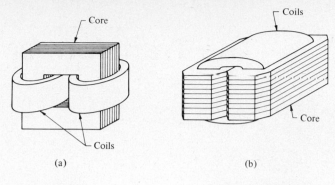

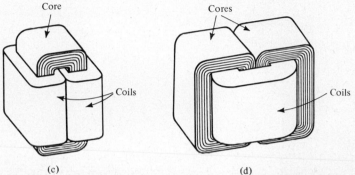

**Figure 3-1** Transformers. (a) Core type with stacked core. (b) Shell type with stacked core. (c) Core type with wound core. (d) Shell type with wound core.

but in the form of flat strips stacked so that the flux path is also in the direction of rolling. A cutaway section of a distribution transformer is shown in Fig. 3-2.

## 3-2 THE IDEAL TWO-WINDING TRANSFORMER

The ideal transformer is one that has no losses, no leakage fluxes, and a core of infinite magnetic permeability and of infinite electrical resistivity. The two-winding shell-type transformer in Fig. 3-3 is assumed to be ideal. Its primary winding has $N_1$ turns and its secondary winding has $N_2$ turns. The dashed lines in the core represent approximately the path of the magnetic flux assumed to be confined entirely within the core.

### 3-2.1 Voltage Ratio and Transformer Polarity

A voltage $v_1$ applied to the primary winding, assumed to have zero resistance, produces a flux in the core that links all $N_1$ turns, since the leakage flux is assumed to be zero. Therefore,

$$v_1 = e_1 = \frac{d\lambda_1}{dt} = N_1 \frac{d\phi}{dt} \tag{3-1}$$

**Figure 3-2** Distribution transformer. (*Source:* Westinghouse Electric Corporation.)

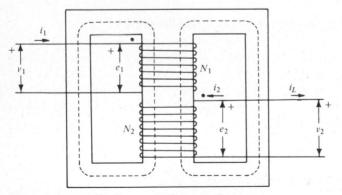

**Figure 3-3** Idealized two-winding transformer.

where $e_1$ is the primary induced voltage, $\lambda_1$ the flux linkage with the primary winding, and $\phi$ the flux in the core.

Since there is no leakage flux, the flux $\phi$ must link all $N_2$ turns of the secondary winding, and since the resistance of the secondary winding is also assumed to be zero, the secondary induced voltage and secondary terminal voltage are equal and are expressed by

$$v_2 = e_2 = N_2 \frac{d\phi}{dt} \tag{3-2}$$

A comparison of Eqs. 3-1 and 3-2 shows that in an ideal transformer the voltage ratio equals the turns ratio;

$$\boxed{\frac{v_1}{v_2} = \frac{e_1}{e_2} = \frac{N_1}{N_2} = a} \tag{3-3}$$

The primary and secondary voltages as shown in Fig. 3-3 have like polarities. The dots near the upper end of each winding are called *polarity marks* and indicate that the upper or marked terminals have like polarities, at a given instant of time when current enters the primary terminal and leaves the secondary terminal.

### 3-2.2 Current Ratio

If the secondary of the ideal transformer in Fig. 3-3 is connected to a load and the instantaneous load current $i_L$ flows in the direction shown, the direction of the instantaneous primary current is as shown. Further, the primary mmf $N_1 i_1$ would produce flux in an upward direction through the windings, whereas the secondary mmf $N_2 i_L$ would produce flux in the downward direction. Since the core has infinite permeability and infinite resistivity, a finite value of flux requires an mmf of zero. Accordingly, the primary and secondary mmfs are equal and opposite and we have

$$N_1 i_1 = N_2 i_L$$

with the result that the current ratio is

$$\boxed{\frac{i_1}{i_L} = \frac{N_2}{N_1} = \frac{1}{a}} \tag{3-4}$$

The polarity marks signify that when positive currents enter both windings at the marked terminals, as in the case of $i_1$ and $i_2$ in Fig. 3-3, the mmfs of the two windings add. It should be apparent that $i_2 = -i_L$. The use of $i_L$ instead of $i_2$ leads to more straightforward phasor diagrams.

### 3-2.3 Impedance Ratio

It is sometimes convenient to refer the impedance connected across one side of a transformer to the other side of the transformer. This is accomplished by multiplying the value of the impedance by the impedance ratio $a^2$ of the transformer, which is derived as follows.

Figure 3-4 shows an impedance $Z_L$ across the secondary of an ideal transformer. The secondary terminal voltage is

$$\mathbf{V}_2 = \mathbf{I}_L Z_L \tag{3-5}$$

Then from Eqs. 3-3, 3-4, and 3-5 it follows that

$$\frac{V_1}{I_1} = \left(\frac{N_1}{N_2}\right)^2 Z_L \tag{3-6}$$

The load impedance viewed from the primary of the transformer is therefore

$$\boxed{Z_1 = \left(\frac{N_1}{N_2}\right)^2 Z_L = a^2 Z_L} \tag{3-7}$$

## 3-3 EXCITING CURRENT, CORE-LOSS CURRENT AND MAGNETIZING CURRENT

The characteristics of iron cores depart from the ideal because of core losses, as discussed in Sec. 2-5, and because the permeability is finite. A current, known as the *exciting current,* therefore flows in the primary when there is no current in the secondary (i.e., at no load). When the secondary delivers current, the exciting current combines with the component $N_2 I_L / N_1$ in the primary to give the total primary current, and the ideal transformer can be modified accordingly, as shown in Fig. 3-5(a), with the corresponding phasor diagram of Fig. 3-5(b).

The exciting current is considered as having two components, the *core-loss current* and the *magnetizing current.* The core-loss current is a real-power component and is due to the core losses. The magnetizing current is, in effect, the component of current that furnishes the mmf to overcome the magnetic reluctance of the core.

### 3-3.1 Core-Loss Current

The core loss $P_c$ is the sum of the hysteresis and eddy-current losses and manifests itself in the form of heat generated in the core. Hence,

$$P_c = P_h + P_e$$

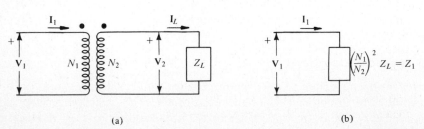

(a)                                         (b)

**Figure 3-4**  Equivalent circuits. (a) Ideal transformer and connected load impedance. (b) Load impedance referred to the primary side.

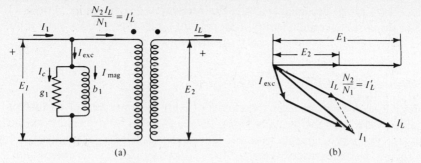

**Figure 3-5** (a) Equivalent circuit of transformer taking exciting current into account. (b) Corresponding phasor diagram.

The core-loss current is in phase with the induced primary voltage and is therefore expressed by

$$I_c = \frac{P_c}{E_1} \qquad (3-8)$$

### 3-3.2 Magnetizing Current

In linear magnetic circuits the magnetizing current can be expressed analytically by

$$i = \frac{\phi \mathcal{R}}{N} \qquad (3-9)$$

where $\mathcal{R}$ is the reluctance of the magnetic circuit. However, in the case of iron circuits the magnetizing current may be calculated from the magnetization curve, such as in Fig. 2-6, for the particular core material. Large power transformers may operate at flux densities as high as 1.4 T, the choice of flux density generally being a compromise of a number of factors such as first cost and performance characteristics. Although the magnetizing current can be calculated to a good degree of approximation from the magnetization curve for a given material, it is generally simpler to rely on curves of exciting volt-amperes per pound of material versus flux density that may be obtained from the manufacturer.†

Nevertheless, the magnetization curve shows important characteristics of the iron. One of these is that the material starts to saturate at a flux density of about 1.1 T, requiring a magnetizing force of about 225 ampere turns/m. The upper limit of 1.4 T does not leave a very large margin for overexcitation. That is, if the voltage applied to the transformer were to exceed rated value by one-third, the flux density would be increased to $\frac{4}{3} \times 1.4$ T = 1.87 T, with a corresponding increase in the mmf by a ratio of about 20,000/225 or 89.

On the other hand, the design of a power transformer for operation with the iron unsaturated calls for excessive size. For example, if the number of turns

† *Non-oriented Electrical Steel Sheets* (ADUSS 31-2626) (Pittsburgh, Pa.: U.S. Steel Corporation).

and the voltage rating are such that the flux density is $B_m = \phi_m/A_c = 0.93$ T, the area of the core $A_c$ would need to be 50 percent larger than that for operation at 1.4 T. This is evident from the result of substituting $A_c B_m$ for $\phi_m$ in Eq. 2-35.

### 3-3.3 Waveform of Exciting Current

The energy loop, which includes the effects of eddy currents as well as that of hysteresis, is wider than the hysteresis loop for a sinusoidal time variation of the flux density. The eddy currents lead the flux density wave by an angle of 90° and are therefore a maximum at $B = \phi = 0$ and are zero at $\phi_m$. The upper half of such a loop is shown in the graphical construction of Fig. 3-6 for obtaining the waveform of the exciting current for a sinusoidal time variation of the flux. The phase relation among flux, exciting current, and induced voltage is shown in Fig. 3-7. It is evident from Fig. 3-7 that the waveform of the exciting current is not sinusoidal. However, it is symmetrical, and since $i(\omega t + \pi) = -i(\omega t)$, the exciting current can therefore be represented by a series of *odd* harmonics as expressed by

$$i_{exc} = \sqrt{2}[I_{1e} \cos{(\omega t - \theta_{1e})} + I_{3e} \cos{(3\omega t - \theta_{3e})}$$
$$+ \cdots + I_{ne} \cos{(n\omega t - \theta_{ne})}] \qquad (3\text{-}10)$$

where $n$ is odd and where $I_{1e}, I_{3e}, \ldots, I_{ne}$ are the rms values of the fundamental and the higher harmonic components. The fundamental component of the exciting current $I_{1e}$ lags the induced voltage $E$ by the angle $\theta_{1e}$ with

$$e = \sqrt{2}E \cos{\omega t} \qquad (3\text{-}11)$$

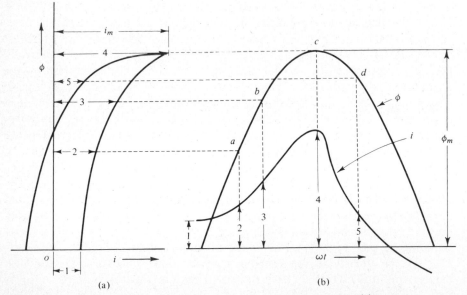

**Figure 3-6**   (a) Upper half of energy loop. (b) Sinusoidal flux and exciting current waves.

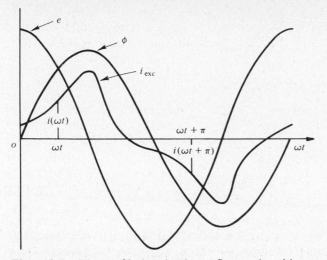

**Figure 3-7**   Waves of induced voltage, flux, and exciting current.

The third harmonic is the most prominent of the higher harmonic components. The amplitude of the harmonics decreases with increasing order for ordinary magnetic materials subjected to a sinusoidal time variation of flux.

### 3-3.4 Core-Loss Current

The core loss is expressed by

$$P_c = EI_{1e} \cos \theta_{1e} = EI_c \qquad (3\text{-}12)$$

where $I_c = I_{1e} \cos \theta_{1e}$ is the core-loss current. The higher harmonics in the exciting current wave make no contribution to the core loss for a sinusoidally varying flux.†

### 3-3.5 Magnetizing Current, Including Harmonics

The rms value of the exciting current is expressed in terms of its harmonic components by

$$I_{exc} = \sqrt{I_{1e}^2 + I_{3e}^2 + \cdots + I_{ne}^2} \qquad (3\text{-}13)$$

If there were no higher harmonics in the exciting current, the magnetizing current would be defined by

$$I_{mag\ 1} = \sqrt{I_{1e}^2 - (I_{1e} \cos \theta_{1e})^2} = \sqrt{I_{1e}^2 - I_c^2}$$
$$= I_{1e} \sin \theta_{1e} \qquad (3\text{-}14)$$

† For a more detailed discussion of exciting current, see L. W. Matsch, *Capacitors, Magnetic Circuits and Transformers* (Englewood Cliffs, N.J.: Prentice-Hall, Inc., 1964), pp. 204–217.

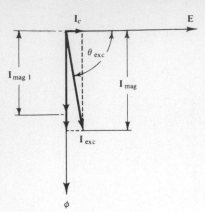

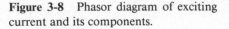

**Figure 3-8** Phasor diagram of exciting current and its components.

However, the total magnetizing current is considered as an equivalent sinusoid, which includes the fundamental component $I_{\text{mag }1}$ and all the higher harmonics and having an rms value of

$$I_{\text{mag}} = \sqrt{(I_{1e} \sin \theta_{1e})^2 + I_{3e}^2 + \cdots + I_{ne}^2} \qquad (3\text{-}15)$$

Since the exciting current can be measured with an ammeter and the core loss by means of a power measurement, the magnetizing current as defined by Eq. 3-15 is calculated simply as

$$\boxed{I_{\text{mag}} = \sqrt{I_{\text{exc}}^2 - I_c^2}} \qquad (3\text{-}16)$$

The relationship among the exciting current, core-loss current, and magnetizing current with respect to the induced voltage is represented by the phasor diagram of Fig. 3-8 in keeping with Eq. 3-16. In Fig. 3-8, $\theta_{\text{exc}} = \cos^{-1} I_c/I_{\text{exc}}$. The exciting current results from the imperfection of the iron core, which is taken into account by the addition of the conductance $g_1 = I_c/E_1$ and the susceptance $b_1 = -I_{\text{mag}}/E_1$ to the ideal transformer circuit in Fig. 3-5(a). The exciting admittance† is

$$\boxed{y_{\text{exc }1} = g_1 + jb_1 = \frac{\mathbf{I}_{\text{exc}}}{\mathbf{E}_1}} \qquad (3\text{-}17)$$

## 3-4 LEAKAGE IMPEDANCE

The windings of the transformer are also imperfect because of resistance and leakage flux (i.e., the component of equivalent flux that links one winding without linking the other winding). Under load, the windings develop sizable mmfs which act on spaces external to the core, thus giving rise to leakage fluxes. The simplified

---

† The variation in the exciting admittance is usually negligible for small departures of the voltage from rated value during normal operation. However, because of saturation, its magnitude may increase far out of proportion to sizable increases in voltage above normal.

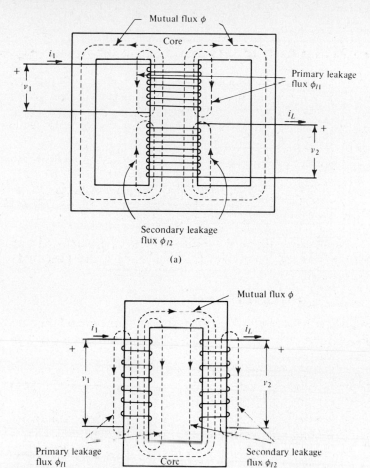

**Figure 3-9** Paths of mutual flux and leakage fluxes. (a) Shell-type transformer. (b) Core-type transformer. Because of the loose magnetic coupling of the windings in both transformers the leakage is high.

diagrams of Fig. 3-9 represent the mutual flux $\phi$, which links both windings, and the primary and secondary leakage fluxes $\phi_{l1}$ and $\phi_{l2}$, which link only their respective windings. The paths taken by the leakage fluxes in transformers are more complex than those indicated in Fig. 3-9. Some of the leakage flux is distributed through its associated winding so that different amounts link different turns, and $\phi_{l1}$ and $\phi_{l2}$ are therefore equivalent leakage fluxes which in linking all the turns of their respective windings account for the total leakage flux linkage.† The coupling between the windings in Fig. 3-9 is loose, giving rise to high magnetic leakage. Winding arrangements used in transformers to make for reduced leakage fluxes are shown in Fig. 3-10.

---

† For more complete descriptions of flux paths, see C. P. Steinmetz, *Theory and Calculation of Electric Circuits* (New York: McGraw-Hill Book Company, 1917), pp. 216–218; also MIT Staff, *Magnetic Circuits and Transformers* (New York: John Wiley & Sons, Inc., 1943), pp. 317–324.

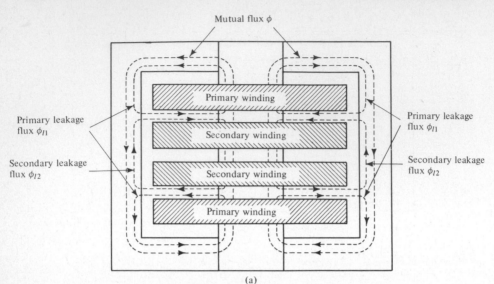

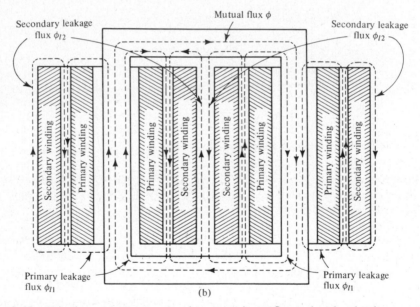

**Figure 3-10** Winding arrangements showing approximate flux paths taken by the mutual flux and the leakage fluxes. (a) Shell-type transformer. (b) Core-type transformer.

### 3-4.1 The Equivalent Circuit

The equivalent circuit can be made to take into account the leakage flux by the addition of an external reactance known as *leakage reactance* in each winding. The addition of external resistances to represent the resistance of each winding completes the equivalent circuit as shown in Fig. 3-11(a). The ideal transformer,

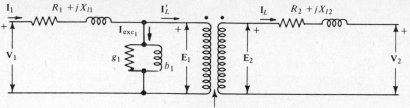

Ideal transformer

(a)

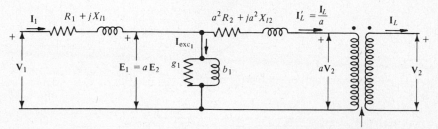

(b)

Ideal transformer

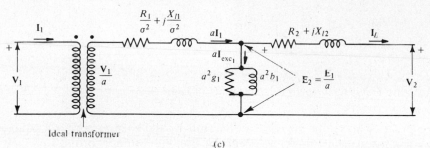

Ideal transformer

(c)

**Figure 3-11** Equivalent circuits of a transformer.

which is assumed to carry only the mutual flux, can also be located as shown in Fig. 3-11(b) and (c). All three of these equivalent circuits lead to identical analytical results. Figure 3-12 shows the phasor diagram for the circuit of Fig. 3-11(b).

## EXAMPLE 3-1

The constants of a 150-kVA 2400/240-V 60-Hz transformer are as follows:

Resistance of 2400-V winding $R_1 = 0.216 \ \Omega$

Resistance of 240-V winding $R_2 = 0.00210 \ \Omega$

Leakage reactance of 2400-V winding $X_{l1} = 0.463 \ \Omega$

Leakage reactance of 240-V winding $X_{l2} = 0.00454 \ \Omega$

Exciting admittance on the 240-V side $y_{exc} = 0.0101 - j0.069 \ \text{S}$

a. Draw the equivalent circuit on the basis of Fig. 3-11(a) showing the numerical values of the leakage impedances and the exciting admittance expressed in complex form.

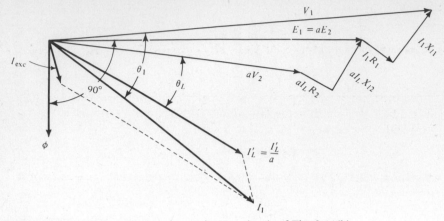

**Figure 3-12**    Phasor diagram for equivalent circuit of Fig. 3.11(b).

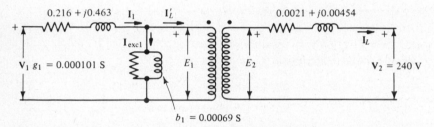

**Figure 3-13**    Equivalent circuit for transformer of Example 3.1.

b.  Determine the values of the emfs $E_1$ and $E_2$ induced by the equivalent mutual flux, the exciting current $I_{exc}$, the primary current $I_1$ at 0.80 power factor, current lagging, and the applied primary voltage when the transformer delivers rated load at rated secondary voltage.

*Solution*
a.  The equivalent circuit is shown in Fig. 3-13.
b.  The rated secondary terminal voltage is 240 V and the rated load (rated output) of the transformer is 150,000 VA at 240 V. The rated load current is therefore

$$I_L = 150,000 \div 240 = 625 \text{ A}$$

Let the phasor that represents the secondary terminal voltage $V_2$ lie on the real axis as shown in Fig. 3-12; then

$$V_2 = 240 + j0 \text{ V}$$

and the load current expressed as a phasor is

$$I_L = 625(0.80 - j0.60) = 500 - j375 \text{ A}$$

The secondary induced voltage $E_2$ in Fig. 3-12 must equal the phasor sum of the secondary terminal voltage plus the voltage drop across the secondary leakage impedance; thus

$$\mathbf{E}_2 = \mathbf{V}_2 + \mathbf{I}_L(R_2 + jX_{l2})$$
$$= 240 + j0 + (500 - j375)(0.00210 + j0.00454)$$
$$= 240 + j0 + 1.05 + 1.70 + j2.27 - j0.79$$
$$= 242.75 + j1.48 = 242.8\underline{/0.35°}$$

Since the primary induced voltage $\mathbf{E}_1$ and the secondary induced voltage $\mathbf{E}_2$ are both produced by the mutual flux, their ratio $\mathbf{E}_1/\mathbf{E}_2$ must equal the turns ratio $a = N_1/N_2 = 10$, and we have

$$\mathbf{E}_1 = a\mathbf{E}_2 = 10 \times 242.8\underline{/0.35°} = 2428\underline{/0.35°} = 2428 + j14.8$$

The load component $\mathbf{I}'_L$ of the primary current expressed in phasor form is

$$\mathbf{I}'_L = \frac{\mathbf{I}_L}{a} = \frac{500 - j375}{10} = 50.0 - j37.5 \text{ A}$$

The primary current is the phasor sum of the load component $\mathbf{I}'_L$ and the exciting current $\mathbf{I}_{exc}$. The exciting current produces the mutual flux, so to speak, and is obtained by multiplying the primary induced voltage by the exciting admittance referred to the primary side:

$$\mathbf{I}_{exc} = \mathbf{E}_1 y_{exc\ 1}$$

The value of the admittance in the given data is as measured on the secondary side. This admittance is referred to the primary side by making use of the impedance ratio $a^2$. It should be remembered that an impedance is transferred from the secondary side to the primary side of a transformer by multiplying its value by the impedance ratio. Further, impedance is the reciprocal of admittance, and on transferring the exciting admittance from the secondary to the primary side, we have

$$\frac{1}{y_{exc\ 1}} = a^2 \frac{1}{y_{exc\ 2}}$$

or
$$y_{exc\ 1} = \frac{y_{exc\ 2}}{a^2} = \frac{0.0101 - j0.069}{(10)^2}$$
$$= (1.01 - j6.9)10^{-4} = 6.98 \times 10^{-4}\underline{/-81.67°}$$

The primary exciting current is therefore

$$\mathbf{I}_{exc} = y_{exc\ 1}\mathbf{E}_1 = 6.98 \times 10^{-4}\underline{/-81.67°} \times 2428\underline{/0.35°}$$
$$= 1.70\underline{/-81.32°} = 0.257 - j1.51$$

From Fig. 3-13 it is evident that the primary current is

$$\mathbf{I}_1 = \mathbf{I}'_L + \mathbf{I}_{exc\ 1}$$
$$= 50.0 - j37.5 + 0.257 - j1.51 = 50.257 - j39.01$$
$$= 63.6\underline{/-37.9°}$$

The primary applied voltage is the sum of the primary induced voltage and the voltage drop across the primary leakage impedance. Hence,

$$\mathbf{V}_1 = \mathbf{E}_1 + \mathbf{I}_1(R_1 + jX_{l1})$$
$$= 2428 + j14.8 + (50.257 - j39.01)(0.216 + j0.463)$$
$$= 2428 + j14.8 + 10.86 + 18.14 + j23.27 - j8.46$$
$$= 2457 + j29.5 = 2457\underline{/0.69°} \text{ V}$$

All three of the equivalent circuits shown in Fig. 3-11 lead to exactly the same values of primary current $\mathbf{I}_1$ and primary applied voltage $\mathbf{V}_1$ as were obtained in the solution of Example 3-1.

## 3-4.2 The Approximate Equivalent Circuit

Calculations of iron-core transformer performance can usually be simplified with negligible sacrifice in accuracy by using the approximate equivalent circuit. Figure 3-14 shows the approximate equivalent circuit, which is a modification of the equivalent circuit of Fig. 3-11(b) and effected by placing the exciting admittance across the primary terminals. The error due to this shift is small in iron-core transformers because the exciting current is only a few percent of the rated current, and even at full-load current the voltage drop across the primary leakage impedance $R_1 + jX_{l1}$ is also only a few percent of the rated voltage.

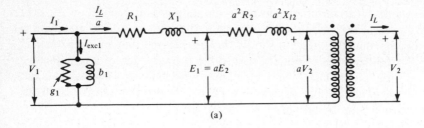

(a)

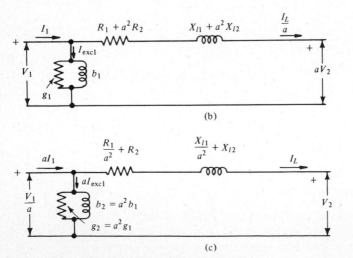

(b)

(c)

**Figure 3-14** Approximate equivalent circuits.

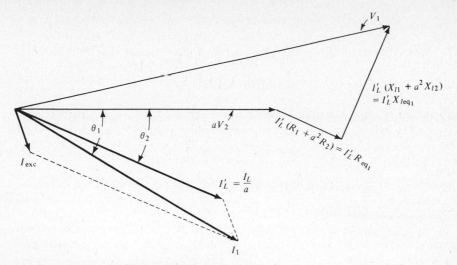

**Figure 3-15** Phasor diagram for approximate equivalent circuit of Fig. 3.14(a).

Similar modifications in the equivalent circuits of Fig. 3-11(a) and (c) will lead to corresponding approximate equivalent circuits. Figure 3-15 shows the phasor diagram for the approximate equivalent circuit of Fig. 3-14(a).

## 3-5 COUPLED-CIRCUIT EQUATIONS

Coupled-circuit equations express the relationship between voltages and currents in two or more inductively or magnetically coupled circuits in terms of resistances and inductances. The equations for two inductively coupled circuits are

$$v_1 = R_1 i_1 + L_{11} \frac{di_1}{dt} + L_{12} \frac{di_2}{dt} \tag{3-18}$$

$$v_2 = R_2 i_2 + L_{22} \frac{di_2}{dt} + L_{21} \frac{di_1}{dt} \tag{3-19}$$

where $R_1$ and $R_2$ are the resistances and $L_{11}$ and $L_{22}$ the self-inductances of the two circuits, and where $L_{12} = L_{21} = M$, the mutual inductance between the two circuits.

### 3-5.1 Leakage Inductance

Although these equations apply strictly only to linear circuits, such as air-core transformers, they afford a basis for clarifying the concept of leakage inductance in transformers with iron cores as well as in air-core transformers. Even in iron-core transformers the leakage inductances are practically independent of normal variations in saturation of the core because the leakage flux paths are largely in air.†

When the secondary current $i_2$ of a transformer is zero, Eq. 3-18 becomes

$$v_1 = R_1 i_1 + L_{11} \frac{di_1}{dt} \tag{3-20}$$

and the voltage induced in the secondary by the equivalent mutual flux $\phi_{21}$ is

$$e_2 = L_{21} \frac{di_1}{dt} = M \frac{di_1}{dt} \tag{3-21}$$

Equations 3-20 and 3-21 can be expressed in terms of equivalent fluxes, as defined in Chap. 2, by

$$v_1 = R_1 i_1 + N_1 \frac{d\phi_{11}}{dt} \tag{3-22}$$

$$e_2 = N_2 \frac{d\phi_{21}}{dt} = k_1 N_2 \frac{d\phi_{11}}{dt} \tag{3-23}$$

The voltage induced in the primary by the mutual flux $\phi_{21} = k_1 \phi_{11}$ is

$$e_1 = k_1 N_1 \frac{d\phi_{11}}{dt} = k_1 L_{11} \frac{di_1}{dt} \tag{3-24}$$

According to Eqs. 3-23 and 3-24,

$$\frac{e_1}{e_2} = \frac{N_1}{N_2} \tag{3-25}$$

and when Eq. 3-21 is substituted in Eq. 3-25, we have

$$e_1 = \frac{N_1}{N_2} M \frac{di_1}{dt} \tag{3-26}$$

Since the primary leakage flux with respect to the secondary is defined by

$$\phi_{l1} = \phi_{11} - \phi_{21} = (1 - k_1)\phi_{11}$$

and the primary leakage inductance is

$$L_{l1} = \frac{N_1 \phi_{l1}}{i_1} = (1 - k_1)L_{11} \tag{3-27}$$

then, by making use of Eqs. 3-24, 3-26, and 3-27, we may rewrite Eq. 3-20 as follows:

$$v_1 = R_1 i_1 + (1 - k_1)L_{11} \frac{di_1}{dt} + \frac{N_1}{N_2} M \frac{di_1}{dt} \tag{3-28}$$

which corresponds to Eq. 3-18 for $i_2 = 0$.

---

† The leakage inductances of transformers designed for high leakage are not linear. Such nonlinearities are treated by H. W. Lord, "An Equivalent Circuit for Transformers in Which Nonlinear Effects are Present," *Trans. AIEE* 78, Part I (1959):580–586.

For a load current $i_L = -i_2$, it is necessary to add to Eq. 3-28 the term $-M(di_L/dt)$ corresponding to $L_{12}(di_2/dt)$ in Eq. 3-18, which results in

$$v_1 = R_1 i_1 + (1 - k_1)L_{11}\frac{di_1}{dt} + M\left(\frac{N_1}{N_2}\frac{di_1}{dt} - \frac{di_L}{dt}\right) \tag{3-29}$$

The primary voltage induced by $i_2$ is

$$e_1 = M\frac{di_2}{dt} = -M\frac{di_L}{dt} \tag{3-30}$$

and Eq. 3-26 shows that the current in the primary which induces the same value of $e_1$ must be

$$\boxed{i'_L = \frac{N_2}{N_1} i_L}$$

on the basis of which Eq. 3-29 becomes

$$v_1 = R_1 i_1 + (1 - k_1)L_{11}\frac{di_1}{dt} + \frac{N_1}{N_2} M\frac{d}{dt}(i_1 - i'_L) \tag{3-31}$$

By the same process, it can be shown that the secondary leakage inductance is $(1 - k_2)L_{22}$, and when the impedance ratio is taken into account and the core losses are neglected, the two coupled circuits can be represented by the equivalent circuit in Fig. 3-16.

## 3-5.2 Magnetizing Inductance

The quantity $(N_1/N_2)M$ in Eq. 3-31 is called the *primary magnetizing* inductance, and it follows from the impedance ratio that the secondary magnetizing inductance must be $(N_2/N_1)M$. Not only do the leakage inductances and the magnetizing inductance influence the performance of transformers, but they also largely determine the characteristics of induction motors and synchronous machines. Iron-core transformers generally have a high value of magnetizing inductance and low leakage inductance. Induction motors have a relatively lower magnetizing inductance because of the air gap between the rotor and stator iron. This air gap is usually made as short as good performance characteristics will permit. Large synchronous machines require a relatively low magnetizing inductance, and therefore have long air gaps (large clearance between rotor and stator iron). The leakage inductances in induction motors and in synchronous machines are somewhat larger than those in transformers.

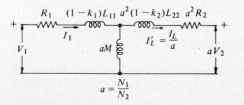

**Figure 3-16** Equivalent circuit for two inductively coupled circuits with negligible core loss.

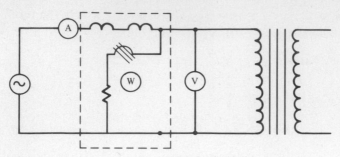

**Figure 3-17** Open-circuit test with indicating instruments.

### 3-5.3 Coefficient of Coupling

The coefficient of coupling between two windings is defined as $k \equiv \sqrt{k_1 k_2}$, also as $M/\sqrt{L_{11}L_{22}}$. It is evident that these relationships are identical from consideration of Fig. 3-16, which shows that $(N_1/N_2)M = k_1 L_{11}$, and on the basis of the impedance ratio $(N_1/N_2)^2$ that $(N_2/N_1)M = k_2 L_{11}$. Accordingly, we have

$$k = \sqrt{k_1 k_2'} = \frac{M}{\sqrt{L_{11}L_{22}}} \tag{3-32}$$

## 3-6 OPEN-CIRCUIT AND SHORT-CIRCUIT TESTS, EXCITING ADMITTANCE, AND EQUIVALENT IMPEDANCE

The open-circuit and short-circuit tests yield the constants for the equivalent circuit. The exciting admittance is obtained from the open-circuit test, which, in the case of power transformers or constant-voltage transformers operating at one specified frequency, consists of the application of the rated voltage at rated frequency usually for reasons of convenience, to the low-voltage winding with the high-voltage winding open-circuited. Measurements of voltage, current, and real power are made with indicating instruments.

Because of the small quantities involved in most communication transformers, ac bridges or other suitable devices are used instead of voltmeters, ammeters, and wattmeters. In Fig. 3-17,

$I_{exc}$ = exciting current as read by ammeter A

$V$ = applied voltage as read by voltmeter V

$P_{oc}$ = power as measured with wattmeter W and corrected for instrument losses

Because the $P_{I^2R}$† $\approx 0$, the exciting admittance is

† In this test because the exciting current is small the $I^2R$ losses of the conductors and the hysteresis and eddy-current losses can be neglected.

$$y \approx \frac{I_{exc}}{V} \qquad (3\text{-}33)$$

and the exciting conductance

$$g \approx \frac{P_{oc}}{V^2} \qquad (3\text{-}34)$$

from which the exciting susceptance is found to be

$$b = \sqrt{y^2 - g^2} \qquad (3\text{-}35)$$

**EXAMPLE 3-2**

An open-circuit test on the 240-V winding of the transformer in Example 3-1 yielded the following data, corrected for instrument losses:

| Volts | Amperes | Watts |
|-------|---------|-------|
| 240   | 16.75   | 580   |

Calculate the exciting admittance, conductance, and susceptance.

*Solution.* The exciting admittance is

$$y \approx \frac{I_{exc}}{V} = \frac{16.75}{240} = 0.0698 \text{ S}$$

$$g \approx \frac{P_{oc}}{V^2} = \frac{580}{(240)^2} = 0.0101 \text{ S}$$

from which the exciting susceptance is found to be

$$b = \sqrt{y^2 - g^2}$$
$$= \sqrt{(0.0698)^2 - (0.0101)^2} = 0.069 \text{ S}$$

The equivalent impedance, equivalent resistance, and equivalent reactance are found from the short-circuit test data.

The equivalent impedance, referred to the primary in Fig. 3-14, is

$$Z_{eq\ 1} = R_{eq\ 1} + jX_{eq\ 1} \qquad (3\text{-}36)$$

in which the equivalent resistance referred to the primary is

$$R_{eq\ 1} = R_1 + a^2 R_2 \qquad (3\text{-}37)$$

and the equivalent leakage reactance referred to the primary is

$$X_{eq\ 1} = X_{l1} + a^2 X_{l2} \qquad (3\text{-}38)$$

The equivalent impedance, equivalent resistance, and equivalent reactance referred to the secondary are

$$Z_{eq\ 2} = \frac{Z_{eq\ 1}}{a^2} = \frac{R_{eq\ 1}}{a^2} + j\frac{X_{eq\ 1}}{a^2} \qquad (3\text{-}39)$$

and
$$R_{eq\ 2} = \frac{R_1}{a^2} + R_2$$

$$X_{eq\ 2} = \frac{X_{l1}}{a^2} + X_{l2} \tag{3-40}$$

The short-circuit test is made by short-circuiting one winding (usually the low-voltage winding as a matter of convenience) and applying voltage at rated frequency such that rated current results. Measurements of input current, power, and voltage are generally made with indicating instruments, except in the case of most communication transformers, where an ac bridge method or a means adapted for measuring the smaller quantities associated with such transformers is used. In Fig. 3-18,

$V_{sc}$ = applied voltage as read by voltmeter V

$I_{sc}$ = input short-circuit current as read by ammeter A

$P_{sc}$ = input power as read by wattmeter W

For conventional transformers the exciting current is small compared with the rated current and can therefore be neglected. The short-circuit impedance $Z_{sc}$ is therefore assumed to equal the equivalent series impedance of the transformer and we have

$$\boxed{Z_{eq} \approx Z_{sc} = \frac{V_{sc}}{I_{sc}}} \tag{3-41}$$

and since the core loss is negligible at the low value of $V_{sc}$, the equivalent series resistance is practically equal to the short-circuit resistance:

$$\boxed{R_{eq} \approx R_{sc} = \frac{P_{sc}}{I_{sc}^2}} \tag{3-42}$$

The equivalent leakage reactance of the transformer is, accordingly,

$$\boxed{X_{eq} \approx X_{sc} = \sqrt{Z_{sc}^2 - R_{sc}^2}} \tag{3-43}$$

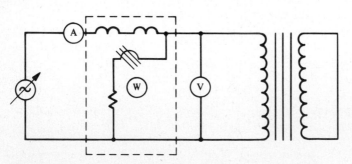

**Figure 3-18**   Short-circuit test with indicating instruments.

**EXAMPLE 3-3**

The following data was obtained in a short-circuit test of the transformer in Example 3-1, with its low-voltage winding short-circuited.

| Volts | Amperes | Watts |
|-------|---------|-------|
| 63.0  | 62.5    | 1600  |

Calculate (a) the equivalent primary impedance, (b) the equivalent primary reactance, and (c) the equivalent primary resistance.

*Solution*

a.   $Z_{eq\ 1} \approx \dfrac{V_{sc}}{I_{sc}} = \dfrac{63.0}{62.5} = 1.008\ \Omega$

b.   $R_{eq\ 1} \approx \dfrac{P_{sc}}{I_{sc}^2} = \dfrac{1660}{(62.5)^2} = 0.425\ \Omega$

c.   $X_{eq\ 1} = \sqrt{Z_{eq\ 1}^2 - R_{eq\ 1}^2} \approx \sqrt{(1.008)^2 - (0.425)^2} = 0.915\ \Omega$

## 3-7 TRANSFORMER LOSSES AND EFFICIENCY

The losses in a transformer are the *core losses,* which for a given voltage and frequency are practically independent of the load; the *copper losses,*[†] due to the resistance of the windings; and the *stray losses,* largely due to eddy currents induced by the leakage fluxes in the tank and other parts of the structure. The sum of the copper losses[‡] and the stray losses is called the *load losses,* being $I^2R_{eq}$, as determined from the short-circuit test. It will be recalled that the core loss is determined from the open-circuit test.

The efficiency of transformers at rated load is quite high. A value of 90 percent is not uncommon for transformers as small as 1 kVA, with greater values of efficiency as the ratings increase. The efficiency is expressed by

$$\text{Efficiency} = 1 - \frac{\text{losses}}{\text{input}} \qquad (3\text{-}44)$$

Methods of testing transformers and of calculating their performance are specified in the ANSI Standards,§ which include corrections and refinements beyond the scope of this text.

---

† The term copper loss is still used to indicate resistance losses of winding materials whether copper or aluminum is used.

‡ The term *copper losses* is still sometimes used instead of *load losses* and when so used is meant to include the stray losses.

§ *ANSI Standard C57, Transformers, Regulators and Reactors* (New York: American National Standards Institute).

While the equivalent circuits, including the approximate equivalent circuit, may be used to calculate the losses for a given output, it is usually more convenient to use the data of the open-circuit and short-circuit tests directly.

**EXAMPLE 3-4**

Calculate the efficiency of the transformer in Examples 3-1, 3-2, and 3-3, (a) at rated load 0.80 power factor and (b) at one-half rated load 0.60 power factor.

*Solution*

a. The real power output is

$$\text{Output} = 150{,}000 \times 0.80 = 120{,}000 \text{ W}$$

The load loss at rated load equals the real power measured in the short-circuit test at rated current:

$$\text{Load loss} = P_{sc} = 1660 \text{ W}$$

The core loss is taken as the no-load loss (i.e., the real power measured in the open-circuit test):

$$\text{Core loss} = P_{oc} = 580 \text{ W}$$

$$\text{Losses} = 1660 + 580 = 2240 \text{ W}$$

$$\text{Real power input} = \text{output} + \text{losses}$$

$$= 120{,}000 + 2240 = 122{,}240 \text{ W}$$

The rated-load efficiency at 0.80 power factor is therefore

$$\text{Efficiency} = 1 - \frac{\text{losses}}{\text{input}} = 1 - \frac{2240}{122{,}240}$$

$$= 1 - 0.0183 = 0.9817$$

b. The real power output at one-half rated load and 0.60 power factor is:

$$\text{Output} = \tfrac{1}{2} \times 150{,}000 \times 0.60 = 45{,}000 \text{ W}$$

The load loss, being $I^2 R_{eq}$, varies as the current squared, and at one-half rated load the rated current is one-half rated value. The load loss is therefore $(\tfrac{1}{2})^2 = \tfrac{1}{4}$ of that at rated current value; hence,

$$\text{Load losses} = \tfrac{1}{4} P_{sc} = \tfrac{1}{4} \times 1660 = 415 \text{ W}$$

The core losses are considered unaffected by the load, as long as the secondary terminal voltage is at its rated value. Therefore,

$$\text{Core losses} = P_{oc} = 580 \text{ W}$$

$$\text{Losses} = 415 + 580 = 995 \text{ W}$$

and

$$\text{Input} = 45{,}000 + 995 = 45{,}995 \text{ W}$$

The efficiency at one-half rated load and 0.60 power factor is

$$\text{Efficiency} = 1 - \frac{995}{45,995} = 0.9784$$

## 3-8 VOLTAGE REGULATION

The voltage regulation is an important measure of transformer performance and is expressed in percent by

$$\boxed{\text{Percent regulation} = \frac{E_{oc\,2} - V_2}{V_2} \times 100} \qquad (3\text{-}45)$$

or by

$$\boxed{\begin{aligned}\text{Percent regulation} &= \frac{aE_{oc\,2} - aV_2}{aV_2} \times 100 \\ &= \frac{V_1 - aV_2}{aV_2} \times 100\end{aligned}}$$

where $V_2$ is the rated secondary voltage at rated load and $E_{oc\,2}$ is the no-load secondary voltage with the same value of primary voltage for both rated load and no load. *The quantities used in Eq. 3-45 are magnitudes, not phasors.*

Calculations of regulation are usually based on the approximate equivalent circuit.

**EXAMPLE 3-5**

Calculate the regulation of the transformer in Example 3-1 for rated load 0.80 power factor, current lagging, on the basis of the data in Example 3-3.

*Solution.* The primary voltage based on the approximate equivalent circuit (Fig. 3-14b) is

$$\mathbf{V}_1 = a\mathbf{V}_2 + (R_{eq\,1} + jX_{eq\,1})\frac{\mathbf{I}_L}{a}$$

$$a\mathbf{V}_2 = 2400 + j0 \text{ V}$$

$$\frac{\mathbf{I}_L}{a} = 62.5(0.8 - j0.6) = 50.0 - j37.5 \text{ A}$$

$$R_{eq\,1} + jX_{eq\,1} = 0.425 + j0.915 \ \Omega$$

$$\mathbf{V}_1 = 2400 + j0 + (0.425 + j0.915)(50.0 - j37.5)$$

$$= 2455.6 + j29.8 = 2456\underline{/0.70°}$$

$$\text{Percent regulation} = \frac{2456 - 2400}{2400} \times 100 = 2.33$$

Loads such as motors and incandescent lamps require operation near rated voltage and frequency. Excessive voltage shortens lamp life, and motors when delivering rated load at subnormal voltage require overcurrents, leading to overheating. Such loads should be supplied by transformers that have small values of regulation (i.e., a few percent). On the other hand, loads such as series lighting systems and welding arcs that operate at nearly constant current are each supplied from its own individual transformer—one with a high value of regulation—designed for high leakage reactance.

## 3-9 AUTOTRANSFORMERS

The two-circuit transformer discussed in the foregoing can be converted to an autotransformer by connecting its windings 1 and 2 in series with each other as shown in Fig. 3-19. The use of an autotransformer in place of the two-circuit transformer effects a great saving in size in applications in which the ratio of transformation does not differ too greatly from unity and which do not require the secondary winding to be isolated from the primary winding. In Fig. 3-19 consider winding 1 to be between points $a$ and $b$ and winding 2 to be between points $b$ and $c$. The high-side voltage $\mathbf{V}_H$ of the autotransformer is the phasor sum of the terminal voltages of windings 1 and 2:

$$\mathbf{V}_H = \mathbf{V}_1 + \mathbf{V}_2 = \mathbf{V}_1 + \mathbf{V}_x \tag{3-46}$$

The phase angle between the primary and secondary voltages of power transformers is small, usually less than $10°$, and the phasor sum is therefore practically equal to the arithmetical sum in magnitude, so

$$V_H \approx V_1 + V_2 = V_x + V_1 \tag{3-47}$$

Also, the ratio of the terminal voltages $\mathbf{V}_1$ and $\mathbf{V}_2$ differs from the turns ratio by only a few percent. Hence,

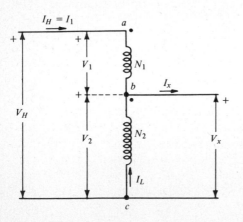

**Figure 3-19** Autotransformer connections.

$$\frac{V_H}{V_x} \approx \frac{V_1 + V_x}{V_x} \approx \frac{N_1 + N_2}{N_2} \tag{3-48}$$

Since the exciting current in iron-core transformers is usually small in terms of the rated current, the mmfs of the two windings are nearly equal. Hence,

$$N_1 I_1 = N_1 I_H \approx N_2 I_L$$

and
$$\mathbf{I}_H = \mathbf{I}_1 \approx \frac{N_2}{N_1} \mathbf{I}_L \tag{3-49}$$

From Fig. 3-19 it is evident that the current in the low-voltage terminals is the phasor sum expressed by

$$\mathbf{I}_x = \mathbf{I}_1 + \mathbf{I}_L \tag{3-50}$$

which, on comparison with Eq. 3-49, shows that

$$\frac{\mathbf{I}_H}{\mathbf{I}_x} \approx \frac{N_2}{N_1 + N_2} \tag{3-51}$$

For ratios near unity the rating as an autotransformer is much higher than for two-circuit operation, as is shown in the following:

$$\text{Autotransformer rating} = V_H I_H \approx (V_1 + V_x) I_1$$
$$\text{Two-circuit transformer rating} = V_1 I_1$$

from which

$$\boxed{\frac{\text{Autotransformer rating}}{\text{Two-circuit transformer rating}} = \frac{V_x}{V_1} + 1} \tag{3-52}$$

Thus, when a two-circuit transformer rated at 2400/240 V is connected as an autotransformer for a voltage rating of 2640/2400, its volt-ampere rating is increased in the ratio of 11:1.

## 3-10 INSTRUMENT TRANSFORMERS

There are two general kinds of instrument transformers, *instrument potential transformers* and *instrument current transformers*. These are used in ac power circuits operating in excess of a few hundred volts to supply instruments, protective relays, and control circuits. Instrument potential transformers usually step high voltages down to about 115 secondary volts for supplying voltmeters and wattmeters, in addition to relays and control devices. Instrument current transformers are connected in series with the line to step high values of current down to a rated value of about 5 A for ammeters and wattmeters. Potential transformers are usually connected line to neutral in three-phase installations, although line-to-line connections are not uncommon. The secondaries of instrument transformers are grounded for reasons of safety, to eliminate the hazard of raising the secondary winding to a high potential through the capacitance coupling with the primary winding. As a matter of safety, the secondary circuit of a current transformer should never be opened under load, because there would then be no secondary mmf to oppose the primary mmf, and all the primary current would become exciting current and thus might induce a very high voltage in the secondary.

## 3-11 THREE-PHASE TRANSFORMER CONNECTIONS

Commercial power is, practically without exception, generated and transmitted as three-phase, involving various three-phase voltage transformations. Three-phase power may be converted from one voltage to another by means of two or three single-phase transformers or by the use of one three-phase transformer. In the case of one three-phase transformer or of three single-phase transformers, several three-phase arrangements can be used. The following are quite common: the delta–delta, the wye–wye, and the wye–delta or delta–wye connections. The equivalent circuits in Figs. 3-11 and 3-14(a) apply to each of the transformers in these various arrangements.

In the case of identical transformers in a given arrangement, each transformer carries one-third of the three-phase load under conditions of balanced load and balanced voltages.

### 3-11.1 Delta-Delta Connection

Three single-phase transformers, assumed to be identical, are shown in Fig. 3-20 with their primaries and secondaries both connected in delta. Figure 3-20 shows that full line-to-line voltage exists across the windings of each transformer. And when the leakage impedances of the transformers are neglected, the secondary line-to-line voltages $V_{ab}$, $V_{bc}$, and $V_{ca}$ are in phase with the primary line-to-line voltages $V_{AB}$, $V_{BC}$, and $V_{CA}$ with the voltage ratios equaling the turns ratio:

$$\frac{V_{AB}}{V_{ab}} = \frac{V_{BC}}{V_{bc}} = \frac{V_{CA}}{V_{ca}} = a$$

Phasor diagrams for an ideal transformer bank connected delta–delta are shown in Fig. 3-21.

Under balanced conditions, the line currents are $\sqrt{3}$ times the currents in the windings when the third harmonics in the exciting current are neglected, as is evident from Figs. 3-20 and 3-21(a); thus,

$$\mathbf{I}_A = \mathbf{I}_{AB} - \mathbf{I}_{CA}$$

and
$$\mathbf{I}_{CA} = \mathbf{I}_{AB}\underline{/120°}$$

from which

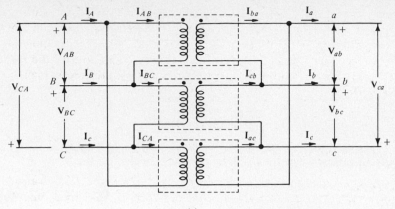

(a)

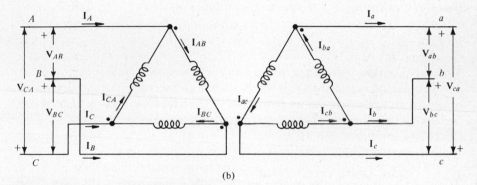

(b)

**Figure 3-20** Delta–delta connection. (a) Common physical arrangement of three single-phase transformers. (b) Schematic diagram.

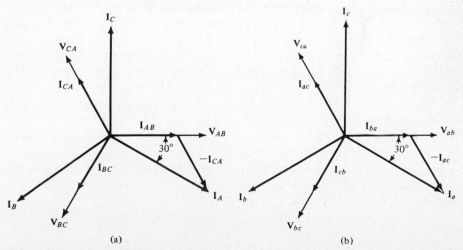

(a)                                                                                       (b)

**Figure 3-21** Phasor diagram for delta–delta bank of ideal transformers supplying balanced noninductive load. (a) Primary. (b) Secondary.

$$\mathbf{I}_A = \mathbf{I}_{AB}(1 - 1\underline{/120°}) = \sqrt{3}\ \mathbf{I}_{AB}\underline{/-30°} \tag{3-53}$$

Similarly,

$$\mathbf{I}_B = \sqrt{3}\ \mathbf{I}_{BC}\underline{/-30°} \tag{3-54}$$

and

$$\mathbf{I}_C = \sqrt{3}\ \mathbf{I}_{CA}\underline{/-30°} \tag{3-55}$$

The current ratios, when the exciting current is neglected, are

$$\frac{I_{AB}}{I_{ab}} = \frac{I_{BC}}{I_{bc}} = \frac{I_{CA}}{I_{ca}} = \frac{I_A}{I_a} = \frac{I_B}{I_b} = \frac{I_C}{I_c} = \frac{1}{a} \tag{3-56}$$

The delta–delta connection is generally used in moderate voltage systems because the windings operate at full line-to-line voltage.

## 3-11.2 Wye–Wye Connection

Figure 3-22 shows three single-phase transformers with their primaries and secondaries each connected in wye (Y). A neutral connection (in many cases consisting of ground) is shown on both sides of the transformer bank. The neutral connection between the primary of the transformers and the source assures balanced line-to-neutral voltage and provides a path for the third-harmonic components in the exciting currents.

Under balanced three-phase conditions the neutral conductor carries no fundamental component of current because the fundamentals in the three phases are equal and 120° apart and their sum is therefore zero. However, the third harmonics are displaced from each other by $3 \times 120°$, or 360°, which means that they are in phase with each other and the neutral conductor carries three times the third harmonic current of one phase. This is also true of all multiples of the third harmonics. In the absence of a neutral connection, the third harmonics and multiples thereof are absent from the exciting current and the corresponding harmonics therefore appear in the flux waveform and consequently in the line-to-neutral voltages. Third harmonics and their multiples are negligible in the line-to-line voltages because the line-to-line voltages are the phasor differences between the line-to-neutral voltages. The third harmonics and their multiples, being in phase, cancel.

It is evident from Fig. 3-22 that the current in the transformer winding, when connected in wye, is the line current. Then for ideal transformers the voltage ratios are

$$\frac{\mathbf{V}_{AN}}{\mathbf{V}_{an}} = \frac{\mathbf{V}_{BN}}{\mathbf{V}_{bn}} = \frac{\mathbf{V}_{CN}}{\mathbf{V}_{cn}} = a \tag{3-57}$$

and the current ratios

$$\frac{\mathbf{I}_A}{\mathbf{I}_a} = \frac{\mathbf{I}_B}{\mathbf{I}_b} = \frac{\mathbf{I}_C}{\mathbf{I}_c} = \frac{1}{a} \tag{3-58}$$

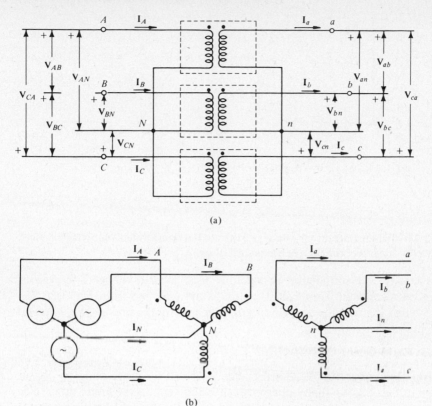

(a)

(b)

**Figure 3-22** Wye–wye connection. (a) Common physical arrangement of three single-phase transformers. (b) Schematic diagram showing primary neutral connected to the source and the secondary neutral going to the neutral of the load. The load is not shown.

Figure 3-23 shows phasor diagrams for the wye–wye connection under balanced conditions, from which the following relationships between line-to-line and line-to-neutral voltages are evident:

$$\mathbf{V}_{AB} = \mathbf{V}_{AN} - \mathbf{V}_{BN}$$

and

$$\mathbf{V}_{BN} = \mathbf{V}_{AN}\underline{/-120°}$$

so that

$$\mathbf{V}_{AB} = \mathbf{V}_{AN}(1 - 1\underline{/-120°}) = \sqrt{3}\,\mathbf{V}_{AN}\underline{/30°} \qquad (3\text{-}59)$$

Similarly,

$$\mathbf{V}_{BC} = \sqrt{3}\,\mathbf{V}_{BN}\underline{/30°} \qquad (3\text{-}60)$$

and

$$\mathbf{V}_{CA} = \sqrt{3}\,\mathbf{V}_{CN}\underline{/30°} \qquad (3\text{-}61)$$

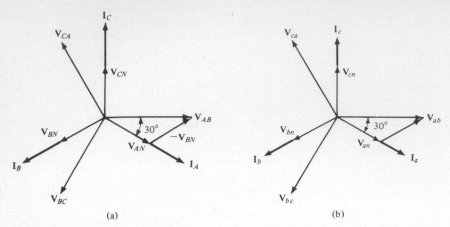

**Figure 3-23** Phasor diagram for wye–wye bank of ideal transformers supplying balanced noninductive load. (a) Primary. (b) Secondary.

The wye–wye connection is generally used in high-voltage applications because, as is apparent from Eqs. 3-59, 3-60, and 3-61 and from Fig. 3-23, the voltage across the transformer winding is only $1/\sqrt{3}$ of the line-to-line voltage.

### 3-11.3 Wye–Delta Connection

The wye–delta connection is shown in Fig. 3-24 and the corresponding phasor diagram in Fig. 3-25. In high-voltage transmission systems the high-voltage side is connected in wye and the low-voltage side in delta. A common arrangement in distribution circuits is the 208/120-V system supplied by the wye connection on the low-voltage side with the high-voltage side of the transformer bank or three-phase transformer connected in delta. In such systems the neutral point of the wye is grounded and single-phase loads are connected line to neutral for 120-V operation, while three-phase equipment such as motors are connected across the three line wires for 208-V operation.

The delta connection assures balanced line-to-neutral voltage on the wye side and provides a path for the circulation of the third harmonics and their multiples without the use of a neutral wire.

When the phases are designated as in Fig. 3-24, there is an angle of 30° between the line-to-line voltages on one side and the corresponding voltages on the other side. With other designations of the phases, angles of 90° and 150° may result.

### 3-11.4 Open-Delta or V–V Connection

In the open-delta (or V–V) connection, two instead of three single-phase transformers are used for three-phase operation, as shown in Fig. 3-26. This connection is sometimes used in the case of instrument transformers for reasons of economy and sometimes initially in load centers, the full growth of which may require several years, at which time a third transformer is added for delta–delta operation.

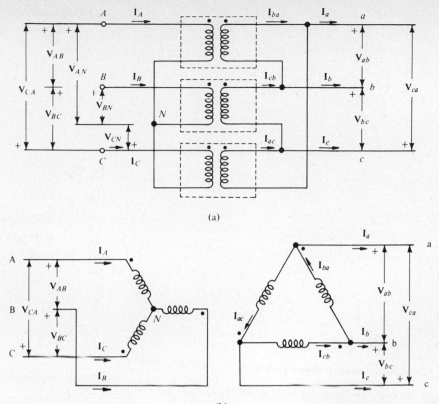

(a)

(b)

**Figure 3-24** Wye–delta connection. (a) Common physical arrangement of three single-phase transformers. (b) Schematic diagram.

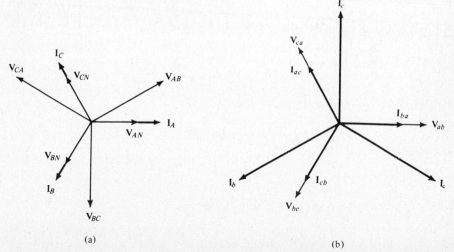

(a)

(b)

**Figure 3-25** Phasor diagram for wye–delta arrangement of Fig. 3.24 for ideal transformers supplying balanced noninductive load. (a) Primary wye–connected. (b) Secondary delta–connected.

**125**

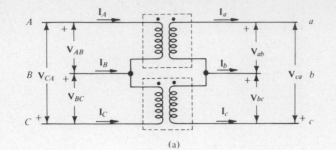

(a)

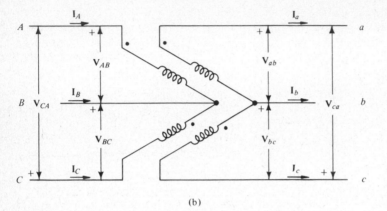

(b)

**Figure 3-26** Open delta (or V–V) connection. (a) Common physical arrangement. (b) Schematic diagram.

Since the line current is also the current in the windings of the transformers, the rating of two identical transformers operating open-delta is only $1/\sqrt{3}$ of three similar transformers connected delta–delta. Balanced three-phase voltage applied to the primary would produce balanced three-phase voltage on the secondary side if the leakage impedance were negligible. Thus, if $\mathbf{V}_{AB}$, $\mathbf{V}_{BC}$, and $\mathbf{V}_{CA}$ are balanced voltages applied to the primary, we have

$$\mathbf{V}_{BC} = \mathbf{V}_{AB}\underline{/-120°}$$

and
$$\mathbf{V}_{CA} = \mathbf{V}_{AB}\underline{/-240°}$$

Then, if the leakage impedances of the transformers are neglected, the secondary voltages are

$$\mathbf{V}_{ab} = \frac{\mathbf{V}_{AB}}{a}$$

and
$$\mathbf{V}_{bc} = \frac{\mathbf{V}_{BC}}{a} = \frac{\mathbf{V}_{AB}}{a}\underline{/-120°}$$

Also, from Kirchhoff's law,

$$V_{ab} + V_{bc} + V_{ca} = 0$$

and
$$V_{ca} = V_{ab}\underline{/-240°}$$

In the case of appreciable leakage reactance, the secondary voltages will be somewhat unbalanced, even with balanced voltage applied to the primaries, because in the open-delta arrangement there are only two, instead of three, leakage impedance voltage drops.

## 3-11.5 Three-Phase Transformers

The three-phase transformer is one in which the cores and windings for all three phases are combined in a single structure—an arrangement that affords a considerable saving in space and initial investment compared with three single-phase transformers of the same total rating. This is extremely important, since modern power transformers are of large size—as, for example, a three-phase autotransformer rated at 500 MVA, 765–345 kV, 60 Hz in operation on the American Electric Power System.† Common arrangements of the core and windings in three-phase transformers are shown schematically in Fig. 3-27.

The core in Fig. 3-27(a) has only three legs and, since the sum of the three fluxes, $\phi_a$, $\phi_b$, and $\phi_c$, is zero for balanced three-phase operation, a fourth leg is unnecessary.

In the shell-type transformer of Fig. 3-27(b), the polarities of both middle windings are reversed with respect to those of the outer windings. With this arrangement the fluxes from the adjacent phases *add,* in the regions of *x*, to that of the middle phase instead of subtracting. For balanced three-phase voltages, the fluxes are 120° apart and the maximum value of the sum equals the maximum value of the flux in one phase. On the other hand, the maximum value of the difference is $\sqrt{3}$ times the maximum value of the flux in one phase. The width of the legs *y* would need to be $\sqrt{3}$ times that of legs *z* if the polarities of the middle windings were the same as those of the outer windings. A three-phase transformer is shown in Fig. 3-28.

## 3-11.6 Three-to-Six-Phase Transformation

Electronic rectifiers convert large amounts of ac power to dc power, and a smoother waveform is more easily obtained on the dc side as the number of phases is increased. Objectionable harmonics in the alternating currents are also reduced with a greater number of phases. Six-phase is therefore preferable to three-phase for rectification, and twelve-phase rectifiers are used in larger installations. Transformers may be connected in several ways to effect three-to-six-phase transformation. One arrangement, known as the six-phase star connection, is shown in Fig. 3-29. The secondaries have center taps which are connected together to form the neutral on the six-phase side. Figure 3-29 shows the primaries connected in delta, but these may also be connected in wye. Three

† "First 765-kV Transformer," *Power Eng.* (May 1968):38–39.

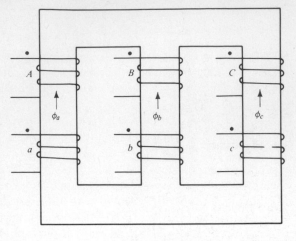

(a)

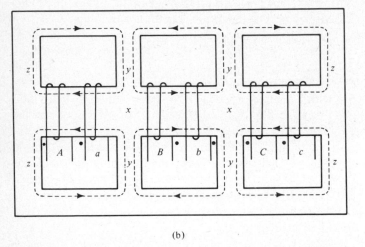

(b)

**Figure 3-27**   Three-phase transformers. (a) Core type. (b) Shell type.

single-phase transformers or one three-phase transformer may be used for three-to-six-phase transformation.

## 3-12 PER-UNIT QUANTITIES OF TRANSFORMERS

In Chap. 1 it was shown that per-unit quantities afford a ready means for comparing the performance of a line of transformers or of a line of rotating machines. In addition, they simplify the analyses of complex power systems involving transformers of different ratios. Per-unit quantities are the ratios of current, voltage, impedance, admittance, and power to selected base values of these respective quantities:

**Figure 3-28**    Three-phase core-type transformer rated 69,000/15,000 V, 60 Hz.

$$\text{Per-unit current} = \frac{\text{current in amperes}}{\text{base amperes}} \qquad (3\text{-}62)$$

$$\text{Per-unit voltage} = \frac{\text{voltage in volts}}{\text{base volts}} \qquad (3\text{-}63)$$

$$\text{Per-unit impedance} = \frac{\text{impedance in ohms}}{\text{base ohms}} \qquad (3\text{-}64)$$

$$\text{Per-unit admittance} = \frac{\text{admittance in mhos}}{\text{base mhos}} \qquad (3\text{-}65)$$

$$\text{Per-unit power} = \frac{\text{power in volt-amperes}}{\text{base volt-amperes}} \qquad (3\text{-}66)$$

The values of base quantities are selected as a matter of convenience. When dealing with a specific piece of equipment by itself, such as a transformer, the rating of the transformer is generally taken as the base.

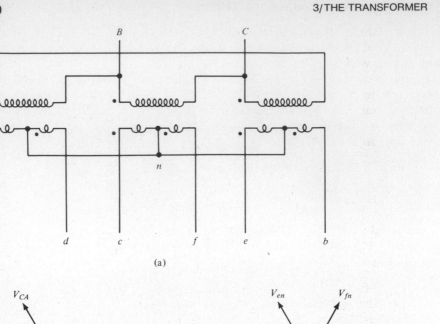

(a)

(b)

**Figure 3-29** (a) Three-phase delta to six-phase star connection. (b) Phasor diagram of primary and secondary voltages.

In studies of power systems involving several pieces of equipment it is usually convenient to use a relatively large MVA base. The per-unit impedance values of generators and transformers are generally known in terms of their individual ratings and may be conveniently converted to any other volt-ampere base. Thus, if $Z_{pu\ 1}$ is the impedance per unit for a volt-ampere base $VA_1$, its per-unit impedance for a volt-ampere base of $VA_2$ is

$$Z_{pu\ 2} = Z_{pu\ 1}\frac{VA_2}{VA_1} \tag{3-67}$$

The per-unit impedance is generally expressed by

$$Z_{pu} = Z_{ohms}\frac{VA_{base}}{(V_{base})^2} \tag{3-68}$$

Equation 3-68 also applies to three-phase systems in which $Z_{ohms}$ is the line-to-neutral impedance in ohms per phase and $VA_{base}$ is the three-phase volt-ampere base with $V_{base}$ the base line-to-line volts.

There are cases when generators and also transformers are operated in power systems at somewhat reduced voltages, as for example, 13.8-kV transformers operating at 13.2 kV and at normal frequency. The change in voltage base must be taken into account when changing the per-unit impedance from one base to another as follows:

$$Z_{\text{pu 2}} = Z_{\text{pu 1}} \frac{VA_{\text{base 2}}}{VA_{\text{base 1}}} \left(\frac{V_{\text{base 1}}}{V_{\text{base 2}}}\right)^2 \tag{3-69}$$

Typical values of power-transformer impedances and their ranges are listed in Tables 3-1 and 3-2. Table 3-3 shows typical values of efficiency for power transformers.

**EXAMPLE 3-6**

Using the transformer of Example 3-5, find the per-unit impedance for the transformer

    a. On its own base.
    b. On a base of 10,000 kVA.

*Solution*

$$\text{Base volt-amperes} = 150,000$$

$$\text{Base volts} = 240$$

$$\begin{aligned} \text{Base amperes} &= \text{base volt-amperes} \div \text{base volts} \\ &= 150,000 \div 240 \\ &= 625 \end{aligned}$$

$$\begin{aligned} \text{Base ohms} &= \text{base volts} \div \text{base amperes} \\ &= 0.384 \end{aligned}$$

$$\begin{aligned} \text{Base mhos} &= \text{base amperes} \div \text{base volts} \\ &= \text{reciprocal of base ohms} \\ &= 2.604 \end{aligned}$$

The base quantities on the high-voltage side are, accordingly, 150,000 VA, 2400 V, 62.5 A, 38.4 $\Omega$, and 0.02604 S.

    a. The leakage impedance of this transformer was found in Example 3-3 to be 1.008 $\Omega$ referred to the high-voltage side, and the per-unit impedance, from Eq. 3-64, is

$$Z_{eq} = \frac{1.008}{38.4} = 0.0263 \text{ per unit}$$

The same result is obtained by dividing the leakage impedance referred to the low-voltage side (0.01008 $\Omega$) by the base impedance (0.384 $\Omega$) on that side. This shows *the per-unit impedance of a transformer to have the same value whether referred to the high-voltage or to the low-voltage side.*

b. The per-unit impedance of the transformer in Example 3-5 for a base of 10,000 kVA would be

$$Z_{pu} = 0.0263 \times \frac{10,000,000}{150,000} = 1.755$$

**TABLE 3-1.** TRANSFORMER IMPEDANCES[a]
Standard Range in Impedances for Two-Winding Power
Transformers Rated at 55°C Rise (Both 25- and
60-Hz Transformers)

| High-voltage winding insulation class, kV | Low-voltage winding insulation class, kV | Class[b] OA OW OA/FA OA/FA/FOA | | Class[b] FOA FOW | |
|---|---|---|---|---|---|
| | | Min. | Max. | Min. | Max. |
| 15 | 15 | 4.5 | 7.0 | 6.75 | 10.5 |
| 25 | 15 | 5.5 | 8.0 | 8.25 | 12.0 |
| 34.5 | 15 | 6.0 | 8.0 | 9.0 | 12.0 |
| | 25 | 6.5 | 9.0 | 9.75 | 13.5 |
| 46 | 25 | 6.5 | 9.0 | 9.75 | 13.5 |
| | 34.5 | 7.0 | 10.0 | 10.5 | 15.0 |
| 69 | 34.5 | 7.0 | 10.0 | 10.5 | 15.0 |
| | 46 | 8.0 | 11.0 | 12.0 | 16.5 |
| 92 | 34.5 | 7.5 | 10.5 | 11.25 | 15.75 |
| | 69 | 8.5 | 12.5 | 12.75 | 18.75 |
| 115 | 34.5 | 8.0 | 12.0 | 12.0 | 18.0 |
| | 69 | 9.0 | 14.0 | 13.5 | 21.0 |
| | 92 | 10.0 | 15.0 | 15.0 | 23.25 |
| 138 | 34.5 | 8.5 | 13.0 | 12.75 | 19.5 |
| | 69 | 9.5 | 15.0 | 14.25 | 22.5 |
| | 115 | 10.5 | 17.0 | 15.75 | 25.5 |
| 161 | 46 | 9.5 | 15.0 | 13.5 | 21.0 |
| | 92 | 10.5 | 16.0 | 15.75 | 24.0 |
| | 138 | 11.5 | 18.0 | 17.25 | 27.0 |
| 196 | 46 | 10 | 15.0 | 15.0 | 22.5 |
| | 92 | 11.5 | 17.0 | 17.25 | 25.5 |
| | 161 | 12.5 | 19.0 | 18.75 | 28.5 |
| 230 | 46 | 11.0 | 16.0 | 16.5 | 24.0 |
| | 92 | 12.5 | 18.0 | 18.75 | 27.0 |
| | 161 | 14.0 | 20.0 | 21.0 | 30.0 |

[a] Reprinted by permission from *Electrical Transmission and Distribution Reference Book*, 4th ed. (East Pittsburgh, Pa., Westinghouse Electric Corporation, 1950).

[b] The impedances are expressed in percent on the self-cooled rating of OA/FA and OA/FA/FOA. *Note:* The through impedance of a two-winding autotransformer can be estimated knowing rated circuit voltages, by multiplying impedance obtained from this table by the factor $(HV - LV)/HV$. Definition of transformer classes: OA—oil-immersed, self-cooled; OW—oil-immersed, water-cooled; OA/FA—oil-immersed, self-cooled/forced-air-cooled; OA/FA/FOA—oil-immersed, self-cooled/forced-air-cooled/forced-oil-cooled; FOA—oil-immersed, forced-oil-cooled with forced air cooler; FOW—oil-immersed-oil-cooled with water cooler.

TABLE 3-2. TRANSFORMER IMPEDANCES[a]
Standard Reactances and Impedances for Ratings 500 kVA and Below (for 60-Hz Transformers)

| Single-phase kVA rating[b] | Rated-voltage class, kV | | | | | | | |
|---|---|---|---|---|---|---|---|---|
| | 2.5 | | 15 | | 25 | | 69 | |
| | Average reactance, % | Average impedance, % | Average reactance, % | Average impedance, % | Average reactance, % | Average impedance, % | Average reactance, % | Average impedance, % |
| 3 | 1.1 | 2.2 | 0.8 | 2.8 | | | | |
| 10 | 1.5 | 2.2 | 1.3 | 2.4 | 4.4 | 5.2 | | |
| 25 | 2.0 | 2.5 | 1.7 | 2.3 | 4.8 | 5.2 | | |
| 50 | 2.1 | 2.4 | 2.1 | 2.5 | 4.9 | 5.2 | 6.3 | 6.5 |
| 100 | 3.1 | 3.3 | 2.9 | 3.2 | 5.0 | 5.2 | 6.3 | 6.5 |
| 500 | 4.7 | 4.8 | 4.9 | 5.0 | 5.1 | 5.2 | 6.4 | 6.5 |

[a] Reprinted by permission from *Electrical Transmission and Distribution Reference Book*, 4th ed. (East Pittsburgh, Pa., Westinghouse Electric Corporation, 1950).
[b] For three-phase transformers use one-third of the three-phase kVA rating, and enter table with rated line-to-line voltages.

**TABLE 3-3.** APPROXIMATE VALUES OF EFFICIENCY FOR 60-HZ
TWO-WINDING, OIL-IMMERSED SELF-COOLED
THREE-PHASE POWER TRANSFORMERS[a,b]
Full load, unity power factor, at 75°C

|        | Voltage class |         |        |         |         |
|--------|--------|---------|--------|---------|---------|
| kVA    | 15 kV  | 34.5 kV | 69 kV  | 138 kV  | 161 kV  |
| 2,000  | 98.97  | 98.89   | 98.83  | 98.56   | 98.47   |
| 10,000 | 99.23  | 99.22   | 99.17  | 99.12   | 99.11   |
| 50,000 |        | 99.47   | 99.45  | 99.44   | 99.44   |

[a] Reprinted by permission from *Electrical Transmission and Distribution Reference Book*,
4th ed. (East Pittsburgh, Pa: Westinghouse Electric Corporation, 1950).

[b] These figures apply also to OA/FA and OA/FA/FOA transformers, at loads corresponding
to their OA ratings.

## 3-13 MULTICIRCUIT TRANSFORMERS

One multicircuit transformer† or multiwinding transformer is frequently used
to interconnect three or more circuits that may have different voltages. This
makes for greater economy than the use of two or more two-circuit transformers
to effect the same interconnection. The distribution transformer is an example
of a three-winding transformer in which the primary winding is usually rated at
2000 V or above and in which there are two 120-V secondaries connected in
series with the common or neutral grounded as shown in Fig. 3-30. The transformer thus supplies a load comprised of 120-V appliances connected from line
to neutral, and divided roughly equally between the two 120-V secondaries.
Appliances, such as electric stoves that operate at 240 V, are connected across
the two windings in series.

Large three-winding power transformers are commonly used in three-phase
power systems. The primaries and secondaries are connected for wye–wye operation and the tertiaries‡ are connected in delta to assure balanced voltages and

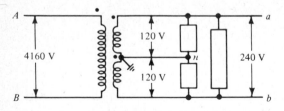

**Figure 3-30**   Distribution transformer.

† For a discussion of transformers with more than three circuits, see MIT Staff, *Magnetic
Circuits and Transformers* (New York: John Wiley & Sons, Inc., 1943), Chap. V, and L. F. Blume
et al., *Transformer Engineering* (New York: John Wiley & Sons, Inc., 1954), pp. 117–133.

‡ Tertiary—the third winding of a transformer using the same core.

to provide a path for the third harmonics and their multiples in the exciting current.

Figure 3-31(a) shows a schematic diagram of the windings in a three-circuit transformer. If the transformer were ideal, we would have

$$\frac{V_2}{V_1} = \frac{N_2}{N_1} \tag{3-70}$$

$$\frac{V_3}{V_1} = \frac{N_3}{N_1} \tag{3-71}$$

$$\boxed{N_1 I_1 = N_2 I_s + N_3 I_T} \tag{3-72}$$

where $V_1$, $V_2$, and $V_3$ are the primary, the secondary, and the tertiary terminal voltages, respectively, and $N_1$, $N_2$, and $N_3$ are the turns in the respective windings. Also, $I_1$, $I_s$, and $I_T$ are the currents in the three windings.

The equivalent circuits of a three-circuit or three-winding transformer that take into account the leakage impedance and exciting admittance are shown in Fig. 3-31(b) and (c).

## 3-13.1 Open-Circuit and Short-Circuit Tests

The open-circuit test is made in the same manner as that for a two-winding transformer, and it yields the data for calculating the exciting admittance. Since

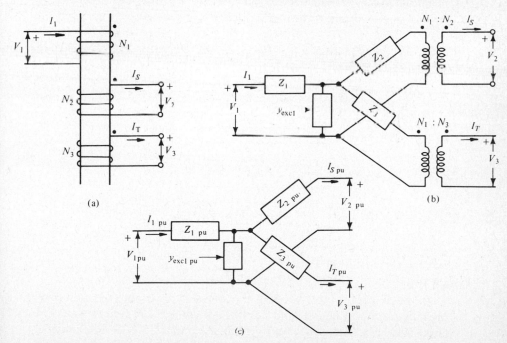

**Figure 3-31** Three-circuit transformer. (a) Circuit diagram. (b) Equivalent circuit in ohms and siemens referred to primary. (c) Equivalent circuit in per-unit.

there are three windings, *three* short-circuit tests are required to obtain the data for the equivalent circuits of Fig. 3-31(b) and (c). Assume the primary to have the highest voltage rating, the tertiary the lowest voltage rating, and the secondary to have the intermediate voltage rating. The short-circuit tests are then usually made in the following manner.

1. Applying voltage to the primary with the secondary short-circuited and the tertiary open-circuited. The exciting current is neglected and the measured impedance is then

$$Z_{12} = Z_1 + Z_2 \tag{3-73}$$

2. Applying voltage to the primary with the tertiary short-circuited and the secondary open-circuited, so that, when the exciting current is neglected, we have

$$Z_{13} = Z_1 + Z_3 \tag{3-74}$$

3. Applying voltage to the secondary with the tertiary short-circuited and the primary open-circuited, whence

$$Z_{23} = Z_2 + Z_3 = \left(\frac{N_1}{N_2}\right)^2 \frac{V_s}{I_{ssc}} \tag{3-75}$$

when the exciting current is again neglected. In Eq. 3-75 $V_s$ is the voltage applied to the secondary and $I_{ssc}$ is the secondary current with the tertiary short-circuited and the primary open.

The leakage impedances $Z_1$, $Z_2$, and $Z_3$, all referred to the primary, are then given by

$$\boxed{Z_1 = \frac{Z_{12} + Z_{13} - Z_{23}}{2}} \tag{3-76}$$

$$\boxed{Z_2 = \frac{Z_{23} + Z_{12} - Z_{13}}{2}} \tag{3-77}$$

$$\boxed{Z_3 = \frac{Z_{13} + Z_{23} - Z_{12}}{2}} \tag{3-78}$$

**EXAMPLE 3-7**

Three single-phase, three-winding transformers are connected for three-phase operation at 66, 11, and 2.2 kV. All voltages are line to line. The 66-kV and 11-kV sides are connected in wye and the 2.2-kV side in delta. Assume the 66-kV side to be the primary, the 11-kV side to be the secondary, and the 2.2-kV side to be the tertiary. The following data was obtained on test:

a. Short-circuit test:

| Winding | | Excited winding | |
|---|---|---|---|
| Excited | Short-circuited | Volts | Amperes |
| Primary | Secondary | 2850 | 788 |
| Primary | Tertiary | 5720 | 657 |
| Secondary | Tertiary | 529 | 1575 |

b. Open-circuit test:

| Winding | Volts | Amperes |
|---|---|---|
| Tertiary | 2200 | 146 |

Determine the values of reactance and admittance to be used in the equivalent circuits of Fig. 3-31(b) and (c). Neglect the resistance of the windings. Use a base of 30,000 kVA per transformer or 90,000 kVA as the three-phase base for determining the per-unit values.

*Solution*

a. The values of reactance are based on the short-circuit test and will be referred to the primary.

$$Z_{12} = j2850 \div 788 = j3.62 \ \Omega$$

$$Z_{13} = j5720 \div 657 = j8.72 \ \Omega$$

$$Z_{23} = \left(\frac{N_1}{N_2}\right)^2 j529 \div 1575 = \left(\frac{66}{11}\right)^2 j529 \div 1575 = j12.10 \ \Omega$$

$$Z_1 = \frac{Z_{12} + Z_{13} - Z_{23}}{2} = \frac{j3.62 + j8.72 - j12.10}{2} = j0.12 \ \Omega$$

$$Z_2 = \frac{Z_{23} + Z_{12} - Z_{13}}{2} = \frac{j12.10 + j3.62 - j8.72}{2} = j3.50 \ \Omega$$

$$Z_3 = \frac{Z_{13} + Z_{23} - Z_{12}}{2} = \frac{j8.72 + j12.10 - j3.62}{2} = j8.60 \ \Omega$$

The exciting admittance, if the core losses are neglected, when referred to the tertiary, is

$$y_3 = -j\frac{146}{2200} = -j0.0664 \ \text{S}$$

and $$\left(\frac{N_3}{N_1}\right)^2 y_3 = \left(\frac{2.2}{66/\sqrt{3}}\right)^2 (-j0.0664) = -j0.000221 \ \text{S}$$

referred to the primary.

b. Since the impedances are all referred to the primary, the base impedance for 30,000 kVA and for the primary voltage of 66 kV/$\sqrt{3}$ will suffice as a base for all three impedances.

$$Z_{\text{base}} = \frac{(\text{volts})^2}{\text{volt-amperes}} = \frac{(66,000/\sqrt{3})^2}{30,000,000} = \frac{(66)^2}{90} = 48.4 \ \Omega$$

Hence,

$$Z_{\text{1pu}} = j0.12 \div 48.4 = j0.00248$$
$$Z_{\text{2pu}} = j3.50 \div 48.4 = j0.0724$$
$$Z_{\text{3pu}} = j8.60 \div 48.4 = j0.178$$
$$Y_{\text{pu}} = -j0.000221 \times 48.4 = -j0.0107$$

## 3-14 THIRD HARMONICS IN THREE-PHASE TRANSFORMER OPERATION

The third harmonic component of the exciting current may produce undesirable effects in three-phase transformer operation, particularly in the case of the wye–wye connection. Figure 3-32 shows a wye–wye connection, assumed for three identical single-phase transformers operating without load and supplied by a three-phase generator. The sum of the instantaneous primary currents including the neutral current must equal zero:

$$i_A + i_B + i_C + i_N = 0 \tag{3-79}$$

The sum of all harmonic components except the third harmonic and its multiples in $i_A$, $i_B$, and $i_C$ is equal to zero, and the neutral current $i_N$ must therefore contain only third harmonics and multiples thereof. Further since the phase displacement between harmonics is $120h$ degrees, where $h$ is the order of the harmonic, the neutral current must equal three times the third-harmonic current and multiples of the third harmonic. The multiples of the third harmonic have relative small amplitudes and the neutral current may be considered one that has three times the frequency of the fundamental.

If the neutral connection between the transformer primaries and the generator is opened, $i_N = 0$ and the third-harmonic currents must be zero. As a result, the flux cannot be sinusoidal—it will contain a third harmonic, which in

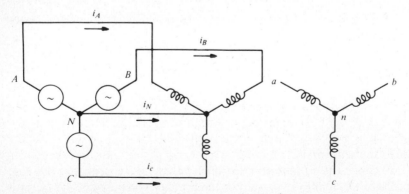

**Figure 3-32**   Exciting currents in wye–wye connection.

turn produces a third-harmonic voltage in the transformer line-to-neutral voltages. Since the transformers are identical, no third harmonics will exist in the line-to-line voltages because

$$v_{AB} = v_{AN} - v_{BN} \tag{3-80}$$

The third harmonic voltages in the three transformers are equal and in phase:

$$v_{AN3} = v_{BN3} = v_{CN3} \tag{3-81}$$

and

$$v_{AB3} = v_{AN3} - v_{BN3} = 0 \tag{3-82}$$

When the primaries of the identical transformers are connected in delta there will be no third-harmonic components in the line currents $i_A$, $i_B$, and $i_C$ because the line currents are the differences between the currents flowing in the delta-connected windings. It is evident from Fig. 3-20 that the instantaneous currents

$$i_A = i_{AB} - i_{CA} \tag{3-83}$$

and

$$i_{AB3} = i_{BC3} = i_{CA3} \tag{3-84}$$

from which

$$i_{A3} = i_{AB3} - i_{CA3} = 0 \tag{3-85}$$

The third-harmonic currents circulate in the delta.

If the primary neutral is disconnected from the source neutral in a wye–delta connection, third-harmonic current circulates in the delta-connected secondary windings so that the flux is sinusoidal. This results from the fact that the third-harmonic voltages, if they exist, in the three phases must be in phase with each other and their sum must therefore be equal to three times that of one phase. However, the sum of the voltages around the closed delta (i.e., that of the secondary line-to-line voltages) must equal zero in accordance with Kirchhoff's voltage law, and no third-harmonic voltage exists in the secondaries of the transformers.

The delta connection, in addition to assuring balanced voltages, provides a path for the third-harmonic currents, which accounts for the popularity of the wye–delta or delta–wye connection. Where wye–wye transformation is required, it is quite common to incorporate a tertiary winding connected in delta.

## 3-15  CURRENT INRUSH

Frequently upon energizing a power transformer, there is an inrush of exciting current which may initially be as high as eight times the rated current of the excited winding even with all other windings open. Because of losses in the winding and the magnetic circuit, this current eventually dwindles to the normal value of the exciting current (i.e., to perhaps 5 percent or less of rated value).

When a linear magnetic circuit is energized from an ac source the current may have a dc transient component having an initial value equal at most to the amplitude of the ac component and which decays at a rate determined by the time constant of the circuit. However, under the same conditions, the transient current or inrush current in a power transformer or iron-core reactor is proportionately much greater due to saturation and more complex due to the nonlinear characteristics of the iron. Figure 3-33 shows the approximate waveform of inrush current for a transformer.

The inrush is most severe when the transformer is energized at the instant the voltage goes through zero immediately following which the polarity of the voltage is such that the flux increases in the direction of the residual flux. For these conditions, the applied voltage

$$e = \sqrt{2}\,E \sin \omega t = \frac{d\lambda}{dt} = N \frac{d\phi}{dt} \tag{3-86}$$

is applied at $t = 0$. The value of the flux is found by integrating Eq. 3-86 thus:

$$\phi = \frac{\sqrt{2}\,E}{N} \int_0^t \sin \omega t \, dt + \phi(0) \tag{3-87}$$

where $\phi(0) = \phi_r$, the residual flux. Hence,

$$\phi = \frac{\sqrt{2}E}{\omega N} (1 - \cos \omega t) + \phi_r$$

$$= -\phi_m \cos \omega t + \phi_m + \phi_r \tag{3-88}$$

If the dc component fluxes $\phi_m + \phi_r$ are assumed to remain constant, then at $\omega t = \pi$ the instantaneous flux must be

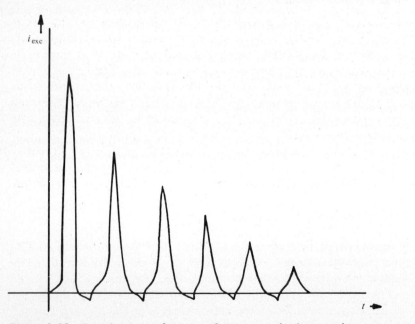

**Figure 3-33**   Inrush current for a transformer energized at zero instantaneous voltage.

$$\phi = 2\phi_m + \phi_r \tag{3-89}$$

Now suppose that under steady-state conditions $B_m = 1.32$ T. Then the peak value of the exciting current would correspond to a magnetizing force of $H_m = 600$ ampere turns/m, according to Fig. 2-6. If $\phi_r$ is assumed to equal $0.6\phi_m$, then at the end of the first half-cycle ($\omega t = \pi$) the flux density would have a peak value of $B_{peak} = 2 \times 1.32 + 0.6 \times 1.32 = 3.42$ T. This value is far beyond the range of the magnetization curve in Fig. 2-6. However, from this curve it is evident that $H_m$ and the peak value of the inrush current would reach extremely high values.

Inrush currents can be troublesome in that they may cause improper operation of such protective devices as relays and fuses.


## 3-16 REACTORS

Reactors are coils that have substantial inductance, and those which operate at or below audio frequencies usually have ferromagnetic cores interrupted by one or more air gaps. The air gap reduces the effect of the nonlinearity of the iron and lowers the sensitivity of the reactor's inductance to variations in the saturation of the iron with changes in flux density. Iron-core reactors are used for many purposes, among which are suppressing the ac flux ripple in rectifier circuits and limiting starting currents in motors, arc furnaces, and arc welders, as well as limiting fault currents in power systems.


### 3-16.1 Volume of Air Gap

The physical size of a reactor is determined by its volt-ampere and frequency ratings. The cores may be of the shell-type or of the core-type configuration. The design of the core and winding must be such as to make for a minimum amount of materials compatible with operation at safe temperature of the core and winding. In addition, the configurations of the core and winding must lend themselves to relative ease of manufacture. While a complete treatment of reactor design† is outside the scope of this book, some relationships basic to such design are discussed in the following.

For effective operation the resistance of the winding must be low compared with the inductive reactance, and the induced voltage may therefore generally be assumed to equal the applied voltage. Then the rating of the reactor is expressed by

$$VI = EI = \omega LI^2 \tag{3-90}$$

If the reluctance of the iron and the leakage are neglected, all the energy in the magnetic circuit is stored in the field of the air gap. According to Eq. 2-54 the instantaneous maximum energy stored in the air gap is

---

† For a thorough treatment of reactor design, see Alexander Kusko and T. Wroblewski, *Computer-Aided Design of Magnetic Circuits,* Research Monograph No. 55 (Cambridge, Mass.: The MIT Press, 1969).

$$W_\phi = \text{Vol}_g \frac{B_m H_m}{2} = \text{Vol}_g \frac{B_m^2}{2\mu_0} \tag{3-91}$$

which may also be expressed in terms of the maximum instantaneous current by Eq. 2-62 as

$$W_\phi = \tfrac{1}{2}L(\sqrt{2}I)^2 = LI^2 \tag{3-92}$$

The volume of the air gap is found from Eqs. 3-90, 3-91, and 3-92 to be

$$\text{Vol}_g = \frac{4EI \times 10^{-7}}{fB_m^2} \tag{3-93}$$

Equation 3-93 shows the volume of one or more of the air gaps in the core to be inversely proportional to $B_m^2$, which seems to indicate the desirability of a high value of $B_m$. However, if the reluctance of the iron is to be small compared with that of the air gap and the core losses are not to be excessive, then $B_m$, as expressed by Eq. 2-34, must not be so high as to produce excessive saturation of the iron. But, if $B_m$ is made too small, the reactor is too large and for that reason the iron is worked in the "knee" of the magnetization curve plotted in rectangular coordinates, which calls for a value of $B_m$ roughly in the neighborhood of 1.2 T.

Although Eq. 3-93 yields the approximate volume of air in series with the iron, it gives no indication as to the configuration (i.e., length and area) of the one or more air gaps. Most reactor cores have rectangular cross sections with the air gaps between plane-parallel surfaces of the iron.

However, the configuration of the air gap is at the disposal of the designer. For a given value of $B_m$ the dimensions of the air gap largely determine the number of turns $N$. Practical reactor dimensions call for an air-gap length which is short compared with the length of the flux path in the iron. An initial value $g_0$ of air-gap length may be chosen, perhaps based on known design of reactors, not necessarily of the same rating, or other empirical information. Since there is some saturation present, the reluctance of the iron should be taken into account by modifying the value of the mechanical length of the air gap as follows.

If $l_i$ is the assumed length of flux path in the iron, the mmf required to overcome the reluctance of the iron is

$$H_i l_i = \frac{B_m l_i}{\mu_i} \tag{3-94}$$

The mmf for the air gap is

$$H_m g_0 = \frac{B_m g_0}{\mu_0} \tag{3-95}$$

so the total mmf must be

$$H_m g_1 = H_m g_0 + H_i l_i$$

where $g_1$ is the length of a fictitious air gap or the modified length of air gap. Hence,

$$g_1 = g_0 + \frac{H_i}{H_m} l_i \tag{3-96}$$

$H_i$ is determined in ampere turns per meter from Fig. 2-6. The magnetizing force for the air gap is

$$H_m = \frac{B_m}{4\pi \times 10^{-7}}$$

$$g_1 = g_0 + \frac{4\pi \times 10^{-7} H_i l_i}{B_m} \tag{3-97}$$

A corresponding value for the number of turns can now be obtained on the basis that for a sinusoidal current,

$$H_m = \frac{N\sqrt{2}\, I}{g_1} = \frac{B_m}{\mu_0}$$

and

$$N = \frac{B_m g_1 \times 10^7}{\sqrt{2} \times 4\pi I} \tag{3-98}$$

The air-gap area can be determined by making use of Eq. 3-93. The area of the air gap, including the effect of fringing, is therefore

$$A_g = \frac{\text{Vol}_g}{g_1} = \frac{4EI \times 10^{-7}}{g_1 f B_m} \tag{3-99}$$

The net cross-sectional area of the iron is smaller than $A_g$ because of fringing at the air gap and the stacking factor of the core. In addition, leakage flux is not taken into account. Nevertheless, Eq. 3-99 is useful for obtaining approximate results.

**EXAMPLE 3-8**

A 200-V, 3-A, 60-Hz shell-type reactor has an air gap 3 mm long in its center leg, which is of square cross section. The length of flux path in the iron is 36 cm. The core material is electrical sheet steel for which the magnetization curve is shown in Fig. 2-6. The maximum flux density in the iron is 1.27 T and the stacking factor of the core is 0.94. The effect of fringing is such as to increase the effective area of the air gap to 1.07 that of the gross area of the core. Calculate (a) the flux density $B_m$ in the air gap, (b) the modified length of the air gap, (c) the number of the turns in the winding, and (d) the cross-sectional area of the center leg of the core.

*Solution*

a. The flux density in the air gap is

$$B_m = \frac{B_{\text{iron}} \times \text{stacking factor}}{\text{correction for fringing}}$$

$$= \frac{1.27 \times 0.94}{1.07} = 1.115 \text{ T}$$

b. From Fig. 2-6, $H_i = 400$, then, from Eq. 3-97,

$$g_1 = g_0 + \frac{4\pi \times 10^{-7} H_i l_i}{B_m}$$

$$= 0.003 + \frac{4\pi \times 10^{-7} \times 400 \times 0.36}{1.115}$$

$$= 0.003163$$

c. From Eq. 3-98,

$$N = \frac{B_m g_1 \times 10^7}{\sqrt{2} \times 4\pi I} = \frac{1.115 \times 0.00317 \times 10^7}{\sqrt{2} \times 4\pi \times 3}$$

$$= 660 \text{ turns}$$

d. From Eq. 3-99,

$$A_g = \frac{4EI \times 10^{-7}}{g_1 f B_m^2} = \frac{4 \times 200 \times 3 \times 10^{-7}}{0.003163 \times 60 \times (1.115)^2}$$

$$= 10.15 \times 10^{-4} \text{ m}^2$$

$$A_{\text{core}} = \frac{A_g}{\text{correction for fringing}}$$

$$= \frac{10.15 \times 10^{-4}}{1.07} = 9.48 \times 10^{-4} \text{ m}^2$$

## 3-16.2 Rating of Reactors and Transformers

For given values of frequency, flux density, and current density, the volts per turn are proportional to the cross-sectional area of the core and the ampere turns are proportional to the area occupied by the winding, so

$$\frac{V}{N} = 4.44 f B_m A_c \tag{3-100}$$

where $A_c$ is the net cross-sectional area of the core and

$$NI = J k_w A_w \tag{3-101}$$

where $J$ is the rated current density, $A_w$ the area of the window occupied by the winding, and $k_w$ a space factor somewhat less than unity because not all the window area is occupied by conductor material of the winding, some of the space being occupied by electrical insulation and voids between turns and between winding and core. The volt-ampere rating of the reactor is the product of Eqs. 3-100 and 3-101:

$$VI = 4.44 f B_m J k_m A_w A_c$$

For a given configuration, an increase in all linear dimensions by a factor $k$ increases the product of the two areas $A_w A_c$ by $k^4$. Accordingly, the volt-ampere rating of the reactor increases as the $\frac{4}{3}$ power of the volume or weight:

$$VI \propto \text{Vol}^{4/3} \quad \text{or} \quad VI \propto \text{weight}^{4/3} \tag{3-102}$$

and the volt-amperes per unit weight are

$$\frac{VI}{\text{weight}} \propto \text{weight}^{1/3} \propto k \tag{3-103}$$

On the basis of Eq. 3-103 the volt-ampere rating per pound varies directly as the linear dimension. For a fixed current density the $I^2R$ losses in the winding vary at a given temperature as the volume of the winding, and when the flux density and frequency are fixed, the core loss varies directly as the volume of the core. Hence, under rated conditions the heat generated in the reactor or transformer is proportional to its volume or $k^3$. The surface from which the heat is radiated varies as $k^2$; so the temperature rise would vary about as the linear dimension or $k$.

Accordingly, if frequency, flux density, current density, and configuration remain fixed, greater provision in proportion to $k$ must be made to maintain the same temperature rise in the core and windings as the size and rating are increased. For small ratings, normal radiation and convection dissipate the generated heat without excessive temperature rise, but beyond a certain range it becomes necessary to facilitate cooling by means of forced air, circulation of oil, or some other method when operating at the same frequency, current density, and flux density. Therefore, Eq. 3-103 serves as a guide not to be interpreted too literally, particularly in a comparison of widely differing ratings.

## STUDY QUESTIONS

1. What is transformer action? What conditions are essential for transformer action to take place?
2. Can transformer action take place in a dc circuit? Explain your answer.
3. What will be the frequency of the induced voltage in a coil coupled to the coil which has an alternating voltage impressed on it?
4. What is a static transformer? Explain how a static transformer differs from a rotating transformer.
5. Are transformers more efficient than rotating electric machines and, if so, why?
6. Describe the difference between a core-type and shell-type transformer.
7. The primary of a transformer is connected to the supply circuit. What is the secondary of a transformer connected to?
8. What is meant by leakage flux?
9. Investigate why the first few turns of high-voltage transformer coils are more highly insulated than the rest of the coils. Also, investigate for three-phase transformer whether or not the last few turns of a transformer coil might not be more insulated than the middle coils.
10. Why do transformers hum? How might this hum be minimized?
11. Why are transformers placed in oil-filled tanks? List several properties that a good transformer oil should possess. Make a list of common insulating materials used in transformers.

12. Some of the tanks of large transformers are corrugated. Explain why.

13. What is meant by the term oil sludging? Explain how oil sludging occurs.

14. When the secondary of a transformer is open-circuited, what current flows in the primary? What function does this current serve?

15. What factors affect the induced voltage in the primary of a transformer? In the secondary?

16. Describe the relationship that exists between primary and secondary voltages and turns.

17. Describe what is meant by the ratio of transformation. How can this ratio of transformation be determined experimentally?

18. Is there a relationship between the primary and secondary currents and turns? If so, what is it?

19. Describe the relationship that exists between primary secondary and secondary voltages and currents.

20. What is the difference between a step-up and a step-down transformer?

21. What is the difference between a power and a distribution transformer?

22. Explain why distribution transformers are tapped.

23. Explain how the primary current increases automatically in direct proportion to the increase in current delivered by the secondary.

24. Explain why the principle of transformer action requires that primary and secondary ampere turns be equal.

25. What is meant by the term "voltage regulation"? What factors affect the regulation of the transformer?

26. What is meant by the term "leakage reactance drop"?

27. Explain why the induced voltage lags behind the flux by 90°.

28. Assume a constant primary voltage. Explain how the secondary terminal voltage changes when the power factor is unity lagging or leading at very low values.

29. Outline completely the procedure for performing a short-circuit test. What information is obtained from the short-circuit test?

30. Outline the procedure for performing an open-circuit test. What information is obtained from the open-circuit test?

31. What components make up the core loss?

32. How is hysteresis loss affected by a change in flux density?

33. Assuming that the source of voltage has constant frequency, explain why voltage change affects the hysteresis loss or the eddy-current loss.

34. Assume that the source is a constant voltage. How does the frequency change affect the hysteresis loss and the eddy-current losses?

35. Losses in a transformer are often called copper losses, even though the transformer may be made of aluminum windings. Explain how the copper losses vary with the load.

36. Make a list of the losses in a power transformer, and explain how these losses can be measured experimentally.

37. Why are the core losses unaffected by load?

38. Describe the condition under which a transformer has maximum efficiency. Would you expect a distribution transformer to be designed to develop maximum efficiency at loads lower than rated value? If so, why?

39. What conditions would be essential for maximum efficiency to occur at rated load?

40. Why and when would it be desirable to have a power transformer operate at maximum efficiency when it is delivering rated load?

41. What is an autotransformer? Describe the advantages and disadvantages of auto-transformers.

42. Define the terms transformed power and conducted power as they refer to an auto-transformer.

43. Show the electrical connections that would be made to convert a two-winding trans-former into a step-up autotransformer, and a step-down autotransformer.

44. What is meant by the term "current transformer"?

45. What is meant by the term "potential transformer"?

46. What special precautions must be taken when using a current transformer?

47. How is the ratio of transformation for current transformers specified, and what is the significance of this notation?

48. Explain the principle of operation for a clip-on ammeter.

49. What does the polarity of a transformer mean?

50. What names are given to the standard polarities of transformers?

51. What tests could be performed to determine the polarity of a transformer connection?

52. When would it be necessary to know the polarity of a transformer?

53. When connecting two transformers in parallel, certain conditions must be fulfilled. List them.

54. When two transformers are operated in parallel at no load, describe the conditions that will exist if there is no circulating current between the transformers.

55. Two transformers with equal ratios of transformation are connected in parallel. How is the load divided between them?

56. If two transformers with unequal ratios of transformation are connected in parallel, how will the total load divide between them?

57. If two transformers have equal values of impedance, but different ratios of equivalent resistance to equivalent reactance, will the total load divide equally between them? Will their power factors be equal?

58. There are four possible ways of connecting a bank of three transformers for three-phase service. List them.

59. How can three-phase service be supplied using a bank of two transformers? What is this connection called?

60. Two transformers are connected in open delta. At what power factor will the individual transformers operate when a balanced three-phase load is delivered at a power factor of 0.707 and at a universal power factor equal to cosine $\theta$?

61. What are the advantages and disadvantages of three-phase transformers?

62. Explain how a Scott T connection is made for a three-phase to three-phase or a three-phase to two-phase operation.

**63.** Find a reference and list several interesting and unique transformer connections other than the standard connections discussed.

## PROBLEMS

**3-1.** This is a schematic diagram of an ideal transformer supplying a noninductive resistance $R = 10\ \Omega$. The primary winding has $N_1 = 100$ turns and the secondary winding has $N_2 = 50$ turns. The instantaneous voltage applied to the primary winding is $v_1 = 150$ V. (a) Calculate $d\lambda_1/dt$, $d\phi_1/dt$, $d\lambda_2/dt$, $d\phi_2/dt$, the instantaneous load current $i_L$ and indicate its direction through $R$, and the instantaneous primary current $i_1$ and indicate its direction (i.e., whether into or out of the marked primary terminal). (b) Specify the directions, whether clockwise or counterclockwise, with respect to the core of the mmf produced by $i_1$ and that produced by $i_L$. (c) Where should the secondary polarity mark be placed with the primary polarity mark located as shown? (d) The transformer and its connected load are replaced by a noninductive resistance which takes the same value of current at 150 V as that in the primary of the transformer. What is the value of the resistance in ohms? (e) What is the ratio of the resistance in part (d) to that of the resistance $R$ in the secondary? What is this ratio called?

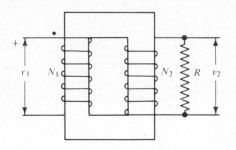

**Prob. 3-1**   Ideal transformer.

**3-2.** A 2300/230-V, 60-Hz distribution transformer has 1200 turns on the high side. If the net cross-sectional area of the iron is 60 cm², calculate: (a) the total flux, (b) the maximum flux density in the core, (c) the number of secondary turns.

**3-3.** A transformer was tested under load and was found to deliver 62 A at 228 V when the primary current measured 3.2 A. Calculate: (a) the primary input voltage, (b) the ratio of transformation.

**3-4.** An audio output transformer operates from a 4000-$\Omega$ resistance source delivering 5 W to a noninductive load of 10 $\Omega$. Assume the transformer to be ideal and determine (a) the turns ratio such that the load impedance referred to the primary has a value of 4000 $\Omega$, (b) the secondary and primary currents, and (c) the primary applied voltage.

**3-5.** A transformer has two 2400-V primary coils and two 240-V secondary coils. Indicate, by sketching, four possible ways to connect the transformer and for each one determine the primary to secondary voltages, and the ratios of transformation.

**3-6.** An ideal transformer is represented schematically. Note the current directions and voltage polarities and draw a phasor diagram on a one-to-one ratio basis for these directions and polarities. Assume the load current to lag the load voltage by an angle of 30°.

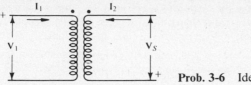

**Prob. 3-6**  Ideal transformer.

**3-7.** A transformer has two 1200-V primary coils and two 220-V secondary coils. Each of the primary coils has a 5 percent tap. Sketch the five possible ways to connect the transformer for three-wire service, and for each one indicate the ratio of the primary to secondary voltages.

**3-8.** A 2400/240-V 60-Hz transformer has a core with a gross cross-sectional area of 200 cm². The maximum flux density is not to exceed 0.93 T. The stacking factor, which takes into account reduction from the gross area due to the laminated construction of the core, is 0.90. Calculate (a) the number of primary turns and (b) the number of secondary turns.

**3-9.** The high-tension winding of a transformer consists of two halves which may be connected in series or in parallel. When these halves are connected in series across 2400 V, 60 Hz, the current is 0.59 A and the power is 121 W at no load. What are the no-load current and power if (a) the two halves of the winding are connected in parallel across a 1200-V 60-Hz source? (b) One of the halves is connected across the 1200-V 60-Hz source? Neglect resistance of the winding.

**3-10.** The illustration shows an iron core, assumed to be ideal, with three air gaps and carrying a primary winding of 200 turns and a secondary winding of 100 turns. Neglect fringing at the air gaps and calculate (a) the self-inductances $L_{11}$ and $L_{22}$ of the primary and secondary windings, (b) the coefficients of coupling $k_1$ and $k_2$, (c) the mutual inductance $M$ between the two windings, (d) the leakage inductances $L_{l1}$ and $L_{l2}$ of the primary and secondary windings, and (e) the magnetizing inductance of the primary winding.

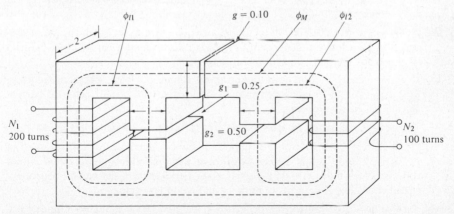

**Prob. 3-10**  Magnetic circuit. Dimensions are in centimeters.

**3-11.** Calculate the exciting current for the circuit shown in Prob. 3-10 at no load for a frequency of 1130 Hz when (a) the applied primary voltage is 100 V and (b) the applied secondary voltage is 50 V. How does the value of the flux that links the primary winding in part (a) compare with the value of the flux that links the secondary winding in part (b)?

**3-12.** To demonstrate the fact that the coefficient of coupling for conventional iron-core transformers is only slightly smaller than unity, calculate the coefficient of coupling of the transformer in Example 3-1. Neglect the effect of core loss.

**3-13.** The results of an open-circuit test and short-circuit test at rated frequency on a 500 kVA 40,000/2400-V 60-Hz transformer are as follows

Open-circuit test—voltage
applied to low-voltage side:

| Volts | Amperes | Watts |
|-------|---------|-------|
| 2400  | 9.1     | 1925  |

Short-circuit test—low-
voltage side short-circuited:

| Volts | Amperes | Watts |
|-------|---------|-------|
| 2300  | 12.5    | 4075  |

(a) Determine the following quantities referred to the low-voltage side of the transformer: (1) the exciting admittance, conductance, and susceptance; (2) the equivalent leakage impedance; and (3) the equivalent resistance. (b) Repeat part (a), referring all quantities to the high-voltage side. (c) Determine the leakage reactance of each winding on the basis that the same amount of equivalent leakage flux links each winding. (d) Determine the resistance of each winding on the basis of equal amounts of copper and equal current densities in the two windings. (e) What is the magnitude of the exciting current when the 40,000-V winding is the primary?

**3-14.** Determine the efficiency of the transformer in Prob. 3-13 when delivering at 2400 secondary volts (a) rated load at unity power factor, (b) $\frac{1}{4}$, $\frac{1}{2}$, $\frac{3}{4}$, 1, and $\frac{5}{4}$ rated load all at 0.80 power factor. Plot efficiency versus per-unit output in part (b). The vertical scale should range from 0.97 to 0.99 instead of from zero to 1.00 and about 2.5 cm corresponding to 0.01 on the efficiency scale.

**3-15.** (a) Show that the efficiency of the transformer is a maximum when the load losses equal the core losses independent of the load power factor. (b) In Prob. 3-14(b), what is the value of the load at maximum efficiency?

**3-16.** Determine the regulation based on the approximate equivalent circuit of the transformer in Prob. 3-13 when delivering rated load at 0.80 power factor (a) current lagging, (b) current leading.

**3-17.** The percent regulation of a 2300/230-V transformer is 2.5 percent. Calculate: (a) the no-load secondary voltage, (b) the turns ratio.

**3-18.** A 100-kVA 2400/240-V 60-Hz transformer has the following equivalent circuit constants: $R_1 = 0.42$; $X_1 = 0.72$, $R_2 = 0.0038$; $X_2 = 0.0068$. Calculate the equivalent circuit $R$, $X$, and $Z$ in terms of primary and secondary values.

**3-19.** A 240/120-V autotransformer delivers a current of 180 A to a load connected to the low side. Neglect the exciting current and determine the current in each of the windings.

**3-20.** Use the data of Prob. 3-18. Calculate the following voltage drops in primary and secondary terms: (a) $IR$ drop; (b) $IX$ drop; and (c) use this information and calculate the percent regulation for unity power factor, for a 0.8 power factor leading, and a 0.8 power factor lagging load.

**3-21.** The eddy-current loss of a 2300-V 60-Hz transformer is 280 W. What will this loss be if the transformer is connected to: (a) A 2300-V 50-Hz source; (b) a 2400-V 60-Hz source; (c) a 2200-V 25-Hz source?

**3-22.** A 2400/240-V transformer has an efficiency of 0.94 at rated load, 0.80 power factor, when operating as a two-circuit or two-winding transformer. This transformer is reconnected to operate as an autotransformer, with the windings carrying their rated currents and feeding a 0.80-power-factor load. Show a diagram of connections and determine the efficiency when the autotransformer is connected for operation (a) 2640/2400 V, (b) 2400/2640 V, (c) 2400/2160 V, and (d) 2640/240 V.

**3-23.** A 5-kVA 2300/460-V distribution transformer is to be connected as an autotransformer to step up the voltage from 2300 to 2760 V. (a) Make a wiring diagram of the connection. (b) When used to transform 5 kVA, calculate the kilovolt-ampere load output.

**3-24.** This autotransformer is assumed to be ideal. (a) Calculate the ratio of the load impedances $Z_b/Z_a$ such that the current in the winding between taps $a$ and $b$ is zero. (b) What is the combined impedance of $Z_a$ and $Z_b$ viewed from the primary in terms of $Z_a$?

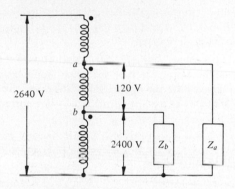

**Prob. 3-24** Autotransformer.

**3-25.** A 2400/2200-V autotransformer delivers a load of 90 kW at a power factor of 0.707. Calculate the current in each winding section and the kVA rating of the autotransformer.

**3-26.** An autotransformer designed for 4160- to 2300-V operation supplies a load of 50 kW at a power factor of 0.8. Calculate: (a) the transformed power, (b) the conducted power.

**3-27.** Three single-phase transformers are to be connected for 4000/440-V three-phase operation. The voltages are line-to-line. The three-phase rating is to be 750 kVA. There are four ways of connecting this transformer bank. Show a diagram of connections for each of the four ways and specify the voltage and current ratings of both windings in each transformer for each of the connections.

**3-28.** A 12,000-V, three-phase bus supplies (1) a bank of three-single-phase transformers connected delta-delta and delivering a balanced three-phase load of 5000 kVA, 0.90 power factor, current lagging, at 4000 V, and (2) a delta-wye-connected, three-phase transformer delivering a balanced three-phase load of 2500 kVA, 0.71 power factor, current lagging, at 2300 V. Determine the total real and reactive load supplied to the two transformer arrangements by the 12,000-V bus. Assume the transformers to be ideal.

**3-29.** The following information is given for two transformers that are connected in parallel. Transformer 1: Rating equals 100 kVA, 4600/230 V, circuit equivalent impedance equals 0.1 Ω in secondary terms. Transformer 2: Rating equals 7.5 kVA, 4160/230 V, equivalent impedance equals 0.2 Ω in secondary terms. Calculate the secondary circulating current between the transformers at no load.

**3-30.** The following information is given for two transformers connected in parallel. Transformer 1: Rating equal 100 kVA, 2400/240 V, transformer equivalent impedance equals 2.2 Ω in secondary terms. Transformer 2: Rating equals 75 kVA, 2400/240 V, equivalent circuit impedance equals 3.1 Ω in secondary terms. Calculate the kVA load carried by each transformer if the total load on both transformers is 150 kVA.

**3-31.** Three 25-kVA single-phase transformers are connected delta/delta: (a) What total rated kVA can the bank deliver? (b) To what kVA would the load be reduced if one of the transformers is removed on each side of the delta so that the bank operates open-delta?

**3-32.** Assume that the open-delta bank described in the previous problem delivers a balanced three-phase load of 60 kVA at 460 V: (a) What current flows in the secondary of each transformer? (b) What kVA load does each transformer carry?

**3-33.** Two identical 4000/240-V single-phase transformers are connected open-delta and have a three-phase rating of 11.53 kVA. What is the three-phase rating if another similar transformer is added for delta-delta operation?

**3-34.** Three single-phase transformers, each rated 2400/240/120 V and 10 kVA, have their primaries connected in delta and the center taps on their secondaries to form a neutral for six-phase operation. Determine the line-to-line voltages on the six-phase side and the rated current in the six-phase line wires.

**3-35.** The distribution transformer in Fig. 3-30 delivers 30 A at 120 V, 1.00 power factor $a$ to $n$; 40 A at 120 V, 0.707 power factor, current lagging $b$ to $n$; 25 A at 240 V, 0.90 power factor, current lagging $a$ to $b$. Neglect transformer impedance and admittance and calculate the primary current and the primary power factor.

**3-36.** The transformer bank in Example 3-7 delivers a load of 60,000 kVA, 0.80 power factor, current lagging to a balanced three-phase load on the 11-kV side. There is no load on the 2.2-kV side. Neglect resistance and determine the voltage on (a) the 66-kV side, (b) the 2.2-kV side, if the load voltage is 11 kV line to line.

**3-37.** Determine for the transformer of Example 3-1 the following base quantities: (a) on the high side, (b) on the low side, using the rating of the transformer for the base power: (1) current, (2) voltage, (3) impedance.

**3-38.** The manufacturers data for a 1000-kVA 4160/480-V transformer is given as follows: No-load losses equal 2160 W; full-load losses equal 16,750 W. Assume that the transformer operates at 80 percent capacity 16 hours/day, 5 days/week, and at 5 percent capacity at other times. Estimate the total annual losses in kilowatt hours. For your area of the country estimate the total cost of annual losses in dollars.

**3-39.** Determine the per-unit values of (a) exciting current, (b) resistance, (c) leakage reactance, and (d) impedance, of the transformer in Example 3-1, using the rating of the transformer as the base power.

**3-40.** Repeat Prob. 3-39 for a base kVA of 450.

**3-41.** A three-phase transformer rated at 150,000 kVA, 138/13.8 kV has an impedance of 0.10 per unit and is operating at 132/13.2 kV. Calculate the per-unit impedance for a 1,000,000-kVA base and 132-kV base.

**3-42.** Calculate the per-unit values of $I_{exc}$, $R_{eq}$, $X_{eq}$, and $Z_{eq}$ using the transformer rating as a base *directly* (i.e., without converting from siemens and ohms) from the data of Prob. 3-13.

**3-43.** Three identical single-phase 10,000-kVA transformers are connected delta on the 22-kV side and wye on the 132-kV side with the neutral on the 132-kV side isolated. Balanced three-phase voltages are applied to the 22-kV side such that

$$v_{AB} = \sqrt{2}(22{,}000 \cos 377t)$$

and the exciting current in the delta between $A$ and $B$ lines is

$$i_{exc\,AB} = \sqrt{2}[3.2 \cos (377t - 81°) + 1.50 \cos (1131t + 87°)$$
$$+ 0.4 \cos (1885t - 94°)]$$

Express as functions of time (a) the voltage $v_{BC}$ and $v_{CA}$ on the 22-kV side, (b) the exciting currents $i_{exc\,BC}$ and $i_{exc\,CA}$ in the delta, and (c) the no-load current $i_{exc\,A}$, $i_{exc\,B}$, and $i_{exc\,C}$ in the 22-kV lines.

**3-44.** The three single-phase transformers of Prob. 3-43 are operating without load and are excited from the 132-kV side with the neutral connected to the source. Assume $a$ phase line-to-neutral voltage on the 132-kV side to be in phase with the line-to-line voltage between $A$ and $B$ phases on the 22-kV side. If the delta connection is opened, (a) how do the exciting currents $i_{exca}$, $i_{excb}$, and $i_{excc}$ vary as functions of time? (b) How does the neutral current $i_n$ vary as a function of time?

**3-45.** The three single-phase transformers of Probs. 3-13 and 3-44 are operating without load and are excited from the 132-kV side with the neutral isolated and the delta closed. Express the exciting currents $i_{exca}$, $i_{excb}$, and $i_{excc}$ as functions of time. Express the currents in the delta, for this condition, as time functions.

## BIBLIOGRAPHY

Blume, L. F., et al. *Transformer Engineering*, 2d ed. New York: John Wiley & Sons, Inc., 1954.

Gibbs, J. B. *Transformer Principles and Practice*, 2d ed. New York: McGraw-Hill Book Company, 1950.

Kuhlman, J. H. *Design of Electrical Apparatus*, 3d ed. New York: John Wiley & Sons, Inc., 1950.

Landee, R. W., et al. *Electronic Designers Handbook*, New York: McGraw-Hill Book Company, 1957.

Lee, R. *Electronic Transformers and Circuits*, 2d ed. New York: John Wiley & Sons, Inc., 1955.

Matsch, Leander W. *Electromagnetic & Electromechanical Machines*, 2d ed., New York: Dun-Donnelley Publishing Co., 1977.

McPherson, George. *An Introduction to Electrical Machines and Transformers*, New York, John Wiley & Sons, Inc., 1981.

MIT Staff. *Magnetic Circuits and Transformers*. New York: John Wiley & Sons, Inc., 1943.

Myers, S. D., J. J. Kelly, and R. H. Parrish. *A Guide to Transformer Maintenance*, Akron, Ohio: Transformer Maintenance Institute, 1981.

*Selection of Armco Electric Steels for Magnetic Cores.* Armco, Inc., Middletown, Ohio, 1984.

Siskind, Charles S. *Electrical Machines: Direct & Alternating Current,* 2d ed. New York: McGraw-Hill Book Company, 1959.

Westinghouse Electric Corporation. *Electrical Transmission and Distribution Reference Book,* 4th ed. East Pittsburgh, Pa.: Westinghouse Electric Corporation, 1950. Chap. 5.

# Chapter 4

# Synchronous Machines

## 4-1 INTRODUCTION

Conventional electric motors and generators, called *electric machines,* convert energy by means of rotational motion. Electric motors are built in sizes from a small fraction of horsepower to thousands of horsepower. Ratings of several hundred thousand kilowatts are not uncommon for present-day electric generators. Simplicity of construction and compactness of design as well as the nature of the connected apparatus (in the case of the generator, the prime mover and in the case of a motor, the driven load) dictate rotary motion for electric generators and most motors.

*Electric power* generated by the power industry is converted from *mechanical power* usually supplied to the electric generators by means of steam and water turbines and in a few cases internal-combustion engines. Because of the large quantities of energy involved in the energy-conversion process, economy of operation as well as reliability of the equipment are extremely important. Efficiency and economics of power demand require the use of very large generators; as a result ac generators rated in excess of 1000 MVA† are in use. Large generators have high efficiency—in fact, at ratings greater than 50,000 kW, the efficiency usually exceeds 98 percent. Generally, the higher the rating of a machine, for a given speed, the greater is the efficiency. Electric motors also have high efficiencies, in some cases greater than 80 percent for ratings as low as 1 hp or even less.

---

† Features of a 1200-MW generator are presented by A. Abolins and F. Richter. "Test Results of the World's Largest Four-Pole Generator with Water-Cooled Stator and Rotor Windings," *Trans. IEEE Power Apparatus and Systems* PAS-94, No. 4 (July–August 1975):1103–1108.

By way of contrast, acoustical devices such as microphones are extremely sensitive and respond to such small amounts of power as a few microwatts, while loudspeakers may have input ratings exceeding 50 W, generally with an efficiency of less than 50 percent. Such values of power are practically insignificant when compared with the ratings of large rotating machines and transformers.

Electric machines consist of a magnetic circuit, one or more electric circuits and mechanical supports, with at least one winding. For a motor such a winding is energized from an electrical source of energy and for a generator such a winding (if there is only one) is a source of electrical energy. The magnetic circuit contains iron interrupted by an air gap between the stationary member or stator and the rotating member or rotor. Magnetic cores subjected to alternating magnetic fluxes or fluxes that undergo rapid time variations are usually laminated to ensure low eddy-current losses and rapid responses. However, smaller devices, with magnetic circuits of complex configurations, generally have laminated structures to reduce manufacturing costs, whether the excitation is ac or dc.

The turns in the windings of small machines consist of round wire. In larger machines the conductor material has rectangular cross section for more compact nesting in the space occupied by the winding. The most common conductor material is copper, although aluminum has come into limited use.

Conventional systems of the power industry are supplied by three-phase synchronous generators which fall into two general classifications—cylindrical-rotor machines and salient-pole machines. The cylindrical-rotor construction is peculiar to synchronous generators driven by steam turbines and which are also known as *turboalternators* or *turbine generators.* The stator core and rotor iron of a four-pole turbine generator are shown in Fig. 4-1, and the stator with its three-phase winding of a smaller machine is shown in Fig. 4-2. Steam turbines operate at relatively high speeds, 1800 and 3600 rpm being common for 60 Hz, accounting for the cylindrical-rotor construction, which, because of its compactness, readily withstands the centrifugal forces developed in the large sizes at those speeds. In addition, the smoothness of the rotor contour makes for reduced windage losses and for quiet operation.

Salient-pole rotors are used in low-speed synchronous generators such as those driven by waterwheels. They are also used in synchronous motors. Because of their low speeds salient-pole generators require a large number of poles as, for example, 72 poles for a 100-rpm 60-Hz generator. This follows from the fact that in one revolution a voltage undergoes $P/2$ cycles and the relationship between frequency and speed is

$$f = \frac{Pn_{\text{syn}}}{120} \tag{4-1a}$$

where $P$ = number of poles and $n_{\text{syn}}$ = synchronous speed in rpm.

In the terminology applied to rotating electric machines, an angle has the same value in electrical measure as in mechanical measure for two-pole machines because the voltage induced in the armature coil by a constant field flux goes through one cycle or $2\pi$ radians per revolution of the armature. However, in a

(a)

(b)

**Figure 4-1**   (a) Stator core for an ac turbine generator. (b) Four-pole turbine generator rotor. (Courtesy of Siemans-Allis, Inc.)

machine that has $P$ poles (i.e., one or more pair of poles), the armature coil generates $P/2$ cycles or $P\pi$ radians in electrical measure while the armature makes one revolution which corresponds to $2\pi$ radians in mechanical measure. Accordingly,

$$\theta = \frac{P}{2}\,\theta_m$$                                              (4-1b)

where $\theta$ is in electrical measure and $\theta_m$ in mechanical measure.

Figure 4-3 shows a hydrogenerator or waterwheel generator being assembled at the site of its installation. A synchronous motor is shown in Fig. 4-4. The salient-pole structure is simpler and more economical to manufacture than would be a cylindrical one with a large number of poles.

In contrast with the dc machine, the field winding, instead of the armature winding of conventional synchronous machines, is carried by the rotor, because the field winding is less massive than the armature winding, operating as it does at lower voltage with smaller current. In addition, the field winding is excited with direct current, requiring it to terminate in only two slip rings as are evident on the rotor shown in Fig. 4-4. If on the rotor, the armature winding would require at least three slip rings and in most cases a fourth for the neutral of the three-phase winding which is generally connected in wye. Slip rings and brushes are eliminated in synchronous machines with brushless excitation systems. These

**Figure 4-2**   Stator with three-phase winding. (Courtesy General Electric Company.)

require the field winding to be on the rotor. For example, a certain 432,000-kVA 22-kV three-phase 1800-rpm 60-Hz generator has a rated armature current of 11,340 A, while the field is rated at 500 V and 1940 A.

## 4-2 WAVEFORM

Conventional three-phase generators deliver practically sinusoidal voltage under normal conditions. Features that contribute to the production of good waveform are the use of distributed armature windings (i.e., among several slots per phase and pole), fractional-pitch armature coils (i.e., coils that span less than 180° in electrical measure), distribution of the field winding among several slots per pole in cylindrical rotors, and by shaping the pole shoes of salient-pole rotors so that the air gap is shortest under the pole center and increasing in length toward the pole tips.

Figure 4-5 illustrates two- and four-pole cylindrical rotors along with a developed view of the field winding for one pair of poles. One pole and its

**Figure 4-3** Salient-pole rotor being lowered into the stator of a hydrogenerator. (Courtesy General Electric Company.)

associated field coil of a salient-pole rotor is shown in Fig. 4-5(d). The stator slots in which the armature winding is embedded are not shown for reasons of simplicity. The approximate path taken by the field flux, not including leakage flux, is indicated by the dashed lines in Fig. 4-5(a), (b), and (d). The field coils in Fig. 4-5(c) are represented by filaments but actually (except for the insulation between turns and between the coil sides and the slot) practically fill the slot more nearly in keeping with Fig. 4-6.

The stepped curve in Fig. 4-6 represents the waveform of the mmf produced by the distributed field winding if the slots are assumed to be completely filled by the copper in the coil sides instead of containing current filaments. The shape of the mmf wave may be verified for this assumption by taking line integrals of **H** around appropriate paths. The sinusoid indicated by the dashed line in Fig. 4-6 represents approximately the fundamental component of the mmf wave.

The air gap in cylindrical-rotor machines is practically of uniform length

**Figure 4-4**   Synchronous motor. (Courtesy Westinghouse Electric Corporation.)

except for the slots in the rotor and in the stator, and when the effect of the slots and the tangential component of **H**†—which is quite small for the low ratio of air-gap length to the arc subtended by one pole in conventional machines—are neglected, the stepped mmf wave in Fig. 4-6 produces a flux-density space wave in which the corners of the steps are rounded due to fringing. The flux-density waveform is therefore more nearly sinusoidal than the mmf waveform when the effect of the slots is neglected. However, saturation of the iron in the region of maximum mmf tends to flatten the top of the flux-density wave.

## 4-3  AC ARMATURE WINDINGS

The armature winding of an ac machine is the source of the induced voltage, and for that reason some of the more elementary aspects of ac windings are treated in this chapter.

Armature windings are generally comprised of one or more turns and are so interconnected that their electric and magnetic effects are cumulative. The coils may have full pitch or fractional pitch. A full-pitch coil spans 180° in

† For a rigorous treatment of **H** due to current filaments in air gaps, see B. Hague, *The Principles of Electromagnetism Applied to Electrical Machines* (New York: Dover Publications, Inc., 1962), Chap. 6.

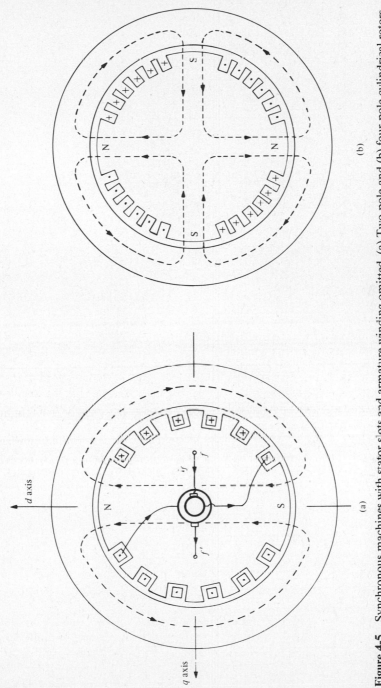
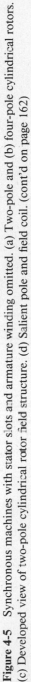

(b)

(a)

**Figure 4-5** Synchronous machines with stator slots and armature winding omitted. (a) Two-pole and (b) four-pole cylindrical rotors. (c) Developed view of two-pole cylindrical rotor field structure. (d) Salient pole and field coil. (cont'd on page 162)

(cont'd on page 162)

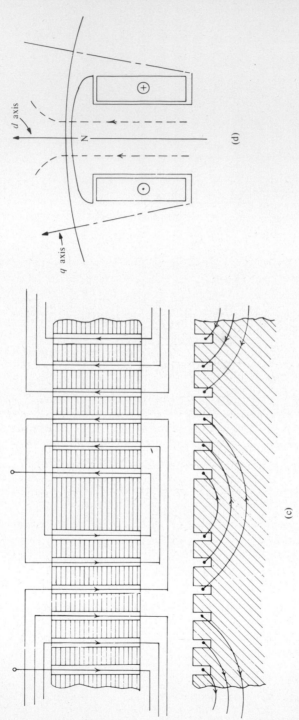

**Figure 4-5** (Continued)

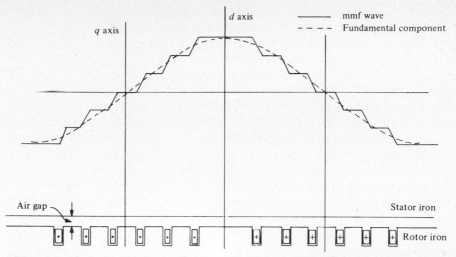

**Figure 4-6**   Cylindrical-rotor mmf wave and its fundamental of a synchronous machine.

electrical measure and a fractional-pitch coil spans less than 180° but seldom less than 120°. Full-pitch and fractional-pitch coils are shown in Figs. 4-7 and 4-8.

   The armature in Fig. 4-7 has three slots per pole, which corresponds to one slot per phase and pole for a three-phase winding. The three coils that are shown belong to one phase arbitrarily designated as $a$ phase, hence the letter designation $a_1$, $a_2$, and $a_3$. These three coils may be connected in series to form a single-circuit winding or they may be connected in parallel, resulting in a three-circuit winding. A developed view of the single-circuit connection is shown in Fig. 4-7(b), and a side view of the coil sides in the slots is shown in Fig. 4-7(c). Only the simplest of a large variety of armature windings used in three-phase machines are treated in this text.† However, the principles underlying the characteristics of these simple windings are basic, with minor modifications also for the more complex arrangements.

   A three-phase winding results from the addition of another two sets of armature coils displaced 120° and 240° in electrical measure from the first phase to produce a system of three voltages equal in magnitude and displaced from each other by 120°. A three-phase full-pitch winding is shown in Fig. 4-9(a), (b), and (c) in which $b$ phase is displaced from $a$ phase by two slots in the direction of rotation, with $c$ phase similarly displaced from $b$ phase. Since each slot corresponds to 60° in electrical measure, the windings are displaced so that $b$-phase and $c$-phase voltages lag $a$-phase voltage by 120° and 240°, respectively, as shown by the phasor diagram in Fig. 4-9(d). This phase sequence ($a$-$b$-$c$) is called *positive*

---

   † For a more complete treatment of armature windings in ac machines, see M. Liwschitz-Garik, *Winding Alternating-Current Machines* (New York: D. Van Nostrand Company, Inc., 1950); C. S. Siskind, *Alternating-Current Armature Windings* (New York: McGraw–Hill Book Company, 1951); A. M. Dudley, *Connecting Induction Motors,* 3d ed. (New York: McGraw-Hill Book Company, 1936).

*phase sequence.* A reversal in the direction of rotation results in *negative-phase sequence* (*a-c-b*) as shown by the phasor diagram in Fig. 4-9(e). The winding in Fig. 4-9 has one-half as many coils as there are slots or one coil side per slot. The more common arrangement of two coil sides per slot is shown in Fig. 4-10, with only one phase shown in Fig. 4-10(a) and (b). A side view of the slots and the coil sides for all three phases is shown in Fig. 4-10(c). A comparison of Fig. 4-10(c) with Fig. 4-9 shows that the former has two layers of coil sides in the slots and the latter has one layer, hence the terms *two-layer* and *single-layer* windings. Although single-layer windings are not common, they are sometimes used in induction motors of 10 hp or less. The chief advantage of the two-layer winding is that of accommodating fractional-pitch coils which have shorter end turns or end connections than full-pitch coils and as a result have lower resistance without a proportionate decrease in their flux linkage. Fractional pitch also assists in improving the waveform of the induced emf and the armature mmf. Three coils of a $\frac{2}{3}$-pitch winding are illustrated in Fig. 4-11, and a $\frac{5}{6}$-pitch winding is shown in Fig. 4-12.

The windings treated in this chapter are called *integral-slot windings,* since they occupy a structure in which the number of slots per pole is an integer. A more common arrangement for ac machines is the fractional-slot winding† for which the number of slots per pole is a fraction. The analysis of integral-slot windings is simpler than that of fractional-slot windings and yet serves to bring out the basic principles regarding the mmfs and inductances of armature windings. Fractional-slot windings have two advantages: (1) it is possible to use the same stator laminations, with resulting lower investment in dies, for salient-pole structures with a variety of a number of poles, and (2) the contribution toward good waveform is equivalent to that of an integral-slot winding with a larger number of slots per pole. Fractional-slot windings are also used to some extent in induction motors.

## 4-4 INDUCED ARMATURE VOLTAGE

A phasor diagram‡ which includes voltage-, current-, and flux-linkage phasors facilitates the analysis of steady-state synchronous machine-behavior. In order to relate the phase of flux linkage to the voltage that results from its time variation, it is necessary to establish conventions regarding the sign of the induced voltage in a generator and in a motor with regard to assumed current direction and the time phase of the flux linkage.§ Therefore, consider the magnetic circuit and

---

† See M. Liwschitz-Garik and C. C. Whipple, *Alternating-Current Machines,* Vol. II (New York: D. Van Nostrand Company, Inc., 1961).

‡ D. B. Harrington, "Recommended Phasor Diagram for Synchronous Machines." *IEEE Paper No. TP 143-PWR.* Presented at IEEE Winter Power Meeting, New York, January 1969.

§ For a more complete discussion, see W. A. Lewis, "Simplicity in Three-Phase System Circuit Conventions and Concepts," *Elec. Eng.* October 1958: 937–939; November 1958: 1038–1040; December 1958: 1126–1128.

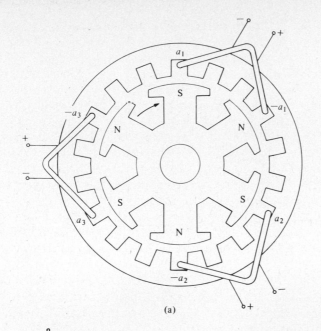

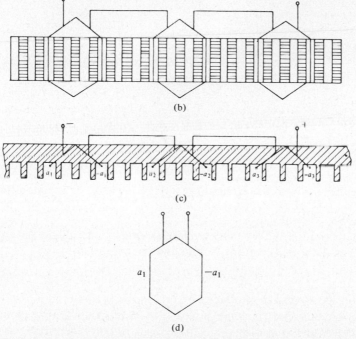

**Figure 4-7** (a) Simplified magnetic structure of a six-pole synchronous machine showing three full-pitch coils for one phase only. (b) Developed view of the three armature coils connected in series. (c) Side view of slots and end connections. (d) Schematic representation of an armature coil.

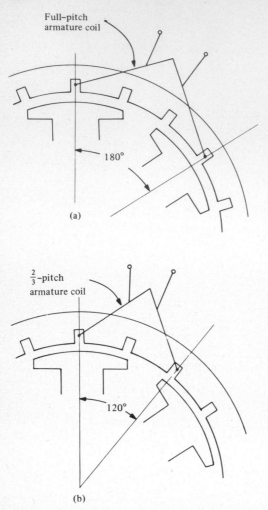

Full–pitch
armature coil

180°

(a)

$\frac{2}{3}$–pitch
armature coil

120°

(b)

**Figure 4-8** Armature coils in multipolar machines. (a) Full pitch. (b) Two-thirds pitch.

resistance $r_a$ in series with an independent voltage source $e_s$ and a load voltage $v_{an}$ as shown in Fig. 4-13(a). According to Kirchhoff's voltage law,

$$e_s = r_a i_a + p\lambda_a + v_{an} \qquad (4\text{-}2)$$

where the flux linkage $\lambda_a$ may be produced by $i_a$ alone or in combination with currents in other circuits (not shown) coupled inductively with the magnetic circuit. The time derivative $p\lambda_a$ may result from single or combined time variations of current, magnetic permeance, and inductive coupling. It is important to note that, when the net mmf is the same direction as that due to $i_a$ alone, the flux is in the direction of $\lambda_a$, as indicated in Fig. 4-13.

The magnetic circuit and the resistance $r_a$ can represent one phase of a synchronous generator if the source $e_s$ is removed, as in Fig. 4-13(b), which reduces Eq. 4-2 to

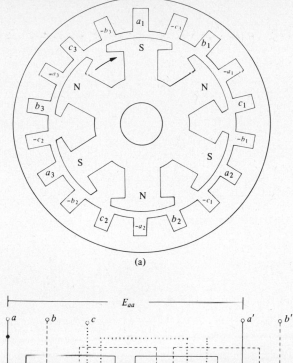

(a)

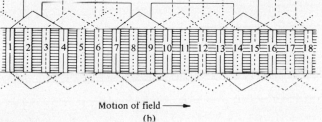

Motion of field ⟶

(b)

**Figure 4-9** Three-phase six-pole machine. (a) Arrangement of coil sides in stator slots. (b) Developed view of armature winding. (c) Developed side view of slots. (d) Positive-sequence voltage phasors. (e) Negative-sequence voltage phasors. (f) Wye and delta connections.

$$v_{an} = -r_a i_a - p\lambda_a \tag{4-3}$$

By similar reasoning, Fig. 4-13(c) shows the same circuit when representing one phase of a motor and for which the applied terminal voltage is expressed by

$$v_{an} = r_a i_a + p\lambda_a \tag{4-4}$$

However, $\lambda_a$ results from both the mmf of the armature current and that of the field current. In the motor the direction of the field mmf relative to that of the armature is opposite that in the generator, which is in agreement with Fig. 4-27(a) and (b). The value of $\lambda_a$ is therefore different in the generator [Fig.

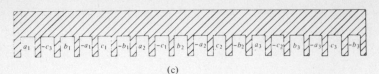

(c)

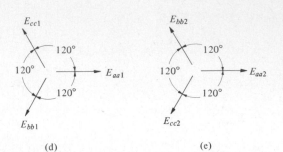

(d)                                          (e)

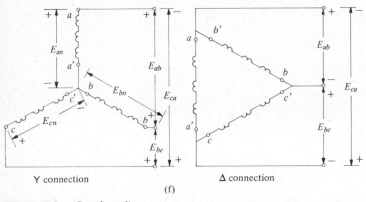

Y connection                         Δ connection

(f)

**Figure 4-9**   (Continued)

4-13(b)] from that in the motor [Fig. 4-13(c)] for the same values of armature current and field current, and Eqs. 4-3 and 4-4 must be interpreted accordingly.

### 4-4.1  Voltage Induced in a Generator Armature Coil

Figure 4-14(a) shows an armature coil of $N_{\text{coil}}$ turns and of pitch $p$ linking a sinusoidally distributed magnetic flux. The direction of the flux density is assumed radial and positive when directed from the stator iron to the rotor iron. This also is the direction of the flux through the armature coil by a current entering the negative terminal of the coil as in the case of a generator.

All angles in Fig. 4-14(a) are in electrical measure, but $\theta_m$ in Fig. 4-14(b) is in mechanical measure. The flux density at $\theta$ in Fig. 4-14(a) is expressed by

$$B_\theta = B_{\text{amp}} \sin \theta \qquad (4\text{-}5)$$

where $B_{amp}$ is the amplitude of the flux density space wave and assumed constant. The flux in the strip subtended by $d\theta$ is

$$d\phi_{coil} = B_\theta \, dA \tag{4-6}$$

where $dA = \frac{1}{2}LD \, d\theta_m$, in which $L$ is the axial length of the stator iron and $D$ the diameter of the inner surface (neglecting slots) of the armature iron as indicated in Fig. 4-14(b). The flux is assumed to be confined to the axial length $L$, and since $\theta_m = 2\theta/P$, Eq. 4-6 can be rewritten as

$$d\phi_{coil} = \frac{LD}{P} B_{amp} \sin \theta \, d\theta$$

and if the coil sides are regarded as filaments, the flux that links the coil is

$$\phi_{coil} = \frac{LD}{P} B_{amp} \int_\alpha^{\alpha+p\pi} \sin \theta \, d\theta$$

$$= \frac{LD}{P} B_{amp}[\cos \alpha - \cos (\alpha + p\pi)]$$

$$= \frac{2LD}{P} B_{amp} \sin p \frac{\pi}{2} \sin \left( \alpha + p \frac{\pi}{2} \right)$$

which results in the flux linkage

$$\lambda_{coil} = N_{coil}\phi_{coil} = \frac{2LD}{P} N_{coil}B_{amp} \sin p \frac{\pi}{2} \sin \left( \alpha + p \frac{\pi}{2} \right)$$

$$= \lambda_{coil\ M} \sin \left( \alpha + p \frac{\pi}{2} \right) \tag{4-7}$$

where

$$\lambda_{coil\ M} = \frac{2LD}{P} N_{coil}B_{amp} \sin p \frac{\pi}{2} \tag{4-8}$$

If the generator is driven at a constant angular velocity $\omega$ in electrical measure and in the direction indicated in Fig. 4-14(a)

$$\alpha = -\omega t$$

and the flux linkage is then expressed as a function of time by

$$\lambda_{coil} = -\lambda_{coil\ M} \sin \left( \omega t - p \frac{\pi}{2} \right) \tag{4-9}$$

which induces a coil voltage of

$$e_{coil} = -p\lambda_{coil} = \omega\lambda_{coil\ M} \cos \left( \omega t - p \frac{\pi}{2} \right) \tag{4-10}$$

Equation 4-8 expresses the maximum flux linkage, and if the coil had full pitch (i.e., $p = 1$), it could then link the entire flux per pole $\phi$ at the armature

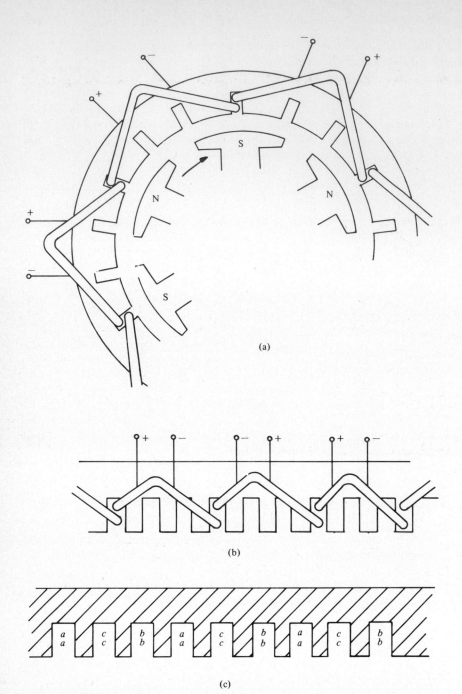

**Figure 4-10** (a) Partial representation of one phase of full-pitch two-layer three-phase winding. (b) Developed view. (c) Arrangement of coil sides in all three phases.

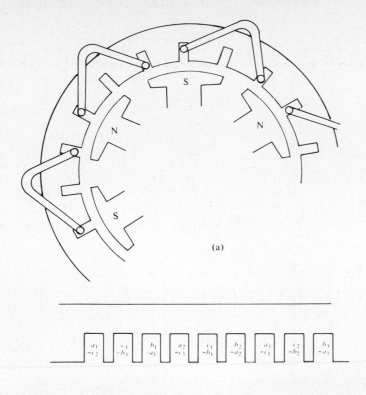

(a)

(b)

**Figure 4-11** (a) Partial representation of one phase of a ⅔-pitch three-phase winding. (b) Arrangement of coil sides for all three phases.

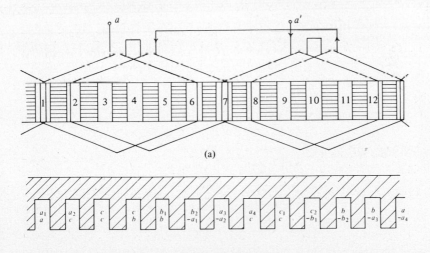

(a)

(b)

**Figure 4-12** (a) Developed view of a phase of a ⅚-pitch three-phase winding for one pair of poles. (b) Side view of coil sides in slots.

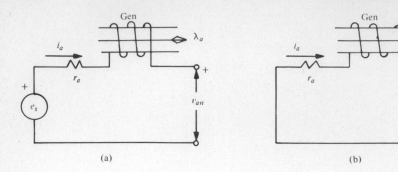

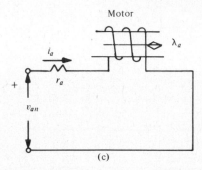

**Figure 4-13** Conventions for voltage polarity, flux linkage, and current directions. (a) One phase of a generator in series with another voltage source $e_s$. (b) The source $e_s$ removed. (c) One phase of a motor.

surface of the air gap. It follows that $\phi = (2LD/P)B_{amp}$ and the induced voltage is

$$e_{coil} = \omega N_{coil}k_p\phi \cos\left(\omega t - p\frac{\pi}{2}\right) \tag{4-11}$$

where

$$k_p = \sin p\frac{\pi}{2} \tag{4-11a}$$

and is called the *pitch factor.*

Since $\omega = 2\pi f$, the rms value of the coil voltage is

$$E_{coil} = \frac{2\pi f N_{coil}k_p\phi}{\sqrt{2}} = 4.44 f k_p N_{coil}\phi \tag{4-12}$$

The voltage induced in a full-pitch coil is found by letting $k_p$ = unity in Eq. 4-12, and the resulting equation is then similar to Eq. 2–36 for a transformer in which a sinusoidally time-varying flux has an instantaneous maximum value of $\phi_m$ equal to the flux per pole $\phi$.

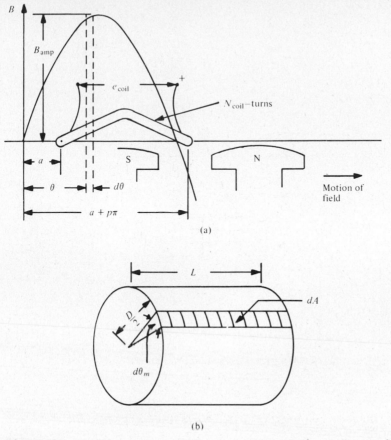

**Figure 4-14**  $N$-turn armature coil in a sinusoidally distributed magnetic field.

The flux linkage leads the resulting voltage induced in a *generator* coil by 90° in time phase as shown by a comparison of Eqs. 4-9 and 4-11 and illustrated graphically in Fig. 4-15. With this convention wattmeter and varmeter indications would be positive when connected to read *generator output* to an inductive load.

## 4-4.2 Voltage Induced in a Distributed Winding

A distributed winding is one in which each phase occupies more than one slot per pole as, for example, the winding in Fig. 4-12. It is evident from Eq. 4-7 that the flux linkages of coils occupying different slots under a given pair of poles differ since the angle $\alpha$ must have different values for the different coils. As a result, the flux linkages are not in time phase with each other. The resultant flux linkage of a series-connected coil group such as in Fig. 4-16(a) may be found by means of phasor addition, as in Fig. 4-16(b), as long as the flux density is distributed sinusoidally. The angle between adjacent slots in Fig. 4-16 is $\gamma$, expressed in electrical measure, and the flux-linkage phasors as well as the coil voltage

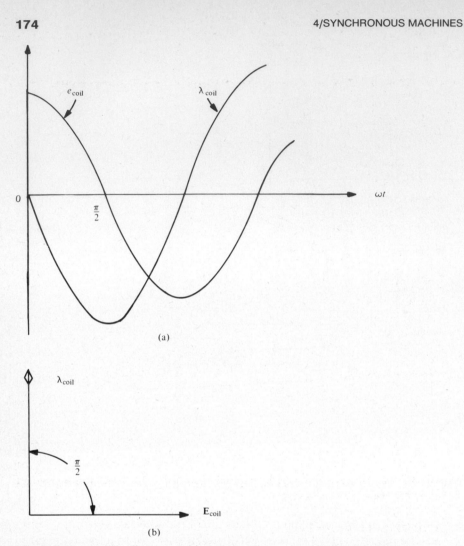

**Figure 4-15** (a) Flux linkage and induced voltage waves. (b) Phasor diagram.

phasors are displaced from each other by the same angle. The sums of these phasors are $\lambda_{group}$ and $\mathbf{E}_{group}$, as shown in Fig. 4-16(b) and (c), with the voltage phasors lagging their corresponding flux linkage phasors by 90°.

The ratio of the phasor sum to the arithmetical sum is called the *breadth factor* and is expressed by

$$k_b = \frac{E_{group}}{nE_{coil}} \tag{4-13}$$

where $n$ is the number of coils in the group. Figure 4-17 shows a phasor diagram for obtaining a general expression for $k_b$. The phasors **AB, BC, CD,** and **DE** represent the coil voltages and phasor **AE** the voltage of the coil group. The

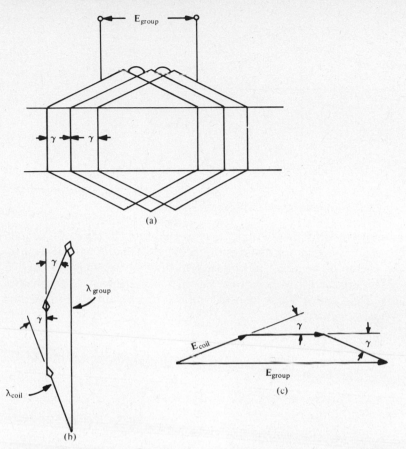

**Figure 4-16** (a) Group of three coils in a distributed winding. (b) and (c) Phasor diagrams of flux linkages and induced voltages.

angle occupied under one pole by one phase is represented by $\beta$ with $\gamma$ as the angle between adjacent slots and $n$ the number of slots per phase and pole. Then, on the basis of Eq. 4-13 and Fig. 4-17, the breadth factor is found to be

$$k_b = \frac{AE}{nAB} = \frac{OA \sin (\beta/2)}{nOA \sin (\gamma/2)} = \frac{\sin (\beta/2)}{n \sin (\gamma/2)} \qquad (4\text{-}14)$$

**EXAMPLE 4-1**

The stator of a three-phase four-pole generator has 36 slots. Calculate $k_b$.

*Solution.* In Eq. 4-14,

$$\beta = 180 \div 3 = 60° \qquad n = \frac{36}{\text{poles} \times \text{phases}} = \frac{36}{4 \times 3} = 3$$

$$k_b = \frac{\sin (60°/2)}{3 \sin (20°/2)} = 0.96$$

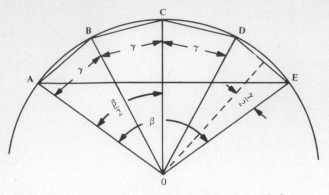

**Figure 4-17**   Phasor diagram for obtaining breadth factor.

The various coil groups of a phase may be connected in series, in parallel, or, in series–parallel, depending on the number of poles and the general arrangement of the armature winding. The connections of all phases must be alike. If the connections are such that there are $a$ paths in each phase (i.e., $a$ armature circuits in parallel), and if $N_{ph}$ is the total number of turns in each phase, the induced voltage per phase is

$$E_{ph} = \frac{4.44 f k_w N_{ph} \phi}{a} \tag{4-15}$$

where $k_w = k_p k_b$. Further, the line-line voltage is $\sqrt{3}\, E_{ph}$ for a wye connection and $E_{ph}$ for a delta connection.

### 4-4.3  Pitch Factor and Breadth Factor for Harmonics

Since the flux density wave in practical machines departs somewhat from a sinusoid, it must contain harmonics which may not be negligible. The $h$ harmonic is expressed by

$$B_{\theta h} = B_{\text{amp } h} \sin{(h\theta + \delta_h)} \tag{4-16}$$

and produces the harmonic in the coil voltage

$$e_{\text{coil } h} = \frac{2LD}{P} \omega N_{\text{coil}} B_{\text{amp } h} \sin hp\, \frac{\pi}{2} \cos\left(h\omega t + \delta_h - hp\, \frac{\pi}{2}\right) \tag{4-17}$$

Equation 4-17 can be verified by following the derivation that led to Eq. 4-11.

The pitch factor for the $h$ harmonic is found from Eq. 4-17 to be

$$k_{ph} = \sin hp\, \frac{\pi}{2}$$

In the case of a $\frac{5}{6}$-pitch coil, $p = \frac{5}{6}$ and $k_p = \sin p(\pi/2) = \sin 75° = 0.965$ for the fundamental, $k_{p3} = \sin 3 \times 75° = -0.707$ for the third harmonic, and

$k_{p5} = \sin 5 \times 75° = -0.259$ for the fifth harmonic. However, $k_{p11} = 0.965$, the same as for the fundamental. From these values it is apparent that fractional pitch suppresses most of the harmonics to a greater extent than the fundamental, which results in a voltage having a better waveform than the flux-density distribution.

Since the angle associated with the $h$ harmonic is $h$ times that of the fundamental, as indicated by Eqs. 4-16 and 4-17, it follows that the breadth factor for the $h$ harmonic may be obtained by modifying Eq. 4-14 as follows:

$$k_{bh} = \frac{\sin h(\beta/2)}{n \sin h(\gamma/2)} \tag{4-18}$$

The $h$ harmonic in the phase voltage is

$$E_{ph-h} = \frac{4.44 h f k_{wh} N_{ph} \phi_h}{a} \tag{4-18a}$$

where $k_{wh} = k_{ph}k_{bh}$ and $\phi_h = (2LD/hP)B_{\text{amp } h}$, the $h$th harmonic flux per $h$th harmonic pole.

For example, a three-phase winding distributed among 12 slots per pole would have $n = 12 : 3 = 4$ slots per phase and pole; $\beta = 180° \div 3 = 60°$ per phase; $\gamma = 180° \div 12 = 15°$ between adjacent slots. When these values are used in Eq. 4-18 the breadth factors would be $k_b = 0.957$, $k_{b3} = 0.653$, $k_{b5} = 0.259$, etc. If, in addition, the coil pitch were $p = \frac{5}{6}$, the product of the pitch factors and their respective breadth factors for the fundamental, the third, and the fifth harmonic would be

$$k_w - k_p k_b = 0.965 \times 0.957 = 0.922$$
$$k_{w3} - k_{p3}k_{b3} = 0.707 \times 0.653 = 0.461$$
$$k_{w5} = k_{p5}k_{b5} = 0.259 \times 0.259 = 0.067$$

Because of the symmetrical magnetic circuit, no appreciable even harmonics are present in the flux-density waveform and they therefore need not be considered in the voltage waveform. The rms value of the voltage including the fundamental and harmonics is then expressed by

$$E = \sqrt{E_1^2 + E_3^2 + E_5^2 + \cdots + E_n^2} \tag{4-19}$$

where $n$ is odd and all subscripts refer to the order of the harmonic.

Although the use of fractional pitch and a distributed winding reduces the fundamental component in the voltage by about 8 percent, the reduction in the harmonic content is much greater, which shows the waveform of the voltage to be better than that of the flux density space wave. The third harmonic and its multiple components in the line-to-line voltage are negligible even when appreciable in the phase voltage. In the case of the wye connection this line-to-line voltage is the phasor difference between the line-neutral voltages, and since the third harmonics are displaced by an angle equal to 3 times the displacement between the fundamentals (i.e., $3 \times 120°$), they are in phase with each other. When the armature is connected in delta, the third harmonic voltages are short-circuited, a condition that may result in objectionable third-harmonic currents circulating in the delta. The wye connection is far more common for generators than is the delta connection, and only such third-harmonic currents can flow in the armature as are present in the load. However, for balanced voltages all harmonics except the third and its multiples are $\sqrt{3}$ times as great in the line-to-line voltage of a wye connection as in their respective line-neutral values.

## 4-5 ARMATURE MMF

The armature current in a polyphase synchronous machine produces an mmf which, under steady balanced conditions, is stationary with respect to the mmf of the field winding. The phase angle between the current and the voltage of the armature fixes direction of the armature mmf in the synchronous machine. For example, if the phase angle $\theta_i$ between the armature current and the voltage induced in the armature by the field current is zero, as indicated by the phasor diagram in Fig. 4-18(a), the armature mmf is in the $q$ axis, as illustrated in Fig. 4-18(b), of an elementary three-phase two-pole generator. The field-winding is not shown in Fig. 4-18(b) for reasons of simplicity; however, the $d$ axis is shown passing through the field poles $N_F$ and $S_F$.

If the voltage generated in $a$ phase by the field flux is expressed by

$$e_{af} = \sqrt{2} \, E_{af} \sin \omega t \qquad (4\text{-}20)$$

then on the basis of the phasor diagram in Fig. 4-18(a) the current in $a$ phase is, for $\theta_i = 0$,

$$i_a = \sqrt{2} \, I_a \sin \omega t \qquad (4\text{-}21)$$

and the instantaneous armature currents are as shown graphically in Fig. 4-18(c). At $\omega t = \pi/2$, the instantaneous voltage and current in $a$ phase are both a positive maximum, while those in $b$ and $c$ phases are negative and one-half of their maximum values in accordance with the current directions indicated in Fig. 4-18(b). The armature mmf then produces an mmf somewhat along the path indicated by the dashed lines in Fig. 4-18(b) with the armature poles indicated by $N_A$ and $S_A$. This mmf pattern rotates at synchronous speed in the same direction as the rotor and is therefore stationary relative to the field poles and in its reaction upon the field poles, producing torque opposite the direction of rotation characteristic of generator operation. Since the direction of the armature mmf is along the $q$ axis when the current is in phase with the induced voltage, it has no magnetizing or demagnetizing effect upon the field as long as the magnetic circuit is unsaturated. The fact that the armature mmf rotates in synchronism with the field mmf can be demonstrated by the condition that when $\omega t$ has increased from $\pi/2$ to $5\pi/6$ (i.e., by an angle of $\pi/3$ rad), the negative value of $i_c$ is a maximum while at the same instant $i_a$ and $i_b$ are at one-half the positive

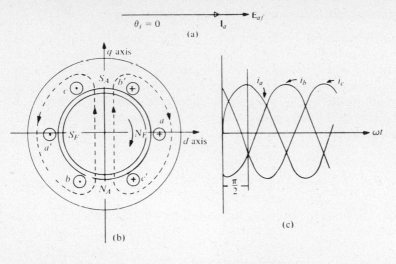

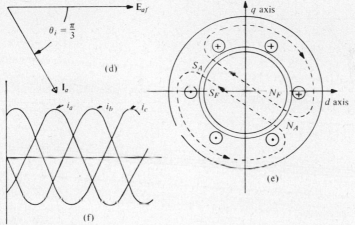

**Figure 4-18** Rotating armature mmf in a synchronous generator. (a) Phasor diagram of current in phase with generated voltage. (b) Elementary two-pole three-phase generator showing approximate path of flux due to armature mmf, current in phase with generated voltage. (c) Instantaneous armature currents. (d) Phasor diagram for current lagging induced voltage 60°. (e) and (f) Armature flux path and instantaneous currents, current lagging 60°.

maximum values. During that interval the positions of the armature flux pattern as well as the rotor have advanced $\pi/3$ rad.

Under practical operating conditions the current usually lags the induced voltage by a sizable phase angle. Figure 4-18(d) shows the phasor diagram for $\theta_i = \pi/3$ with the resulting relationships indicated in Fig. 4-18(e) and (f). The armature mmf now lags the direct axis by $\pi/3$ rad and has components in both the direct and quadrature axis, thus exerting a demagnetizing effect on the field. By the same token, when the armature current leads the generated voltage it strengthens the field.

## 4-5.1 Fundamental Component of mmf Space Wave

Figure 4-19(a) shows a full-pitch armature coil which may be in a two-pole machine or in a $P$-pole machine which has one such coil for each pair of poles. A uniform air gap $g$ is assumed. If the coil has $N_{coil}$ turns and its sides are assumed to be filaments, a current of $i_a$ amperes in the coil produces a rectangular mmf wave of amplitude $N_{coil}i_a/2$ ampere turns per pole, since the total mmf of the coils is $N_{coil}i_a$ ampere turns per pair of poles. If $\theta$ is the space angle in electrical measure from a coil side, the mmf waveform can be represented by the Fourier series.

$$\mathscr{F}_\theta = \frac{N_{coil}i_a}{2} \frac{4}{\pi} \left( \sin \theta + \frac{1}{3} \sin 3\theta + \cdots + \frac{1}{n} \sin n\theta \right) \tag{4-22}$$

where $n$ is odd. The fundamental is the largest and the only desirable component being expressed by

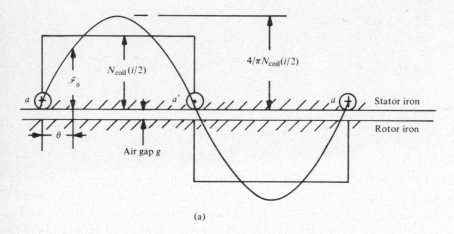

(a)

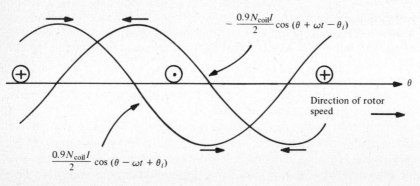

(b)

**Figure 4-19** (a) Mmf wave for a full-pitch coil of $N_{coil}$ turns per pair of poles. (b) Traveling-wave components of mmf.

$$\mathcal{F}_{\theta 1} = \frac{N_{\text{coil}} i_a}{2} \frac{4}{\pi} \sin \theta \qquad (4\text{-}23)$$

The effect of fractional pitch and distribution of the armature winding suppresses the harmonics in the mmf wave in a manner similar to that in which it improves the voltage wave. If the induced voltage is expressed by Eq. 4-20 and the current lags that voltage by $\theta_i$, then

$$i_a = \sqrt{2}\, I \sin (\omega t - \theta_i) \qquad (4\text{-}24)$$

and the amplitude of the mmf varies sinusoidally with time. The *amplitude* of the fundamental is therefore expressed as a *time function* by

$$\mathcal{F}_{\text{amp 1}} = \frac{\sqrt{2}\, N_{\text{coil}} I}{2} \frac{4}{\pi} \sin (\omega t - \theta_i) = 0.9 N_{\text{coil}} I \sin (\omega t - \theta_i) \qquad (4\text{-}25)$$

and the instantaneous value of the fundamental at the *space* angle $\theta$ is

$$\mathcal{F}_{\theta 1} = 0.9 N_{\text{coil}} I \sin \theta \sin (\omega t - \theta_i) \qquad (4\text{-}26)$$

It is important not to confuse the *space angle* $\theta$ with the *time-phase angle* $\theta_i$ in Eq. 4-26.

By making use of the identity

$$2 \sin x \sin y = \cos (x - y) - \cos (x + y)$$

Eq. 4-26 may be rewritten as

$$\boxed{\mathcal{F}_{\theta 1} = 0.9 \frac{N_{\text{coil}} I}{2} [\cos (\theta - \omega t + \theta_i) - \cos (\theta + \omega t - \theta_i)]} \qquad (4\text{-}27)$$

which represents two mmf waves, each having an amplitude of $0.9 N_{\text{coil}} I/2$ and of which the one associated with the first cosine term travels at synchronous speed in the same direction as the rotor; the second cosine term applies to the wave traveling at synchronous speed in the opposite direction. Thus, the component $0.9(N_{\text{coil}} I/2) \cos (\theta - \omega t + \theta_i)$ is stationary with respect to the rotor while the component $-0.9(N_{\text{coil}} I/2) \cos (\theta + \omega t - \theta_i)$ travels at twice the rotor speed, as illustrated in Fig. 4-19(b).

The three-phase armature currents are

$$i_a = \sqrt{2}\, I \sin (\omega t - \theta_i) \qquad \text{for } a \text{ phase}$$

$$i_b = \sqrt{2}\, I \sin \left( \omega t - \theta_i - \frac{2\pi}{3} \right) \qquad \text{for } b \text{ phase}$$

$$i_c = \sqrt{2}\, I \sin \left( \omega t - \theta_i - \frac{4\pi}{3} \right) \qquad \text{for } c \text{ phase} \qquad (4\text{-}28)$$

when the load is balanced and when the phase sequence is positive (i.e., $E_a = E$, $E_b = E\underline{/-2\pi/3}$, $E_c = E\underline{/-4\pi/3}$). To satisfy the requirement for positive-sequence voltages the phases of the armature winding must be displaced from each other in accordance with the arrangement illustrated in Fig. 4-20 for an

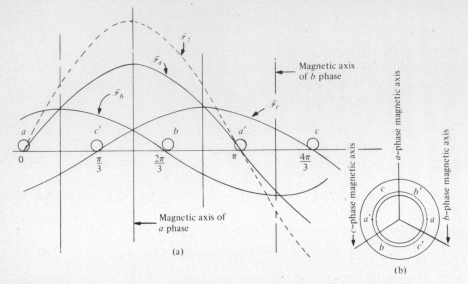

**Figure 4-20**    (a) Fundamental mmf waves in each phase and the total for $(\omega t - \theta_i) =$ $\pi/2$ or $i_a = \sqrt{2}\,I$. (b) Elementary two-pole three-phase machine.

elementary three-phase generator such as in Fig. 4-18 or any multipolar generator in which there is one armature slot per phase and pole, as, for example, in Fig. 4-20, which shows the three mmf waves and their resultant or total when $\omega t - \theta_i = \pi/2$ or when $i_a$ is a maximum. It is important that when $i_a$ is a maximum, the total armature mmf $A$ coincides with the magnetic axis of $a$ phase and the resulting flux linkage with $a$ phase is then a maximum. It may therefore be concluded that the total armature mmf related to any one phase of the armature is in time phase with the current in that phase of the armature winding, verified analytically in the following.

The magnetic axes of $b$ and $c$ phases are behind that of $a$ phase by $2\pi/3$ and $4\pi/3$ rad in electrical measure with respect to the direction of rotation. Hence, the contribution each phase makes to the mmf at the space angle $\theta$, considering only the fundamental and simplifying the notation by dropping the subscript 1, is expressed by

$$\mathscr{F}_{\theta a} = 0.45 N_{\text{coil}} I [\cos (\theta - \omega t + \theta_i) - \cos (\theta + \omega t - \theta_i)]$$

$$\mathscr{F}_{\theta b} = 0.45 N_{\text{coil}} I \left[ \cos (\theta - \omega t + \theta_i) - \cos \left( \theta + \omega t - \theta_i - \frac{4\pi}{3} \right) \right]$$

$$\mathscr{F}_{\theta c} = 0.45 N_{\text{coil}} I \left[ \cos (\theta - \omega t + \theta_i) - \cos \left( \theta + \omega t - \theta_i - \frac{2\pi}{3} \right) \right] \quad (4\text{-}29)$$

The total mmf is

$$\mathscr{F}_{\theta A} = \mathscr{F}_{\theta a} + \mathscr{F}_{\theta b} + \mathscr{F}_{\theta c} \quad (4\text{-}30)$$

The sum of the second cosine terms in Eq. 4-29 is zero, since these may be represented by three phasors of equal magnitude displaced from each other by $2\pi/3$ rad. Hence, for the three-phase armature under discussion,

$$\mathcal{F}_{\theta a} = 1.35 N_{\text{coil}} I \cos{(\theta - \omega t + \theta_i)} \qquad (4\text{-}31)$$

If the number of phases were $m$ instead of 3, it is found from an extension of the preceding derivation that

$$\mathcal{F}_{\theta A} = 0.45 m N_{\text{coil}} I \cos{(\theta - \omega t + \theta_i)} \qquad (4\text{-}32)$$

Equations 4-31 and 4-32 can be reduced to

$$\mathcal{F}_{\theta A} = A \cos{(\theta - \omega t + \theta_i)} \qquad (4\text{-}33)$$

and since the magnetic axis of $a$ phase is at $\theta = \pi/2$, the mmf in that axis due to the armature current is

$$A \cos{\left(\frac{\pi}{2} - \omega t + \theta_i\right)} = A \sin{(\omega t - \theta_i)} \qquad (4\text{-}34)$$

A comparison of Eq. 4-34 with the first of Eqs. 4-28 *shows the armature mmf along the magnetic axis of a-phase to be in time phase with* $i_a$.

As mentioned earlier, the mmf wave produced by balanced armature currents has the important property of rotating at synchronous speed in the same direction as the rotation of the rotor. Then, for a given steady-state condition, the armature mmf is stationary relative to the field poles. This also follows from Eq. 4-33, which shows the amplitude of the armature mmf to coincide with $\theta$ such that $\theta = \omega t - \theta_i$, signifying that the wave travels at $p\theta = \omega$ rad/sec or at synchronous speed.

### 4-5.2 Angular Displacement between mmf Waves

Since the flux linkage leads the induced voltage by 90°, the voltage expressed by Eq. 4-20 results from the flux linkage

$$\lambda_{af} = \lambda_{afm} \sin{\left(\omega t + \frac{\pi}{2}\right)} = \lambda_{afm} \cos{\omega t} \qquad (4\text{-}35)$$

If the magnetic circuit is assumed to be linear, the maximum flux linkage with the coil is proportional to the amplitude of the sinusoidal mmf wave that produces it [i.e., $\lambda_{afm} \propto F$, where $F$ is the amplitude of the mmf wave as indicated in Fig. 4-21(a), in which only one phase of the armature winding is shown]. Then the mmf in the magnetic axis of the coil due to the field current is

$$\mathcal{F}_F = F \cos{\omega t} \qquad (4\text{-}36)$$

and that due to the armature current is, from Eq. 4-34,

$$\mathcal{F}_A = A \sin{(\omega t - \theta_i)} = A \cos{\left(\omega t - \frac{\pi}{2} - \theta_i\right)} \qquad (4\text{-}37)$$

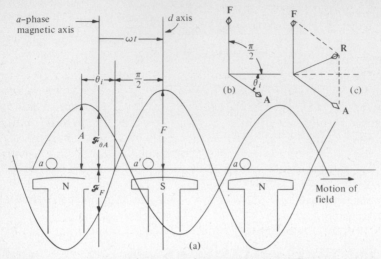

**Figure 4-21** (a) Fundamental mmf waves. (b) and (c) Phasor representation for a poly-phase synchronous generator (only $a$ phase indicated).

The relationships expressed by Eqs. 4-36 and 4-37 are shown graphically in Fig. 4-21(a) and (b), from which it is seen that the armature mmf lags the field mmf by the angle $\pi/2 + \theta_i$, both waves traveling at synchronous speed in the direction of rotation.

The same angular relationships exist between the magnetic axes of the field and armature windings in multipolar machines which are distributed among more than one slot per phase and pole and which in addition may have fractional pitch. This follows from the fact that the sinusoidal flux linkages of several coils may be obtained from the phasor sum of the indivdual coil flux linkages as in Fig. 4-16(b). The magnetic axis of a coil group passes through the geometric center of the group, just as the magnetic axis of one coil passes through its center. The resultant mmf may be obtained by adding the **A** and **F** waves, the amplitude **R** being the phasor sum of the amplitudes **A** and **F** of the armature and field mmfs, as shown in Fig. 4-21(c).

The effect of distributing and winding and of fractional pitch on the fundamental of the mmf wave is the same as that on the fundamental of the induced voltage and can be taken into account by means of the winding factor $k_w$. Accordingly, Eq. 4-32 can be modified to include the general case of a $P$-pole $m$-phase machine with $a$ current paths in the armature to yield

$$A = \frac{0.9mk_wN_{ph}I}{Pa} \quad \text{ampere turns/pole} \tag{4-38}$$

and for a three-phase winding $m = 3$, for which

$$A = \frac{2.7k_wN_{ph}I}{Pa} \tag{4-39}$$

# 4-6 UNSATURATED INDUCTANCES OF A CYLINDRICAL-ROTOR MACHINE

The magnetic reluctance of the iron in conventional machines is negligible when unsaturated. The effect of slots in the stator and in the rotor on the reluctance of the air gap, which is assumed to consume the entire mmf, may be taken into account by increasing the length of the air gap from the actual clearance $g$ between the stator and rotor iron to an effective length $g_e$. The ratio $g_e/g$ is nearly unity for large synchronous machines because of their relatively long air gaps; it is generally less than 1.1 for dc machines and large induction motors and usually between 1.1 and 1.25 for small induction motors in which $g$ ranges from 0.025 to 0.076 cm.†

## 4-6.1 Inductance of the Field

If the curvature of the air gap is neglected and $H$ is assumed normal to the iron surfaces, the amplitude of the $H$ wave due to the field mmf is

$$H_{\text{amp F}} = \frac{F}{g_e} \tag{4-40}$$

Since phasor operations are valid for sinusoidal space functions just as for sinusoidal time functions, the winding factor $k_{wf}$ applies to the mmf of a distributed and fractional-pitch winding and the amplitude of the fundamental in the field mmf is expressed by

$$F - \frac{4}{\pi} k_{wf} \frac{N_f i_f}{P} \quad \text{ampere turns/pole} \tag{4-41}$$

where $N_f$ is the number of series turns in the field winding (total field turns ÷ paths in field winding), and $i_f$ is the field current.

The amplitude of the $B$ wave due to $i_f$ is, from Eqs. 4-40 and 4-41,

$$B_{\text{amp F}} = \mu_0 H_{\text{amp F}} = \frac{4\mu_0 K_{wf} N_f i_f}{\pi P g_e} \tag{4-42}$$

Since the average value of a sinusoid is $2/\pi$ times its amplitude, the flux per pole must be

$$\phi_F = \frac{2}{\pi} B_{\text{amp F}} \times \text{area/pole}$$

If the mean diameter at the air gap is $D_g$, the area per pole is $\pi D_g L/P$, where $L$ is the effective axial length of the iron. Then

---

† For methods of correcting air-gap lengths, see P. L. Alger, *Induction Machines* (New York: Gordon and Breach, Inc., 1970), pp. 182–185, and A. E. Knowlton (ed.), *Standard Handbook for Electrical Engineers,* 9th ed. (New York: McGraw–Hill Book Company, 1957), Sec. 7–39.

$$\phi_F = \frac{8\mu_0 D_g L k_{wf} N_f i_f}{\pi P^2 g_e} \tag{4-43}$$

with a resulting field flux linkage of

$$\lambda_{ff} = k_{wf} N_f \phi_F = \frac{8\mu_0 D_g L (k_{wf} N_f)^2 i_f}{\pi P^2 g_e} \tag{4-44}$$

Equation 4-44 does not take into account the field leakage flux, which, however, is small compared with the radial field flux in cylindrical rotor machines. The self-inductance of the field winding is therefore slightly larger than expressed by

$$L_{ff} \cong \frac{\lambda_{ff}}{i_f} = \frac{8\mu_0 D_g L}{\pi g_e} \left(\frac{k_{wf} N_f}{P}\right)^2 \quad \text{H} \tag{4-45}$$

Since $\mu_0 = 4\pi \times 10^{-7}$,

$$L_{ff} = \frac{3.2 D_g L}{g_e} \left(\frac{k_{wf} N_f}{P}\right)^2 \times 10^{-6} \tag{4-46}$$

**EXAMPLE 4-2**

The following data are for a three-phase 13,800-V wye-connected 60,000-kVA 60-Hz synchronous generator:

P = 2; stator slots = 36, stator coils = 36, turns in each stator coil = 2

Stator coil pitch = $\frac{3}{2}$, rotor slots = 28

Spacing between rotor slots = $\frac{1}{37}$ of circumference

Rotor coils = 14

Turns in each rotor coil = 15

i.d. of stator iron = 0.948 m

o.d. of rotor iron = 0.865 m

Net axial length of stator iron = 3.365 m, stator coil connection is two-circuit ($a = 2$) series rotor-coil connection

Assume that $g_e = 1.08 g$ and calculate the unsaturated self-inductance of the field winding based on the fundamental component of the air-gap flux.

*Solution.* From Eq. 4-46,

$$L_{ff} \cong \frac{3.2 D_g L}{g_e} \left(\frac{k_{wf} N_f}{P}\right)^2 \times 10^{-6}$$

$$D_g = \frac{0.948 + 0.865}{2} = 0.9065 \text{ m}$$

$$L = 3.365 \text{ m}$$

$$g_e = \frac{1.08(0.948 - 0.865)}{2} = 0.0449 \text{ m}$$

$$N_f = 14 \times 15 = 210 \text{ turns}$$

$$P = 2$$

The field winding is equivalent to a full-pitch distributed winding, and the winding factor $k_{wf}$ is found from Eq. 4-14 as

$$k_{wf} = \frac{\sin (\beta_f/2)}{n \sin (\gamma_f/2)}$$

where $\beta_f = \dfrac{28\pi}{37} = 2.38$ rad or $136.3°$, the angle occupied by the field winding under one pole

$\gamma_f = \dfrac{2\pi}{37} = 0.0541$ rad or $9.73°$, the angle between adjacent slots

$n = 14$ slots per pole

$$k_{wf} = \frac{\sin (136.3/2)}{14 \sin (9.73/2)} = \frac{0.928}{14 \times 0.085} = 0.78$$

Hence,

$$L_{ff} = \frac{3.2 \times 0.9065 \times 3.365}{0.0449} \left(\frac{0.78 \times 210}{2}\right)^2 \times 10^{-6} = 1.46 \text{ H}$$

## 4-6.2 Magnetizing Inductance (Inductance of Armature Reaction)

The radial flux component due to the armature mmf $A$ resulting from balanced armature current can be determined on the same basis as $\phi_F$ in terms of $F$ and is then found to be

$$\phi_A = \frac{2\mu_0 A D_g L}{P g_e} \tag{4-47}$$

Substitution of Eq. 4-38 in Eq. 4-47 yields

$$\phi_A = \frac{1.8\mu_0 m D_g L k_w N_{ph} I}{P^2 a g_e} \tag{4-48}$$

This flux in rotating at synchronous speed induces a component of voltage in each armature phase in accordance with Eq. 4-15 which is expressed by

$$E_A = \frac{8 f \mu_0 m D_g L}{g_e} \left(\frac{k_w N_{ph}}{P a}\right)^2 I \tag{4-49}$$

The *magnetizing reactance or reactance of armature reaction* is defined by

$$X_{ad} = \frac{E_A}{I} = \frac{8 f \mu_0 m D_g L}{g_e} \left( \frac{k_w N_{ph}}{Pa} \right)^2 \qquad (4\text{-}50)$$

from which it follows that magnetizing inductance is

$$L_{ad} = \frac{X_{ad}}{2\pi f} = \frac{4\mu_0 m D_g L}{\pi g_e} \left( \frac{k_w N_{ph}}{Pa} \right)^2$$

$$= 1.6 \frac{m D_g L}{g_e} \left( \frac{k_w N_{ph}}{Pa} \right)^2 \times 10^{-6} \qquad (4\text{-}51)$$

and for a three-phase winding, $m = 3$:

$$L_{ad} = \frac{4.8 D_g L}{g_e} \left( \frac{k_w N_{ph}}{Pa} \right)^2 \times 10^{-6} \text{ H/phase} \qquad (4\text{-}52)$$

### 4-6.3  Self- and Mutual-Inductance Components of Magnetizing Inductance in Three-Phase Windings

The self-inductance of one phase of the armature resulting from radial flux can be found by adapting Eq. 4-46, with the result that

$$L_{aaM} = \frac{3.2 D_g L}{g_e} \left( \frac{k_w N_{ph}}{Pa} \right)^2 \times 10^{-6} \text{ H/phase} \qquad (4\text{-}53)$$

The mutual inductance component between any two phases is one-half that expressed by Eq. 4-53 and is negative:

$$L_{abM} = - \frac{1.6 D_g L}{g_e} \left( \frac{k_w N_{ph}}{Pa} \right)^2 \times 10^{-6} \text{ H/phase} \qquad (4\text{-}54)$$

which follows because of the 120° displacement between the magnetic axes the flux linkage produced in one phase by the current in another phase is one-half (cos 120°) that of the current-carrying phase.

The angular displacement between the magnetic axis of the field and that of a given armature phase determines the mutual inductance between the field and that phase of the armature. When the magnetic axes are in alignment, the mutual inductance is a maximum. Since the self-inductances $L_{ff}$ and $L_{aaM}$ as expressed by Eqs. 4–46 and 4-53 do not include leakage fluxes, the coefficient of coupling $k$ is unity when the magnetic axes are in alignment. Accordingly, the maximum is on the basis of Eq. 2-74.

$$L_{afM} = 1.00\sqrt{L_{ff}L_{aaM}}$$

$$= \frac{3.2D_g L k_{wf} N_f k_{wa}(N_{ph}/a)}{g_e P^2} \times 10^{-6} \qquad (4\text{-}55)$$

**EXAMPLE 4-3**

Calculate the magnetizing reactance of the generator in Example 4-2.

*Solution.* From Eq. 4-52,

$$L_{ad} = \frac{4.8 D_g L}{g_e}\left(\frac{k_w N_{ph}}{Pa}\right)^2 \times 10^{-6}$$

$N_{ph}$ = stator coils per phase $\times$ turns per stator coil

$$= \frac{36}{3} \times 2 = 24 \text{ turns/phase}$$

$a = 2$      (number of paths in the armature)

$k_w = k_p k_b$

$k_p = \sin p\dfrac{\pi}{2}$      (from Eq. 4-11a)

$p = \frac{2}{3}$ so   $k_p = \sin(\pi/3) = 0.866$

$k_b = \dfrac{\sin(\beta/2)}{n \sin(\gamma/2)}$      (from Eq. 4-14)

$\beta = \dfrac{\pi}{m} = \dfrac{\pi}{3}$     $n = \dfrac{\text{slots per pole}}{m} = \dfrac{36}{2 \times 3} = 6$

$\gamma = \dfrac{\pi}{\text{slots per pole}} = \dfrac{\pi}{18}$

$k_b = \dfrac{\sin(\pi/6)}{6 \sin(\pi/36)} = 0.955$      $k_w = 0.826$

When these values are substituted in Eq. 4-52, the result is

$$L_{ad} = \frac{4.8 D_g L}{g_e}\left(\frac{k_w N_{ph}}{Pa}\right)^2 \times 10^{-6}$$

$$= 0.00804 \text{ H/phase}$$

The magnetizing reactance is therefore

$$x_{ad} = 2\pi f L_{ad} = 2\pi \times 60 \times 0.00804 = 3.03 \ \Omega/\text{phase}$$

## 4-7 PHASOR DIAGRAM OF CYLINDRICAL-ROTOR SYNCHRONOUS GENERATOR

The fundamental component fluxes $\phi_F$ and $\phi_A$ produce flux linkages with the armature which induce corresponding components of armature voltage. The flux linkage with the armature winding due to the field current is

$$\lambda_{af} = \frac{k_w N_{ph}}{a} \phi_F \tag{4-56}$$

and that due to the balanced armature current is

$$\lambda_a = \frac{k_w N_{ph}}{a} \phi_A \tag{4-57}$$

The resultant **R** of the field and the armature mmfs produce the net air-gap flux

$$\phi_R = \phi_F + \phi_A \tag{4-58}$$

which results in the net armature flux linkage

$$\lambda_{ag} = \lambda_{af} + \lambda_a \tag{4-59}$$

where all quantities in Eqs. 4-58 and 4-59 are phasors, and which generates the corresponding components of armature voltage

$$\mathbf{E}_{ag} = \mathbf{E}_{af} + \mathbf{E}_A \tag{4-60}$$

These relationships are shown in the phasor diagram in Fig. 4-22(a).

## 4-7.1 Leakage Flux

Although the configurations of the windings and of the magnetic circuits in rotating machines are more complex than those in conventional transformers, the nature of leakage fluxes in both are quite similar. Analytical approaches to the calculation of leakage inductance are not nearly as straightforward as those of magnetizing reactance and are therefore not included in this text.† However, the ratio of leakage inductance to magnetizing inductance ranges from about 0.09 to 0.20 for cylindrical-rotor machines.

## 4-7.2 Synchronous Reactance

The equivalent leakage flux $\phi_l$ and the flux $\phi_A$ of armature reaction or of magnetizing inductances are both in phase with the current in a given phase of the armature winding. The magnetizing inductance and leakage inductance may therefore be added to yield the synchronous inductance:

$$L_d = L_{ad} + L_l \qquad \text{H/phase} \tag{4-61}$$

with the corresponding synchronous reactance

$$\omega L_d = \omega L_{ad} + \omega L_l$$

---

† For more complete treatment of leakage reactance, see C. A. Adams, "The Leakage Reactance of Induction Motors," *Trans Int. Elec. Congress,* St. Louis, 1904, Vol. 1 (1905), pp. 706–724; P. L. Alger, "Calculation of Armature Reactance in Synchronous Machines," *Trans. AIEE* 47 (1928): 493–512; M. Liwschitz-Garik and C. C. Whipple, *Alternating-Current Machines* (Princeton, N.J.: D. Van Nostrand Company, Inc., 1961), Chap. 57; P. L. Alger, *Induction Machines* (New York: Gordon and Breach, Inc., 1970), Chap. 7.

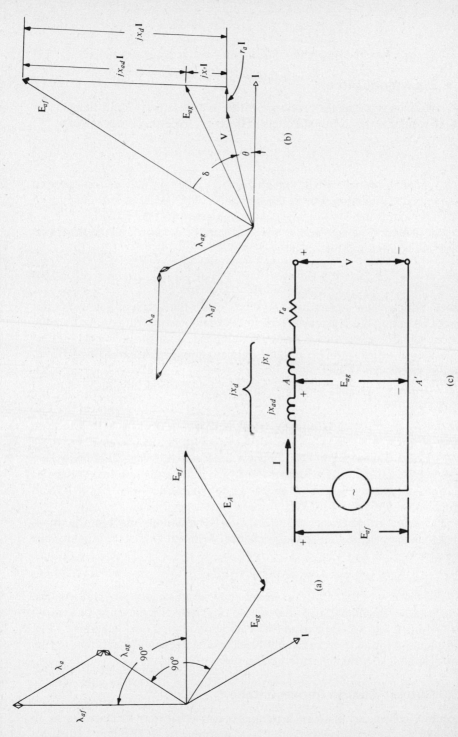

**Figure 4-22** (a) Phasor diagram for cylindrical-rotor generator showing flux linkages and corresponding induced voltages. (b) Phasor diagram including leakage impedance voltage drop. (c) Equivalent circuit.

or
$$x_d = x_{ad} + x_l \qquad \text{ohms/phase}$$
(4-62)

where $x_{ad}$ and $x_l$ are the magnetizing reactance and leakage reactance.

### 4-7.3 Equivalent Circuit

If the resistance and leakage reactance of the armature were zero, then $\mathbf{E}_{ag}$ in Eq. 4-60 would be the terminal voltage. However, since this is not the case,

$$\mathbf{E}_{ag} = \mathbf{V} + (r_a + jx_l)\mathbf{I}$$
(4-63)

where $\mathbf{V}$ is the terminal voltage per phase and $r_a$ is the armature resistance in ohms per phase. Equation 4-63 is taken into account by the phasor diagram in Fig. 4-22(b) and is the basis for the equivalent circuit in Fig. 4-22(c).

The armature voltage due to the field current is known as the generated voltage and is expressed by

$$\mathbf{E}_{af} = \mathbf{V} + (r_a + jx_d)\mathbf{I}$$
(4-64)

for a cylindrical-rotor generator. The relationship for a salient-pole generator includes an additional reactance term to take into account the nonuniformity of the air gap. Equation 4-64 is sometimes applied to salient-pole machines for calculations in which the effect of pole saliency is not important. Although synchronous motors have, practically without exception, salient poles, they are often treated as cylindrical-rotor machines and Eq. 4-64 is applicable if the sign of $\mathbf{I}$ is made negative to yield

$$\mathbf{V} = \mathbf{E}_{af} + (r_a + jx_d)\mathbf{I}$$
(4-65)

Figure 4-23 shows a phasor diagram for a synchronous motor based on Eqs. 4-4, 4-62, and 4-65. The current $\mathbf{I}$ is shown leading the terminal voltage $\mathbf{V}$, a relationship which requires the motor to be overexcited (i.e., $E_{ag} > V$) and which is common practice for synchronous motor operation.

The armature resistance of three-phase synchronous machines is much smaller than the synchronous reactance; hence, the magnitude of the synchronous impedance is

$$z_d = \sqrt{r_a^2 + x_d^2} \cong x_d$$
(4-66)

For that reason $r_a$ is omitted from many analyses of steady-state performance of synchronous machines. Such analyses are then based on the simplified equivalent circuit in Fig. 4-24(a) and on the corresponding phasor diagrams in Fig. 4-24(b) and (c) for an overexcited and underexcited generator, and that in Fig. 4-24(d) for an overexcited synchronous motor.

### 4-7.4 Current-Source Representation

Since the synchronous machine generates a voltage, it may be classified as an *active network*. An active network may be represented by its Thévenin equivalent circuit, which consists of a voltage source in series with an impedance, and by its Norton equivalent circuit, in which a current source is shunted by an imped-

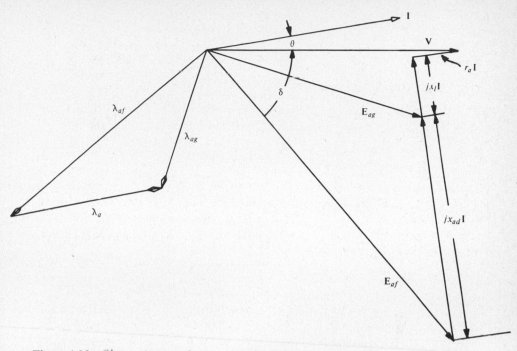

**Figure 4-23**   Phasor diagram for a synchronous motor based on cylindrical-rotor theory.

ance, the impedance having the same value in both equivalents.† The Norton equivalent makes for a ready comparison of synchronous machines with induction motors.

The Thévenin equivalent viewed from the terminals of a synchronous machine is as shown in Fig. 4-22(c). However, when the Thévenin equivalent to the left of points $A$ and $A'$ in Fig. 4-22(c) is converted to its Norton equivalent, the circuit in Fig. 4-25 is the result. The current $I_F$ is an equivalent balanced polyphase armature current which when applied to the armature with the dc field current, $i_f = 0$, produces the same value of air-gap flux as that produced by a value of dc field current such that $E_{af} = I_F x_{ad}$ when the armature current is zero and the generator is driven at synchronous speed.

**EXAMPLE 4-4**

The generator in Examples 4-2 and 4-3 is delivering rated load at 0.80 power factor, current lagging. The leakage reactance is 0.12 times the magnetizing reactance, and the armature resistance is negligible for this problem.

Neglecting saturation calculate (a) the synchronous reactance in ohms per phase and in per unit; (b) $E_{ag}$, the voltage behind the leakage impedance; (c) $E_{af}$, the voltage due to the field current; (d) the resultant flux linking the armature; (e) the flux that links the armature due to the field current; and (f) the flux produced by the armature current.

Show a phasor diagram of the current, voltage, and flux phasors.

---

† Thévenin and Norton equivalent circuits are treated in most modern textbooks on electric circuits. See, for instance, R. E. Lueg, *Basic Electronics for Engineers* (New York: IEP, A. Dun-Donnelley Publisher, 1963), pp. 26–30.

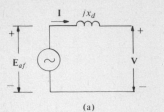

(a)

(b)

(c)

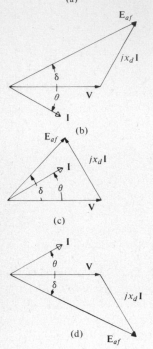

**Figure 4-24** (a) Simplified equivalent circuit of a synchronous machine. Phasor diagrams for overexcited generator. (c) For underexcited generator. (d) For overexcited motor.

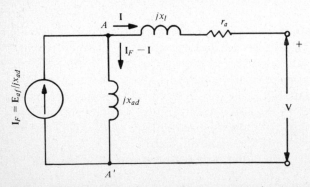

**Figure 4-25** Synchronous generator represented as a current source.

*Solution*

a.    $x_{ad} = 3.03\ \Omega/\text{phase}$    (from Example 4-3)

$$x_l = 0.12x_{ad} = 0.12 \times 3.03 = 0.36\ \Omega/\text{phase}$$

$$x_d = x_{ad} + x_l \quad \text{(from Eq. 4-62)}$$

$$= 3.03 + 0.36 = 3.39\ \Omega/\text{phase}$$

$$x_{d(\text{per unit})} = x_{d(\text{ohms})} \frac{VA_{\text{base}}}{(\text{volts}_{\text{base}})^2}$$

$$= \frac{3.39 \times 60 \times 10^6}{(13.8 \times 10^3)^2} = 1.07$$

b. $\mathbf{E}_{ag} = \mathbf{V} + (r_a + jx_l)\mathbf{I}$    (from Eq. 4-63)

For convenience let $\mathbf{V}$ lie on the axis of reals:

$$\mathbf{V} = \frac{13,800}{\sqrt{3}}(1 + j0) = 7960 + j0\ \text{V/phase}$$

The rated armature current is

$$I = \frac{VA}{\sqrt{3}\ V_{L-L}} = \frac{60 \times 10^6}{\sqrt{3} \times 13.8 \times 10^3} = 2510\ \text{A/phase}$$

and when expressed as a phasor,

$$\mathbf{I} = 2510(0.80 - j0.60) = 2010 - j1510$$

Since $r_a \cong 0$,

$$\mathbf{E}_{ag} = 7960 + j0.36(2010 - j1510)$$

$$= 7960 + 544 + j724 - 8504 + j724$$

$$= 8530\underline{/4.86°}$$

c. $\mathbf{E}_{af} = \mathbf{V} + (r_a + jx_d)\mathbf{I}$

$$= 7960 + j3.39(2010 - j1510)$$

$$= 7960 + 5120 + j6820 = 13,080 + j6820$$

$$= 14,750\underline{/27.5°}$$

d. $\phi_R = \dfrac{aE_{ag}}{4.44\,f\,k_w N_{ph}}$    (from Eq. 4-15, where $k_w = k_p k_b$)

$$k_w = 0.826 \quad \text{(from Example 4-3)}$$

$$\phi_R = \frac{2 \times 8530}{4.44 \times 60 \times 0.826 \times 24} = 3.23\ \text{Wb/pole}$$

$\phi_R$ leads $\mathbf{E}_{ag}$ by 90°; hence,

$$\phi_R - 3.23\underline{/90°} + 4.8° = 3.23\underline{/94.8°}$$

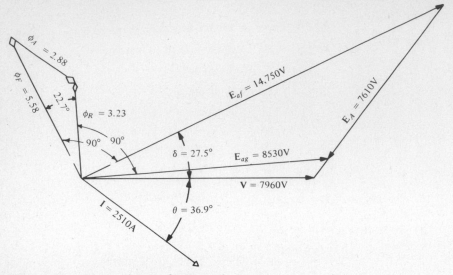

**Figure 4-26** Phasor diagram for Example 4.4.

e. $\phi_F = \dfrac{E_{af}}{E_{ag}} \phi_R = \dfrac{14{,}750}{8{,}530} \times 3.23 = 5.58$ Wb/pole

$\phi_F = 5.58\underline{/90° + 27.5°} = 5.58\underline{/117.5°}$

f. $E_A = X_{ad}I = 3.03 \times 2510 = 7610$ V/phase

$\phi_A = \dfrac{E_a}{E_{ag}} \phi_R = \dfrac{7610}{8530} \times 3.23 = 2.88$ Wb/pole

Also                          $\mathbf{I} = 2510(0.80 - j0.60) = 2510\underline{/-36.9°}$

and since the armature flux is in phase with armature current,

$$\phi_A = 2.89\underline{/-36.9°}$$

The phasor diagram is shown in Fig. 4-26.

## 4-8 IDEALIZED THREE-PHASE GENERATOR— GENERAL RELATIONSHIP IN TERMS OF INDUCTANCES

Figure 4-27(a) shows a schematic diagram of a three-phase Y-connected synchronous machine with voltage polarities and current directions for motor (load) operation and Fig. 4-27(b) for operation as a generator or source. The only difference between Fig. 4-27(a) and (b) is that the direction of the armature currents $i_a$, $i_b$, and $i_c$ in one are opposite to those in the other.

Let $r_a$ = resistance of each phase of the armature
   $L_{aa}$ = self-inductance of each phase of the armature
   $r_f$ = resistance of the field winding

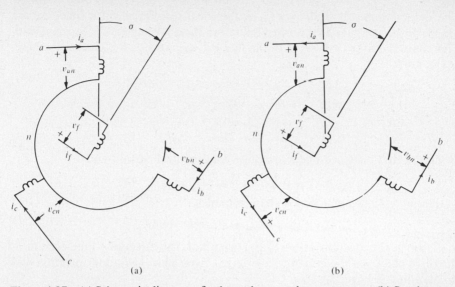

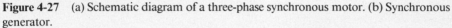

**Figure 4-27**    (a) Schematic diagram of a three-phase synchronous motor. (b) Synchronous generator.

$L_{ff}$ = self-inductance of the field winding

$L_{af}$ = mutual inductance between $a$ phase and the field winding

$L_{bf}$ = mutual inductance between $b$ phase and the field winding

$L_{cf}$ = mutual inductance between $c$ phase and the field winding

$L_{afm}$ = maximum value of mutual inductance between any phase and field winding

$L_{ab}$ = mutual inductance between any two phases of the armature (the mutual inductance between phases was shown to be negative in Eq. 4-54)

The applied voltages in Fig. 4-27(a) for motor operation are

$$v_{an} = r_a i_a + p\lambda_a$$
$$v_{bn} = r_a i_b + p\lambda_b$$
$$v_{cn} = r_a i_c + p\lambda_c$$
$$v_f = r_f i_f + p\lambda_f \tag{4-67}$$

where $v_{an}$, $v_{bn}$, and $v_{cn}$ are the terminal voltages of phases $a$, $b$, and $c$, respectively, and $v_f$ is the voltage applied to the field. The quantities $\lambda_a$, $\lambda_b$, and $\lambda_c$, and $\lambda_f$ are the flux linkages with the respective phases and with the field winding. The relationship between the flux linkages and the inductances can be represented in matrix form by

$$
\begin{bmatrix} \lambda_a \\ \lambda_b \\ \lambda_c \\ \lambda_f \end{bmatrix}
=
\begin{bmatrix}
L_{aa} & -L_{ab} & -L_{ab} & -L_{af} \\
-L_{ab} & L_{aa} & -L_{ab} & -L_{bf} \\
-L_{ab} & -L_{ab} & L_{aa} & -L_{cf} \\
-L_{af} & -L_{bf} & -L_{cf} & L_{ff}
\end{bmatrix}
\begin{bmatrix} i_a \\ i_b \\ i_c \\ i_f \end{bmatrix}
\tag{4-68}
$$

The inductances $L_{aa}$, $L_{ab}$, and $L_{ff}$ in the ideal cylindrical-rotor machine are practically constant and for normal operation the field current $i_f$ is constant also. Therefore, the time derivatives of the flux linkages in Eq. 4-67 are on the basis of Eq. 4-68:

$$p\lambda_a = L_{aa}pi_a - L_{ab}p(i_b + i_c) - i_f pL_{af}$$
$$p\lambda_b = L_{aa}pi_b - L_{ab}p(i_c + i_a) - i_f pL_{bf}$$
$$p\lambda_c = L_{aa}pi_c - L_{ab}p(i_a + i_b) - i_f pL_{cf}$$
$$p\lambda_f = -p(L_{af}i_a + L_{bf}i_b + L_{cf}i_c) \tag{4-69}$$

Under normal balanced operation there is no neutral current and

$$i_a + i_b + i_c = 0$$

so that the voltage applied to $a$ phase becomes

$$v_{an} = [r_a + (L_{aa} + L_{ab})p]i_a - i_f pL_{af} \tag{4-70}$$

When the machine operates as a generator, the direction of the current in the three phases of the armature is reversed from that for motor operation and the terminal voltages are then

$$v_{an} = -[r_a + (L_{aa} + L_{ab})p]i_a - i_f pL_{af}$$
$$v_{bn} = -[r_a + (L_{aa} + L_{ab})p]i_b - i_f pL_{bf}$$
$$v_{cn} = -[r_a + (L_{aa} + L_{ab})p]i_c - i_f pL_{cf}$$
$$v_f = r_f i_f + p(L_{af}i_a + L_{bf}i_b + L_{cf}i_c) \tag{4-71}$$

## 4-9 GENERATOR DELIVERING BALANCED LOAD

At no load $i_a = i_b = i_c = 0$ and the terminal voltages are also the generated voltages,

$$v_{an} = e_{an} = -i_f pL_{af}$$
$$v_{bn} = e_{bn} = -i_f pL_{bf}$$
$$v_{cn} = e_{cn} = -i_f pL_{cf}$$
$$v_f = r_f i_f \tag{4-72}$$

Since sinusoidal mmf and flux-density distribution waves are assumed for the idealized polyphase machine, the mutual inductance between the armature winding and the field winding is

$$
\boxed{
\begin{aligned}
L_{af} &= L_{afm} \cos \sigma \\
L_{bf} &= L_{afm} \cos \left( \sigma - \frac{2\pi}{3} \right) \\
L_{cf} &= L_{afm} \cos \left( \sigma - \frac{4\pi}{3} \right)
\end{aligned}
}
\tag{4-73}
$$

where $\sigma$ is the angle in electrical measure between the magnetic axis of the field and that of $a$ phase as in Fig. 4-27.

For a constant rotor angular velocity of $\omega$ electrical radians per second,

$$\sigma = \omega t + \sigma_0 \tag{4-74}$$

and the no-load voltage is found from Eqs. 4-72 and 4-73 to be

$$e_{an} = \omega i_f L_{afm} \sin (\omega t + \sigma_0)$$

$$e_{bn} = \omega i_f L_{afm} \sin \left( \omega t + \sigma_0 - \frac{2\pi}{3} \right)$$

$$e_{cn} = \omega i_f L_{afm} \sin \left( \omega t + \sigma_0 - \frac{4\pi}{3} \right) \tag{4-75}$$

which, by letting $\omega L_{afm} i_f = \sqrt{2}\, E_{af}$, can be reduced to

$$e_{an} = \sqrt{2}\, E_{af} \sin (\omega t + \sigma_0)$$

$$e_{bn} = \sqrt{2}\, E_{af} \sin \left( \omega t + \sigma_0 - \frac{2\pi}{3} \right)$$

$$e_{cn} = \sqrt{2}\, E_{af} \sin \left( \omega t + \sigma_0 - \frac{4\pi}{3} \right) \tag{4-76}$$

The no-load voltages of the three phases are equal in magnitude to $E_{af}$ rms and are displaced from each other by $2\pi/3$ radians or 120 degrees with positive phase sequence and can therefore be represented by phasors as follows:

$$\mathbf{E}_{an} = \mathbf{E}_{af}$$

$$\mathbf{E}_{bn} = \mathbf{E}_{af} \underline{/-120°}$$

$$\mathbf{E}_{cn} = \mathbf{E}_{af} \underline{/-240°} \tag{4-77}$$

If the generator supplies a balanced three-phase load such that the current in each phase lags the induced or no-load emf by an angle $\theta_i$, the instantaneous currents are

$$i_a = \sqrt{2}\, I \sin (\omega t + \sigma_0 - \theta_i)$$

$$i_b = \sqrt{2}\, I \sin \left( \omega t + \sigma_0 - \theta_i - \frac{2\pi}{3} \right)$$

$$i_c = \sqrt{2}\, I \sin \left( \omega t + \sigma_0 - \theta_i - \frac{4\pi}{3} \right) \tag{4-78}$$

When Eqs. 4-76 and 4-78 are substituted in Eq. 4-71, the instantaneous terminal voltage of $a$ phase is found to be

$$v_{an} = -r_a \sqrt{2}\, I \sin (\omega t + \sigma_0 - \theta_i) - \omega (L_{aa} + L_{ab}) \sqrt{2}\, I \cos (\omega t + \sigma_0 - \theta_i)$$
$$+ \sqrt{2}\, E_{af} \sin (\omega t + \sigma_0) \tag{4-79}$$

where the synchronous reactance is

$$\omega (L_{aa} + L_{ab}) = x_d$$

$$v_{an} = -r_a \sqrt{2}\, I \sin (\omega t + \sigma_0 - \theta_i) - x_d \sqrt{2}\, I \cos (\omega t + \sigma_0 - \theta_i)$$
$$+ \sqrt{2}\, E_{af} \sin (\omega t + \sigma_0) \tag{4-80}$$

which can be abbreviated further by means of phasor representation as follows:

$$\mathbf{V} = \mathbf{V}_{an} = -(r_a + jx_d)\mathbf{I} + \mathbf{E}_{af}$$ (4-81)

where $\mathbf{V}$ is the terminal voltage of $a$ phase.

It follows from Eqs. 4-77, 4-78, and 4-79 that

$$\mathbf{V}_{bn} = \mathbf{V}_{an}\underline{/-120°}$$
$$\mathbf{V}_{cn} = \mathbf{V}_{an}\underline{/-240°}$$

The flux linkages expressed by Eq. 4-68 are for motor operation. The relationship for generator action is obtained by reversing the sign of the armature current, which results in the field flux linkage

$$\lambda_f = L_{ff}i_f + L_{af}i_a + L_{bf}i_b + L_{cf}i_c$$ (4-82)

Substitution of Eqs. 4-73 and 4-78 in Eq. 4-82 yields

$$\lambda_f = L_{ff}i_f + L_{afm}\sqrt{2}I\left[ \sin (\omega t + \sigma_0 - \theta_i) \cos (\omega t + \sigma_0) \right.$$
$$+ \sin \left( \omega t + \sigma_0 - \theta_i - \frac{2\pi}{3} \right) \cos \left( \omega t + \sigma_0 - \frac{2\pi}{3} \right)$$
$$\left. + \sin \left( \omega t + \sigma_0 - \theta_i - \frac{4\pi}{3} \right) \cos \left( \omega t + \sigma_0 - \frac{4\pi}{3} \right)\right]$$ (4-83)

When the relationship

$$\sin x \cos y = \tfrac{1}{2}[\sin (x - y) + \sin (x + y)]$$

is applied to Eq. 4-83, the result is

$$\lambda_f = L_{ff}i_f + \frac{L_{afm}\sqrt{2}\,I}{2}\left\{ -3 \sin \theta_i + \sin [2(\omega t + \sigma_0) - \theta_i] \right.$$
$$\left. + \sin \left[ 2\left( \omega t + \sigma_0 - \frac{2\pi}{3} \right) - \theta_i \right] + \sin \left[ 2\left( \omega t + \sigma_0 - \frac{4\pi}{3} \right) - \theta_i \right]\right\}$$

in which

$$\sin [2(\omega t + \sigma_0) - \theta_i] + \sin \left[ 2\left( \omega t + \sigma_0 - \frac{2\pi}{3} \right) - \theta_i \right]$$
$$+ \sin \left[ 2\left( \omega t + \sigma_0 - \frac{4\pi}{3} \right) - \theta_i \right] = 0$$

so that

$$\lambda_f = L_{ff}i_f - \tfrac{3}{2}L_{afm}\sqrt{2}I \sin \theta_i$$ (4-84)

Equation 4-84 shows that the flux linkage with the field due to steady balanced three-phase armature currents is constant, signifying that the armature flux or armature mmf rotates at synchronous speed. This agrees with the principles discussed in Sec. 4-5, in which it was shown that the armature mmf rotates at synchronous speed. When the armature current in a generator lags the induced emf, it has a demagnetizing effect on the field while a leading current magnetizes the field.

Since the flux linkage with the field winding is constant under steady balanced three-phase load, $\lambda_f$ is constant and $p\lambda_f = 0$; so the voltage applied to the field winding is constant, being

$$v_f = r_f i_f$$

It is a simple matter to show that the equivalent armature current $I_F$ in Fig. 4-25 is

$$I_F = \frac{L_{afm}}{\sqrt{2}\,L_{ad}}\,i_f \tag{4-85}$$

## 4-10 TORQUE

Equation 2-110 shows the torque to equal the change, with angular displacement, of the energy stored in the magnetic field for a given set of currents:

$$T = \frac{\partial W_\phi}{\partial \sigma_m}\bigg|_{\text{all currents constant}}$$

The energy stored in the magnetic field of a three-phase cylindrical-rotor synchronous machine is

$$W_\phi = \tfrac{1}{2}[L_a(i_a^2 + i_b^2 + i_c^2) + 2L_{ab}(i_a i_b + i_b i_c + i_c i_a) \\ + 2i_f(L_{af}i_a + L_{bf}i_b + L_{cf}i_c) + \tfrac{1}{2}L_{ff}i_f^2] \tag{4-86}$$

The inductances $L_{aa}$, $L_{ab}$, and $L_{ff}$ are independent of angular position in a cylindrical-rotor machine and are therefore constant. Hence,

$$T = \frac{\partial W_\phi}{\partial \sigma_m}\bigg|_{\text{all currents constant}} = i_f\left[i_a \frac{dL_{af}}{d\sigma_m} + i_b \frac{dL_{bf}}{d\sigma_m} + i_c \frac{dL_{cf}}{d\sigma_m}\right]$$

$$= -i_f L_{afm} \frac{P}{2}\left[i_a \sin \sigma + i_b \sin\left(\sigma - \frac{2\pi}{3}\right) + i_c \sin\left(\sigma - \frac{4\pi}{3}\right)\right] \tag{4-87}$$

When the currents are balanced as expressed by Eq. 4-78 and substituted in Eq. 4-87, the torque is found to be

$$T = -i_f L_{afm} \frac{P}{2}\sqrt{2}\,I\left[\sin \sigma \sin(\sigma - \theta_i) + \sin\left(\sigma - \frac{2\pi}{3}\right)\right.$$

$$\times \sin\left(\sigma - \theta_i - \frac{2\pi}{3}\right)$$

$$\left. + \sin\left(\sigma - \frac{4\pi}{3}\right)\sin\left(\sigma - \theta_i - \frac{4\pi}{3}\right)\right] \tag{4-88}$$

which can be reduced by making use of the relationship

$$\sin x \sin y = \tfrac{1}{2}[\cos (x - y) + \cos (x + y)]$$

to

$$T = -\frac{3}{2}\frac{P}{2}\, i_f L_{afm}\sqrt{2}\, I \cos \theta_i \qquad (4\text{-}89)$$

Equation 4-89 shows that for constant field current and constant balanced three-phase armature current, the instantaneous torque is constant. The torque developed in a generator opposes the direction of rotation and is therefore negative. However, when the armature currents are unbalanced, the torque is not constant even when the field current is constant. An extreme case is that in which only one phase carries current and for which the torque would have a large double-frequency component, as may be verified by including only the first term in the brackets in Eq. 4-88. There would then also be a double-frequency voltage induced in the field winding, which follows from Eq. 4-83. In fact, unbalanced armature currents induce not only an emf in the field winding, which contains double frequency and higher-order harmonics, but also eddy currents in the solid iron structure of the rotor, which may cause the rotor iron to overheat.

It should be remembered that the inductances treated in this chapter are for unsaturated machines. Economic considerations, however, require the magnetization of the iron to be carried somewhat into the region of saturation. The maximum flux density in the armature teeth is generally held to values not exceeding 1.55 T, being lower in the other parts of the iron. High saturation of the teeth produces high iron losses and requires a large number of ampere turns on the field winding. Operating the teeth at low flux densities requires an uneconomically large cross section of iron. The flux density for the air gaps of 60-Hz synchronous machines is generally between 0.54 and 0.85 T. The relationship between the flux per pole and its determining factors is therefore of importance to an understanding of design considerations.

## 4-11 OPEN-CIRCUIT AND SHORT-CIRCUIT TESTS

The effect of saturation on the performance of synchronous machines is taken into account by means of the magnetization curve and other data obtained by tests on an existing machine or from design data. Calculations based on design are beyond the scope of this textbook, and only some basic test methods are considered.† The unsaturated synchronous impedance and an approximate value of the saturated synchronous impedance can be obtained from the open-circuit and short-circuit tests.

In the case of a constant voltage source having a constant impedance, the impedance can be found by dividing the open-circuit terminal voltage by the

† The various tests on synchronous machines are described in *IEEE Test Procedures for Synchronous Machines No. 115* (New York: Institute of Electrical and Electronic Engineers).

short-circuit current. However, when the impedance is a function of the open-circuit voltage, as it is when the machine is saturated, the open-circuit characteristic or magnetization curve in addition to the short-circuit characteristic is required.

The unsaturated synchronous reactance is constant because the reluctance of the unsaturated iron is negligible. The equivalent circuit of one phase of a polyphase synchronous machine is shown in Fig. 4-28(a) for the open-circuit condition and in Fig. 4-28(b) for the short-circuit condition. Now $\mathbf{E}_{af}$ is the same in both cases when the impedance $z = \mathbf{E}_{af}/\mathbf{I}_{sc}$, where $\mathbf{E}_{af}$ is the open-circuit volts per phase and $\mathbf{I}_{sc}$ is the short-circuit current per phase.

### 4-11.1 Open-Circuit Characteristic

To obtain the open-circuit characteristic the machine is driven at its rated speed without load. Readings of line-to-line voltage are taken for various values of field current. The voltage except in very low-voltage machines is stepped down by means of instrument potential transformers. Figure 4-29(a) shows the open-circuit characteristic or saturation curve. Two sets of scales are shown; one, line-

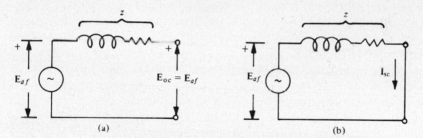

**Figure 4-28** Synchronous generator. (a) Open circuit. (b) Short circuit.

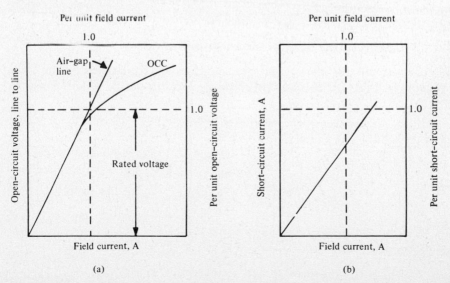

**Figure 4-29** (a) Open-circuit characteristic. (b) Short-circuit characteristic.

to-line volts versus field current in amperes and the other per-unit open-circuit voltage versus per-unit field current. If it were not for the magnetic saturation of the iron, the open-circuit characteristic would be linear as represented by the air-gap line in Fig. 4-29(a). It is important to note that 1.0 per unit field current corresponds to the value of field current that would produce rated voltage if there were no saturation. On the basis of this convention, the per-unit representation is such as to make the air-gap lines of all synchronous machines identical.

### 4-11.2 Short-Circuit Test

The three terminals of the armature are short-circuited each through a current-measuring circuit, which except for small machines is an instrument current transformer with an ammeter in its secondary. A diagram of connections in which the current transformers are omitted is shown in Fig. 4-30. The machine is driven at approximately synchronous (rated) speed and measurements of armature short-circuit current are made for various values of field current, usually up to and somewhat above rated armature current. The short-circuit characteristic (i.e., armature short-circuit current versus field current) is shown in Fig. 4-29(b). In conventional synchronous machines the short-circuit characteristic is practically linear because the iron is unsaturated up to rated armature current and somewhat beyond, because the magnetic axes of the armature and the field practically coincide (if the armature had zero resistance the magnetic axes would be in exact alignment), and the field and armature mmfs oppose each other. A phasor diagram for a synchronous machine under a three-phase short circuit is shown in Fig. 4-31. The flux linkage $\lambda_{af}$ is of an order that would produce an open-circuit voltage $\mathbf{E}_{af}$ of about rated value. The resultant flux linkage $\lambda_{ag}$ is just enough to overcome armature leakage reactance and armature resistance and is of the order of about one-fourth or less of $\lambda_{af}$ and is therefore insufficient to produce saturation.

### 4-11.3 Unsaturated Synchronous Impedance

The open-circuit and short-circuit characteristics are represented on the same graph in Fig. 4-32. The field current $oa$ produces a line-to-line voltage $oc$ on the

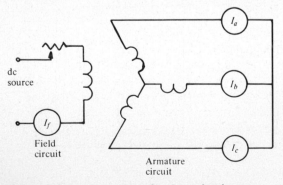

**Figure 4-30**   Connections for short-circuit test.

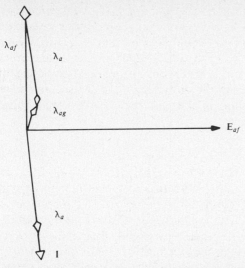

**Figure 4-31**    Phasor diagram for short-circuit test condition.

air-gap line, which would be the open-circuit voltage if there were no saturation. The same value of field current produces the armature current $o'd$ and the unsaturated synchronous reactance is

$$z_{du} = \frac{oc}{\sqrt{3}\; o'd} \qquad \Omega/\text{phase} \tag{4-90}$$

When the open-circuit characteristic, air-gap line, and the short-circuit characteristic are plotted in per-unit, then the per-unit value of unsaturated synchronous reactance equals the per-unit voltage on the air-gap line which results from the same value of field current as that which produces rated short-circuit (one per unit) armature current. In Fig. 4-32 this would be the per-unit value on the air-gap line corresponding to the field current $og$.

### 4-11.4  Approximation of the Saturated Synchronous Reactance

As mentioned previously, economical size requires the magnetic circuit to be somewhat saturated under normal operating conditions. However, the machine is unsaturated in the short-circuit test, and the synchronous reactance based on short-circuit and open-circuit test data is only an approximation at best. Nevertheless, there are many studies in which a value based on rated open-circuit voltage and the short-circuit current suffices. Hence, in Fig. 4-32, if $oc$ is rated voltage, $ob$ is the required no-load field current, which also produces the armature current $o'e$ on short circuit. The synchronous impedance is, accordingly,

$$z_d = \frac{oc}{\sqrt{3}\; o'e} \tag{4-91}$$

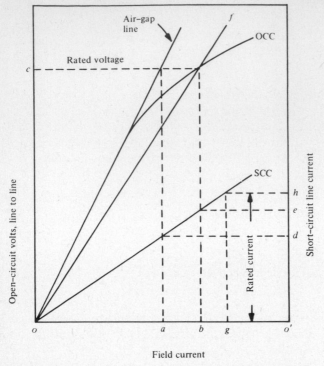

**Figure 4-32** Open-circuit and short-circuit characteristics.

Except in very small machines, the synchronous reactance is much greater than the resistance $r_a$ of the armature, and the saturated value as well as the unsaturated value of the synchronous reactance is considered equal to the magnitude of the synchronous impedance:

$$x_d = \sqrt{z_d^2 - r_a^2} \simeq z_d \qquad (4\text{-}92)$$

The line *of* in Fig. 4-32 is more nearly representative of the saturated machine than is the air-gap line. On the basis of this line, an estimate of the field current can be obtained for a given terminal voltage, load current, and power factor. This is done by calculating $\mathbf{E}_{af}$ and making use of the saturated synchronous reactance as follows:

$$\boxed{\mathbf{E}_{af} = \mathbf{V} + z_d\,\mathbf{I}}$$

The field current is that required to produce $E_{af}$ on the line *of*.

## 4-12 VOLTAGE REGULATION

If $V$ is the rated terminal voltage at a given load and power factor and rated speed, $E_{oc}$ is the open-circuit voltage at rated speed when the load is removed without changing the field current, the regulation is

$$\boxed{\text{Regulation} = \frac{E_{oc} - V}{V}} \qquad (4\text{-}93)$$

**EXAMPLE 4-5**

The open-circuit and short-circuit characteristics for a 133,689-kVA three-phase 13.8-kV 60-Hz ac generator are shown in Fig. 4-33. Find (a) the unsaturated synchronous reactance; (b) the approximate saturated synchronous reactance; (c) the estimated field current for rated voltage, rated current at 0.80 power factor lagging; and (d) the voltage regulation.

*Solution*

a.  The field current of about 485 A required for rated voltage 13.8 kV line-to-line on the air-gap line produces a short-circuit current of 3400 A per phase; hence,

$$x_{du} = \frac{13,800}{\sqrt{3} \times 3400} = 2.34 \ \Omega/\text{phase}$$

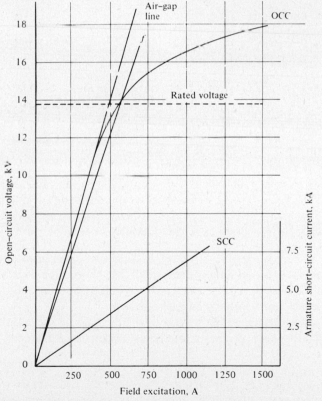

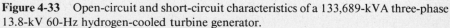

**Figure 4-33**   Open-circuit and short-circuit characteristics of a 133,689-kVA three-phase 13.8-kV 60-Hz hydrogen-cooled turbine generator.

b. A field current of about 550 A produces rated voltage on the open-circuit characteristic and a short-circuit current of 3900 A, from which the approximate synchronous reactance is found to be

$$x_d = \frac{13,800}{\sqrt{3} \times 3900} = 2.04 \ \Omega/\text{phase}$$

c. In Fig. 4-33 the induced emf (line-to-line) on line *of* yields approximately the value of the field current. The induced emf per phase on that basis is

$$\mathbf{E}_{af} = \mathbf{V} + jx_d \mathbf{I}$$

The rated current is

$$I = \frac{133,689}{\sqrt{3} \times 13.8} = 5594 \ \text{A/phase}$$

and

$$\mathbf{E}_{af} = \frac{13,800}{\sqrt{3}} + j2.04 \times 5594(0.80 - j0.60)$$

$$= 7970 + 6840 + j9120 = 17,350\underline{/31.7°}$$

and the line-to-line magnitude of the induced emf is

$$\sqrt{3} \ E_{af} = \sqrt{3} \times 17,350 = 30,100 \ \text{V}$$

The field current required to produce this voltage on line *of* in Fig. 4-33 is 1200 A.

d. A field current of 1200 A produces a no-load line-to-line voltage of 17,300 V ($\sqrt{3} \ E_{oc}$) on the open-circuit characteristic, and the regulation is

$$\text{Regulation} = \frac{17,300 - 13,800}{13,800} = 0.254$$

## 4-13 SHORT-CIRCUIT RATIO

The short-circuit ratio (SCR) is a measure of the physical size of a synchronous machine rated at a given kVA, power factor, and speed. The short-circuit ratio is defined as the ratio of the field current for rated no-load voltage at rated speed to the field current for rated short-circuit armature current.

In Fig. 4-32, *ob* is the field current that produces rated voltage on the open-circuit characteristic and *og* the field current for rated short-circuit current. Hence, the short-circuit ratio is

$$\text{SCR} = \frac{ob}{og} \tag{4-94}$$

To gain an idea as to the influence of physical size on the short-circuit ratio, consider a synchronous machine in which the length *g* of the air gap is doubled while the armature winding and all dimensions of the stator iron remain unchanged. If the reluctance of the iron were negligible, the no-load field current would need to be just about doubled to produce the same voltage as before. Doubling the length of the air gap reduces the unsaturated value of the reactance of armature reaction $x_{ad}$ to one-half its original value, so that only about one-half of the original resultant flux is necessary to produce rated short-circuit current. Since this one-half value of flux now traverses twice the original length of air gap, the value of field mmf required for rated short-circuit armature current is practically unchanged. However, about twice the field current or field mmf in ampere turns is required to produce rated no-load voltage, since the mutual inductance $L_{afm}$ is reduced to one-half, and the field winding must be increased in size if the heating is to remain the same. As a consequence, the machine must be made larger to accommodate the larger field winding.

## 4-14  REAL AND REACTIVE POWER VERSUS POWER ANGLE

The real and reactive power delivered by a synchronous generator can be expressed as a function of the terminal voltage, generated voltage, synchronous impedance, and the power angle or torque angle $\delta$. This is also true for the real and reactive power taken by a synchronous motor. If the angle $\delta$ is increased gradually, the real power output increases, reaching a maximum when $\delta = \tan^{-1} (x_s/r_a)$ or practically $\pi/2$. This is known as the steady-state power limit. The maximum torque or pull-out torque of a synchronous motor occurs at $\delta \simeq \pi/2$ on the basis of cylindrical-rotor theory if the armature resistance $r_a$ is neglected. On that basis any increase in the mechanical power to the synchronous generator or in the mechanical output of the synchronous motor after $\delta$ has reached 90° produces a decrease in real electrical power, and the generator accelerates while the motor decelerates—resulting in a loss of synchronism.

Consider a cylindrical-rotor synchronous generator driven at synchronous speed and let

$\mathbf{V}$ = terminal voltage or bus voltage

$\mathbf{E}_{af}$ = generated voltage

$z_d = r_a + jx_d$, the synchronous impedance

The quantities $\mathbf{V}$, $\mathbf{E}_{af}$, and $z_d$ may be expressed in volts and ohms per phase, in which case the real and reactive power are also per-phase, or they may be expressed in per-unit, with the real and reactive power also in per-unit.

The effects of saturation should be included in evaluating $E_{af1}$ and $x_d$, and several methods are available. One method is to make use of the approximation discussion in Sec. 4-11 and illustrated in Fig. 4-32, in which the magnetic characteristic is assumed to be represented by the line $of$. When per-unit values are used, this method is also known as the *short-circuit ratio method*.† The complex

---

† These methods are discussed in *Electrical Transmission and Distribution Reference Book*, 4th ed. (East Pittsburgh, Pa: Westinghouse Electric Corporation, 1950), pp. 446–453.

power output of the generator in volt-amperes per phase or in per-unit is, from Chap. 1,

$$S = P + jQ = \mathbf{VI}^* \tag{4-95}$$

In Fig. 4-34, the terminal voltage is

$$\mathbf{V} = V + j0$$

and the generated emf is

$$\mathbf{E}_{af} = E_{af}(\cos \delta + j \sin \delta)$$

from which and from Fig. 4-34, it follows that the current is

$$\mathbf{I} = \frac{\mathbf{E}_{af} - \mathbf{V}}{z_d} = \frac{E_{af} \cos \delta - V + jE_{af} \sin \delta}{z_d}$$

and its conjugate

$$\mathbf{I}^* = \frac{E_{af} \cos \delta - V - jE_{af} \sin \delta}{z_d^*} \tag{4-96}$$

where $z_d^* = r_a - jx_d$, the conjugate of the synchronous impedance.

When the numerator and denominator in Eq. 4-96 are multiplied by $z_d$, the conjugate current is expressed by

$$\mathbf{I}^* = \frac{z_d(E_{af} \cos \delta - V - jE_{af} \sin \delta)}{z_d^2}$$

$$= \frac{r_a(E_{af} \cos \delta - V) + x_d(E_{af} \sin \delta)}{z_d^2}$$

$$+ \frac{j[x_d(E_{af} \cos \delta - V) - r_a(E_{af} \sin \delta)]}{z_d^2} \tag{4-97}$$

From Eqs. 4-95 and 4-97

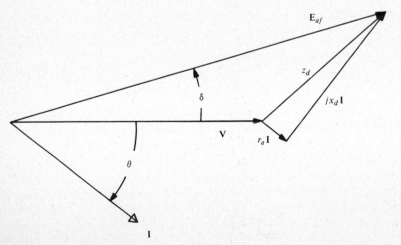

**Figure 4-34**   Phasor diagram of cylindrical-rotor synchronous generator, including the effect of armature resistance.

$$S = \frac{r_a(VE_{af}\cos\delta - V^2) + x_d(VE_{af}\sin\delta)}{z_d^2}$$

$$+ \frac{jx_d(VE_{af}\cos\delta - V^2) - r_a(VE_{af}\sin\delta)}{z_d^2} \tag{4-98}$$

the real and reactive power being

$$P = \frac{r_a(VE_{af}\cos\delta - V^2) + x_d(VE_{af}\sin\delta)}{z_d^2} \tag{4-99}$$

$$Q = \frac{x_d(VE_{af}\cos\delta - V^2) - r_a(VE_{af}\sin\delta)}{z_d^2} \tag{4-100}$$

In practical polyphase synchronous machines $r_a \ll x_d$ and $r_a$ can be neglected in the power equation, so $z_d \simeq x_d$ and

$$P \simeq \frac{VE_{af}}{x_d}\sin\delta \tag{4-101}$$

$$Q \simeq \frac{VE_{af}\cos\delta - V^2}{x_d} \tag{4-102}$$

To obtain the total power for a three-phase generator, Eqs. 4-101 and 4-102 must be multiplied by 3 when the voltages are line-to-neutral. However, if the line-to-line values (magnitudes) are used, these equations express the total three-phase power. The maximum real-power output per phase of the generator for a given terminal voltage and a given induced emf is

$$P_{max} \simeq \frac{VE_{af}}{x_d} \tag{4-103}$$

Any further increase in the prime-mover input to the generator causes the real-power output to decrease, the excess power going into acceleration, causing the generator to increase its speed and to pull out of synchronism. Hence, the steady-state stability limit is reached when $\delta \simeq \pi/2$. The transient stability limit is generally of greater interest in power-system operation.

The power-angle or torque-angle characteristic is shown graphically in Fig. 4-35. For normal steady operating conditions, the torque angle is well under 90°. In Example 4-5 the torque angle is calculated to be 31.7°, a practical value.

Although polyphase synchronous motors are of the salient-pole rather than cylindrical-rotor type, cylindrical-rotor theory applied to the motor yields results to a good degree of approximation for certain kinds of analysis. Accordingly, the maximum torque that a three-phase synchronous motor can develop for a gradually applied load is

$$T_{max} = \frac{P_{max}}{\omega_m} \simeq \frac{3E_{af}V}{2\pi(n_{syn}/60)x_d} = \frac{90E_{af}V}{\pi n_{syn}x_d} \tag{4-104}$$

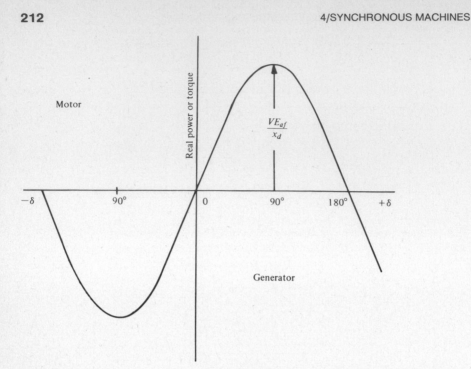

**Figure 4-35** Power- or torque-angle characteristic of a cylindrical-rotor synchronous machine.

The maximum torque approximates the "pull-out torque," which, according to the IEEE Dictionary,† is "the maximum sustained torque which a synchronous motor will develop at synchronous speed with rated voltage applied at rated frequency and with normal excitation."

## 4-15 SYNCHRONOUS-MOTOR V CURVES

When a synchronous motor delivers constant mechanical power while energized from a source of constant voltage and constant frequency, the armature current is a function of the field excitation being a minimum when the power factor of the motor is unity. If the armature resistance of the motor is neglected and cylindrical-rotor theory is applied, relationships are obtained that are quite similar to those developed for parallel operation of synchronous generators. When the armature current of a synchronous motor is plotted against field excitation for a given value of mechanical power, the result is a V curve. The solid lines in Fig. 4-36 represent a family of V curves for an unsaturated synchronous motor having an unsaturated synchronous reactance of 1.00 per unit. The dashed lines are loci for constant power factor and are called *compounding curves*.

† *IEEE Standard Dictionary of Electrical and Electronics Terms,* ANSI/IEEE Std 100-1984 (New York: Institute of Electrical and Electronic Engineers).

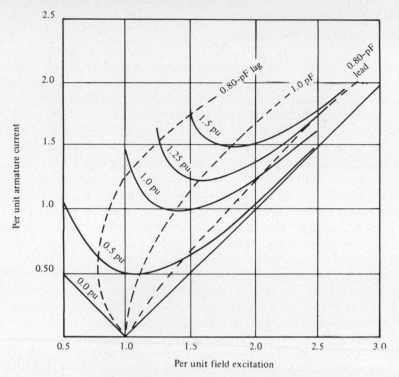

**Figure 4-36** Calculated V curves of unsaturated synchronous motor for 0, 0.5, 1.0, 1.25, and 1.50 per unit developed mechanical power. $X_d = 1.00$ per unit.

The excitation of synchronous motors is usually adjusted so that the motor draws leading current, thus generating reactive power. In fact, an overexcited synchronous machine generates reactive power whether operating as a motor or as a generator. Large electric power systems make use of synchronous condensers to generate some of the required reactive power.[†] A *synchronous condenser* is a synchronous machine, usually of the salient-pole type, that has neither a prime mover nor a mechanical load; it is, in effect, a synchronous motor running idle with an overexcited field. The V curve for zero mechanical power in Fig. 4-36 approximately represents synchronous condenser operation. When rotational losses and armature resistance are neglected, the real power of a synchronous condenser is zero, and consequently the torque angle $\delta = 0$. The reactive power output according to Eq. 4-103 is therefore

$$Q \simeq \frac{VE_{af} - V^2}{X_d} \qquad (4\text{-}105)$$

[†] A synchronous condenser, with an output of 345,000 kVA, was installed by American Electric Power Co. on its 765-kV power system near South Bend, Indiana. The synchronous condenser will supply reactive power to stabilize system voltage and thus improve the power-transmission capacity of the AEP system. It will be connected to an ASEA transformer with a rating of 1.5 million kVA. The condenser's stator and rotor have direct cooling. The stator has a transport weight of 167 tons.

## 4-16  EXCITATION SYSTEMS FOR SYNCHRONOUS MACHINES

A number of arrangements for supplying direct current to the fields of synchronous machines have come into use. Adjustments in the field current may be automatic or manual depending upon the complexity and the requirements of the power system to which the generator is connected.

Excitation systems are usually 125 V up to ratings of 50 kW with higher voltages for the larger ratings. The usual source of power is a direct-connected exciter, motor-generator set, rectifier, or battery. A common excitation system in which a conventional dc shunt generator mounted on the shaft of the synchronous machine furnishes the field excitation is shown in Fig. 4-37(a) and (b). The output of the exciter (i.e., the field current of the synchronous machine) is varied by adjusting the exciter-field rheostat. A somewhat more complex system that makes use of a pilot exciter, a compound dc generator also mounted on the generator shaft which in turn excites the field of the main exciter, is shown in Fig. 4-37(c) and (d). This arrangement makes for greater rapidity of response, a feature that is important in the case of synchronous generators when there are disturbances on the system to which the generator is connected. In some installations a separate motor-driven exciter furnishes the excitation. An induction motor is used instead of a synchronous motor because in a severe system disturbance a synchronous motor may pull out of synchronism with the system. In addition, a large flywheel is used to carry the exciter through short periods of severely reduced system voltage.

### 4-16.1  Brushless Excitation System

The brushless excitation system† eliminates the usual commutator, collector rings, and brushes. One arrangement in which a permanent-magnet pilot exciter, an ac main exciter, and a rotating rectifier are mounted on the same shaft as the field of the ac turbogenerator is shown in Fig. 4-38. The permanent-magnet pilot exciter has a stationary armature and a rotating permanent magnet field. It feeds 420-Hz, three-phase power to a regulator, which in turn supplies regulated dc power to the stationary field of a rotating-armature ac exciter. The output of the ac exciter is rectified by diodes and delivered to the field of the turbogenerator.

Brushless excitation systems have been also used extensively in the much smaller generators employed in aircraft applications where reduced atmospheric pressure intensifies problems of brush deterioration. Because of their mechanical simplicity, such systems lend themselves to military and other applications that involve moderate amounts of power.‡

---

† For a more complete discussion of brushless excitation systems, see E. H. Myers, "Rotating Rectifier Exciters for Large Turbine A-C Generators," paper presented before the American Power Conference, Illinois Institute of Technology, Chicago, 27th Annual Meeting, April, 1965, and D. B. Hoover, "The Brushless Excitation System for Large A-C Generators," *Westinghouse Engineer*, September 1964.

‡ See, for instance, D. H. Miller and A. S. Rubenstein, "Excitation Systems for Small Generators," *Elec. Eng.*, June 1962: 434–440.

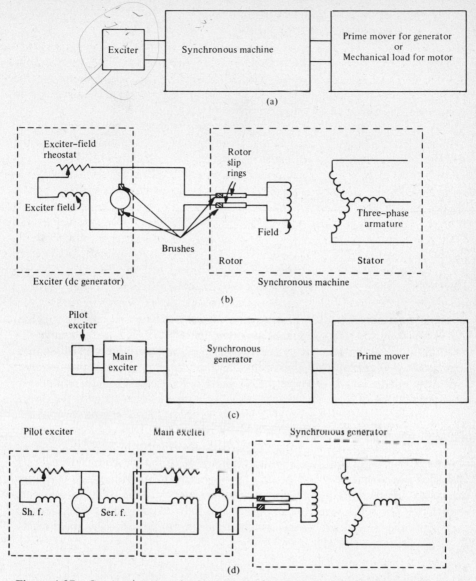

**Figure 4-37** Conventional excitation systems for synchronous machines. (a) Physical arrangement. (b) Circuit diagram for shaft-mounted exciter. (c) Physical arrangement. (d) Circuit diagram for shaft-mounted exciter and pilot exciter.

## 4-17 DIRECT-AXIS AND QUADRATURE-AXIS SYNCHRONOUS REACTANCE IN SALIENT-POLE MACHINES— TWO-REACTANCE THEORY

While the air gap in synchronous machines of the cylindrical-rotor construction is practically of uniform length, that of the salient-pole machine is much longer

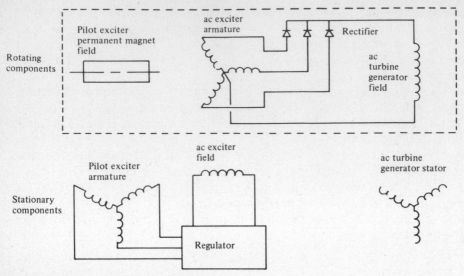

**Figure 4-38**   Brushless excitation system.

in the quadrature axis (i.e., in the region midway between poles) than in the direct axis or at the pole centers, as is evident from Fig. 4-7. Since the air gap is of minimum length in the direct axis, a given armature mmf directed along that axis produces a maximum value of flux, and the same armature mmf directed along the quadrature axis where the air gap has its greatest length produces a minimum value of flux. The synchronous reactance associated with the direct axis is therefore a maximum and is known as the *direct-axis synchronous reactance, $x_d$*. The minimum synchronous reactance $x_q$ is called the *quadrature-axis synchronous reactance*. In addition, because of the nonuniform length of air gap, a sinusoidal mmf wave with its amplitude in the direct axis produces a distorted flux-density wave somewhat as shown in Fig. 4-39(a), while the same sinusoidal mmf will produce a flux density wave of a different shape, about as shown in Fig. 4-39(b), when the amplitude is in the quadrature axis. Flux-density waves of other shapes are produced when the sinusoidal armature mmf reacts

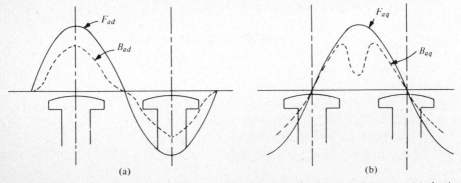

**Figure 4-39**   Sinusoidal armature mmf wave and resulting flux-density wave. (a) *d* axis. (b) *q* axis.

along an axis that lies between the direct and quadrature axes. These complications of variable reactance and of waveform for different locations of the magnetic axis of armature mmf relative to that of the field poles make a rigorous
treatment of the salient-pole machine along the lines of cylindrical-rotor theory
too cumbersome to be practical.

Cylindrical-rotor theory when modified to take waveform into account
could be applied to the salient-pole machine if the armature current were 90°
out of phase with the generated emf $E_{af}$ or if it were in phase with $E_{af}$. In the
first case the armature mmf would react along the direct axis and the direct-axis
synchronous reactance $x_d$ would apply. A phasor diagram for a salient-pole generator with the current lagging the generated emf by 90° is shown in Fig. 4-39(a).
Since the mmf produced by the armature, in this case, reacts entirely along the
direct axis, the armature current is designated as $I_d$, the direct-axis current. And
for this condition the phasor relation between generated emf terminal voltage,
and impedance is expressed by

$$E_{af} = V + (r_a + jx_d)I_d \qquad (4\text{-}106)$$

In the second case the armature mmf reacts along the quadrature axis and the
quadrature-axis synchronous reactance $x_q$ is used as shown in Fig. 5-46(b), and
the armature current is therefore designated as $I_q$, the quadrature-axis current.
The synchronous reactance associated with this current is $x_q$ and the voltages
are related to each other in accordance with

$$E_{af} = V + (r_a + jx_q)I_q \qquad (4\text{-}107)$$

The phasor relations for Eqs. 4-106 and 4-107 are illustrated in Fig. 4-40,
in which $\lambda_{af}$ is the fundamental component of flux linkage due to the field
current and $\lambda_{ad}$ is produced by $I_d$ while $\lambda_{aq}$ is the fundamental quadrature-axis
flux linkage due to $I_q$.

The armature current in synchronous generators and in synchronous motors is normally displaced from the generated voltage $F_{af}$ by some angle lying
between 0° and 90° and may then be divided into the two components $I_d$ and
$I_q$ as shown in Fig. 4-40(a). Then in keeping with Eqs. 4-106 and 4-107, it follows
that the generated voltage of the salient-pole generator is

$$E_{af} = V + (r_a + jx_q)I_q + (r_a + jx_d)I_d$$

and since

$$\boxed{\begin{aligned} I_d + I_q &= I \\ E_{af} = V + r_a I &+ jx_q I_q + jx_d I_d \end{aligned}} \qquad (4\text{-}108)$$

The components $I_q$ and $I_d$ are usually not given, since the only known quantities
are $V$, $I$, the load power-factor angle $\theta$, $r_a$, $x_d$, and $x_q$. The current components
$I_d$ and $I_q$ may, however, be readily obtained by making use of the known quantities
to establish the angle $\delta$ in Fig. 4-41(b), and we then have

$$I_d = I \sin (\theta + \delta) = I \sin \theta_i$$

and

$$I_q = I \cos (\theta + \delta) = I \cos \theta_i \qquad (4\text{-}109)$$

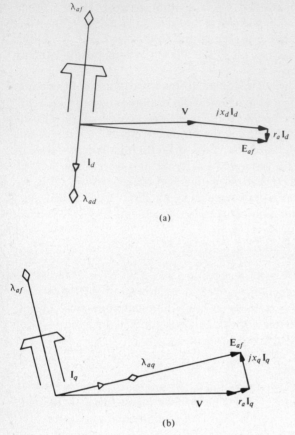

(a)

(b)

**Figure 4-40** Phasor diagram for salient-pole generator with the armature mmf. (a) $d$ axis. (b) $q$ axis.

The phasor diagram in Fig. 4-42 affords a basis for determining the value of $\delta$. Assume the current components $\mathbf{I}_d$ and $\mathbf{I}_q$ to be known. Then if the currents $\mathbf{I}_d$, $\mathbf{I}_q$, and $\mathbf{I}$ are multiplied by $jx_q$, the voltage triangle $ABC$ that is similar to the current triangle $abc$ is obtained and the phasor $jx_q\mathbf{I} = \mathbf{AC}$ terminates at point $C$ on the phasor $\mathbf{E}_{af}$, so

$$\mathbf{aC} = \mathbf{V} + (r_a + jx_q)\mathbf{I}$$

and

$$\tan \delta = \frac{\text{Im } \mathbf{aC}}{\text{Re } \mathbf{aC}}$$

$$= \frac{AC \cos \theta - r_a I \sin \theta}{V + r_a I \cos \theta + AC \sin \theta}$$

$$= \frac{x_q I \cos \theta - r_a I \sin \theta}{V + r_a I \cos \theta + x_q I \sin \theta} \tag{4-110}$$

Hence,

$$\mathbf{E}_{af} = [aC + (x_d - x_q)I_d]\underline{/\delta} \tag{4-111}$$

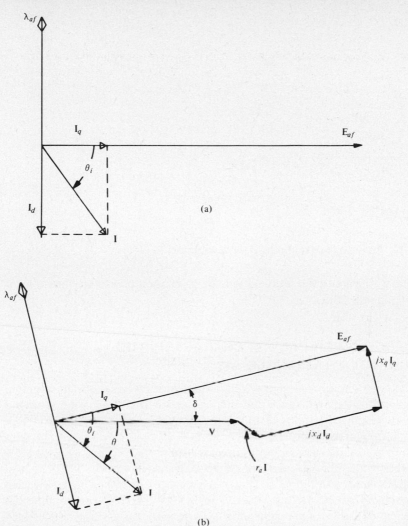

**Figure 4-41** Phasor diagrams for salient-pole generator. (a) Angular relationships of currents and induced voltage. (b) Terminal voltage and impedence drops included.

The armature resistance can generally be neglected and

$$\tan \delta = \frac{x_q I \cos \theta}{V + x_q I \sin \theta} \qquad (4\text{-}112)$$

A phasor diagram of flux-linkage components for an overexcited salient-pole generator is shown in Fig. 4-43. The expressions for inductance derived in Sec. 4-6 can be applied to salient-pole machines by use of multiplying factors that take into account the effects of saliency.† However, the self-inductance of

---

† See L. A. Kilgore, "Calculations of Synchronous Machine Constants," *Trans. AIEE* 5, No. 4 (1931): 1201–1213.

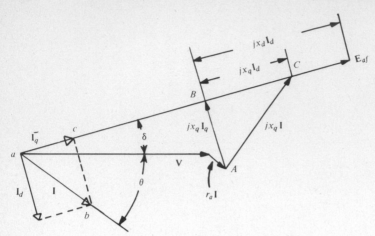

**Figure 4-42** Phasor diagram as a basis for determining $\delta$.

each phase by itself and the mutual inductance between phases are functions of rotor position (i.e., of $\sigma$).

## 4-18 ZERO-POWER-FACTOR CHARACTERISTIC AND POTIER TRIANGLE

While the open-circuit and short-circuit characteristics yield the unsaturated value of synchronous reactance and a rough approximation of the saturated values, closer approximations of the saturated synchronous reactance can be obtained for cylindrical-rotor as well as for salient-pole machines from their zero-power-factor and open-circuit characteristics.

In the zero-power-factor test the generator is loaded with an inductive load of low power factor. An unloaded synchronous motor of about the same rating as the generator can be used as a load. The field of the synchronous motor is underexcited and that of the generator is overexcited so as to produce rated current at various values of terminal voltage. A diagram of connections is shown in Fig. 4-44(a).

It is generally not necessary to obtain a complete zero-power-factor curve, as two points usually suffice for most practical purposes—one is near rated terminal voltage and the other at zero terminal voltage (i.e., short circuit). In that case, the first point may be obtained by connecting the generator or motor to supply a three-phase bus that operates at or somewhat above the rated voltage to assure saturation. The real-power output of the generator is made approximately zero by adjusting the output of its prime mover, or the machine may be operated as an unloaded synchronous motor. The field is overexcited so that the machine supplies its rated value of current to the bus. The zero-terminal voltage point is obtained from the short-circuit characteristic.

Two results are achieved by means of zero-power-factor load. One is that the magnetic circuit is saturated in the presence of armature current, whereas

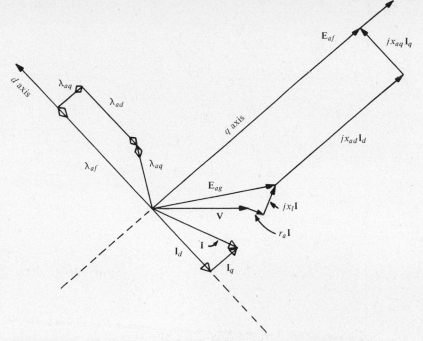

**Figure 4-43**   Phasor diagram for overexcited salient-pole generator.

the iron is unsaturated at rated short-circuit armature current in conventional machines. The other is that the synchronous reactance drop is practically in phase with the terminal voltage $V$ and with the generated voltage $E_{af}$ as shown in the phasor diagram of Fig. 4-44(b), so that the terminal voltage $V$, the synchronous-reactance drop $jx_dI$, and $E_{af}$ are all practically in phase, and $V$ and $x_dI$ add arithmetically. Under the idealized conditions of $r_a \cong 0$ and of 90° lagging current with zero field-leakage flux, the phasor diagram for the generator can be reduced to that of Fig. 4-44(c), in which the synchronous reactance is divided into two components, the leakage reactance $x_l$ and the magnetizing reactance $x_{ad}$. The zero-power-factor and open-circuit characteristics of a synchronous machine are shown in Fig. 4-44(d) for line-neutral voltage.

### 4-18.1  Graphical Determination of the Potier Triangle

The leakage reactance $x_l$ is assumed to be constant since it is relatively unaffected by saturation in the normal range of current as the paths of the armature leakage fluxes are largely in air. In Fig. 4-44(d), $F$ is the field current which produces $\lambda_{af}$, $A$ is the component of field current required to overcome the mmf of armature reaction which produces $\lambda_a$, and $R$ is the component of field current which produces the resultant flux linkage $\lambda_{ag}$, which in turn accounts for the air-gap voltage,

$$E_{ag} = V + x_lI \tag{4-113}$$

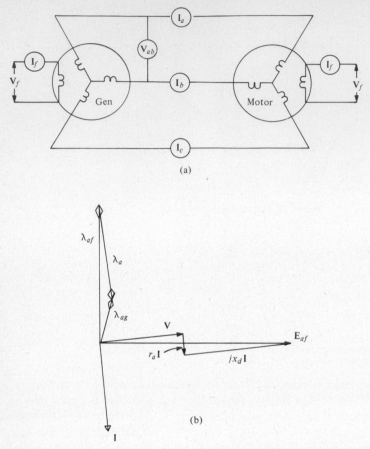

(a)

(b)

**Figure 4-44** Zero-power-factor test. (a) Circuit diagram. (b) Phasor diagram including armature resistance drop and the effect of small real power output. (c) Phasor diagram for zero real power, neglecting $r_a$. (d) Open-circuit and zero-power-factor characteristics.

The mmf of armature reaction then is represented by the base $ab$ and the leakage reactance drop by the altitude $ac$ of the triangle $abc$ known as the *Potier triangle.* For a given armature current the base and altitude of this triangle are practically constant when the effect of field-leakage flux is small, and if vertex $c$ is moved along the open-circuit characteristic, vertex $b$ traces the zero-power-factor characteristic. At zero terminal voltage the Potier triangle has its base on the abscissa, as shown by the triangle $a'b'c'$ in Fig. 4-44(d), where $ob'$ is the field mmf or field current required to produce rated short-circuit current, $a'b'$ is again the mmf of armature reaction, and $a'c'$ is the leakage reactance voltage drop $Ix_l$. The tests for determining the Potier triangle are summarized as follows:

1. Open-circuit characteristic.
2. Field current required to give rated short-circuit current or a known fraction thereof.

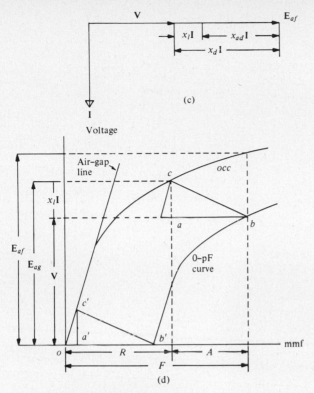

**Figure 4-44** (Continued)

3. Field current for rated current or the same fraction thereof as in (2) at zero power factor and terminal voltage equal to or greater than rated value. (The voltage must be sufficient to require appreciable magnetic saturation.)

This data is represented in Fig. 4-45, where $of$ is the field current and $fb = V$, the terminal voltage at rated current at zero power factor. The field current to produce rated short-circuit current is $ob'$. For the assumed conditions the sides of triangle $o'b'c'$ in Fig. 4-44(d) are fixed regardless of the terminal voltage for given value of zero-power-factor armature current, and when this triangle is moved upward on the open-circuit characteristic beyond the air-gap line, the side $oc$ remains parallel to the air-gap line. Accordingly, in Fig. 4-45 if $o'b$ is taken equal to $ob'$ and the line $o'c$ is drawn parallel to the air-gap line, the Potier triangle is established. The vertical projection of $o'c$ is the leakage reactance voltage drop and equals $\sqrt{3}\ ac$ in Fig. 4-44(d).

When full field current $of$ produces less than rated armature current at zero power factor, the same procedure as for rated armature current is followed except that the field current $ob'$ for the short-circuit test is reduced correspondingly.

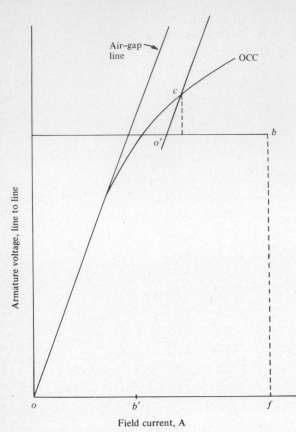

**Figure 4-45**   Graphical construction for Potier triangle.

## 4-18.2  Potier Reactance

The nature of the leakage fluxes associated with the field winding in cylindrical-rotor machines is somewhat similar to that of the leakage fluxes in the armature because the field winding is distributed among slots in much the same fashion as is the armature winding. However, the field winding is not as large, the end connections enclose smaller areas, and the field leakage flux in cylindrical-rotor machines is therefore correspondingly less than the armature leakage flux. The effect of the field leakage flux in combination with the armature leakage flux gives rise to an equivalent leakage reactance $x_p$ known as the *Potier reactance,* which is greater than the armature leakage reactance $x_l$. In the case of cylindrical-rotor machines the Potier reactance is not much greater than the leakage reactance, a typical ratio† of $x_p/x_l$ being about 1.3 and in many cases‡ $x_p$ is assumed to equal $x_l$.

† L. A. March and S. B. Crary, "Armature Leakage Reactance of Synchronous Machines," *Trans. AIEE* 54 (April 1935): 378–381.

‡ See, for example, Westinghouse Electric Corporation, *Electrical Transmission and Distribution Reference Book,* 4th ed. (East Pittsburgh, Pa.: Westinghouse Electric Corporation, 1950), Chap. 6, Secs. 1, 2, 3.

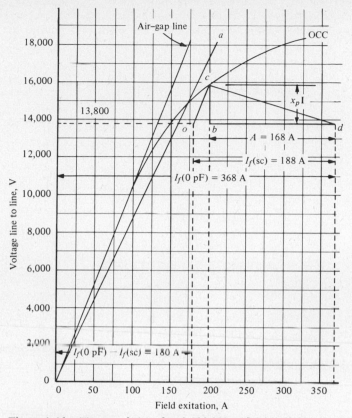

**Figure 4-46** Characteristics of a 13,529-kVA 13.8-kV 60-Hz three-phase two-pole turbine generator.

However, the field-leakage flux in salient-pole machines is sizable, particularly in those that have long slender poles, and the Potier reactance may be as much as 3 times the armature leakage reactance.[†]

Although the Potier reactance decreases with increasing saturation, the value obtained from the zero-power-factor test at about rated voltage is assumed constant for calculations in the normal range of balanced loads under steady-state conditions.

### EXAMPLE 4-6

The open-circuit characteristic of a 13,529-kVA 13.8-kV 60-Hz three-phase two-pole turbine generator is shown in Fig. 4-46.

The zero-power-factor test data follows:

| Line to line (V) | Field current (A) |
| --- | --- |
| 0 | 188 |
| 13,800 | 368 |

[†] Sterling Beckwith, "Approximating Potier Reactance," *Trans. AIEE* 56 (July 1937): 813–818.

Draw the Potier triangle and determine the Potier reactance $x_p$ and the component of field current $A$ to overcome the mmf of armature reaction.

*Solution*

$$I = \text{rated current of 566 A} = \frac{13,529}{\sqrt{3} \times 13.8}$$

Distance $bd$ is 168 A $= A$, the component of field current to overcome the mmf of armature reaction.

$$\text{Potier reactance } x_p = \frac{2000}{\sqrt{3} \times 566} = 2.04 \ \Omega,$$

$$x_p \text{ in per-unit} = \frac{2000}{13,800} = 0.145$$

## 4-19 USE OF POTIER REACTANCE TO ACCOUNT FOR SATURATION

Several methods use Potier reactance to take saturation of synchronous machines into account. The results obtained with the different methods are in good agreement and make it possible to calculate the field current and regulation under load with satisfactory accuracy. Only one of these methods, known as the saturation-factor method, is presented in the following.

### 4-19.1 Saturation-Factor Method

This method, introduced by Kingsley,[†] makes use of the Potier voltage expressed by

$$\boxed{\mathbf{E}_p = \mathbf{V} + (r_a + jx_p)\mathbf{I}} \qquad (4\text{-}114)$$

$r_a$ can be neglected for machines rated at 1000 kVA or above when computing $E_p$.

Saturation is taken into account by means of a *saturation factor k*, which is the ratio of the resultant mmf for the air gap plus iron to that for the air gap alone under the magnetic conditions determined by the voltage $E_p$.

In Fig. 4-47 the field mmf required to overcome the reluctance of the air gap for the resultant flux linkage $\lambda_{ag}$ is $R_{ag}$ and that to overcome both the reluctance of the air gap and the iron is $R$. The saturation factor $k$ is given by the ratio

$$\boxed{k = \frac{R}{R_{ag}}} \qquad (4\text{-}115)$$

† Charles Kingsley, Jr., "Saturated Synchronous Reactance," *Trans. AIEE* 54, No. 3 (March 1935): 300–305.

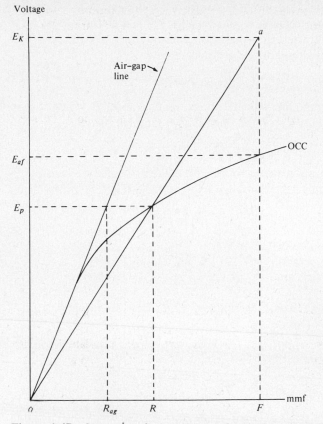

**Figure 4-47** Saturation factor method for determining regulation of a synchronous machine.

The Potier reactance includes the effect of field leakage flux and is taken as the leakage reactance so that Eq. 4-62 for the unsaturated synchronous reactance is modified to

$$x_{du} = x_p + x_{ad} \qquad (4\text{-}116)$$

Since a large part of the path for the leakage fluxes is in air, saturation has less effect on the Potier reactance than on the magnetizing reactance, as the path for the flux associated with $x_{ad}$ is mostly through iron and the only air path is the relatively short length of the air gap. Further, the component $x_{ad}$ is several times larger than $x_p$. The saturated synchronous reactance is therefore taken as

$$\boxed{x_d = x_p + \frac{x_{ad}}{k}} \qquad (4\text{-}117)$$

Then from Eqs. 4-116 and 4-117 the saturated synchronous reactance is expressed in terms of $x_{du}$ and $x_p$ by

$$x_d = x_p + \frac{x_{du} - x_p}{k} \qquad (4\text{-}118)$$

If the magnetic state of the iron remained fixed as defined by the saturation factor $k$, the synchronous reactance would have the value specified by Eq. 4-118 and the open-circuit characteristic would be represented by the line $oa$ in Fig. 4-47. The generated emf would be, accordingly,

$$\mathbf{E}_k = \mathbf{V} + (r_a + jx_d)\mathbf{I} \tag{4-119}$$

The magnitude of $\mathbf{E}_k$ on the line $oa$ determines the field mmf $F$, which in turn produces the generated emf $\mathbf{E}_{af}$ on the open-circuit characteristic.

**EXAMPLE 4-7**

Calculate the saturation factor $k$ and the regulation of the synchronous generator in Example 4-6, when delivering rated load at 0.85 power factor current lagging.

*Solution*

$$\text{Terminal volts per phase, } V = 13,800 \div \sqrt{3} = 7960$$

$$\text{Rated current amp per phase, } I = \frac{13,529}{\sqrt{3} \times 13.8} = 566$$

Potier reactance ohms per phase = 2.04 from Example 4-6

$$\mathbf{E}_p \cong \mathbf{V} + jx_p\mathbf{I}$$

$$\mathbf{E}_p = 7960 + j2.04 \times 566(0.85 - j0.527)$$

$$= 8570 + j981 = 8630\underline{/6.65°} \text{ V/phase}$$

The line-to-line voltage is

$$\sqrt{3}\, E_p = 14,950$$

which requires a field current of $R = 173$ A as determined on the *OCC* characteristic in Fig. 4-46. The same value of voltage requires an mmf of $R_{ag} = 142$ A on the air-gap line in Fig. 4-46. Hence,

$$k = \frac{R}{R_{ag}} = \frac{173}{142} = 1.22$$

The unsaturated synchronous reactance is found from the voltage on the air-gap line produced by a field current of 188 A, which is required for the rated short-circuit armature current of 566 A. This voltage is found to be 19,800 V line to line when the air-gap line is extended. So

$$x_{du} = \frac{19,800}{\sqrt{3} \times 566} = 20.2\ \Omega$$

and since $x_p = 2.04\ \Omega$, the saturated synchronous reactance, from Eq. 4-118, is

$$x_d = x_p + \frac{x_{du} - x_p}{k} = 2.04 + \frac{20.2 - 2.04}{1.22} = 16.9\ \Omega$$

and the induced voltage $E_k$ when $r_a$ is neglected is found to be, from Eq. 4-119,

$$\mathbf{E}_k \cong \mathbf{V} + jx_d\mathbf{I} = 7960 + j16.9 \times 566(0.85 - j0.527)$$

$$- 13,000 + j8130 = 15,320\underline{/32.0°}$$

The line-to-line magnitude of this voltage is

$$\sqrt{3}\ E_k = 26,600 \text{ V}$$

The field current obtained for this value on the line $oa$ in Fig. 4-46 is

$$F = 315 \text{ A}$$

that produces a line-to-line voltage on the open-circuit characteristic of 18,300 V. The regulation is therefore

$$\text{Regulation} = \frac{18.3 - 13.8}{13.8} = 0.326$$

## 4-20  SLIP TEST FOR DETERMINING $x_d$ AND $x_q$

Since the armature mmf reacts entirely on the direct axis when the armature current lags the generated voltage by 90°, the zero-power-factor test yields the value of direct-axis synchronous reactance $x_d$. However, it is impractical to attempt loading a generator so that its armature mmf reacts solely along the quadrature axis in order to obtain the quadrature-axis synchronous reactance $x_q$.

The slip test is made by applying reduced balanced three-phase voltage at rated frequency to the stator, while the rotor is made to rotate slightly above or below synchronous speed with the field circuit open. The direct axis and quadrature axis of the rotor thus alternately slip past the axis of the armature mmf, causing the armature mmf to react alternately along the direct and quadrature axes. The phase sequence of the applied voltage must be such that the armature mmf and the rotor rotate in the same direction. Oscillograms are taken of the armature terminal voltage, armature current, and the voltage across the open-field winding.† The oscillograms are somewhat as shown in Fig. 4-48. The slip indicated in Fig. 4-48 is too high for accurate values of $x_d$ and $x_q$ because of eddy currents induced in the pole faces and damper windings, when such are in the pole faces. The slip should be made as low as possible without the rotor pulling into synchronism as a result of the reluctance torque.

This is also a reason for applying reduced voltage.

The direct-axis and quadrature-axis reactances are found from the oscillograms as follows:

$x_d =$ ratio of applied volts per phase to the
              armature amperes per phase for the direct-
              axis position                                 (4-120)

† *IEEE Test Procedure for Synchronous Machines No. 115* (New York: Institute of Electronic Engineers). See also S. H. Wright, "Determination of Synchronous Machine Constants by Test," *Trans. AIEE* 50, No. 4 (December 1931): 1331–1350.

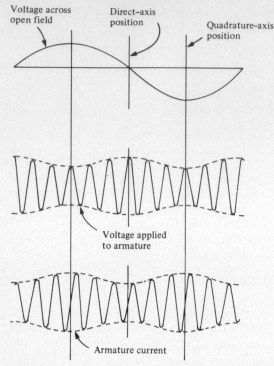

**Figure 4-48**   Oscillograms for slip test. (In practice the slip should be much less than indicated here.)

The voltage across the open-field winding is zero for the direct-axis position.

$x_q$ = ratio of applied volts per phase to the
armature amperes per phase for the quadrature-
axis position                                              (4-121)

Approximate values of $x_d$ and $x_q$ can be obtained from the readings of voltmeters and ammeters. The ammeter indicates a minimum value of current for the direct-axis position and a maximum value for the quadrature-axis position, its deflections thus oscillating between a minimum and a maximum. If the source has appreciable impedance, the oscillations in the ammeter readings are accompanied by oscillations in the voltmeter readings, with the maximum voltage $V_{max}$ occurring when the current is a minimum $I_{min}$. This is indicated by the oscillograms in Fig. 4-48. When armature resistance is neglected.

$$x_d = \frac{V_{max}}{I_{min}}$$                          (4-122)

and

$$x_q = \frac{V_{min}}{I_{max}}$$                          (4-123)

The pointers of the indicating instruments are subject to overswing, with a resulting error in their readings. Oscillographic measurements are therefore preferred. $x_q$ is assumed to be unaffected by saturation, and the saturation factor $k$ need be applied only to $x_d$.

## 4-21 TORQUE-ANGLE CHARACTERISTIC OF SALIENT-POLE MACHINES

The resistance $r_a$ of the armature has negligible effect on the relationship between the power output of a synchronous machine and its torque angle $\delta$ and is therefore neglected in the following. $r_a$ is neglected in the phasor diagram of Fig. 4-49 which is used as a basis for this derivation.

**Power Associated with $I_q$** The complex power associated with quadrature-axis current $\mathbf{I}_q$ is

$$\mathbf{S}_q = \mathbf{VI}_q^* \tag{4-124}$$

From Fig. 4-49

$$\mathbf{V} = V + j0 \tag{4-125}$$

$$\mathbf{I}_q = I_q(\cos \delta + j \sin \delta) \tag{4-126}$$

$$\mathbf{I}_q^* = I_q(\cos \delta - j \sin \delta)$$

$$I_q = \frac{V \sin \delta}{x_q} \tag{4-127}$$

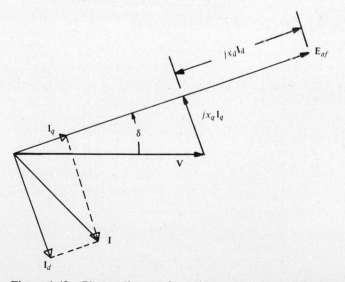

**Figure 4-49**  Phasor diagram for salient-pole generator, neglecting $r_a$.

Substitution of Eqs. 4-125, 4-126, and 4-127 in Eq. 4-124 yields

$$\mathbf{S}_q = \frac{V^2}{x_q} (\sin \delta \cos \delta - j \sin^2 \delta)$$

**Power Associated with $I_d$**  The complex power associated with the direct-axis current $\mathbf{I}_d$ is

$$\mathbf{S}_d = \mathbf{VI}_d^* \tag{4-128}$$

in Fig. 4-49,

$$\mathbf{I}_d = I_d(\sin \delta - j \cos \delta)$$

and $\qquad\qquad \mathbf{I}_d^* = I_d(\sin \delta + j \cos \delta) \tag{4-129}$

$$I_d = \frac{E_{af} - V \cos \delta}{x_d} \tag{4-130}$$

When Eqs. 4-129 and 4-130 are substituted in Eq. 4-128, the result is

$$\mathbf{S}_d = \frac{VE_{af} - V^2 \cos \delta}{x_d} (\sin \delta + j \cos \delta) \tag{4-131}$$

**Total Complex Power**  The complex power output per phase

$$\mathbf{S} = \mathbf{S}_q + \mathbf{S}_d$$
$$= \frac{VE_{af} \sin \delta}{x_d} + \left(\frac{1}{x_q} - \frac{1}{x_d}\right) V^2 \sin \delta \cos \delta$$
$$+ j\left(\frac{VE_{af} \cos \delta - V^2 \cos^2 \delta}{x_d} - \frac{V^2 \sin \delta}{x_q}\right)$$

which can be reduced to

$$\boxed{\begin{aligned}\mathbf{S} &= \frac{VE_{af}}{x_d} \sin \delta + \frac{x_d - x_q}{2x_d x_q} V^2 \sin 2\delta \\ &+ j\left\{\frac{VE_{af}}{x_d} \cos \delta - \frac{V^2}{2x_d x_q} [(x_d + x_q) - (x_d - x_q) \cos 2\delta]\right\}\end{aligned}} \tag{4-132}$$

Equation 4-132 expresses the power per phase unless it is per-unit. Hence, the real power output of a three-phase generator when expressed in watts is

$$\boxed{3P = 3 \operatorname{Re} \mathbf{S} = 3\left(\frac{VE_{af}}{x_d} \sin \delta + \frac{x_d - x_q}{2x_d x_q} V^2 \sin 2\delta\right)} \tag{4-133}$$

and the reactive power in vars is

$$3Q = 3 \text{ Im } \mathbf{S}$$

$$= 3\left\{\frac{VE_{af}}{x_d}\cos\delta - \frac{V^2}{2x_dx_q}[(x_d + x_q) - (x_d - x_q)\cos 2\delta]\right\} \qquad (4\text{-}134)$$

The power-angle characteristic of a salient-pole machine is shown in Fig. 4-50. Motor action results when $\delta$ is negative. The reactive power *output* is independent of the sign of $\delta$, which signifies that, when $r_a$ is neglected, an overexcited synchronous machine *delivers* reactive power, whether operating as a generator or as a motor. Synchronous motors are normally overexcited, so that in addition to delivering mechanical load, they also furnish reactive power. Industrial loads normally operate with lagging current, and when an overexcited synchronous motor is part of such a load it tends to improve the power factor of the combined load, a feature known as *power-factor correction*.

The electromagnetic torque or developed torque for a three-phase synchronous machine is

$$T_{em} = \frac{3P}{\omega_m} = \frac{3 \times 60}{2\pi n_{syn}}\left(\frac{VE_{af}}{x_d}\sin\delta + \frac{x_d - x_q}{2x_dx_q}V^2\sin 2\delta\right) \qquad (4\text{-}135)$$

The component

$$\frac{3 \times 60}{2\pi n_{syn}}\left(\frac{x_d - x_q}{2x_dx_q}V^2\sin 2\delta\right)$$

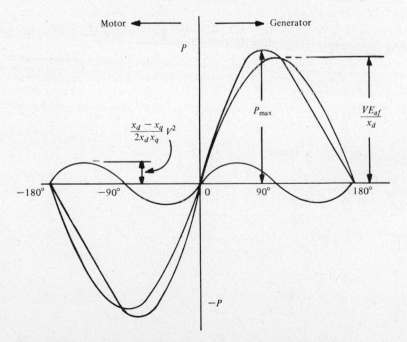

**Figure 4-50**  Power-angle characteristic of salient-pole machine.

is known as the *reluctance torque*. It is independent of the excitation and exists only if the machine is connected to a system receiving reactive power from other synchronous machines that maintain the terminal voltage $V$. The reluctance torque is due to the saliency of the field poles which tends to align the direct axis with that of the armature mmf. This feature causes a salient-pole synchronous motor to develop some torque without field current. Small single-phase synchronous motors such as are used in electric clocks depend entirely on the reluctance torque to maintain synchronism as these motors have no field winding or permanent-magnet rotors.

In addition, integral horsepower synchronous motors, known as reluctance motors, operate without dc field excitation and are finding increasing use despite their relatively large size and their requirement of large reactive power input.

It is impractical, however, to operate a synchronous generator without field excitation on a power system because it could then deliver only about 25 percent or less of its real power rating. In addition, it would absorb an excessive amount of reactive power.

## 4-22 SYNCHRONOUS-MOTOR STARTING

While the rotating field resulting from the armature current produces eddy currents in the rotor iron and hysteresis loss, the net torque due to these effects is insufficient to start synchronous motors except very small motors in some cases. The instantaneous torque over and above the small amount developed through eddy-current and hysteresis effects, however, is alternating, and because of the inertia of the rotor and its connected load, it cannot produce acceleration.

A common method is to start the synchronous motor as an induction motor by means of a starting winding, which in its simplest form is comprised of damper bars embedded in the pole faces and terminating in short-circuiting end rings as shown in Fig. 4-51(a). A typical field-pole lamination designed to accommodate seven damper bars is shown in Fig. 4-51(b). The starting winding

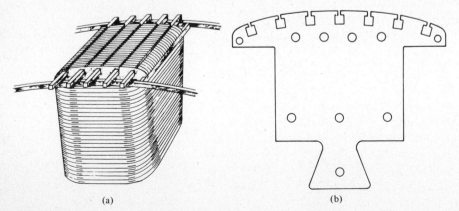

(a)                                                          (b)

**Figure 4-51**     (a) Salient-field pole with five damper bars. (b) Salient-pole lamination for seven damper bars.

may also be a double squirrel-cage winding, and in some cases when high starting torque is required it may be a wound-rotor winding. Squirrel-cage and wound-rotor windings are discussed in Chap. 5. The starting winding also serves to damp out speed oscillations due to pulsating load torques.

Synchronous motors are generally started and synchronized by means of automatic control equipment. The field winding is short-circuited through a discharge resistor during the starting period and is connected to the dc source as the motor approaches synchronous speed. A polarized relay is commonly used to provide for energizing the field with direct current at a point in the ac wave for smooth synchronizing. During starting, the field winding is generally short-circuited, and as the motor approaches the synchronous speed (which it cannot actually attain as an induction motor), the short circuit is removed from the field winding and direct current applied to the field. If the torque requirements of the load and its inertia do not exceed the *pull-in* torque of the motor, synchronism will result. The pull-in torque of a synchronous motor is that developed while operating as an induction motor at the speed from which it will pull into synchronism. It is important to distinguish this from the *pull-out* torque, which is the maximum torque the motor can deliver and stay in synchronism.

In the case of very high ratings, as in frequency changers, for example, where a synchronous motor of several thousand horsepower drives a synchronous generator, an auxiliary motor is generally used to bring the motor-generator up to speed. The synchronous motor is then synchronized to the system in the same manner as for the large synchronous generators described in Sec. 4-14. The exciter is used as a starting motor, when mounted on the shaft of the synchronous motor, in some installations of moderate size where a suitable dc source is available.

To reverse the direction of rotation of a three-phase synchronous motor or three-phase induction motor, it is necessary to reverse the direction in which the armature mmf rotates. This is accomplished by reversing the phase sequence of the applied voltage by interchanging any two of the three-phase leads to the armature.

## 4-23 FEATURES AND APPLICATION OF SYNCHRONOUS MOTORS

The rotors of conventional polyphase synchronous motors are of the salient-pole construction. The synchronous motor is particularly economical from the standpoint of cost and efficiency for slow-speed applications. Some synchronous motors have power-factor ratings of 0.80, current leading—which means that at rated load and rated excitation the motor supplies reactive power equal to about three-fourths its mechanical power rating. Other synchronous motors are rated to operate at unity power factor. Since torque is a function of terminal voltages $V$ and generated voltage $E_{af}$, the 0.80-power-factor motor is larger and has greater pull-in and pull-out torque than the unity-power-factor motor of the same horsepower and speed rating, and is therefore capable of meeting larger peak loads, which are characteristic of such devices as ball mills and crushers.

Synchronous motors are particularly suitable for speeds below 500 rpm for direct-connected loads (without reduction gears), such as compressors, grinders, and mixers, especially in ratings of 100 hp or more. At these low speed ratings, the synchronous motor is less expensive than an induction motor of the same rating, and has the additional advantage of being a source of reactive power.

## STUDY QUESTIONS

1. In what respect are dc and ac generators similar?
2. Would it be possible to convert a dc generator into an ac generator?
3. What is meant by the term alternator?
4. What is synchronous speed and how is it calculated?
5. List at least five reasons that ac generators are built with a revolving field instead of a revolving armature.
6. Why is it possible to construct ac synchronous generators of much larger capacities than dc generators?
7. What is an exciter? What type of machine is it and where is it located?
8. Describe two general types of generator fields. List advantages and disadvantages to the two construction types.
9. The speed of an ac synchronous generator is maintained at a constant value at all times. Why is this essential?
10. What two frequencies are generally used throughout the world?
11. What range of voltages are generally used for field excitation?
12. What are collector rings and what purpose do they serve?
13. Ac generators normally have special ventilation and cooling systems. What are the different methods used to cool ac synchronous generators? Explain the purpose of each method.
14. What special kind of insulating material is used in large ac generators?
15. Assuming that your generator develops a sine wave, describe the three factors that affect the generated voltage of your generator.
16. What is meant by the term "group of armature coils"? How are the coils in such a group connected?
17. In what two ways can the three phases of a three-phase generator be connected?
18. What is meant by a full-pitch winding? A fractional-pitch winding?
19. Why is it that fractional-pitch windings are generally used in generators?
20. What is meant by the term pitch factor?
21. What effect does the pitch factor have on the generated voltage of a generator?
22. Why are distributed windings generally in use in ac generators?
23. Describe what is meant by the term distribution factor.
24. Describe the effect that the distribution factor has on the generated voltage of a generator.
25. What is meant by the term regulation when applied to an ac generator?
26. What factors affect the regulation of a generator?

27. In any ac generator there are basically three voltage drops. What are they?

28. How are these three voltage drops used in making calculations for the regulation of an ac generator?

29. Why does the armature winding of a generator have reactance? What factors affect the reactance of the winding?

30. Describe a loading condition wherein it is possible for the generated emf of the generator to be greater at full load than at no load.

31. Describe the principles of operation of a voltage regulator.

32. Explain how a zero-lagging power factor load current tends to demagnetize the main field of a synchronous generator.

33. Explain how a zero-leading power factor current tends to aid the main field magnetization.

34. Define the following terms: synchronous reactance; synchronous reactance drop; synchronous impedance; synchronous impedance drop; armature resistance; armature reactance; armature resistance voltage drop; armature reactance voltage drop.

35. Explain how a resistance test would be performed on an ac generator.

36. What is meant by the effective resistance of the armature winding?

37. Explain why the effective resistance of an armature winding would be larger than the dc resistance.

38. What relationship exists between the resistance between terminals and the resistance per phase if the windings are connected delta or wye?

39. Describe how the open-circuit test is performed.

40. What information is obtained from the open-circuit test?

41. Describe the short-circuit test for an ac synchronous machine.

42. What information is obtained from a short-circuit test?

43. How is the synchronous impedance determined?

44. Various losses occur in a synchronous machine. List them.

45. How may the friction and windage loss in a synchronous machine be determined.

46. Hydrogen cooling is used in large ac synchronous generators. Explain why hydrogen is used as the cooling medium.

47. How is the core loss of an ac synchronous machine determined?

48. How is the copper loss in the field winding of a synchronous machine found?

49. How is the copper loss in the armature winding of a synchronous machine found?

50. Describe what is meant by stray-load losses. Where do they occur?

51. How can stray-load losses be included in the efficiency calculation of an ac synchronous machine?

52. What factors determine the speed of the synchronous motor? How is the speed related to each of these factors?

53. Explain how a synchronous motor differs from an induction motor with regard to its excitation and the power factor with which it operates.

54. It is necessary to employ special starting equipment in order to start a synchronous motor. Explain why.

55. Two important characteristics are possessed only by the synchronous motor. What are they?

56. Describe how the power factor of the input power to a synchronous motor can be controlled.

57. What is meant by the term synchronous condenser? What function is served when such a machine is installed in an electrical system?

58. Describe the special construction features of the stator core and winding of the synchronous motor.

59. What factor determines the number of poles for which a synchronous motor stator is wound?

60. What is meant by class A or class B insulation?

61. What kinds of insulation materials are used in synchronous-motor stator windings?

62. Describe the general construction of a rotor of a synchronous motor.

63. Why would special bracing be necessary in the rotor construction for synchronous machines operating at high speed?

64. How is the dc excitation supplied to the revolving poles?

65. Explain how a synchronous motor is started.

66. Describe methods that may be used to drive an exciter.

67. A small induction motor is mounted on the shaft of a synchronous motor for starting purposes. Why does it have two poles fewer than the synchronous machine?

68. How might a squirrel-cage winding be designed to facilitate the starting of a synchronous motor rotor?

69. Draw the diagram of an automatic starter and explain the sequence of operation when the starter would be used to start a synchronous motor.

70. Does the resistance of a squirrel cage affect the starting torque of a synchronous motor? If so, how?

71. If the squirrel cage is used in the rotor of a synchronous motor, how is the starting current kept to reasonably low values?

72. When starting a synchronous motor, why is it desirable to short-circuit the dc field?

73. What methods can be used to help develop high starting torque in synchronous motors?

74. It is not possible for the rotor of a synchronous motor to operate at other than synchronous speeds. Why?

75. What will determine the angular displacement between the centers of the stator revolving field and the rotator rotor poles? What is this angle called?

76. Describe the conditions under which a synchronous motor will stall.

77. Why is the speed of a synchronous motor called the average constant speed, and not the instantaneous speed?

78. Is it possible for the counter emf of a synchronous motor to be equal to or greater than the impressed stator voltage per phase?

79. What is meant by hunting in a synchronous motor? In what conditions may hunting become particularly serious?

80. List as many synchronous-motor applications as possible.

81. In selecting a motor for a given application, what factors must be taken into consideration?

## PROBLEMS

**4-1.** The nameplate data of a 60-Hz hydrogenerator shows a rated speed of 72 rpm. Calculate the number of poles for this generator.

**4-2.** A 36-pole hydrogenerator is operated at 200 rpm. What frequency is generated?

**4-3.** An arrangement in which a 60-Hz synchronous machine and a 25-Hz synchronous machine have their shafts connected directly (i.e., without an intervening gear) is known as a *frequency changer*. Calculate the smallest possible number of poles in each machine.

**4-4.** At what speed will a synchronous motor operate if is has 10 poles and is connected to a 50-Hz source, a 25-Hz source, a 60-Hz source?

**4-5.** At what speed should a six-pole generator be driven to develop 40-Hz?

**4-6.** How many poles does a synchronous motor have if it operates at 200 rpm when connected to a 60-Hz source?

**4-7.** A three-phase generator has a rating of 5000 kVA at 13,200 V. Calculate the full-load line current. If this generator delivers a load of 3600 kW at a power factor of 0.8, calculate the line current.

**4-8.** The voltage of a generator rises from 4160 V at full load to 5350 V at no load. Calculate the percent regulation.

**4-9.** The illustration shows a developed view of teeth and slots in the field structure. The widths of the teeth and of the slots are represented by $T$ and $S$. Each field coil has $N_{coil}$ turns and carries a current of $I_f$ A. Express in terms of $x_1$, $x_2$, $x_3$, $T$, $S$, $N_{coil}$, and $I_f$ the line integral of **H** taken around each of the paths 1, 2, and 3 and plot the mmf waveform from the $q$ axis out to the distance $A$. The direction of the current is the same in all three slots.

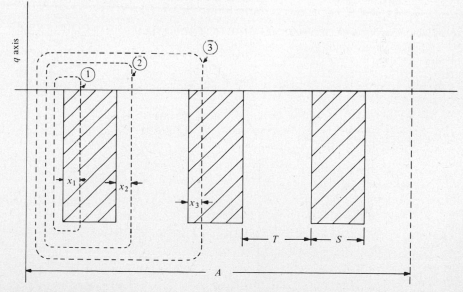

**Prob. 4-9**   Developed view of rotor slots and teeth.

**4-10.** A three-phase 60-Hz synchronous generator has a $\frac{7}{9}$-pitch armature in which each coil has three turns. The flux per pole is 0.02 Wb, sinusoidally distributed. Calculate the induced coil voltage.

**4-11.** The generator in Prob. 4-10 has four poles. The axial length of the iron is 12.5 cm and the diameter of the inner surface of the armature is 30 cm. Calculate the amplitude of the flux-density wave.

**4-12.** The generator in Probs. 4-10 and 4-11 has 36 stator slots. The armature is wye-connected for single-circuit operation. (a) Calculate: (1) the phase voltage, (2) the line-to-line voltage. (b) Repeat parts (1) and (2) for a two-circuit wye connection. (c) Repeat parts (1) and (2) for a single-circuit delta connection.

**4-13.** A generator similar to that in Probs. 4-10, 4-11, and 4-12 has the same fundamental component in the flux density and has in addition third-, fifth-, and seventh-harmonic components in the flux density having amplitudes 0.20, 0.10, and 0.05 that of the fundamental. The armature winding is single-circuit connected in wye. Calculate the rms values of the fundamental and of each harmonic in (a) the line-to-neutral voltage, (b) the line-to-line voltage.

**4-14.** Repeat Prob. 4-13 for a full-pitch armature winding. How does the waveform of the voltage compare with that in Prob. 4-13?

**4-15.** The line-to-line voltage of a three-phase generator is 173.2 V and the line-to-line neutral voltage is 104.4 V. Calculate the third-harmonic component, including its multiples in the phase voltage.

**4-16.** The purpose of this problem is to demonstrate that the error introduced into Eq. 4-40 by neglecting the effect of air-gap curvature on $H$ is small for the air gaps present in conventional machines. (a) If the amplitude of the magnetic field intensity midway between the inner and outer radii of the air gap in the accompanying illustration is $H_{amp}$ and if $\phi$ is the flux per pole in a $P$-pole structure, show that according to Eq. 4-40 the amplitude of the mmf across the air gap is

$$F = \frac{P\phi(r_o - r_i)}{2\mu_0 L(r_o + r_i)}$$

(b) Show that, when curvature is taken into account, the amplitude of the magnetic field intensity at the radius $x$ within the air gap is

$$H_{amp\ x} = \frac{P\phi}{4\mu_0 Lx}$$

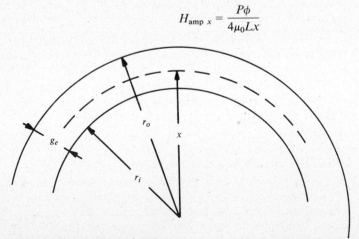

**Prob. 4-16**   Annular air gap.

where $L$ is the axial length of the iron; and that the correct value of the mmf amplitude is, accordingly,

$$F_{corr} = \frac{P\phi}{4\mu_0 L} \ln \frac{r_o}{r_i}$$

(c) Calculate the error by comparing $F$ in part (a) with $F_{corr}$ in part (b) for a two-pole machine in which $r_i = 86.0$ cm and $r_0 = 95.0$ cm. Is the ratio $F/F_{corr}$ a function of $P$?

**4-17.** A three-phase wye-connected generator delivers a unity-power-factor load at 2300 V. If the synchronous reactance voltage drop is 60 V per phase, and the resistance voltage drop per phase is 5 V, what is the percent regulation?

**4-18.** A three-phase 1000-kVA, 2300-V generator is short-circuited and is operated at rated speed to produce rated current. When the short-circuit is removed and the excitation is held constant, the voltage between stator terminals is 1300 V. The effective resistance between stator terminals is 2 Ω. Determine the percent regulation of the alternator at a power factor of 0.8 lagging. What would be the percent regulation of this alternator at 0.8 leading power factor?

**4-19.** A 1000-kVA 4600-V three-phase generator is short-circuited and operated at rated speed. Field excitation is adjusted so that rated armature current occurs. After the short-circuit is removed, the field current and speed are held constant. The open-circuit voltage between stator terminals is found to be 1744 V. The effective ac resistance per phase of the armature winding is 1.1 Ω. Calculate the percent regulation of the generator at a 0.8 lagging power factor.

**4-20.** A 25-kVA generator has a total loss of 2 kW when it delivers rated kVA to a load at a power factor of 0.85. Calculate its percent efficiency.

**4-21** A 25-kVA 220-V three-phase generator delivers rated kVA at a power factor of 0.9. The ac resistance between armature winding terminals is 0.18 Ω. The field takes 9.3 A at 115 V. The friction and windage loss is 460 W, and the core loss is 610 W. Compute the percent efficiency of the generator.

**4-22.** A 50-hp synchronous motor has a full-load efficiency of 90 percent and operates at a leading power factor of 0.8. Calculate the power input to the motor and the kVA input to the motor.

**4-23.** Starting torque of a 25-hp 720-rpm synchronous motor is 150 percent of its rated torque. Calculate the starting torque in newton-meters and in foot-pounds.

**4-24.** The inrush current to a synchronous motor is 235 A when it is started from a rated voltage source of 2300 V. What percent tap on a compensator should be used if the starting current on the line side is to be limited to 100 A? What voltage will be impressed upon the motor at the starting instant?

**4-25.** The full-load losses in a 5000-kVA three-phase 5900-V synchronous condenser are 150 kW. Calculate the full-load current and power factor.

**4-26.** It is desired to correct the 2400-kVA, 0.707 lagging power factor load in a plant to unity power factor, using a synchronous condenser. Calculate the required kVA rating. Assume the condenser is 100 percent efficient.

**4-27.** An industrial plant has a load of 1000 kW at a power factor of 0.707 lagging. It is desired to purchase a synchronous motor of sufficient capacity to deliver a load of 200 kW, and serve to correct the overall plant power factor to 0.95. Assume that the synchronous motor has a 91 percent efficiency. Determine its kVA input rating and the power factor at which it will operate to provide the above correction.

**4-28.** Calculate (a) the values of the flux linkages $\lambda_{ag}$, $\lambda_{af}$, and $\lambda_a$ for the generator in Examples 4-2, 4-3, and 4-4 using the values of $\phi_R$, $\phi_F$, and $\phi_A$ in Example 5-4; (b) the voltages $E_{ag}$ and $E_{af}$ on the basis that $E = 4.44f\lambda$ and compare with the values in Example 4-4.

**4-29.** The generator in Examples 4-2, 4-3, and 4-4 is operating at no load, at synchronous speed, and with $\phi_F = 5.56$ Wb/pole. Calculate (a) the average flux density at the mean radius of the air gap, (b) the amplitude of the flux-density wave, (c) the amplitude of the H wave, and (d) the field mmf and the field current.

**4-30.** A generator has an air gap 1.25 times the length of that in the generator in Examples 4-2, 4-3, and 4-4 but is similar in all other respects, including the voltage, power, and frequency ratings. Neglect saturation and calculate the field current when this generator delivers rated load at 0.80 power factor, current lagging.

**4-31.** (a) Two synchronous generators $A$ and $B$ have identical voltage, frequency, speed, and power-factor ratings. All their corresponding dimensions are identical except the effective axial length $L_B$ of generator $B$ is $1.20L_A$. The armature windings in the two machines have identical pitch and breadth factors and have the same rated value of current density. The magnetic flux densities in the air gaps are also identical. If the turns per phase, rated kVA, and synchronous reactance in ohms of generator $A$ are $N_A$, $S_A$, and $X_A$, what are the ratios $N_B/N_A$, $S_B/S_A$, and $X_B/X_A$, where $N_B$, $S_B$, and $X_B$ are the corresponding quantities for generator $B$? (Assume the ratio of leakage reactance to be the same as that of the magnetizing reactances.) What is the ratio of the per-unit synchronous reactances? (b) Repeat part (a) but for the condition that all the dimensions $L$, $g_e$, $D_e$, and those of the slots in generator $B$ are 1.20 times those in generator $A$. (c) Show that, in a cylindrical-rotor machine of a given number of poles and a given winding factor, if the effective axial length and the effective mean diameter of the air gap are both increased by a factor $k$, the effective air gap must be increased by a factor $k^2$ if the per-unit magnetizing reactance is to remain unchanged.

**4-32.** A 6250-kVA three-phase 2400-V 60-Hz two-pole synchronous generator with a Y-connected armature has the following constants:

$$L_{aa} = 0.00176 \text{ H} \qquad L_{afm} = 0.0315 \text{ H}$$
$$L_{ab} = 0.000809 \text{ H} \qquad L_{ff} = 0.670 \text{ H}$$
$$r_a = 0.00250 \text{ } \Omega \qquad r_f = 0.280 \text{ } \Omega$$

The field current is a direct current having a constant value of 275 A. Calculate for $\sigma_0 = 0$ at synchronous speed (a) the flux linkages expressed as functions of time for each phase when the armature current is zero, (b) the generated voltage in each phase, (c) the voltage across the field winding.

**4-33.** The rotor of the generator in Prob. 4-32 is driven at synchronous speed with the field circuit open. Balanced positive-phase sequence currents (i.e., $\mathbf{I}_a = \mathbf{I}$, $\mathbf{I}_b = \mathbf{I}\underline{/-120°}$, $\mathbf{I}_c = \mathbf{I}\underline{/-240°}$) are applied to the armature with $a$ phase current $i_a = 2121 \sin(\omega t + \sigma_0 - 58.7°)$. Neglect the resistance of the armature winding and calculate (a) the flux linkage with (1) $a$ phase expressed as a function of time, (2) the field winding expressed as a function of time, and (b) the rms value of the voltage applied to the armature expressed in volts per phase and in line-to-line volts.

**4-34.** (a) Repeat Prob. 4-33 but for the rotor at standstill and neglecting eddy currents in the rotor iron as well as the resistance of the armature. (b) Calculate the voltage induced in the field winding with the rotor at standstill.

**4-35.** The generator in Prob. 4-32 carries a constant field current of 275 A and positive-sequence currents with $a$ phase current $i_a = 2121 \sin(\omega t + \sigma_0 - 58.7°)$ while being driven at synchronous speed. $\sigma_0 = 0$. Determine as functions of time the (a) flux linkage (1) with each phase of the armature winding and (2) with the field winding, (b) the voltage applied to the field winding, (c) the instantaneous terminal voltage of each phase of the armature neglecting armature resistance, and (d) the electromagnetic torque.

**4-36.** (a) Show a phasor diagram for the generator of Prob. 4-32 and for the load condition in Prob. 4-35. Include the armature induced emf $\mathbf{E}_{af}$ due to the field current alone and $\mathbf{E}_A$ induced in the armature by the armature current itself if $x_{ad} = 0.85x_d$. What is the value of the terminal voltage? (b) Show a diagram of the equivalent circuit with a current source as in Fig. 4-25 and calculate the equivalent armature current $\mathbf{I}_F$.

**4-37.** The following data applies to the synchronous generator of Example 4-6.

| kVA | 13,259 | Rotor slots | 16 |
|---|---|---|---|
| Volts | 13,800 | Rotor coils | 8 |
| | | (single circuit) | |
| Hz | 60 | Turns per rotor coil | 36 |
| Phases | 3 | Stator slots | 42 |
| Connection | Two-circuit | Stator coils | 42 |
| Speed | 3600 | Turns per stator coil | 5 |
| Poles | 2 | Field resistance | 0.38 Ω |
| | | Stator coil pitch slots | 17 |
| | | Armature resistance per phase | 0.072 Ω |

The rotor slots cover an angle $\beta$ of 125.2 degrees per pole.

Open Circuit Characteristic

| Field current, A | Armature voltage, line-to-line |
|---|---|
| 40 | 4,180 |
| 80 | 8,350 |
| 120 | 11,900 |
| 160 | 14,270 |
| 200 | 15,900 |
| 280 | 17,820 |
| 380 | 18,850 |

Short-Circuit Characteristic

| Field current, A | Armature current, A |
|---|---|
| 186 | 566 |

(a) Calulate the unsaturated synchronous reactance in ohms and per-unit. (b) Calculate the unsaturated values of $L_{afm}$, $L_{ff}$, and $L_{aaM}$. (c) Compute the synchronous reactance of armature reaction $x_{ad}$ in ohms and per-unit. (d) On the basis of the short-circuit data and the value of $x_{ad}$ in part (c), compute the leakage reactance $x_l$ in ohms and per-unit.

**4-38.** (a) Calculate the short-circuit ratio SCR for the generator of Prob. 4-37. (b) Assume the leakage reactance to remain unaffected by changes in the air gap and calculate the short-circuit ratio if the length of the air gap is increased by 25 percent.

**4-39.** The open-circuit data on a 133,689-kVA 13.8-kV 60-Hz 3600-rpm cylindrical-rotor generator is given in the table.

| Field current, A | Line-line voltage, kV |
|---|---|
| 0 | 0 |
| 385 | 11.05 |
| 545 | 13.80 |
| 700 | 15.20 |
| 980 | 16.50 |
| 1190 | 17.25 |
| 1470 | 17.95 |
| 1650 | 18.20 |

The zero-power-factor data of the same generator follow:

| Field current, A | Armature current, A | Line-line voltage, kV |
|---|---|---|
| 795 | 5594 | 0 |
| 1615 | 5594 | 13.8 |

Find (a) the unsaturated synchronous reactance, (b) the Potier reactance, (c) the reactance of armature reaction, and (d) the field current to give one-half rated current at zero power factor and at rated voltage.

**4-40.** Use the saturation-factor method to determine (a) the field current and (b) the regulation of the generator in Prob. 4-39 when delivering rated load at rated terminal voltage, rated frequency and at 0.85 power factor, current lagging.

**4-41.** In a slip test on a 5-kVA 240-V 60-Hz synchronous motor, the following readings were taken:

Line-to-line voltage, V = 200 maximum, 180 minimum

Line current, A = 11.25 maximum, 7.20 minimum

Neglect the effect of instrument overswing and calculate the unsaturated values of $x_d$ and $x_q$ in ohms and per-unit on a line-to-neutral basis. What would the per-phase values be if the armature were delta-connected?

## BIBLIOGRAPHY

Adkins, B. *The General Theory of Electrical Machines.* New York: John Wiley & Sons, Inc., 1957.

Brown, David, and E. P. Hamilton III. *Electromechanical Energy Conversion.* New York: The Macmillan Company, 1984.

Concordia, C. *Synchronous Machines.* New York: John Wiley & Sons, Inc., 1951.

Crosno, C. Donald. *Fundamentals of Electromechanical Conversion.* New York: Harcourt, Brace and World, Inc., 1968.

Fitzgerald, A. E., C. Kingsley, Jr., and Alexander Kusko. *Electric Machinery,* 3d ed. New York: McGraw-Hill Book Company, 1971.

Gaurishankar, V., and D. H. Kelly. *Electromechanical Energy Conversion,* 2d ed. New York: IEP, A Dun-Donnelley Publisher, 1973.

Knowlton, A. E. *Standard Handbook for Electrical Engineers,* 8th ed. New York: McGraw-Hill Book Company, 1949, Sec. 7.

Lawrence, R. R., and H. E. Richards. *Principles of Alternating-Current Machinery,* 4th ed. New York: McGraw-Hill Book Company, 1953.

Lewis, W. A. *The Principles of Synchronous Machines.* Chicago: Illinois Institute of Technology, 1954.

Liwschitz-Garik, M., and C. C. Whipple. *Alternating-Current Machines.* New York: D. Van Nostrand Company, Inc., 1961.

Matsch, Leander W., *Electromagnetic & Electromechanical Machines.* 2d ed., New York, Dun-Donnelley Publishing Co., 1977.

McPherson, George. *An Introduction to Electrical Machines and Transformers,* New York, John Wiley & Sons, Inc., 1981.

Meisel, J. *Principles of Electro-mechanical Energy Conversion.* New York: McGraw-Hill Book Company, 1966.

Nasar, S. A. *Electromagnetic Energy Conversion Devices and Systems.* Englewood Cliffs, N.J., Prentice-Hall, Inc., 1970.

Puchstein, A. F., et al. *Alternating-Current Machines.* New York: John Wiley & Sons, Inc., 1954.

Siskind, Charles S., *Electrical Machines: Direct & Alternating Current,* 2d ed. New York: McGraw-Hill Book Company, 1959.

Slemon, G. R. *Magnetoelectric Devices.* New York: John Wiley & Sons, Inc., 1966.

# Chapter 5

# The Induction Motor

## 5-1 THE POLYPHASE INDUCTION MOTOR

The polyphase induction motors used in industrial applications are practically without exception three-phase, thus corresponding to the number of phases in commercial power systems. In conventional induction motors the stator winding is connected to the source and the rotor winding is short-circuited for many applications, or it may be closed through external resistances. While the synchronous motor has certain advantages—such as constant speed, the ability to generate reactive power with an overexcited field, and low cost of slow-speed motors—it has the disadvantages of requiring a dc source (exciter) for its field excitation and relatively higher cost for high-speed motors. The polyphase induction motor, however, requires no means for its excitation other than the ac line. It is economical to build for the higher speeds, and one type (i.e., the wound-rotor motor) lends itself to a fair degree of speed control. The induction motor runs below synchronous speed and is known as an *asynchronous machine.* Its speed decreases with increasing load torque. The full-load speed of polyphase induction motors is, in most cases, within 7 percent of synchronous speed, although full-load speeds of 1 percent below synchronous speed are not unusual. The induction motor has no inherent means for producing its excitation; it requires reactive power and draws a lagging current. While the power factor at rated load is generally above 80 percent, it is low at light loads, which has the disadvantage of incurring a less favorable price rate for electric power when the power factor (current lagging) falls below a certain value in commercial and industrial installations. In order to limit the reactive power, the magnetizing reactance should be high, and the air gap is therefore shorter than in synchronous motors of the same size and rating (except for small motors). Mechanical considerations limit the minimum length of air gap; other factors are motor noise and magnetic losses in the tooth faces. Too short an air gap may prevent the motor from accelerating to rated speed on starting and cause it to run at a fraction of its rated speed.

The stator windings of polyphase induction motors are fundamentally the same as the stator windings of polyphase synchronous machines. Figures 5-1(a) shows a partially wound stator of a three-phase induction motor. However, polyphase induction motors fall into two general categories, depending on the kind of rotor used—the wound-rotor and the squirrel-cage rotor. The stator iron as well as the rotor iron is laminated and slotted to contain the windings. The wound-rotor has a three-phase winding similar to that in the stator and is wound for the same number of poles as the stator winding. The rotor winding terminates in slip rings mounted on the rotor shaft, as shown in Fig. 5-1(b). Brushes ride on the slip rings of the wound-rotor motor and during starting are connected externally to three equal resistances—one in each phase, connected in wye—that are short-circuited simultaneously in one or more steps as the motor comes up to speed. The external resistance is sometimes used for speed control in such applications as cranes and hoists. Instead of containing a winding, the slots in the squirrel-cage rotor are occupied by bars of copper or of aluminum, known as *rotor bars,* short-circuited in two end rings of the same material as the rotor bars. There is one ring at each end of the stack of rotor laminations, as shown in Fig. 5-1(c). A schematic diagram of a squirrel cage in Fig. 5-2 shows eleven rotor bars and the two end rings. The rotating magnetic field produced by the polyphase voltages applied to the stator winding induces currents in the squirrel-cage rotor circuit that develop the same number of rotor poles as stator poles. The rotor poles react upon the stator flux, thus developing torque in the same direction of rotation as that of the stator flux. As long as the rotor rotates below or above synchronous speed, there is relative motion between it and the rotating stator flux, and voltage is induced in the rotor circuits. At synchronous speed there is no motion of the rotating field relative to the rotor, and no emf is induced in the rotor by the fundamental component of the flux. The induction motor must therefore operate below synchronous speed. It can operate as an induction generator if driven above synchronous speed while connected to a system furnished with reactive power from other sources, such as synchronous machines.

## 5-2 MAGNETIZING REACTANCE AND LEAKAGE REACTANCE

### 5-2.1 Magnetizing Reactance

Since the induction motor as shown in Fig. 5-3 has an air gap that is uniform except for the presence of the slots, Eq. 4-50 for the magnetizing reactance of the cylindrical-rotor machine applies to the induction motor. Hence, for an *m*-phase stator winding if the iron is unsaturated,

$$x_{Mu} = \frac{8f\mu_0 m D_g L}{g_e} \left(\frac{k_{w1}N_{ph1}}{Pa_1}\right)^2 \tag{5-1}$$

(a)

(b)

(c)

**Figure 5-1** (a) Stator for three-phase motor. (b) Wound-rotor. (c) Squirrel-cage rotor. (Courtesy General Electric Company.)

248

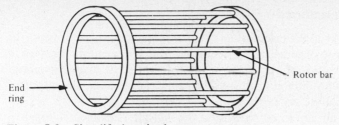

**Figure 5-2**   Simplified squirrel cage.

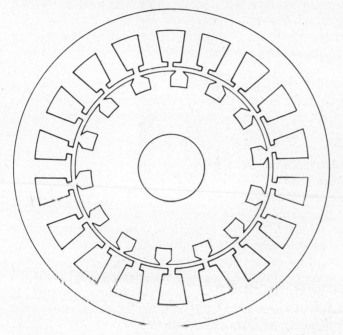

**Figure 5-3**   Section of the magnetic circuit in an induction motor.

where the subscript 1 refers to the stator. However, the actual magnetizing reactance is lower because of saturation, which may be taken into account by the factor $k_i$, which usually lies between 1.15 and 1.40 for conventional induction motors. The magnetizing reactance is then expressed by

$$x_M = \frac{8f\mu_0 m D_g L}{k_i g_e} \left(\frac{k_{w1} N_{ph1}}{Pa_1}\right)^2 \tag{5-2}$$

For a three-phase winding $m = 3$, and since $\mu_0 = 4\pi \times 10^{-7}$,

$$x_M = \frac{3.02 f D_g L}{k_i g_e} \left(\frac{k_{w1} N_{ph1}}{Pa_1}\right)^2 \times 10^{-5} \tag{5-3}$$

## 5-2.2 Leakage Reactance

Analytical approaches† to the calculation of leakage reactance are not nearly as straightforward as those of magnetizing reactance and are therefore omitted in this text. While the configurations of the windings and of the magnetic circuit are more complex in an induction motor than in a transformer, the effects of leakage reactance are quite similar in both.

Because of the air gap in ac machines, the leakage reactance is larger in proportion to the magnetizing reactance than in transformers. A typical ratio for a 25-hp induction motor is about 0.05, as compared with a range between about 0.09 and 0.20 for cylindrical-rotor synchronous machines. By way of comparison, the transformer in Example 3-1 has a primary leakage reactance of 0.463 Ω and a primary magnetizing reactance of $10^2/0.069$, resulting in a ratio of 0.0003.

Although there is physically only one flux, of which the pattern changes from instant to instant with the instantaneous time variations of the currents in the windings, the leakage fluxes may be divided into the following components,‡ as shown in Fig. 5-4:

1. Slot-leakage and tooth-top leakage flux
2. Coil-end leakage flux
3. Differential or air-gap leakage flux

The slot-leakage fluxes (1) and (2) in Fig. 5-4(a) leak across the slots and link only the coil sides that produce these fluxes. The coil-end leakage flux in Fig. 5-4(b) links the portions of the coils of one winding extending beyond the iron, known as the *end connections,* without linking the other winding. The differential leakage flux results from the higher harmonics in the air-gap flux. The effect of these harmonics produced by one winding is to induce parasitic currents in the other winding. Such flux is not a useful flux and is therefore considered a leakage flux. The differential leakage flux is small in windings that occupy several slots per phase and pole and in fractional-slot windings.

## 5-3 ROTOR CURRENT AND SLIP

Although the basic relationships are the same for the squirrel-cage induction motor as for the wound-rotor motor, the latter will be introduced first because these relationships are perhaps more easily visualized. Figure 5-5(a) shows a schematic diagram for a three-phase wound-rotor motor with external resistance

---

† For more complete treatment of leakage reactance, see M. Liwschitz-Garik and C. C. Whipple, *Alternating-Current Machines* (Princeton, N.J.: D. Van Nostrand Company, Inc., 1961), Chap. 57; P. L. Alger, *The Nature of Induction Machines* (New York: Gordon and Breach, Inc., 1965). Chap. 7.

‡ Alger (ibid. 7) divides the leakage reactances into seven components.

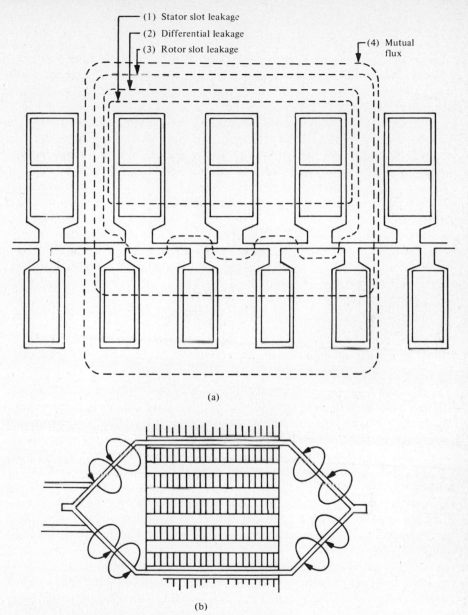

(a)

(b)

**Figure 5-4**  Paths of (a) air-gap fluxes and (b) end-connection leakage flux in an induction motor. [Adapted from P. L. Alger, *The Nature of Induction Machines* (New York: Gordon and Breach, Inc., 1965), p. 200.]

in the rotor circuit for starting or for running conditions, in which variable speed is obtained by varying the amount of external rotor resistance. In the schematic diagram of Fig. 5-5(b) the rotor is shown short-circuited, which is the normal condition for operation under load at rated speed.

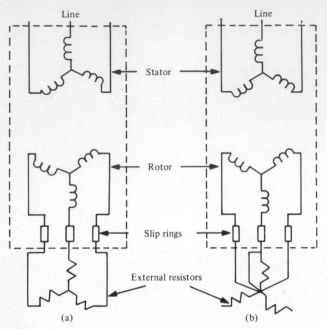

**Figure 5-5** Schematic diagram of wound-rotor induction motor. (a) With external resistance in rotor circuit. (b) With rotor short-circuited.

Consider the motor at standstill with the rotor open-circuited (i.e., the external resistance of each rotor phase infinite). Balanced three-phase voltage applied to the stator produces balanced three-phase stator current, which in turn produces a fundamental component of mmf, resulting in a practically sinusoidally distributed flux wave rotating at synchronous speed with respect to the stator winding. This fundamental flux can be divided into two component fluxes—mutual flux and leakage flux, just as in the case of the transformer. The mutual flux corresponds to the air-gap flux. When the rotor circuit is open, the fundamental component of the air-gap flux would be expressed by Eq. 4-48 if there were no saturation. Assume both the stator and the rotor windings to be connected in wye.

If $\phi_M$ = the fundamental component of air-gap flux or mutual flux, $E_2$ = the *stator* volts per phase induced by the air-gap flux, and $E_r$ = the volts induced by the air-gap flux in each phase of the rotor winding with the rotor at standstill, from Eq. 4-15,

$$E_2 = \frac{4.44 f k_{w1} N_{ph1} \phi_M}{a_1} \tag{5-4}$$

and

$$E_r = \frac{4.44 f k_{w2} N_{ph2} \phi_M}{a_2} \tag{5-5}$$

where $N_{ph1}$ and $N_{ph2}$ are the number of turns in the stator and rotor windings and $a_1$ and $a_2$ are the corresponding numbers of current paths. Also, $k_{w1}$ and

$k_{w2}$ are, correspondingly, combined pitch and breadth factors. The transformation ratio is

$$b = \frac{E_2}{E_r} = \frac{k_{w1}N_{ph1}a_2}{k_{w2}N_{ph2}a_1} \tag{5-6}$$

If the rotor slip rings are connected to three equal resistances as shown in Fig. 5-5(a) and the rotor is prevented from rotating, the operation is that of a three-phase transformer. With the rotor at standstill the flux rotates at synchronous speed relative to both the stator and the rotor. Consequently, the frequency of the rotor-induced emf is the same as the stator frequency at standstill.

## 5-3.1 Induction Motor Slip

Suppose that the rotor circuit is open and that the rotor is made to rotate at a speed $n$ rpm by some external means in the direction of the rotating flux $\phi_M$. If $n_{syn}$ is the synchronous speed in rpm (i.e., the rotational speed of $\phi_M$), the slip is defined by

$$s = \frac{n_{syn} - n}{n_{syn}} \tag{5-7}$$

When the rotor is rotating at a slip $s$, the speed of the stator flux relative to the rotor no longer equals the synchronous speed but is the slip speed, $sn_{syn}$. The rotor frequency must therefore be

$$f_2 = sf$$

In addition, there is a reduction in the magnitude of the rotor voltage from its standstill value to $sE_r$, which follows from substitution of $sf$ for $f$ in Eq. 5-5.

## 5-3.2 Rotor Current

If $r_{22}$ is the resistance of the wound rotor in ohms per phase (line to neutral or one-half the resistance between slip rings) and $L_{22}$ the wound-rotor *leakage inductance* in henries per phase† (line to neutral), then rotor leakage reactance is

$$x_{22} = 2\pi f L_{22} \qquad \Omega/\text{phase at standstill}$$

and at slip $s$ it is

$$sx_{22} = 2\pi sf L_{22} \qquad \Omega/\text{phase}$$

When the slip rings are short-circuited, the rotor current is given by

---

† While the symbol $L_{22}$ is generally used in coupled circuit theory to represent the total self-inductance (leakage inductance plus mutual inductance) of circuit 2, it is used here as the symbol for rotor leakage inductance.

$$I_{22} = \frac{sE_r}{\sqrt{(r_{22})^2 + (sx_{22})^2}} = \frac{E_r}{\sqrt{(r_{22})^2/s + (x_{22})^2}} \tag{5-8}$$

Equation 5-8 expresses the current in terms of the standstill voltage, which in combination with Eqs. 5-4 and 5-5 suggests an ideal transformer supplying a load comprised of a resistance $r_{22}/s$ Ω in series with an inductive reactance $x_{22}$, as shown in Fig. 5-6(a). The rotor current reacts on the stator winding at stator frequency regardless of the value of the slip under balanced steady-state operation and thus induces a voltage in the stator or primary winding at stator or primary frequency regardless of the slip $s$. This follows from the fact that the polyphase rotor currents at a slip $s$ have a frequency of $sf$ and therefore produce a rotor mmf that rotates at $sn_{syn}$ rpm relative to the rotor in the same rotational direction as the stator flux while the rotor is rotating at a speed of $(1 - s)n_{syn}$ rpm in the same direction. The resultant speed of the rotor mmf relative to the stator is the sum of these two speeds [i.e., $sn_{syn} + (1 - s)n_{syn} = n_{syn}$], which is the synchronous speed, also that of the mmf produced by the stator current at the stator frequency $f$. The ideal transformer may then be used in the equivalent circuit of the induction motor.

   The ideal transformer can be removed from the equivalent circuit by making use of the ratio $b$ and the impedance ratio $b^2$ just as in the case of the static

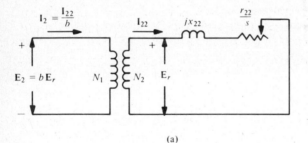

(a)

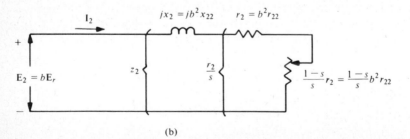

(b)

**Figure 5-6** Equivalent rotor circuit of a wound-rotor induction motor. (a) Rotor quantities shown in the secondary of an ideal transformer. (b) Rotor quantities referred to the stator.

transformer. This is shown in Fig. 5-6(b), where the equivalent rotor resistance $r_2/s$ is shown in two parts, the rotor resistance $r_2$ and the dynamic resistance $[(1 - s)/s]r_2$, which represents the mechanical load, where $r_2$ is the rotor resistance referred to the stator.

## 5-4 ROTOR COPPER LOSS AND SLIP

From Fig. 5-6 it is evident that the real power input to the rotor is

$$P_{2(\text{in})} = I_2^2 \frac{r_2}{s} \qquad \text{W/phase} \tag{5-9}$$

The power expended in heating the rotor winding is

$$P_{2(\text{loss})} = I_2^2 r_2 \qquad \text{W/phase} \tag{5-10}$$

When Eq. 5-10 is divided by Eq. 5-9, the result equals the slip:

$$s = \frac{\text{rotor copper loss}}{\text{rotor real power input}} \tag{5-11}$$

The developed mechanical power must be the difference between the power input and the power that produces heat in the rotor winding. Accordingly,

$$P_{em} = P_{2(\text{in})} - P_{2(\text{loss})} = I_2^2 \frac{1 - s}{s} r_2 \tag{5-12}$$

Equation 5.12 expresses watts per phase and includes the windage and friction loss $P_{fw}$ as well as the stray-load loss $P_{\text{stray}}$. The net mechanical power output is

$$P_{\text{mech}} = P_{em} - (P_{fw} + P_{\text{stray}}) \tag{5-13}$$

Equation 5-8 shows the developed mechanical power to be zero at zero slip or at synchronous speed, indicating that the induction motor cannot attain or exceed synchronous speed. The slip is negative when the rotor speed is greater than synchronous speed according to Eq. 5-7, and the developed mechanical power is therefore negative, meaning that the rotor must be driven mechanically to attain or exceed synchronous speed. The component of rotor current $I_{22}$ in phase with $E_r$ reverses its direction when the slip $s$ goes from a positive to a negative value, as shown when Eq. 5-8 is expressed in complex form by

$$\mathbf{I}_{22} = \frac{s\mathbf{E}_r}{r_{22} + jsx_{22}} \tag{5-14}$$

and the result is generator action. Generator action, however, is possible only if reactive power is supplied to the motor from the bus to which it is connected.

## 5-5 EQUIVALENT CIRCUIT OF THE POLYPHASE WOUND-ROTOR INDUCTION MOTOR

At standstill ($s = 1$) the mechanical power is zero and all the real power transferred across the air gap from the stator to the rotor (i.e., the real power input to the rotor) is converted to heat. The standstill condition of the wound-rotor motor, with the slip rings short-circuited, corresponds to that of a transformer with its secondary short-circuited. At zero slip the rotor represents an open circuit (although the slip rings may be short-circuited), because with $s = 0$, $r_2/s$ is infinite and the rotor current must therefore be zero when the effects of harmonics are neglected. This condition of zero slip corresponds to operating a transformer without load. Whether the rotor is actually open-circuited or whether it is short-circuited and running at zero slip, part of the voltage applied to the stator winding, $E_2$, produces the mutual flux that requires an exciting current, just as in the case of the transformer.

The magnetizing component of the current is given by

$$\mathbf{I}_e = \frac{\mathbf{E}_2}{jx_M}$$

The core-loss component of the exciting current or iron-loss current $\mathbf{I}_{fe}$ corresponds to $\mathbf{I}_{CL}$ in the case of the transformer. The exciting current $\mathbf{I}_M$ is the sum of these components†; thus,

$$\mathbf{I}_M = \mathbf{I}_e + \mathbf{I}_{fe}$$

The circuit in Fig. 5-7(a) may also be compared with that in Fig. 4-25 for the cylindrical-rotor synchronous generator in which the core-loss branch $r_{fe}$ is omitted. If the direction of the armature current $\mathbf{I}$ in Fig. 4-25 is reversed to represent motor operation, it corresponds to the stator current $\mathbf{I}_1$ in Fig. 5-7(a) and the equivalent field current $\mathbf{I}_f$ corresponds to $-\mathbf{I}_2$ in Fig. 5-7(a). In fact, it is possible to operate the wound-rotor induction motor as a synchronous machine by applying direct current to the rotor through the slip rings. However, in the induction motor the magnetizing reactance $x_M$ is made high to keep the magnetizing current low, while a high value of $x_M$ or $x_{ad}$ in the synchronous machine leads to poor regulation and a low stability limit, as indicated by the power angle characteristic in Fig. 4-35.

The voltage applied to the stator, just as in the case of that applied to the primary of a transformer, must not only be sufficient to produce the voltage $\mathbf{E}_2$ but must, in addition, overcome the stator leakage impedance. So if

$\mathbf{I}_1$ = stator current, A/phase

$r_1$ = stator resistance, $\Omega$/phase

$x_1$ = stator leakage reactance, $\Omega$/phase

---

† Although $\mathbf{I}_M$ is generally used to represent magnetizing current in transformers, it is used here to represent the exciting current of the induction motor.

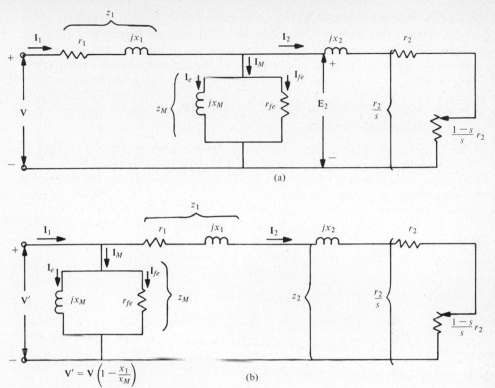

**Figure 5-7** (a) Equivalent circuit. (b) Approximate equivalent circuit of polyphase induction motor.

the applied stator voltage must be in V/phase:

$$V_1 = (r_1 + jx_1)I_1 + E_2 \tag{5-15}$$

Equation 5-15 applies to the transformer, and it follows that the equivalent circuit of the transformer also serves as that of one phase of the polyphase induction motor, whether the motor has a wound rotor or whether it has a squirrel-cage rotor.

Figure 5-7(a) shows the equivalent circuit and Fig. 5-7(b) the approximate equivalent circuit of the polyphase induction motor for steady-state operation with applied balanced voltages. In both circuits the noninductive resistance $[(1 - s)/s]r_2$ represents the mechanical load, which corresponds to a noninductive load on a transformer.

While the equivalent circuit in Fig. 5-7(a) affords a convenient basis for the performance calculations when the numerical values of the motor constants are known, it leads to general derivations in which the effect of the different parameters on the characteristics of the motor become somewhat obscured. The approximate equivalent circuit in Fig. 5-7(b) yields good accuracies for most iron-core transformers and is sometimes used for approximate numerical calculations of induction motor performance. However, because of the relatively large magnetizing current required by the air gap of the induction motor, the approximate equivalent circuit gives rise to values of rotor current that are appreciably higher than the actual values when the slip is at or above rated value.

### 5-5.1 Approximate Equivalent Circuit with Adjusted Voltage

The errors in the relationships based on the approximate equivalent circuit can be reduced to a small amount by reducing the actual value of the applied voltage, as shown in the following. Starting with the equivalent circuit of Fig. 5-7(a), the terminal voltage is expressed by

$$\mathbf{V} = \mathbf{E}_2 + \mathbf{I}_1 z_1 \tag{5-16}$$

and since

$$\mathbf{I}_1 = \mathbf{I}_2 = \mathbf{I}_M \tag{5-17}$$

Equation 5-16 can be rewritten in the form

$$\mathbf{V} = \mathbf{E}_2 + z_1(\mathbf{I}_2 + \mathbf{I}_M) = \mathbf{E}_2 + z_1\mathbf{I}_2 + z_1\mathbf{I}_M$$

$$= (z_1 + z_2)\mathbf{I}_2 + z_1\mathbf{I}_M$$

$$\mathbf{I}_M = \frac{\mathbf{E}_2}{z_M}$$

and we have

$$\mathbf{V} = (z_1 + z_2)\mathbf{I}_2 + \frac{z_1}{z_M}\mathbf{E}_2 \tag{5-18}$$

Since the reactive component in both impedances $z_1$ and $z_M$ is several times greater than the resistive component in conventional induction motors,

$$\frac{z_1}{z_M} \simeq \frac{x_1}{x_M}$$

In addition, $\mathbf{E}_2$ is only a little smaller and nearly in phase with $\mathbf{V}$ in the normal running range; so Eq. 5-18 can be reduced to

$$\mathbf{V} \simeq (z_1 + z_2)\mathbf{I}_2 + \frac{x_1}{x_M}\mathbf{V}$$

or

$$\boxed{\mathbf{V}' = \mathbf{V}\left(1 - \frac{x_1}{x_M}\right) = \left[r_1 + \frac{r_2}{s} + j(x_1 + x_2)\right]\mathbf{I}_2} \tag{5-19}$$

as indicated in Fig. 5-7(b).

The phasor power input to the motor is taken as $m\mathbf{V}\mathbf{I}_1$ rather than $m\mathbf{V}'\mathbf{I}_1$, because $\mathbf{V}$ is the actual voltage applied to the stator winding. The adjusted value $\mathbf{V}'$ is used only to obtain more realistic values of $\mathbf{I}_1$ and $\mathbf{I}_2$.

## 5-5.2 Mechanical Power and Torque

The developed mechanical power in an $m$-phase induction motor is found from Eq. 5-12 to be

$$P_{em} = mI_2^2 \frac{1-s}{s} r_2 \tag{5-20}$$

The developed torque is the developed mechanical power divided by the mechanical angular velocity of the rotor:

$$T_{em} = \frac{P_{em}}{\omega_m} \qquad \text{newton-meters/radian} \tag{5-21}$$

where

$$\omega_m = \frac{2\pi n}{60} = \frac{2\pi(1-s)n_{syn}}{60} \tag{5-22}$$

When Eqs. 5-20 and 5-22 are substituted in Eq. 5-21, the torque is expressed in terms of slip and rotor current by

$$T_{em} = \frac{m60I_2^2}{2\pi n_{syn}} \frac{r_2}{s} \tag{5-23}$$

The quantity $mI_2^2(r_2/s)$ in Eq. 5-23 is the real power input to the rotor (i.e., the power transferred across the air gap into the rotor winding).

Example 5-1 compares the use of the equivalent circuit with that of the approximate equivalent circuit, for calculating the steady-state performance of a three-phase induction motor. It shows close agreement between the results obtained with the two circuits.

**EXAMPLE 5-1**

A 15-hp 440-V three-phase 60-Hz eight-pole wound-rotor induction motor has its stator and rotor both connected in wye. The ratio of effective stator turns to effective rotor turns is $b = 2.4:1$. The windage and friction losses are 220 W at rated speed and may be assumed constant from no load to full load. The stator and rotor have the following constants per phase:

| Stator, $\Omega$ | Rotor, $\Omega$ |
|---|---|
| $r_1 = 0.52$ | $r_{22} = 0.110$ |
| $x_1 = 1.15$ | $x_{22} = 0.20$ |
| $x_M = 40.0$ | |
| $r_{fe} = 360$ | |

The stray-load loss is 120 W. (a) Use the equivalent circuit in Fig. 5-7(a) to calculate the following for a slip $s = 0.045$ with balanced rated voltage and rated frequency applied to the stator and with the rotor slip rings short-circuited: (1) stator current, (2) power factor, (3) current in the rotor winding, (4) output in horsepower, (5) efficiency, and (6) torque. (b) Repeat part (a) using the approximate equivalent circuit, Fig. 5-7(b). (c) Compare the results of parts (a) and (b) in tabulated form.

*Solution*

a. In both equivalent circuits of Fig. 5-7, the rotor impedance is referred to the stator by use of the impedance ratio:

$$b^2 = (2.4)^2 = 5.76$$

so that

$$r_2 = b^2 r_{22} = 5.76 \times 0.110 = 0.634 \ \Omega/\text{phase}$$

$$x_2 = b^2 x_{22} = 5.76 \times 0.20 = 1.15 \ \Omega/\text{phase}$$

For a slip $s = 0.045$, the impedance of the rotor referred to the stator is

$$z_2 = \frac{r_2}{s} + jx_2 = \frac{0.634}{0.045} + j1.15$$

$$= 14.10 + j1.15 = 14.13\underline{/4.7^\circ} \ \Omega/\text{phase}$$

The leakage impedance of the stator is

$$z_1 = r_1 + jx_1 = 0.52 + j1.15 = 1.26\underline{/65.6^\circ}$$

and the exciting impedance referred to the stator is

$$z_M = \frac{r_{fe}\, jx_M}{r_{fe} + jx_M} = \frac{(360)(j40)}{360 + j40}$$

$$= 39.8\underline{/83.65^\circ} = 4.40 + j39.5 \ \Omega/\text{phase}$$

1. The stator current can now be determined simply by dividing the voltage applied to one phase of the stator by the impedance of the circuit. The impedance is, from Fig. 5-7(a),

$$Z = z_1 + \frac{z_2 z_M}{z_2 + z_M}$$

$$= 1.26\underline{/65.6^\circ} + \frac{(14.13\underline{/4.7^\circ})(39.8\underline{/83.65^\circ})}{14.13\underline{/4.7^\circ} + 39.8\underline{/83.65^\circ}}$$

$$= 0.52 + j1.15 + \frac{562\underline{/88.35^\circ}}{44.8\underline{/65.6^\circ}}$$

$$= 12.09 + j6.00 = 13.47\underline{/26.4^\circ} \ \Omega/\text{phase}$$

The voltage per phase is

$$V = \frac{440}{\sqrt{3}} = 254$$

which produces the stator current

$$I_1 = \frac{V}{Z} = \frac{254\underline{/-26.4^\circ}}{13.47} = 18.85\underline{/-26.4^\circ} \ \text{A/phase}$$

2. The power factor of the motor is

$$PF = \cos \theta = \cos 26.4° = 0.895$$

3. $$I_2 = \frac{E_2}{z_2} = \frac{I_1 z_2 z_M}{(z_2 + z_M)z_2} = \frac{I_1 z_M}{z_2 + z_M}$$

$$= \frac{(18.85/-26.4°)(39.8/83.65°)}{44.8/65.6°} = 16.75/-8.35°$$

$$I_{22} = bI_2 = 2.4 \times 16.75/-8.35°$$

$$= 40.2/-8.35° \text{ A/phase}$$

4. $$P_{em} = mI_2^2 \frac{1 - s}{s} r_2 = 3(16.75)^2 \frac{1 - 0.045}{0.045} 0.634$$

$$= 11,300 \text{ W}$$

$$P_{mech} = P_{em} - (P_{fw} + P_{stray})$$

$$= 11,300 - (220 + 120) = 10,960 \text{ W}$$

$$= 10,960 \div 746 = 14.7 \text{ hp}$$

5. The efficiency is the ratio of output to input. The real power input is

$$P_{in} = mVI_1 \cos \theta$$

$$= 3 \times 254 \times 18.85 \times 0.895 = 12,880 \text{ W}$$

$$\text{Efficiency} = \frac{10,960}{12,880} = 1 - \frac{1920}{12,880} = 1 - 0.149 = 0.851$$

6. Torque is the ratio of mechanical power to the mechanical angular velocity of rotation. The angular velocity, from Eq. 5-7, is

$$\omega_m = \frac{2\pi n}{60} = \frac{2\pi(1 - s)n_{syn}}{60} = \frac{2\pi(1 - 0.045)[(120 \times 60)/8]}{60} = 90.0 \text{ rad/s}$$

and a torque of

$$T = \frac{10,960}{90} = 121.8 \text{ N-m/rad}$$

b. 1. On the basis of the approximate equivalent circuit of Fig. 5-7(b) and Eq. 5-19, the rotor current referred to the stator is found to be

$$I_2 = \frac{V(1 - x_1/x_M)}{z_1 + z_2} = \frac{V(1 - x_1/x_M)}{r_1 + (r_2/s) + j(x_1 + x_2)}$$

$$= \frac{254(1 - 1.15/40)}{0.52 + 14.1 + j2.30} = \frac{246.7}{14.62 + j2.30} = \frac{246.7}{14.80/8.95°}$$

$$= 16.7/-8.95° = 16.5 - j2.60 \text{ A/phase}$$

The exciting current is

$$\mathbf{I}_M = \frac{\mathbf{V}(1 - x_1/x_m)}{z_M} = \frac{246.7}{360} + \frac{246.7}{j40} = 0.69 - j6.17$$

and the stator current is

$$\mathbf{I}_1 = \mathbf{I}_2 + \mathbf{I}_M = 16.5 - j2.60 + 0.69 - j6.17$$
$$= 17.19 - j8.77 = 19.30\underline{/-27.0°}$$

2. The power factor is

$$PF = \cos \theta = \cos 27.0° = 0.890$$

The actual current in the rotor winding is

$$I_{22} = bI_2 = 2.4 \times 16.7 = 40.1 \text{ A/phase or per slip ring}$$

3. The developed mechanical power, from Eq. 5-12, is

$$P_{em} = mI_2^2 \frac{1 - s}{s} r_2 = 3(16.7)^2 \frac{1 - 0.045}{0.045} 0.634 = 11,260 \text{ W}$$

4. The mechanical power, from Eq. 5-13, is

$$P_{mech} = P_{em} - (P_{fw} + P_{stray}) = 11,260 - (220 + 120) = 10,920 \text{ W}$$

with a horsepower output of

$$hp = 10,920 \div 746 = 14.65$$

5. The power input is

$$P_{in} = 3VI_1 \cos \theta$$
$$= 3(254)(19.30)(0.890) = 13,080 \text{ W}$$

and the efficiency is

$$\text{Efficiency} = \frac{P_{mech}}{P_{in}} = \frac{10,920}{13,080} = 1 - \frac{2160}{13,080} = 0.835$$

6.   $$\text{Torque} = \frac{P_{mech}}{\omega_m} = \frac{11,040}{90.0} = 122.7 \text{ N-m/rad}$$

c. *Tabulation of results:*

| | Circuit | |
|---|---|---|
| | Equivalent (a) | Approximate equivalent (b) |
| Slip, $s$ | 0.045 | 0.045 |
| Stator current, $I_1$ | 18.85 | 19.3 |
| Rotor current, $I_{22}$ | 40.2 | 40.1 |
| Power factor | 0.895 | 0.890 |
| Power input (kW) | 12.88 | 13.08 |
| Power output (kW) | 10.96 | 10.92 |
| Efficiency | 0.851 | 0.835 |

The preceding table shows that the calculations based on the approximate equivalent circuit in Fig. 5-7(b) and with the correction applied to stator voltage are in good agreement with calculations based on the more exact equivalent circuit in Fig. 5-7(a). However, in small motors, such as two-phase servomotors used in control systems, the resistance may be several times as great as the leakage reactance. In addition, because of physical limitations, the air gap is proportionally greater than in larger motors, which results in a comparatively low magnetizing reactance. For such small motors the more exact equivalent circuit in Fig. 5-7(a), or as modified in Fig. 5-17(d), should therefore be used as a basis. In the latter case, the core losses are subtracted along with the windage and friction losses from the developed mechanical power.

### 5-5.3 Phasor Diagram of the Polyphase Wound-Rotor Induction Motor

The equivalent circuits in Fig. 5-7 represent one phase and apply to polyphase induction motors with squirrel-cage rotors as well as to polyphase induction motors with wound rotors, since all quantities are referred to the stator winding. The phasor diagram for the polyphase induction motor because of its transformer features is the same as that of the transformer. Figure 5-8 shows the phasor diagram based on the equivalent circuit in Fig. 5-7(a).

### 5-6 POLYPHASE SQUIRREL-CAGE INDUCTION MOTOR

The equivalent circuits and the phasor diagram in Sec. 5.5, although derived on the basis of the wound-rotor motor, apply equally well to the squirrel-cage motor, since all quantities are referred to the stator winding. The characteristics of the squirrel-cage motor are quite similar to those of the wound-rotor motor with its rotor slip rings short-circuited. However, for a given stator construction the

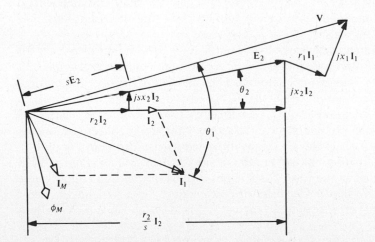

**Figure 5-8**    Phasor diagram for polyphase induction motor.

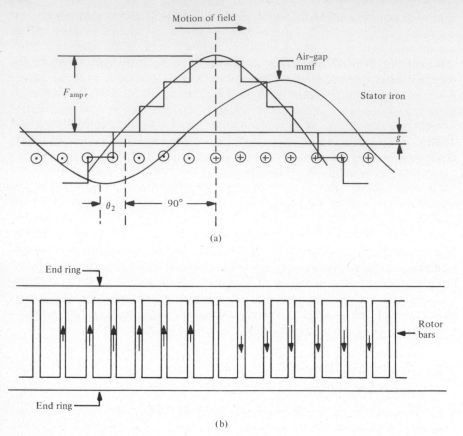

**Figure 5-9** Developed views of squirrel-cage induction motor. (a) Side view of rotor bars. (b) View of rotor cage showing directions of rotor-bar currents with cage assumed between observer and stator.

squirrel-cage rotor generally has lower resistance than the short-circuited wound rotor, largely because of the greater length of the end connections and resistance of slip rings and the brushes in the wound rotor.

Figure 5-9 shows a developed view of the rotor bars and the rotor cage for two poles of a squirrel-cage induction motor, a representation that applies to multipolar as well as to two-pole motors. The end rings of the rotor are omitted in Fig. 5-9(a) for reasons of simplicity. The air-gap flux-density wave, of which only the fundamental is indicated, is produced by the *resultant* of the stator and rotor mmfs and is shown moving from left to right relative to the rotor, inducing emfs in the rotor bars in the directions indicated by the arrows in Fig. 5-9(b). The lengths of the arrows indicate the relative instantaneous values of the induced emfs, being greatest in the rotor bar lying at a given instant in a region where the flux density is a maximum since the emf,† induced in a straight conductor of length **l** moving at a velocity **u** relative to a field which has a uniform flux

† See R. P. Winch, *Electricity and Magnetism* (Englewood Cliffs, N.J.: Prentice-Hall, Inc., 1963).

density **B** along the length of the conductor, is given in vector notation by

$$e = l \cdot (u \times B) \tag{5-24}$$

and since, in the case under discussion, the velocity is at right angles to the magnetic field, the magnitude of the emf is

$$e = Blv$$

The emfs as represented by this arrow pattern are stationary with respect to the rotating flux (i.e., the arrow pattern also rotates at synchronous speed regardless of the rotor speed).

At very small slip, the frequency of the rotor-induced emfs is very low and the phase angle $\theta_2 = \arctan sx_2/r_2$ by which the current in a given rotor bar lags the voltage induced in that same rotor bar is nearly zero, so that the rotor-bar currents are practically in phase with the rotor-bar emfs. The arrow pattern that represents the induced voltages in the rotor bars then also represents the currents in the rotor bars. At appreciable values of slip, the phase angle $\theta_2$ is no longer negligible and the currents in the rotor bars lag the induced emfs by $\theta_2$, as indicated in Fig. 5-9(a). The arrow pattern of the rotor currents is displaced to the left of the arrow pattern of the rotor voltages. The current pattern of the rotor bars shows that the rotor currents distribute themselves in a manner so as to produce a number of rotor poles equal to the number of poles in the stator.

## 5-6.1 Transformation Ratio of the Squirrel-Cage Induction Motor

Since the squirrel cage is a short-circuited winding, the squirrel-cage induction motor has transformer features similar to those of the wound-rotor induction motor with its rotor short-circuited. In the case of the latter, the transformation ratio is given by Eq. 5-6, which is quite straightforward since the rotor as well as the stator have discrete numbers of turns. However, the currents in the rotor bars covering a pair of poles in the squirrel-cage motor are out of phase with each other, as indicated by the differences in the lengths of the arrows in Fig. 5-9(b), which represent the instantaneous values of current. In a balanced $m$-phase system the angle between phases is $360° \div m$. Accordingly, the number of phases in the rotor is the number of rotor bars per pair of poles. Thus, if the total number of rotor slots is $S_2$, the number of phases in the rotor is

$$m_2 = \frac{2S_2}{P}$$

One turn of a full-pitch winding must have two active conductors displaced from each other by $180°$ in electrical measure, and a rotor bar therefore corresponds to one-half a turn per phase (i.e., $N_2 = \frac{1}{2}$). The total effective turns in the stator is $m_1 k_{w1} N_{ph1}/a_1$ and the total turns in the rotor $S_2/2$; so the transformation ratio is

$$b = \frac{m_1 k_{w1} N_{ph1}}{(S_2/2)a_1} = \frac{2m_1 k_{w1} N_{ph1}}{S_2 a_1} \tag{5-25}$$

If $r_{22}$ = resistance of a rotor bar, including that of the end ring sections associated with that bar, and $I_{22}$ = rotor-bar current, then the total rotor-cage copper losses are

$$m_1 I_2^2 r_2 = S_2 I_{22}^2 r_{22} \text{ W}$$

Hence,

$$r_2 = \frac{I_{22}^2}{I_2^2} \frac{S_2 r_{22}}{m_1} \ \Omega/\text{stator phase} \tag{5-26}$$

But

$$\frac{I_{22}}{I_2} = b, \qquad \text{the ratio of transformation} \tag{5-27}$$

When Eqs. 5-25 and 5-27 are substituted in Eq. 5-26, the expression for the rotor resistance referred to the stator becomes

$$\boxed{r_2 = \frac{4m_1 k_{w1}^2 N_{ph1}^2}{S_2^2 a_1^2} r_{22} \ \Omega/\text{phase}} \tag{5-28}$$

At low values of slip the resistance $r_{22}$ is based on dc values.[†] However, at a standstill (i.e., starting), the ac resistance of the rotor is considerably greater than the dc resistance.

## 5-6.2 Double-Squirrel-Cage and Deep-Bar Motors[‡]

While the wound-rotor induction motor has the advantage of some flexibility, it also has the disadvantage of complexity (i.e., the external resistor, short-circuiting arrangement, and slip rings) which results in initial cost and maintenance costs that are higher than for the squirrel-cage motor. High starting torque and good running performance are achieved in some squirrel-cage motors by incorporating two squirrel cages in the rotor, as illustrated in Fig. 5-10(a) and (b), or by the use of deep-bar rotors, as shown in Fig. 5-10(c) and (d). One of these squirrel cages in the double-cage rotor has high resistance with low leakage inductance and predominates during starting, while the other squirrel cage has low resistance with high leakage inductance and predominates at low values of slip. A rough plot of slot leakage flux is shown in Fig. 5-10(b), which shows the rotor bar in the cage nearest the air gap to have a smaller cross section and therefore higher resistance than the rotor bar beneath it. In addition, the larger bar links a greater amount of leakage flux and therefore has higher leakage inductance than the smaller bar above it.

At starting, the frequency of the rotor currents is relatively high, equaling stator frequency, and the leakage reactance of the cage comprised of the larger

[†] A. F. Puchstein, T. C. Lloyd, and A. G. Conrad, *Alternating-Current Machines* (New York: John Wiley & Sons, Inc., 1954), pp. 306–309.

[‡] For a more complete discussion of this subject, the reader is referred to a series of seven papers presented in "Symposium on Design of Double-Cage Induction Motors," *Trans. AIEE* 72, Part III (1953): 621–662.

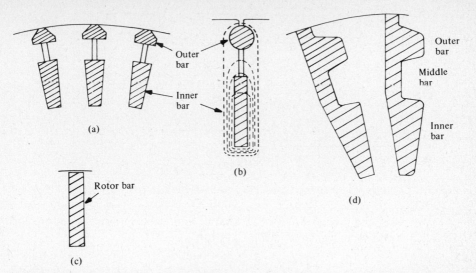

**Figure 5-10**   Slot shapes and rotor bars. (a) and (b) Double-cage rotors. (c) Deep-bar rotor. (d) Triple-cage rotor.

(inner) rotor bars in Fig. 5-10(a) and (b) is high, tending to suppress the current in that cage. The outer cage with the smaller (outer) bars, because of its higher resistance and lower leakage inductance, predominates during starting, thus producing high starting torque. When the motor is operating at normal speed (i.e., at a slip of about 7 percent or less), the rotor frequency is so low that the leakage reactance of the low-resistance cage is considerably lower than its resistance, and the current densities in the two cages are practically equal. As a result, the effective resistance of the rotor is now low, being nearly equal to the dc resistance of the two cages in parallel, making for low slip in the running range from no load to full load. The leakage flux between the two cages is a function of the gap width in the magnetic bridge between the two cages. The current distribution in the deep-bar rotor undergoes a similar change from starting to running, producing high rotor resistance on starting and low resistance at normal speed.

### 5-6.3 Equivalent Circuits for Multiple-Cage Polyphase Induction Motors

The equivalent circuit for the double-cage polyphase induction motor is similar to that for the three-winding transformer and is shown in Fig. 5-11. In that representation the cages are numbered consecutively away from the air gap, starting with 3. If the motor has common end rings, all individual impedance quantities, $r_3$, $r_4$, $x_3$, and $x_4$, refer to the resistances and leakage reactances of the bars in cages 3 and 4, respectively. If the cages have independent end rings, these quantities include the resistance and reactance of the end rings for cages 3 and 4, and $r_e$ and $x_e$, which are the resistance and reactance of the common end rings, are taken as zero. All quantities are referred to the stator.

In the equivalent circuit of Fig. 5-11, $x_{34}$ is the mutual reactance between

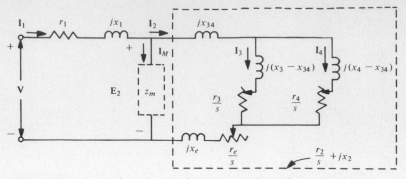

**Figure 5-11**  Equivalent circuit of double-cage induction motor.

the two cages, and $r_e$ and $x_e$ are the resistance and leakage reactance of the end rings when the end rings are common to both cages. Hence,

$$\mathbf{E}_2 = \mathbf{I}_3\left(\frac{r_3}{s} + jx_3\right) + \mathbf{I}_4\, jx_{34} + \mathbf{I}_2\left(\frac{r_e}{s} + jx_e\right) \tag{5-29}$$

in which

$$\mathbf{I}_2 = \mathbf{I}_3 + \mathbf{I}_4$$

so Eq. 5-29 can be rewritten as

$$\mathbf{E}_2 = \mathbf{I}_3\left[\frac{r_3}{s} + j(x_3 - x_{34})\right] + \mathbf{I}_2\left(jx_{34} + \frac{r_e}{s} + jx_e\right) \tag{5-30}$$

Similarly,

$$\mathbf{E}_2 = \mathbf{I}_4\left[\frac{r_4}{s} + j(x_4 - x_{34})\right] + \mathbf{I}_2\left(jx_{34} + \frac{r_e}{s} + jx_e\right) \tag{5-31}$$

For independent end rings, $x_{34}$ includes the mutual reactance between the end rings and $r_e$ and $x_e$ are zero.

    While the double-cage and deep-bar rotors make for high starting resistance and consequently high starting torque along with low running resistance and correspondingly low full-load slip, they do not achieve the flexibility of the wound rotor as far as speed control is concerned. In addition, frequent starting and stopping will overheat the double-cage and deep-bar rotors, as all the heat energy due to increased resistance is generated in the rotor itself. In the case of the wound-rotor motor, the external resistance, which can be made of large physical size to withstand high temperature, dissipates this heat externally to the rotor, while the rotor heating is that associated with the low value of the rotor resistance itself.

## 5-6.4  Skewing

The rotor slots in many squirrel-cage induction motors are skewed or spiraled at an angle of about one slot pitch, as indicated in Fig. 5-12(a) and (b). Skewing

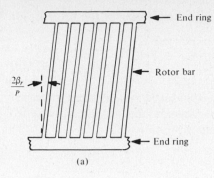

(a)

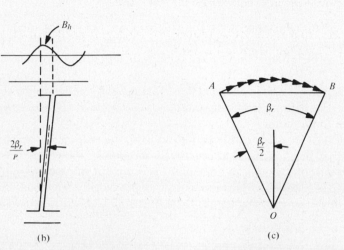

(b)

(c)

**Figure 5-12** (a) Skewed rotor bars and end rings. (b) Skewed rotor bar in harmonic flux. (c) Phasor diagram for skew factor.

reduces torque variations and prevents the rotor from locking in with a harmonic during starting and thus running at a much reduced speed.

A skewed rotor bar lying in the magnetic field of a harmonic of the flux-density wave is shown in Fig. 5-12(b) with the lower end of the bar lying in a weak field and the upper end in a relatively strong field. The instantaneous voltages induced across equal lengths of bar are smallest near the lower end and greatest toward the upper end. Since these voltages are sinusoidal, they all have the same rms values and can therefore be represented by phasors of infinitesimal length and displaced from each other by infinitesimal phase angles. These voltage phasors lie on the arc $AB$ in Fig. 5-12(c), and their phasor sum is represented by the chord $AB$, which corresponds to a winding distributed among an infinite number of straight slots covering the electrical angle of skew $\beta_r$. Hence, the fundamental skew factor is

$$k_r = \frac{\text{chord } AB}{\text{arc } AB} = \frac{2OA \sin (\beta_r/2)}{OA\beta_r} = \frac{\sin (\beta_r/2)}{\beta_r/2} \tag{5-32}$$

where $\beta_r$ is measured in electrical radians. The skew factor of the rotor is, for the $h$ harmonic,

$$k_{rh} = \frac{\sin h(\beta_r/2)}{h(\beta_r/2)} \tag{5-33}$$

The disadvantages of skewing are reduced induced rotor voltage, increased rotor resistance, and increased rotor leakage reactance, but these are outweighed by the better starting performance and quieter operation.

## 5-7 NO-LOAD AND LOCKED-ROTOR TESTS

The equivalent circuit constants of the induction motor can be obtained from a test made with the motor running idle and from the blocked-rotor test.† These tests correspond to the no-load and short-circuit tests on the transformer.

**No-Load Test** Rated balanced voltage at rated frequency is applied to the stator while the motor runs without load. Measurements are made of voltage, current, and power input to the stator.

Because of the low value of the slip at no load, the equivalent resistance $r_2/s$ is so high that the no-load rotor current is negligible. However, a small amount of rotor current, which can be neglected, is present in practical motors even at zero slip, as a result of harmonics in the flux-density wave and slight nonuniformity of the air gap. For the no-load test on a *three-phase motor,* let

$V_0$ = rated line-to-line voltage
$I_0$ = line current
$P_0$ = power input
$r_1$ = stator resistance in ohms per phase on the basis of a Y connection

Since the no-load rotor current is negligible, the rotor circuit may be omitted from the equivalent circuit in Fig. 5-7(a), resulting in that of Fig. 5-13(a) or that of Fig. 5-13(b), in which $z_M$ is represented by an equivalent series impedance to facilitate the evaluation of the motor reactances and resistances. The resistance $r_{fe}$ in Fig. 5-13(a) takes into account not only the stator core losses, but the windage and friction losses as well. Because of the low value of rotor frequency, the rotor core loss is negligible at no load, as shown by Eqs. 2-33 and 2-41. From Fig. 5-13(b),

$$V = \frac{V_0}{\sqrt{3}} \quad \text{V/phase} \tag{5-34}$$

$$z_0 = \frac{V}{I_0} \quad \Omega/\text{phase} \tag{5-35}$$

† For a more complete discussion of tests on induction motors, see *IEEE Test Code for Polyphase Induction Motors and Generators No. 112A* (New York: Institute of Electrical and Electronic Engineers, 1964).

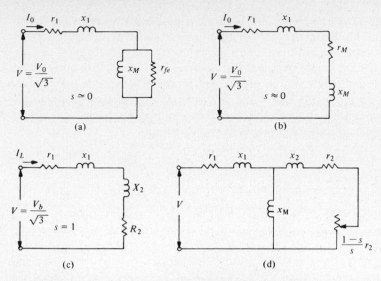

**Figure 5-13** Equivalent circuits for three-phase induction motor. (a) (b) No-load test. (c) Locked-rotor test. (d) Simplified equivalent circuit for motor under load.

$$P_0 = 3I_0^2 r_0 \tag{5-36}$$

where

$$r_0 = r_1 + r_M$$

in which $r_1$ is the stator resistance taken as the dc value. The series resistance $r_M \ll x_M'$ and therefore $x_M' \cong x_M$. The no-load reactance is accordingly very nearly equal to the stator reactance, and if the iron were unsaturated it would correspond to the synchronous reactance of a cylindrical-rotor synchronous machine. Then

$$x_0 \simeq x_1 + x_M \tag{5-37}$$

and
$$x_0 = \sqrt{z_0^2 - r_0^2} \tag{5-38}$$

The rotational losses (i.e., the sum of the windage, friction, and core losses) are found by subtracting the stator copper loss from the no-load power input; thus

$$P_{ro} = P_0 - 3I_0^2 r_1 \tag{5-39}$$

**Locked-Rotor Test** The quantities $x_1$, $x_2$, and $r_2$ can now be determined from the locked-rotor test data, which are obtained with the rotor blocked to prevent it from rotating (i.e., for $s = 1$). The locked-rotor test is also known as the *blocked-rotor test*.

For ordinary single-cage motors of less than 25-hp rating, not including deep-bar machines, reduced balanced three-phase voltage at rated frequency is applied to the stator. In order to obtain constants for the normal running, the voltage is adjusted to produce approximately rated current. Rated voltage would result in excessive currents that would saturate the leakage flux paths across the stator and rotor teeth, giving rise to lower-than-normal values of leakage reactance for the running range. In addition, unless sustained for only a very short period, the excessive currents would overheat the windings. However, if in addition to the running characteristics the full-voltage starting performance is to be determined, a locked-rotor test is also made by applying rated voltage at rated frequency, while measurements of voltage, current, and power, as well as torque, are made as rapidly as possible to prevent overheating.

In the locked-rotor test on a three-phase motor, let

$$V_L = \text{line-to-line voltage}$$
$$I_L = \text{line current}$$
$$P_L = \text{power input}$$

Then the impedance of the motor is

$$z_L = \frac{V_L}{\sqrt{3}I_L} \tag{5-40}$$

its equivalent resistance being

$$r_L = r_1 + R_2 = \frac{P_L}{3I_L^2} \tag{5-41}$$

and its equivalent reactance,

$$x_L = x_1 + X_2 = \sqrt{z_L^2 - r_L^2} \tag{5-42}$$

These parameters are shown in the equivalent circuit in Fig. 5-13(c). When the rotor is locked, the exciting current is small relative to the stator current, the rotor leakage reactance is only slightly greater than $X_2$, and

$$x_1 + X_2 \simeq X_L \tag{5-43}$$

The IEEE Test Code 112A lists the empirical proportions given in Table 5-1 for stator and rotor leakage reactance in three-phase induction motors.

The magnetizing reactance $x_M$ can now be evaluated from Eqs. 5-37 and 5-42 with the aid of Table 5-1. When the classification of the motor is not known, it is assumed that $x_1 = x_2 = 0.5x_L$. The same test code recommends that the net mechanical power output be taken as the developed power minus the sum of rotational losses as expressed by Eq. 5-13 plus the stray-load loss, which makes it possible to eliminate the resistance $r_{fe}$, which results in the simplified equivalent circuit in Fig. 5-13(d). The rotational losses are assumed constant and equal to the no-load value for the running range. The error due to this assumption is

**TABLE 5-1.** EMPIRICAL PROPORTIONS OF INDUCTION-MOTOR LEAKAGE REACTANCES

| Type of motor | Class A | Class B | Class C | Class D | Wound rotor |
|---|---|---|---|---|---|
| $x_1$ | $0.5x_L$ | $0.4x_L$ | $0.3x_L$ | $0.5x_L$ | $0.5x_L$ |
| $x_2$ | $0.5x_L$ | $0.6x_L$ | $0.7x_L$ | $0.5x_L$ | $0.5x_L$ |

small, since the decrease in windage and friction loss with decrease in speed is at least partly offset by an increase in the rotor-core loss. The stray-load tests are described in the IEEE Test Code.

The equivalent resistance $R_2$ in Fig. 5-13(c) is somewhat smaller than the rotor resistance $r_2$ and, since $r_2/s \gg x_1 + x_2$ in the running range, it has a correspondingly greater effect on the performance of the motor within that range. The value of $r_2$ therefore requires a closer approximation than that of $x_2$, which is made as follows. The equivalent circuit of Fig. 5-13(d) applies to locked-rotor conditions by making $s = 1$, and it is then equivalent to the circuit in Fig. 5-13(c). Accordingly,

$$R_2 + jX_2 = \frac{(r_2 + jx_2)jx_M}{r_2 + j(x_2 + x_M)} \tag{5-44}$$

and when the right-hand side of Eq. 5-44 is reduced to a single complex quantity expressed in rectangular form, its real term must equal $R_2$:

$$R_2 = \frac{r_2 x_M^2}{r_2^2 + (x_2 + x_M)^2} \tag{5-45}$$

Since $r_2 \ll (x_2 + x_M)$, Eq. 5-45 can be approximated to

$$R_2 \simeq \frac{r_2 x_M^2}{(x_2 + x_M)^2} \tag{5-46}$$

and the rotor resistance referred to the stator is then found from Eqs. 5-41 and 5-46 to be

$$r_2 = (r_L - r_1)\left(\frac{x_2 + x_M}{x_M}\right)^2 \tag{5-47}$$

**EXAMPLE 5-2**

The following test results were obtained on a 10-hp three-phase 400-V 14-A 60-Hz eight-pole induction motor with a single squirrel-cage rotor (design A).

*No-Load Test.* 440 V line to line, 5.95 A line current, 350 W three-phase power.

*Locked-Rotor Test at 60 Hz.* 94.5 V line to line, 13.85 A line current, 890 W three-phase power.

The dc resistance of the stator was measured immediately after the blocked-rotor test with an average value of 0.77 $\Omega$/phase. Calculate the no-load rotational losses and the constants for the equivalent circuit of Fig. 5-13(d).

*Solution.* From the no-load test data and Eqs. 5-34 and 5-35,

$$z_0 = \frac{440}{\sqrt{3}(5.95)} = 42.8 \ \Omega$$

and, from Eq. 5-36,

$$r_0 = \frac{350}{3(5.95)^2} = 3.30 \ \Omega$$

Further, from Eq. 5-38,

$$x_0 = \sqrt{(42.8)^2 - (3.30)^2} = 42.6 \ \Omega$$

From the locked-rotor test and Eqs. 5-40 and 5-41

$$z_L = \frac{94.5}{\sqrt{3}(13.85)} = 3.94 \ \Omega \qquad r_L = \frac{890}{3(13.85)^2} = 1.545 \ \Omega$$

from which

$$x_L = \sqrt{(3.94)^2 - (1.545)^2} = 3.62 \ \Omega$$

According to Table 5-1,

$$x_1 = x_2 = 0.5 \times 3.62 = 1.81 \ \Omega$$

and, from Eq. 5-37,

$$x_M = 42.6 - 1.81 = 40.79 \ \Omega$$

In Eq. 5-41,

$$r_L - r_1 = 1.545 - 0.77 = 0.775 \ \Omega$$

and

$$r_2 = 0.775 \left(\frac{42.6}{40.77}\right)^2 = 0.845 \ \Omega$$

*Locked-Rotor Test at Reduced Frequency (Double-Cage and Large Deep-Bar Machines).* The value of rotor resistance obtained with full frequency from a blocked-rotor test on double-cage and deep-bar machines is the value for starting and is therefore present in the normal running range. Also, the leakage reactance is lower on starting than in the normal running range. Both of these effects are due to the relatively high frequency of the rotor current at standstill. Therefore, the IEEE Test Code† recommends that the locked-rotor test be made at rated three-phase current at a frequency of 15 Hz to obtain values of rotor resistance and leakage reactance that apply to the normal running range.

**EXAMPLE 5-3**

Tests on a 75-hp three-phase 60-Hz 440-V 88-A six-pole 1170-rpm deep-bar induction motor (class B) yielded results as follows:

*No Load.* 440 V line to line, 24.0 A line current, 2.56 kW three-phase power.

*Locked-Rotor Test at 15 Hz.* 28.5 V line to line, 90.0 A line current, 2.77 kW three-phase power.

† Ibid.

*Average Value of DC Resistance between Stator Terminals.* 0.0966 $\Omega$ terminal to terminal.

*Locked-Rotor Test at Rated Voltage, 60 Hz.* 440 V line to line, 503 A line current, 150.0 kW three-phase power.

Calculate the constants of the equivalent circuit in Fig. 5-13(d) for the running range.

*Solution.* From the no-load test and Eqs. 5-34, 5-35, and 5-38,

$$z_0 = \frac{440}{\sqrt{3}(24.0)} = 10.57 \ \Omega \qquad r_0 = \frac{2560}{3(24)^2} = 1.48 \ \Omega$$

$$x_0 = \sqrt{(10.57)^2 - (1.48)^2} = 10.45 \ \Omega$$

From the dc test,

$$r_1 = 0.0966 \div 2 = 0.0483 \ \Omega$$

From the blocked-rotor test at 15 Hz and Eqs. 5-40, 5-41, and 5-42,

$$z_{L15} = \frac{28.5}{\sqrt{3}(90.0)} = 0.183 \ \Omega \qquad r_{L15} = \frac{2770}{3(90.0)^2} = 0.114 \ \Omega$$

$$x_{L15} = \sqrt{(0.183)^2 - (0.114)^2} = 0.1435 \ \Omega$$

The corresponding 60-Hz value is

$$x_L = \frac{60}{15} x_{L15} = \frac{60}{15} \times 0.1435 = 0.574 \ \Omega$$

The leakage reactances are determined from Table 5-1 to be

$$x_1 = 0.4x_L = 0.4(0.574) = 0.230 \ \Omega$$

$$x_2 = 0.6x_L = 0.6(0.574) = 0.344 \ \Omega$$

The magnetizing reactance, from Eq. 5-37 is

$$x_M = x_0 - x_1 = 10.45 - 0.23 = 10.22 \ \Omega$$

and the rotor resistance referred to the stator, from Eq. 5-47, is

$$r_2 = (r_{L15} - r_1)\left(\frac{x_2 + x_M}{x_M}\right)^2 = 0.0657\left(\frac{0.343 + 10.22}{10.22}\right)^2$$

$$= 0.0697 \ \Omega$$

## 5-8 POLYPHASE-INDUCTION MOTOR–SLIP-TORQUE RELATIONSHIP BASED ON APPROXIMATE EQUIVALENT CIRCUIT

The approximate equivalent circuit in Fig. 5-7(b) along with a modification of the applied voltage is used in the following. The rotor current referred to the stator for a given value of slip $s$ is

$$I_2 = \frac{V(1 - x_1/x_M)}{\sqrt{(r_1 + r_2/s)^2 + (x_1 + x_2)^2}} \qquad (5\text{-}48)$$

The developed mechanical power for all $m_1$ phases of the stator is

$$P_{em} = m_1 I_2^2 \frac{1 - s}{s} r_2$$

and the developed torque is

$$T_{em} = \frac{P_{em}}{\omega_m} = \frac{m_1 I_2^2 [(1 - s)/s] r_2}{2\pi(1 - s)(n_{syn}/60)}$$

$$= \frac{m_1 I_2^2 (r_2/s)}{2\pi n_{syn}/60} \qquad (5\text{-}49)$$

The quantity $m_1 I_2^2 (r_2/s)$ in Eq. 5-49 is the power transferred across the air gap from the stator to the rotor (i.e., the rotor input), from which it follows that the developed torque is directly proportional to the power input to the rotor.

The developed torque is expressed in terms of the stator voltage on the basis of Eqs. 5-48 and 5-49 by

$$T_{em} = \frac{m_1 [V(1 - x_1/x_M)]^2 (r_2/s)}{(2\pi n_{syn}/60)[(r_1 + r_2/s)^2 + (x_1 + x_2)^2]} \qquad (5\text{-}50)$$

At rated voltage and at rated frequency, the voltage $V$ and the synchronous speed $n_{syn}$ in Eq. 5-50 are both constant. Further, in the normal running range (i.e., from no load to full load), the stator and rotor winding resistances $r_1$ and $r_2$, as well as the leakage reactances $x_1$ and $x_2$, are practically constant, and the torque for a given applied voltage at rated frequency is a function only of the slip $s$. Figure 5-14 shows the relationship among torque, mechanical power, and speed based on the approximate equivalent circuit. Motor action is confined to speeds ranging from zero to just below synchronous speed, and generator action occurs when the machine is driven above synchronous speed. When driven backward, the slip is greater than unity, and the rotor circuit absorbs power not only from the stator through the air gap but also mechanical power that is converted into electric power. One way of braking an induction motor or producing a high decelerating torque is to reverse the phase sequence of the applied voltage while the motor is running in the forward direction. This is accomplished by reversing any two of the three leads to the stator, which reverses the direction of the rotating field. All three of these regions of operation are indicated in Fig. 5-14.

## 5-8.1 Starting Torque

On starting, the rotor is stationary and the slip is unity. The expression for the starting torque is obtained by letting $s = 1$ in Eq. 5-50, which results in

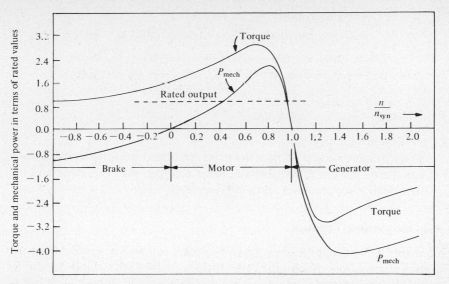

**Figure 5-14** Calculated torque-speed characteristic of a polyphase induction using unsaturated values of $x_1$ and $x_2$ throughout.

$$T_{start} = \frac{m_1[V(1 - x_1/x_M)]^2 r_2}{(2\pi n_{syn}/60)[(r_1 + r_2)^2 + (x_1 + x_2)^2]} \qquad (5\text{-}51)$$

Equation 5-51 shows the starting torque to vary as the square of the applied voltage. While the magnetic circuit is practically linear under rated running conditions, the relative high rotor and stator currents during starting may saturate portions of the iron—the rotor and stator teeth—in the leakage-flux paths. This makes the values of the leakage reactance $x_1$ and $x_2$ somewhat lower at starting than those at normal speed.† Nevertheless, Eq. 5-51 makes for a good approximation of the starting current in wound-rotor induction motors and in squirrel-cage induction motors with a single cage.

**EXAMPLE 5-4**

Using the data given in Example 5-3, calculate the electromagnetic torque for full-voltage starting.

*Full-Voltage Starting Torque.* Equation 5-49 shows the electromagnetic torque to equal the result of dividing the real-power input to the rotor by the synchronous angular velocity. The power input to the rotor is taken as the difference between the real-power input to the stator and the stator copper loss. At full-voltage starting,

† S. S. L. Chang and T. C. Lloyd, "Saturation Effect of Leakage Reactance," *Trans. AIEE* 68, Part III (1949): 1144–1148. See also P. D. Agarval and P. L. Alger, "Saturation Factors for Leakage Reactance of Induction Motors," *Trans. AIEE* 79, Part III (1960): 1037–1042.

Power input to stator = 150,000 W

Stator copper loss = $3I_1^2r_1 = 3(503)^2(0.0483) = 36,700$ W

Power input to rotor = $150,000 - 36,700 = 113,300$ W

$$T_{em} = \frac{113,300}{2\pi n_{syn}/60} \quad \text{and} \quad n_{syn} = 1200 \text{ rpm}$$

$$= \frac{113,300}{40\pi} = 900 \text{ N-m/rad}$$

The measured torque is lower than the calculated value above because in computing the power input to the rotor winding, the core losses and the stray load losses were not taken into account.

## 5-8.2 Maximum Torque

The maximum torque for a given applied voltage can be determined from the laws of maximum and minimum. The constant multipliers in Eq. 5-50 can be replaced by one constant $K$ so that

$$T_{em} = \frac{K(r_2/s)}{(r_1 + r_2/s)^2 + (x_1 + x_2)^2} \tag{5-52}$$

To obtain the maximum, differentiate both sides of Eq. 5-52 with respect to $r_2/s$ and equate the result to zero, as follows:

$$\frac{dT_{em}}{d(r_2/s)} = K\left\{\frac{1}{(r_1 + r_2/s)^2 + (x_1 + x_2)^2} - \frac{2(r_2/s)(r_1 + r_2/s)}{[(r_1 + r_2/s)^2 + (x_1 + x_2)^2]^2}\right\} = 0$$

from which

$$\frac{r_2}{s} = \sqrt{r_1^2 + (x_1 + x_2)^2} \tag{5-53}$$

When Eq. 5-53 is substituted in Eq. 5-50, the maximum torque is found to be

$$T_{em(max)} = \frac{m_1[V(1 - x_1/x_2)]^2}{(4\pi n_{syn}/60)[r_1 + \sqrt{r_1^2 + (x_1 + x_2)^2}]} \tag{5-54}$$

Equation 5-54 shows the maximum torque to be independent of rotor resistance. The rotor resistance, however, determines the slip at which the maximum torque occurs. The maximum torque in conventional induction motors is generally developed at a value of slip several times the rated load slip for which the stator and rotor currents may be large enough to produce some saturation of the iron in the leakage flux paths. The maximum torque would then exceed the value calculated on the basis of unsaturated values of leakage reactance.

## 5-8.3 Influence of Rotor Resistance on Slip

Equation 5-50 shows that the quantity $r_2/s$ must be constant for constant torque with a given applied voltage of normal frequency. The effect of rotor resistance on the torque-slip characteristic of a polyphase induction motor is shown in Fig. 5-15, in which the maximum torque has the same value for all three values of resistance and which shows that the slip at which the maximum torque occurs is directly proportional to the rotor resistance.

### 5-8.4 Influence of Reactances on Motor Performance

It is evident from Eqs. 5-51 and 5-54 that the leakage reactance $(x_1 + x_2)$ must be low—about 0.15 per unit—to assure good starting torque as well as adequate maximum torque. On the other hand, the magnetizing reactance should be high to hold the magnetizing current to a permissible value (i.e., about 0.8 of rated value for small motors and about 0.4 for the largest motors). A large value of magnetizing current results in a poor power factor, particularly at light loads.

## 5-9 WOUND-ROTOR MOTOR STARTING AND SPEED CONTROL

A motor needs to develop only moderate starting torque for such applications as fans and blowers. However, some loads require high starting torque, such as

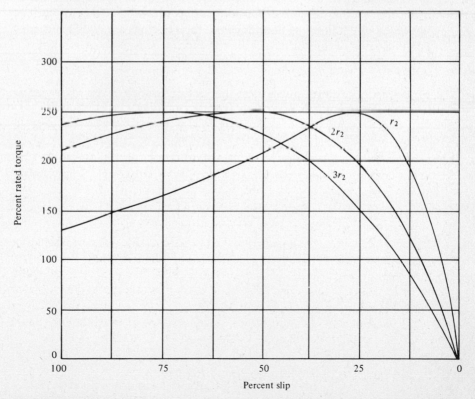

**Figure 5-15** Effect of rotor resistance on torque-slip characteristic of a polyphase induction motor.

conveyors, in which it is necessary to overcome high static torque and loads having high inertia. The starting torque is a function of rotor resistance. As shown by Eq. 5-51, very low rotor resistance results in low starting torque. The motor develops its maximum torque on starting, when the rotor resistance is

$$r_2 = \sqrt{r_1^2 + (x_1 + x_2)^2}$$

as shown by Eq. 5-53. While this large value of resistance produces optimum starting conditions, it causes the motor to develop excessive slip at rated torque, which results in poor efficiency and poor speed regulation. Thus, a fixed value of rotor resistance that promotes high starting torque interferes with good performance in the normal speed range. This difficulty is circumvented in the wound-rotor motor by means of series resistance external to the rotor, which is short-circuited as the motor comes up to rated speed. The speed of the motor may also be controlled by varying the value of the external resistance.

**EXAMPLE 5-5**

The wound-rotor induction motor of Example 5-1 delivers its rated load of 15 hp at rated voltage and frequency with the slip rings short-circuited at a slip of 0.042. Three tapped resistors, one in each phase, are to be connected in wye to the rotor slip rings. Figure 5-16 shows an arrangement in which the external resistance can be cut out in three steps. The resistance at starting should produce maximum torque, and when the rotor current has fallen to 2.0 times its normal value, the slip rings are switched to taps giving a value of resistance such that the torque is again a maximum.

*Solution*

*First Step.* For maximum torque the total resistance of the rotor circuit referred to the stator is, from Eq. 5-53,

$$\frac{r_2 + r_x}{s} = \sqrt{r_1^2 + (x_1 + x_2)^2} = \sqrt{(0.52)^2 + (2.30)^2} = 2.36$$

where $r_x$ is the resistance external to the rotor in ohms per phase referred to the stator. For $s = 1$,

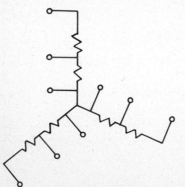

**Figure 5-16**  Tapped external resistors for a three-phase wound rotor.

$$r_x = 2.36 - 0.634 = 1.726 \ \Omega/\text{phase}$$

The actual resistance external to the rotor is

$$r_{2x} = \frac{r_x}{b^2} = \frac{1.726}{(2.4)^2} = 0.300 \ \Omega/\text{phase}$$

The initial rotor current at starting ($s = 1$) is, from Eq. 5-48,

$$I_2 = \frac{V(1 - x_1/x_M)}{\sqrt{(r_1 + [(r_2 + r_x)/s])^2 + (x_1 + x_2)^2}}$$

$$= \frac{254(1 - 1.15/40)}{\sqrt{(0.52 + 2.36)^2 + (2.30)^2}}$$

$$= 66.7 \ \text{A/phase (referred to the stator)}$$

*Rated Rotor Current.*   The rotor current referred to the stator at $s = 0.042$ is

$$I_2 = \frac{254(1 - 1.15/40)}{\sqrt{(0.52 + 0.63/0.042)^2 + (2.30)^2}}$$

$$= \frac{246.7}{15.7} = 15.72 \ \text{A/phase}$$

*Slip at Which Rotor Current Equals 2.0 Times Rated Value.*   At 2 times rated current the impedance encountered by $I_2$ is

$$\sqrt{\left(r_1 + \frac{r_2 + r_x}{s}\right)^2 + (x_1 + x_2)^2} = \frac{15.7}{2.0} = 7.85 \ \Omega$$

$$0.52 + \frac{2.36}{s} = \sqrt{(7.85)^2 - (2.30)^2} = 7.52 \ \Omega/\text{phase}$$

from which

$$s = 0.337$$

*Second Step.*   Resistance for maximum torque at slip $s = 0.337$:

$$\frac{r_2 + r_x}{s} = \sqrt{r_1^2 + (x_1 + x_2)^2} = 2.36$$

$$r_2 + r_x = 2.36 \times 0.337 = 0.796 \ \Omega/\text{phase}$$

$$r_{2x} = \frac{0.796 - 0.634}{(2.4)^2} = \frac{0.16}{(2.4)^2} = 0.028 \ \Omega/\text{phase}$$

The initial value of $I_2$ at $s = 0.337$ is again 66.7 A.

*Slip at Which Rotor Current Has Fallen to 2.0 Times Rated Value*

$$0.52 + \frac{0.796}{s} = 7.52$$

$$s = 0.114$$

*Third Step.* The external resistance required for maximum torque at $s = 0.114$ would have to be negative. Therefore, the third step would simply short-circuit the remaining 0.028 $\Omega$ in the external resistance for which the rotor current referred to the stator would be 39.6 A. Since this value does not exceed twice rated value, the operation is practical.

## 5-10  SPEED CONTROL OF POLYPHASE INDUCTION MOTORS

While the speed of the wound-rotor motor can be controlled by adjusting the external rotor resistance, this method has the disadvantage of low efficiency and poor speed regulation at high values of slip. For large induction motors, complex arrangements that feed the power normally consumed by the external rotor resistance back into the line can be justified. Several of these are described in the literature† but are not taken up in this text.

The speed may also be changed by changing the number of poles, a method better adapted to squirrel-cage motors than to wound-rotor motors because the squirrel-cage automatically assumes the same number of poles as the stator. For the single polyphase stator winding, a speed range of 2 to 1 can be obtained by doubling the number of poles as illustrated in Fig. 5-17, in which the current to terminals $a_2$–$a_2'$ is reversed. Several speed ranges having synchronous values, for example, 3600, 1800, 1200, and 600 rpm, may be obtained by using two separate stator windings for each phase. One of these is 2–4 pole and the other is 6–12 pole. When one of these is in use, the other is idle. Only a limited number of coils for one phase are shown in Fig. 5-17, for reasons of simplicity. To prevent a large difference in the flux densities for the two speeds, the windings are connected in series delta for the larger number of poles and in parallel wye for the smaller number.

### 5-10.1  Variable Frequency

An efficient method for controlling the speed of induction motors and synchronous motors is that of applying variable frequency to the stator, since the field in such motors rotates at synchronous speed. The air-gap field in these motors is nearly constant for a given ratio of voltage to frequency. Unless the voltage and frequency are decreased in about the same proportion, saturation may become excessive. Operation above rated speed, however, is usually at rated voltage.

The rotor current of an induction motor is practically proportional to the slip speed for a given value of flux, regardless of the stator frequency, as shown by Eqs. 5-6 and 5-9. According to Eq. 5-23, the electromagnetic torque is then constant for a given slip speed $sn_{\text{syn}}$. For example, if a 240-V four-pole 60-Hz induction motor is driving a constant-torque load and the speed is 1764 rpm at 60 Hz, then at 40 Hz the voltage should be 160 V, the slip speed would still be

---

† See P. L. Alger, *Induction Machines* (New York: Gordon and Breach, Inc., 1970), pp. 302–305; and A. E. Knowlton (ed.), *Standard Handbook for Electrical Engineers,* 8th ed. (New York: McGraw-Hill Book Company, 1949), pp. 772–775.

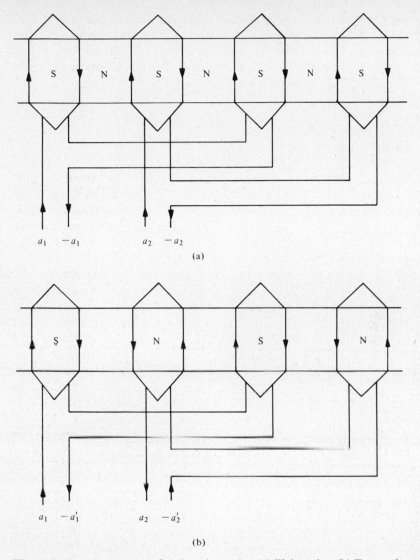

**Figure 5-17**  Arrangement for changing poles, (a) Eight poles. (b) Four poles.

36 rpm, and the motor would run at 1164 rpm. The power output would be 1164/1764 = 0.66 of that at 60 Hz.

Because conventional power systems normally operate at constant frequency, a means for varying the frequency must be interposed between the line and the motor. The most economical arrangement is a SCR rectifier supplying dc voltage to a static inverter, incorporating solid-state components, which in turn supplies the variable frequency to the motor† somewhat as indicated in Fig. 5-18. This arrangement is called a cycloconverter.

† See Alexander Kusko, *Solid-State A.C. Motor Drives* (Cambridge, Mass.: The MIT Press, 1971).

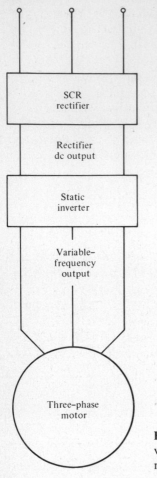

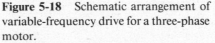

**Figure 5-18**   Schematic arrangement of variable-frequency drive for a three-phase motor.

The three-phase inverter is a circuit arrangement of thyristors, diodes, inductors, and capacitors.† The waveform of the voltage applied to the motor is generally so distorted as to produce above-normal losses and motor heating. Although the frequency range of the inverter is wide, its overload capacity is limited, which introduces difficulties on starting.

## 5-10.2  Line-Voltage Control

The speed of an induction motor can be adjusted by varying the voltage of constant frequency applied to the stator. This method is limited to certain drives, because the developed torque for a given speed is directly proportional to the square of the applied voltage and, in the case of constant-load torque, for example,

---

† B. D. Bedford and R. G. Hoft, *Principles of Inverter Circuits* (New York: John Wiley & Sons, Inc., 1964); see also J. M. D. Murphy, *Thyristor Control of A.C. Motors* (New York: Pergamon Press, 1973).

the current must increase with decreasing voltage, since the flux is nearly proportional to the voltage. However, this method finds considerable application where the load torque decreases with a reduction in speed, as in the case of fans and pumps.

The stator voltage may be varied by means of variable impedance in series with the motor, but the use of thyristors between the line and the motor seems to offer the most economical and effective operation. One arrangement using thyristors is illustrated in Fig. 5-19 for a three-phase induction motor. A similar scheme with one pair of back-to-back connected thyristors instead of three pairs can be used to control the speed of single-phase capacitor motors.

## 5-11 APPLICATIONS OF POLYPHASE INDUCTION MOTORS

Wound-rotor motors are suitable for loads requiring high starting torque and for applications where the starting current must be low; also for loads having high inertia, which results in extremely large rotor energy losses during acceleration. Wound-rotor motors are also used for loads that require a gradual buildup of torque or soft start and for loads that require some speed control.

The maximum torque is usually well above 200 percent of full-load value while the full-load slip may be as low as 3 percent, which makes for a high full-

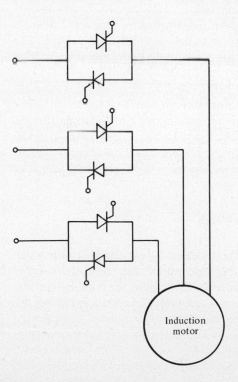

Induction motor

**Figure 5-19** Schematic arrangement for line-voltage control of a three-phase induction motor.

load efficiency, approaching 90 percent. Typical load applications are conveyors, crushers, plunger pumps, hoists, cranes, elevators, and compressors.

Squirrel-cage induction motors are classified by the National Electric Manufacturers Association† as designs A, B, C, D, and F. Design A motors usually have low-resistance single-cage rotors, which make for good running characteristics at the expense of high starting current and moderate starting torque. Because of the high starting current, a reduced-voltage starter (see Sec. 5-12) may be required. Examples of loads are blowers, fans, machine tools, and centrifugal pumps.

Design B motors, the most popular, are of the double-cage or deep-bar design and used for full-voltage starting. They have about the same starting torque as design A with only about 75 percent of the starting current, and their applications are the same as those for the design A.

Design C motors are of the double-cage and deep-bar construction with higher rotor resistance than design B, making for higher starting torque but lower efficiency and somewhat greater slip than for the design B. The application is for practically constant-speed loads requiring fairly high starting torque while drawing relatively low starting current. Typical loads are compressors, conveyors, crushers, and reciprocating pumps.

Design D motors have the highest starting torque of all squirrel-cage induction motors. They generally have a single-cage high-resistance rotor, which results in high starting torque but also high slip with correspondingly low efficiency. These motors are used for high-inertia loads such as bulldozers, die-stamping machines, punch presses, and shears.

Design F motors are usually high-speed motors directly connected to loads that require only low starting torques, such as fans or centrifugal pumps. The rotor has low resistance, which makes for low slip and correspondingly high efficiency but also low starting torque.

Induction motors are better suited for high-speed applications. This is largely because the magnetizing reactance $x_M$ is inversely proportional to the square of the number of poles for a given length of air gap, number of turns, and frame size, as indicated by Eq. 5-2. On that basis, lower-speed motors have a proportionately greater magnetizing current, with a correspondingly lower power factor.

Synchronous motors are generally used for applications requiring constant speeds below about 500 rpm.

## 5-12 REDUCED-VOLTAGE STARTING

Design A squirrel-cage induction motors, because of their relatively low value of leakage reactance, take large starting current, which may place an excessive load on the supply circuit to the motor. In such installations the required starting torque is usually low, and the starting current is lessened by reducing the voltage below normal during starting. As the motor approaches normal speed, rated voltage is applied. The most common method of reduced-voltage starting makes

---

† *NEMA Motor and Generator Standards, Publication MGI-1978* (New York: National Electrical Manufacturers Association).

use of an autotransformer, sometimes called a *starting compensator,* which generally has 65 and 80 percent voltage taps for motors up to 50 hp and 50, 65, and 80 percent taps for larger motors. Usually the lowest-voltage tap that will allow the motor to accelerate is selected. The autotransformer may be three-phase with a winding for each phase, or it may be a three-phase open-delta transformer as shown in Fig. 5-20.

The current drawn from the line at starting and the starting torque are both approximately inversely proportional to the square of the autotransformer ratio. Since the autotransformer is energized only during the brief starting period, it is not designed for continuous operation and is therefore of correspondingly small physical size.

The transfer of the motor from the autotransformer to the line is accompanied by a peak current of a few cycles duration, which may be as high as or higher than the normal starting current at full voltage. Upon disconnecting the motor from the autotransformer, the flux in the motor does not decrease to zero, being sustained by the transient rotor current. In addition, the motor slows down so that the voltage induced in the stator winding by the rotor current is lower than and out of phase with the line voltage upon reclosing. This peak current may produce sufficient momentary torque to damage the shaft or coupling. However, there are many installations in which the effects of the peak current

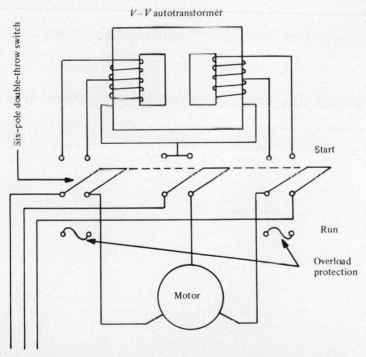

**Figure 5-20** Open-delta autotransformer for reduced-voltage starting of induction motor. (a) Circuit diagram manual operation. (b) Circuit diagram automatic operation. (c) Operation and contactor sequence for automatic starter.

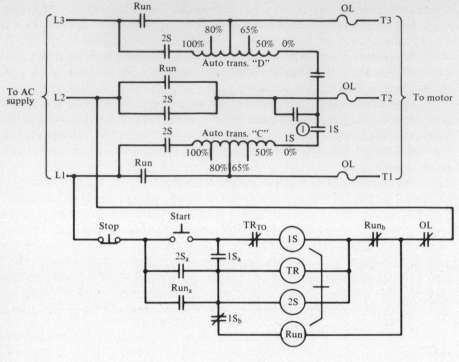

(b)

Closing the start button or other pilot device energizes the start contactor (1S). The interlock (1S$_a$) closes, energizing the timing relay (TR) and contactor (2S), which seal in through the interlock (2S$_a$). With the (1S) and (2S) contactor main contacts closed, the motor is connected through the autotransformer for reduced voltage start. After a preset time interval, the (TR$_{TO}$) contacts time open, deenergizing contactor (1S) and connecting the autotransformer as a reactor in series with the motor. Interlock (1S$_b$) immediately energizes the run contactor (RUN), which seals in through its interlock (RUN$_a$). The run contacts are now closed, and the motor is running at full voltage. Start contactor (2S) and relay (TR) are de-energized when interlock (RUN$_b$) opens.

An overload, opening the stop pushbutton or other pilot device de-energizes the (RUN) contactor, removing the motor from the line.

Contactor Sequence

| Contactor | Start | | Transition | Run |
|-----------|-------|---|------------|-----|
| 1S | • | • | | |
| 2S | | • | • | • |
| RUN | | | • | • |

(c)

**Figure 5-20**   (Continued)

288

can be tolerated. In others, a transition impedance (not shown in Fig. 5-20) between the autotransformer and the line maintains current to the stator during the transfer.

There are numerous other methods for starting polyphase induction motors.† One of these starts the motor at full voltage with the stator winding connected in wye, with this winding connected in delta as the motor approaches normal speed. Part-winding starting is another method in which one of two identical stator windings is used during starting, both windings being connected in parallel when the motor is nearly up to speed. The impedance of one winding by itself is almost twice that of the two windings in parallel.

Wound-rotor motors are always started with full voltage, the starting current being controlled by the resistance external to the rotor.

## 5-13 ASYNCHRONOUS GENERATOR

An induction motor can be made to generate real power by driving it above synchronous speed while receiving ac excitation from the line. Because of this departure from synchronous speed, the machine is called an *asynchronous generator*. When driven above synchronous speed, the slip is negative and the equivalent resistance $r_2/s$ of the rotor circuit is negative, which means that the rotor absorbs negative real power from the stator (i.e., it furnishes the stator with real power). The asynchronous generator cannot generate reactive power; in fact, it requires reactive power from the line to furnish its excitation, since unlike the synchronous generator, it has no means for establishing an air-gap field with the stator open-circuited. Operation of the asynchronous generator requires synchronous machines, whether generators or motors, on the line to supply the asynchronous machine with its needed reactive power. It is this limitation of reactive-power requirements which restricts the use of asynchronous generators to a few rather unusual applications.

## 5-14 SINGLE-PHASE INDUCTION MOTORS

Single-phase induction motors have ratings from a small fraction of a horsepower up to about 10 hp. It is generally better to use three-phase induction motors than single-phase induction motors when the requirements exceed 1 hp, assuming that three-phase supplies are available. However, in many installations such as for domestic loads, only single-phase supplies are available, which precludes the use of conventional three-phase motors. Single-phase induction motors, like polyphase induction motors, have a nearly constant speed characteristic and are

† For other methods of starting induction motors, see D. R. Shoults et al., *Electric Motors in Industry* (New York: John Wiley & Sons Inc., 1952), Chap. V, and I. L. Kosow, *Electrical Machinery and Control* (Englewood Cliffs, N.J.: Prentice-Hall, Inc., 1964), pp. 318–325; P. L. Alger, *Induction Machines* (New York: Gordon and Breach, Inc., 1970), Chap. 8.

used for driving such household equipment as fans, refrigerators, washing machines, and oil burners.

A three-phase induction motor can operate as a single-phase induction motor once it is running by opening one of the stator phases. In order to make for effective utilization of materials in terms of rating, only about two-thirds of the slots per pole are normally occupied by the stator winding. If, instead, all the slots were filled, the rating of the stator winding would be increased in the ratio 1/0.866 or 1.15, while the amount of stator winding and copper losses would each be increased by 50 percent. The fact is that in filling the remaining one-half of the slots, the number of stator turns increases from $N$ to $3N/2$ and the breadth factor decreases according to Eq. 4-14 from

$$\frac{\sin (\pi/3)}{n \sin (\gamma/2)} \quad \text{to} \quad \frac{\sin (\pi/2)}{(3n/2) \sin (\gamma/2)}$$

Then by making use of Eq. 4-15, it is found that the stator voltage is increased in the ratio

$$\frac{E'}{E} = \frac{(3N/2) \sin (\pi/2)n \sin (\gamma/2)}{N \sin (\pi/3)(3n/2) \sin (\gamma/2)} = \frac{1}{\sin (\pi/3)} = 1.15$$

## 5-15 METHODS OF STARTING SINGLE-PHASE INDUCTION MOTORS

Single-phase induction motors generally have squirrel-cage rotors similar to those in polyphase induction motors. A single-phase induction motor cannot start as such, but once started in a given direction by mechanical or other means will develop torque in that direction and approach synchronous speed if the load torque is not excessive.

The most widely used methods for starting single-phase induction motors incorporate features for producing a rotating magnetic field at standstill and are classified as follows:

1. Split-phase motor.
2. Capacitor-start motor.
3. Shaded-pole motor.

The split-phase and capacitor-start motors have two stator windings (i.e., a main and an auxiliary winding) displaced from each other by 90° in electrical measure, thus to some degree simulating a two-phase winding. The auxiliary winding, which is used for starting only, generally has a smaller amount of wire than the main winding.

1. The main and auxiliary windings of the split-phase motor are connected in parallel during starting, and when the motor attains about 75 percent of rated speed, a centrifugal switch disconnects the auxiliary winding.

Figure 5-21(a) shows a schematic diagram. The auxiliary winding is of smaller wire size and usually of fewer turns than the main winding, as a result of which the auxiliary winding has a greater ratio of resistance to leakage reactance than the main winding. The current in the main winding therefore lags that in the auxiliary winding, as shown in Fig. 5-21(b), thus resulting in two stator mmfs displaced from each other in space phase as well as in time phase, both conditions being necessary for the production of a rotating magnetic field required for the production of torque.

2. The capacitor-start motor is similar to the split-phase motor except that it has a capacitor in series with the auxiliary winding, as shown in Fig. 5-21(c). The capacitor, generally a dry-type ac electrolytic capacitor,

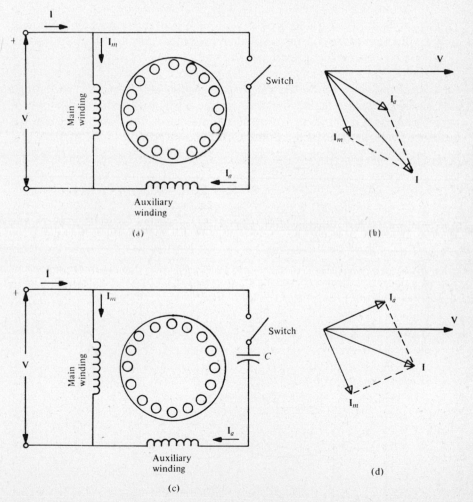

**Figure 5-21** (a) Split-phase motor connections. (b) Phasor diagram for split-phase starting. (c) Capacitor-start motor connections. (d) Phasor diagram for capacitor starting.

has a large value of capacitance ranging from about 70 to 400 $\mu$F for 115-V motors from $\frac{1}{8}$ to 1 hp, and makes it possible for the currents in the main and auxiliary windings to be displaced from each other by about 90° at starting, as shown by the phasor diagram in Fig. 5-21(d). When the windings are displaced 90° in electrical measure and their mmfs are equal in magnitude but 90° apart in time phase, the action is that of a polyphase motor, which represents optimum performance. However, electrolytic capacitors are suitable only for intermittent duty and for periods of a few seconds' duration, and the auxiliary winding must be opened as the motor approaches rated speed.

3. Figure 5-22 shows a schematic diagram of a four-pole shaded-pole induction motor with a concentrated stator winding. Each pole is recessed to accommodate a short-circuited coil usually of one turn and embracing about one-third of the pole. In some motors, two or even three shading coils are used on each pole, with each coil encompassing a different fraction of the pole face. Although the salient pole is the usual construction, other structures are used as well.

The short-circuit current induced in the shading coil causes the flux through the shaded portion to lag the flux through the unshaded portion of the pole in time phase. As a result there is a small component of flux sweeping across the pole face from the unshaded portion to the shaded portion in the manner of a

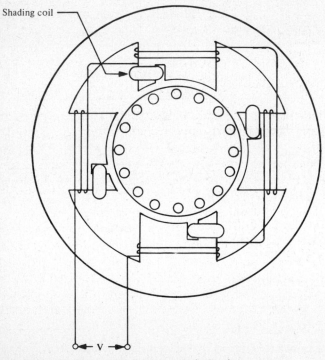

**Figure 5-22**   Four-pole shaded-pole induction motor.

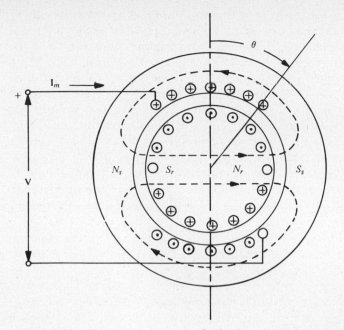

**Figure 5-23** Elementary two-pole single-phase induction motor at standstill.

rotating magnetic field. The losses in the shading coils at rated speed are sizable, resulting in performance inferior to that of other types of single-phase induction motor. The rating of shaded-pole motors is about 50 W or less.

## 5-16 TWO-REVOLVING-FIELD THEORY

The *two-revolving-field theory* is one of the two most common theories used for analyzing the steady-state behavior of the single-phase induction motor. The other, the *cross-field theory,* leads to substantially the same results as the two-revolving-field theory, but because the latter may be regarded as an extension of the concepts applied to the polyphase induction motor, only the two-revolving-field theory will be discussed in this text.†

Consider the elementary two-pole single-phase induction motor in Fig. 5-23 at standstill with its main winding excited only. Since the squirrel cage is a short-circuited winding, this motor at standstill is, in effect, a short-circuited transformer for which the equivalent circuit is shown in Fig. 5-24(a), in which the effect of the iron losses is omitted, to be taken into account subsequently by including them in the rotational losses. The dashed lines in Fig. 5-23 indicate the approximate flux path. When the directions of the instantaneous stator and rotor currents are as indicated, with the stator mmf greater than the rotor mmf,

---

† For a discussion of the cross-field theory, see A. F. Puchstein et al., *Alternating-Current Machines* (New York: John Wiley & Sons, Inc., 1954), Chap. 30.

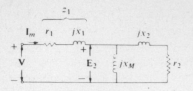

(a)

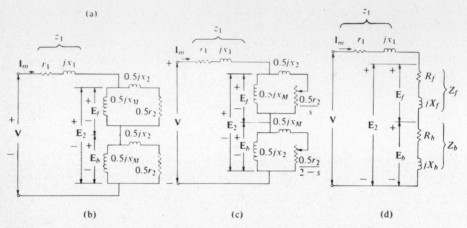

(b)                          (c)                          (d)

**Figure 5-24** Equivalent circuit of single-phase induction (main winding only). (a) and (b) Standstill. (c) Running, detailed circuit. (d) Simplified circuit.

the stator poles are as indicated by $N_s$ and $S_s$ and the rotor poles as indicated by $N_r$ and $S_r$. The magnetic axes of the rotor and the stator therefore coincide, resulting in zero torque. If the main stator winding has $N_m$ turns and a winding factor $k_{wm}$ and Eq. 4-23 is modified accordingly, the fundamental component of the stator mmf at the space angle $\theta$ in Fig. 5-23 is

$$\mathcal{F}_\theta = \frac{4}{\pi} \frac{k_{wm} N_m}{2} i_m \sin \theta \tag{5-55}$$

where $i_m$ is the instantaneous current in the stator winding, which may be expressed in terms of the effective stator current by

$$i_m = \sqrt{2} I_m \sin (\omega t - \theta_M) \tag{5-56}$$

where $\theta_M$ is the angle by which $I_m$ lags the applied stator voltage $v = \sqrt{2} V \sin \omega t$. Substitution of Eq. 5-56 in Eq. 5-55 followed by the procedure in Sec. 4-5 which led to Eq. 4-27 results in a similar expression for two equal mmfs rotating at synchronous speed in opposite directions:

$$\boxed{\mathcal{F}_\theta = \tfrac{1}{2} F_{\text{amp}} \cos (\theta - \omega t + \theta_M) + \tfrac{1}{2} F_{\text{amp}} \cos (\theta + \omega t - \theta_M)} \tag{5-57}$$

where $F_{\text{amp}} = 0.9 k_{wm} N_m I_m$. In the case of a $P$-pole machine with $a$ current paths, the mmf is found by letting $m = 1$ in Eq. 4-38. The first term on the right-hand side of Eq. 5-57 represents the forward-rotating mmf ($\theta$ increasing) and the second term represents the backward-rotating mmf.

At standstill these two mmfs produce equal fluxes rotating in opposite directions at synchronous speed, each of the two fluxes producing rotor currents in a manner similar to that in which the rotating flux in a polyphase induction motor produces rotor current. Hence, two equal components of torque opposing each other are produced, with a resultant torque of zero. The equivalent circuit in Fig. 5-24(a) is rearranged in Fig. 5-24(b) to take into account the effect of the forward and backward rotating fluxes, which produce the voltages $E_f$ and $E_b$ and which equal each other at standstill.

Now assume that the rotor be made to rotate at a speed of $n$ rpm in the forward direction. Then the slip relative to the forward rotating flux is the same as that in the polyphase motor:

$$s = \frac{n_{\text{syn}} - n}{n_{\text{syn}}} \tag{5-58}$$

However, because the direction of rotation is opposite that of the backward-rotating flux, the sign of $n$ must be changed in Eq. 5-58 to obtain the backward slip:

$$s_b = \frac{n_{\text{syn}} + n}{n_{\text{syn}}} \tag{5-59}$$

which can be expressed in terms of the forward slip by taking the sum of Eqs. 5-58 and 5-59 to yield

$$s_b = 2 - s \tag{5-60}$$

When the slip as expressed by Eqs. 5-58 and 5-60 is taken into account, the result is the equivalent circuit in Fig. 5-24(c), which represents the motor running on the main winding, which may be simplified as shown in Fig. 5-24(d).

## 5-16.1 Torque

The resultant electromagnetic torque $T_{em}$ is the difference between the torques developed by the forward and backward rotating torques $T_f$ and $T_b$:

$$T_{em} = T_f - T_b \tag{5-61}$$

As in the case of the polyphase induction motor, the forward torque $T_f$ equals the real power input of the forward rotating field to the rotor divided by the synchronous angular velocity. This is also true of the backward torque $T_b$. Then, on the basis of the equivalent circuit in Fig. 5-24(d),

$$T_{em} = \frac{I_m^2(R_f - R_b)}{\omega_{\text{syn}}} \tag{5-62}$$

and the electromagnetic power is

$$P_{em} = (1 - s)\omega_{syn}T_{em} = (1 - s)I_m^2(R_f - R_b) \qquad (5\text{-}63)$$

The net power output is obtained by subtracting the rotational losses, which include the core losses, from the electromagnetic power as follows:

$$\boxed{P_{mech} = P_{em} - P_{rot}} \qquad (5\text{-}64)$$

**EXAMPLE 5-6**

The following constants are for a $\frac{1}{4}$-hp 60-Hz 115-V four-pole capacitor start motor:

$$r_1 = 2.15 \ \Omega \qquad r_2 = 4.45 \ \Omega$$

$$x_1 = 3.01 \ \Omega \qquad x_2 = 2.35 \ \Omega$$

$$x_M = 70.5 \ \Omega$$

Core loss = 26.0 W, windage and friction loss = 14.0 W. Calculate for a slip of 0.05 the (a) current, (b) power factor, (c) output, (d) torque, and (e) efficiency.

*Solution*

a. The current based on the equivalent circuit in Fig. 5-24(d) is

$$\mathbf{I}_m = \frac{\mathbf{V}}{z_1 + Z_f + Z_b} \text{ A}$$

where $z_1 = r_1 + jx_1 = 2.15 + j3.01$, and where $Z_f$ and $Z_b$ are found by making use of Fig. 5-24(c) as follows:

$$\begin{aligned}
Z_f = R_f + jX_f &= \frac{[(0.5r_2/s) + j0.5x_2)]\,j0.5x_M}{(0.5r_2/s) + j(0.5x_2 + 0.5x_M)} \\
&= \frac{(44.5 + j1.175)\,j35.25}{44.5 + j36.43} = 27.2\underline{/52.1^\circ} \\
&= 16.72 + j21.45 \ \Omega \\
Z_b = R_b + jX_b &= \frac{\{[0.5r_2/(2 - s)] + j0.5x_2\}\,j0.5x_M}{[0.5r_2/(2 - s)] + j(0.5x_2 + 0.5x_M)} \\
&= \frac{(1.14 + j1.175)\,j35.25}{1.14 + j36.43} = 1.58\underline{/47.6^\circ} \\
&= 1.06 + j1.17 \ \Omega
\end{aligned}$$

and $\quad z_1 + Z_f + Z_b = 2.15 + j3.01 + 16.72 + j21.45 + 1.06 + j1.17$

$$= 19.93 + j25.63 = 32.5\underline{/52.1^\circ} \ \Omega$$

$$\mathbf{I}_m = \frac{115}{32.5\underline{/52.1^\circ}} = 3.54\underline{/-52.1^\circ} \text{ A}$$

b. The power factor is cos 52.1° = 0.615.
c. The developed power, from Eq. 5-63, is

$$P_{em} = (1 - s)I_m^2(R_f - R_b)$$

$$= (0.95)(3.54)^2(16.72 - 1.06) = 186 \text{ W}$$

The mechanical power output is

$$P_{\text{out}} = P_{em} - P_{\text{rot}}$$

and the rotational losses are

$$P_{\text{rot}} = 26 + 14 = 40 \text{ W}$$

Hence,

$$P_{\text{out}} = 186 - 40 = 146 \text{ W} \quad \text{or} \quad 0.196 \text{ hp}$$

d.     $\text{Torque} = \dfrac{P_{\text{out}}}{(1 - s)\omega_{\text{syn}}} = \dfrac{146}{0.95[(2\pi \times 1800)/60]}$

$$= 0.816 \text{ N-m/rad}$$

e.     $\text{Efficiency} = \dfrac{\text{output}}{\text{input}}$

$$\text{Input} = VI_m \cos \theta = 115 \times 3.54 \times 0.615 = 250 \text{ W}$$

$$\text{Efficiency} = \frac{146}{250} = 0.584 = 58.4 \text{ percent}$$

## 5-16.2 Double-Frequency Torque

While the equivalent circuit in Fig. 5-24 serves as a convenient basis for determining the operating characteristics of the single-phase induction motor as discussed in the foregoing, it fails to account for the double-frequency-torque component resulting from the interaction of the two revolving fields. The equivalent circuit represents two polyphase induction motors with their stator windings connected in series in a manner such that positive-sequence voltage is applied to one and negative-sequence voltage to the other, and with their rotors coupled mechanically, each of the two rotors thus reacting only to its own air-gap field. The rotor of the single-phase induction motor rotating in the presence of both revolving air-gap fields produces a double-frequency torque component in addition to the steady torque expressed by Eq. 5-62.

Figure 5-9 for the polyphase induction motor can also represent the air-gap flux and rotor current associated with *one* or the *other* rotating fields on the basis of which the torque is expressed by

$$T_{em} = k\phi_{\text{amp}} I_{2\text{amp}} \cos \theta_2$$

where $\phi_{\text{amp}}$ and $I_{2\text{amp}}$ are the amplitudes of the flux wave and pattern of the rotor current displaced from each other by the angle $\theta_2$. However, if $\phi_f$ and $I_b$ are the amplitudes of the forward-rotating flux and backward-rotating rotor-current pattern, then

$$\theta_2 = 2\omega t + \alpha$$

and the resulting component of torque is

$$T_{fb} = k\phi_f I_b \cos (2\omega t + \alpha)$$

where $\omega = 2\pi f$.

The interaction of the backward-rotating flux and forward-rotating rotor-current pattern produces another double-frequency component in the same manner.

The double-frequency torque has an average value of zero but is objectionable because it produces vibration and noise, both of which can be reduced by means of flexible mountings. Flexible couplings are sometimes used to reduce the vibration transmitted to the load.

The utility of the equivalent circuit goes beyond that of making performance calculations as in Example 5-6. The equivalent circuit also serves to bring into focus the physical concepts basic to induction machines, which, in fact, lead to the equivalent circuit itself.

## 5-17 NO-LOAD AND LOCKED-ROTOR TESTS ON THE SINGLE-PHASE INDUCTION MOTOR

The no-load and locked-rotor tests are similar to those made on the polyphase induction motor. However, except for the capacitor-run motor, these tests are made with the auxiliary winding kept open.

### 5-17.1 No-Load Test

The no-load test is made by running the motor without load at rated voltage and rated frequency. Since the no-load slip is small, the resistance $0.5r_2/s$ is correspondingly large and is assumed to be infinite. On the other hand, the resistance $0.5r_2/(2 - s)$ associated with the backward-rotating field is small enough so that the backward magnetizing current may be neglected, which results in the equivalent circuit of Fig. 5-25(a) and for which the total series reactance is

$$x_0 = x_1 + 0.5x_M + 0.5x_2 \qquad (5\text{-}65)$$

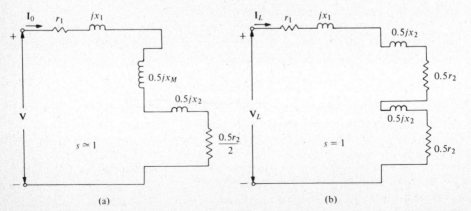

**Figure 5-25**  Equivalent circuit of a single-phase induction motor. (a) No-load test. (b) Locked-rotor test.

### 5-17.2  Locked-Rotor Test

The calculations based on the locked-rotor data can be greatly simplified by neglecting the magnetizing current, which in the case of the motor in Example 5-6 is less than 3.5 percent of the equivalent rotor current at a standstill. Then the equivalent circuit for the locked-rotor test is reduced to that shown in Fig. 5-25(b). It is assumed that $x_1 = x_2$, from which

$$x_1 = x_2 = \frac{x_L}{2} \tag{5-66}$$

where $x_L$ is the locked-rotor reactance.

The magnetizing reactance is found to be, from Eqs. 5-65 and 5-66,

$$x_M = 2x_0 - 1.5x_L \tag{5-67}$$

### 5-17.3  Winding Resistance Test

The resistance $r_1$ of the main winding is measured with dc. The rotor resistance $r_2$ is determined in somewhat the same manner as for the polyphase induction motor, as follows. The equivalent series resistance of the motor with the rotor locked is

$$r_L = \frac{P_L}{I_L^2} \tag{5-68}$$

where $P_L$ and $I_L$ are the power and current input to the motor when the rotor is locked by making use of the equivalent circuit in Fig. 5-24(a). Following the procedure in Sec. 5-7 the same relationship as expressed by Eq. 5-54 for the polyphase motor is found to apply to the main winding of the single-phase induction motor:

$$r_2 = (r_L - r_1)\left(\frac{x_2 + x_M}{x_M}\right)^2 \tag{5-69}$$

The rotational losses are obtained by subtracting the stator and rotor copper losses from the no-load power input and neglecting the backward magnetizing current. The copper losses are then expressed by

$$P_c = I_0^2[r_1 + 0.5(r_L - r_1)]$$
$$= 0.5I_0^2(r_1 + r_L) \tag{5-70}$$

and the rotational losses by

$$P_{rot} = P_0 - P_c \tag{5-71}$$

## 5-18  THE CAPACITOR MOTOR

There are three types of capacitor motors. One of these, the capacitor-start motor treated in Sec. 5-15, uses its auxiliary winding and associated capacitor during starting only. The other two types, the permanent-split or single-value capacitor motor and the two-value capacitor motor, have their auxiliary windings energized not only during starting but throughout the entire operation of the motor, which of course is true of the main winding also.

A permanent-split capacitor motor has the same value of capacitance for both running and starting conditions. Such a motor requires an oil-impregnated-paper type of capacitor, since the less expensive electrolytic capacitor is not suited for continuous operation. Although such motors have good running characteristics, their starting torque is low and they are used principally for fans or blowers.

A two-value capacitor motor uses one value of capacitance for starting and a lower value for running. For example, a $\frac{1}{2}$-hp motor has an electrolytic capacitor of 250 $\mu$F for starting and an oil-impregnated-paper capacitor of 15 $\mu$F for running. This feature enables the motor to develop good starting torque without sacrificing running performance. Common arrangements for these two types of motors are shown in Fig. 5-26.

### 5-18.1 Equivalent Circuit of the Capacitor Motor Based on the Two-Revolving-Field Theory

The current in the auxiliary winding as well as that in the main winding produces equal and oppositely rotating mmfs. If the mmf of the main winding, known as the *main phase,* had the same magnitude as the auxiliary winding, known as the *auxiliary phase,* and if the current in the two windings were 90° out of time phase, two-phase operation would result, since the two backward-rotating fields would cancel each other, just as the backward-rotating mmfs in the three-phase motor cancel each other. The main and auxiliary windings in capacitor motors are displaced from each other by 90° in electrical measure, and since the air gap is practically uniform, there is no appreciable mutual inductance between these two windings. The voltage equations for the main and auxiliary phases can be written as

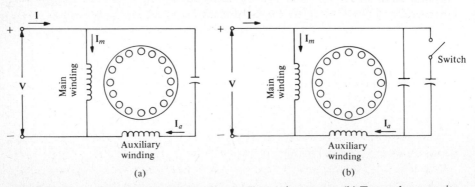

(a)                                                                  (b)

**Figure 5-26**    (a) Permanent split (single-value) capacitor motor. (b) Two-value capacitor motor.

$$\mathbf{V}_m = (r_1 + jx_1)\mathbf{I}_m + \mathbf{E}_{2m} \tag{5-72}$$

$$\mathbf{V}_a = (r_{1a} + jx_{1a} + R_c + jX_c)\mathbf{I}_a + \mathbf{E}_{2a} \tag{5-73}$$

where $r_1 + jx_1 = z_1$, the leakage impedance of the main winding; $r_{1a} + jx_{1a} = z_{1a}$, the leakage impedance of the auxiliary winding; and $R_c + jX_c = Z_c$, the equivalent series impedance of the capacitor. The voltages $\mathbf{E}_{2m}$ and $\mathbf{E}_{2a}$ are induced by the rotating air-gap fluxes. The voltage $\mathbf{E}_{2m}$ not only results from the two revolving fields produced by the main winding but also contains components due to the two revolving fluxes produced by the auxiliary winding. This can be readily understood by considering the motor as running in the forward direction on its auxiliary phase with the main phase open. The forward direction of rotation in a capacitor motor is from the magnetic axis of its auxiliary phase toward that of the main phase, because the current in the auxiliary phase leads that in the main phase. With the main phase open $\mathbf{I}_m = 0$, and for that condition Eqs. 5-72 and 5-73 become

$$\mathbf{V}_m = \mathbf{E}_{2m} \tag{5-74}$$

$$\mathbf{V}_a = (r_{1a} + jx_{1a} + R_c + jX_c)\mathbf{I}_a + \mathbf{E}_{2a} \tag{5-75}$$

where

$$\mathbf{E}_{2a} = \mathbf{E}_{fa} + \mathbf{E}_{ba} \tag{5-76}$$

in which $\mathbf{E}_{fa}$ and $\mathbf{E}_{ba}$ are the voltages induced in the auxiliary phase due to its own forward- and backward-rotating fluxes. Since the main phase is displaced in the direction of rotation from the auxiliary phase by 90° in electrical measure, the voltage induced in it by the forward-rotating flux must lag by 90° that voltage induced in the auxiliary phase by the same flux. The opposite is true for the effect of the backward-rotating flux. Hence,

$$\mathbf{E}_{2m} = -j\frac{\mathbf{E}_{fa}}{a} + j\frac{\mathbf{E}_{ba}}{a} \tag{5-77}$$

where $a$ is the turns ratio of the auxiliary to main phase, taking into account winding factors and the number of current paths. When both stator windings are energized, the effect of all fluxes must be taken into account, which results in

$$\mathbf{V}_m = z_1\mathbf{I}_m + \mathbf{E}_{fm} - j\frac{\mathbf{E}_{fa}}{a} + \mathbf{E}_{bm} + j\frac{\mathbf{E}_{ba}}{a} \tag{5-78}$$

$$\mathbf{V}_a = (z_{1a} + Z_c)\mathbf{I}_a + \mathbf{E}_{fa} + ja\mathbf{E}_{fm} + \mathbf{E}_{ba} - ja\mathbf{E}_{bm} \tag{5-79}$$

Figure 5-27 shows an equivalent circuit for the main and auxiliary phases based on Eqs. 5-78 and 5-79. It should be noted that the circuit in Fig. 5-27(a) is that of Fig. 5-24(b) modified to include the voltages induced in the main phase by the fluxes due to the auxiliary phase. In both cases the rotor quantities $R_f$, $X_f$, $R_b$, and $X_b$ are all referred to the main phase. These same quantities are

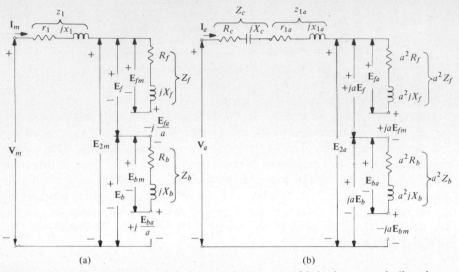

(a)                                           (b)

**Figure 5-27**  Equivalent circuits of a capacitor motor with both stator windings in operation. (a) Main phase. (b) Auxiliary phase.

incorporated in the equivalent circuit in Fig. 5-27(b) for the auxiliary phase by making use of the impedance ratio $a^2$. On the basis of these equivalent circuits, Eqs. 5-78 and 5-79 can be reduced to the following more convenient forms:

$$\mathbf{V}_m = (z_1 + Z_f + Z_b)\mathbf{I}_m - ja(Z_f - Z_b)\mathbf{I}_a \tag{5-80}$$

$$\mathbf{V}_a = ja(Z_f - Z_b)\mathbf{I}_m + [Z_c + z_{1a} + a^2(Z_f + Z_b)]\mathbf{I}_a \tag{5-81}$$

The two phases of the capacitor motor are normally connected in parallel and then

$$\mathbf{V}_a = \mathbf{V}_m = \mathbf{V} \tag{5-82}$$

and the total current supplied to the motor is

$$\mathbf{I} = \mathbf{I}_m + \mathbf{I}_a \tag{5-83}$$

## 5-18.2  Torque

Equation 5-62 shows that the torque due to one stator phase is equal to the angular synchronous velocity divided into the difference between the amount of power transferred across the air gap to the rotor by the oppositely rotating fields. This difference is for both phases:

$$P_{gf} - P_{gb} = \text{Re}\,[(\mathbf{E}_f - \mathbf{E}_b)\mathbf{I}_m^* + j(\mathbf{E}_f - \mathbf{E}_b)a\mathbf{I}_a^*] \tag{5-84}$$

where the terms in the brackets on the right-hand side of Eq. 5-84 represent the complex power in accordance with Eq. 1-36. Equation 5-84 can be rewritten on the basis of Fig. 5-27 as

$$P_{gf} - P_{gb} = (R_f - R_b)[I_m^2 + (aI_a)^2] + \text{Re} \left( -j\frac{E_{fa}}{a} - j\frac{E_{ba}}{a} \right) I_m^*$$

$$+ \text{Re} \, (jaE_{fm} + jaE_{bm})I_a^* \tag{5-85}$$

The second and third terms on the right-hand side of Eq. 5-85 can be combined by means of the following procedure:

$$\frac{-j}{a} (E_{fa} + E_{ba})I_m^* = -ja[R_f + R_b + j(X_f + X_b)]I_a I_m^*$$

$$= [(X_f + X_b) - j(R_f + R_b)]aI_a I_m^* \tag{5-86}$$

and $\qquad ja(E_{fm} + E_{bm})I_a^* = ja[R_f + R_b + j(X_f + X_b)]I_a^* I_m$

$$= -[(X_f + X_b) - j(R_f + R_b)]aI_a^* I_m \tag{5-87}$$

The *real* of a phasor sum is equal to the sum of the reals of the phasors, and the sum of Eqs. 5-86 and 5-87 is

$$\frac{-j}{a} (E_{fa} + E_{ba})I_m^* + ja(E_{fm} + E_{bm})I_a^*$$

$$= [(X_f + X_b) - j(R_f + R_b)]a(I_a I_m^* - I_a^* I_m) \tag{5-88}$$

Now if

$$I_a = I_a \varepsilon^{j\theta_a} \quad \text{and} \quad I_m = I_m \varepsilon^{j\theta_m}$$

then

$$I_a I_m^* - I_a^* I_m = I_a I_m (\varepsilon^{j(\theta_a - \theta_m)} - \varepsilon^{-j(\theta_a - \theta_m)})$$

$$= j2I_a I_m \sin (\theta_a - \theta_m)$$

Hence,

$$\text{Re} \left( -j\frac{E_{fa}}{a} - j\frac{E_{ba}}{a} \right) I_m^* + \text{Re} \, (jaE_{fm} + jaE_{bm})I_a^*$$

in Eq. 5-85 can be reduced to

$$2aI_a I_m (R_f + R_b) \sin (\theta_a - \theta_m)$$

and Eq. 5-85 can be rewritten as

$$\boxed{\begin{aligned} P_{gf} - P_{gb} = & [I_m^2 + (aI_a)^2](R_f - R_b) \\ & + 2aI_a I_m (R_f + R_b) \sin (\theta_a - \theta_m) \end{aligned}} \tag{5-89}$$

The developed torque is

$$T_{em} = \frac{P_{gf} - P_{gb}}{\omega_{syn}} \tag{5-90}$$

It should be emphasized that, since Eq. 5-89 is expressed in terms of the currents $I_m$ and $I_a$, it is valid regardless of whether the main and auxiliary circuits are connected in parallel or whether they are excited from two independent sources.

Indeed, they apply also whether the auxiliary winding is in series with a capacitor or not.

**EXAMPLE 5-7**

The constants of a $\frac{1}{4}$-hp 115-V 60-Hz four-pole two-value capacitor motor are as follows:

Ratio of series turns in auxiliary winding to main winding, $a = 1.18$

Main winding: $r_1 = 2.20 \ \Omega$     $x_1 = 3.05 \ \Omega$     $x_M = 73.0 \ \Omega$

Auxiliary winding: $r_{1a} = 7.80 \ \Omega$     $x_{1a} = 3.52 \ \Omega$

Rotor referred to main winding: $r_2 = 4.50 \ \Omega$     $x_2 = 2.32 \ \Omega$

Starting capacitor: $R_c = 3.10 \ \Omega$     $X_c = -14.7 \ \Omega$

Running capacitor: $R_c = 2.00 \ \Omega$     $X_c = -180 \ \Omega$

No-load core loss = 23 W

No-load windage and friction = 15 W

With the running capacitor in the circuit and for a slip of 0.06, calculate (a) the current in each stator winding, (b) the line current, (c) the power factor of the motor, (d) the voltage across the capacitor, (e) the developed torque, and (f) the mechanical power output.

*Solution.* The stator currents are found by solving Eqs. 5-80 and 5-81:

$$\mathbf{V}_m = (z_1 + Z_f + Z_b)\mathbf{I}_m - ja(Z_f - Z_b)\mathbf{I}_a \text{ V}$$

$$\mathbf{V}_a = ja(Z_f - Z_b)\mathbf{I}_m + [Z_c + z_{1a} + a^2(Z_f + Z_b)]\mathbf{I}_a \text{ V}$$

$$z_1 = 2.20 + j3.04 \ \Omega$$

$$Z_f = \frac{j0.5x_M[(r_2/s) + jx_2]}{(r_2/s) + j(x_2 + x_M)} \quad \text{[from Fig. 5-24(c)]}$$

$$= \frac{j0.5 \times 73.0[(4.50/0.06) + j2.32]}{(4.50/0.06) + j(2.32 + 73.0)} = \frac{j36.5(2.32 + j2.32)}{2.32 + j75.32}$$

$$= 25.7\underline{/46.6°} = 17.67 + j18.7 \ \Omega$$

$$Z_b = \frac{j0.5x_M\{[r_2/(2 - s)] + jx_2\}}{[r_2/(2 - s)] + j(x_2 + x_M)} = \frac{j36.5(2.32 + j2.32)}{2.32 + j75.32}$$

$$= 1.59\underline{/46.8°} = 1.09 + j1.16 \ \Omega$$

$$z_1 + Z_f + Z_b = 20.96 + j22.91 = 31.1\underline{/47.55°} \ \Omega$$

$$ja(Z_f - Z_b) = j1.18(16.58 + j17.54)$$

$$= -20.67 + j19.55 = -28.5\underline{/-43.4°}$$

$$Z_c + z_{1a} + a^2(Z_f + Z_b) = 2.00 - j180 + 7.80 + j3.52$$

$$+ (1.18)^2(18.76 + j19.86)$$

$$= 35.95 - j148.8 = 153.0\underline{/-76.4°}$$

a. The currents in Eqs. 5-80 and 5-81 are expressed in determinant form by

$$
I_m = \frac{\begin{vmatrix} V_m & -ja(Z_f - Z_b) \\ V_a & Z_c + z_{1a} + a^2(Z_f + Z_b) \end{vmatrix}}{\begin{vmatrix} z_1 + Z_f + Z_b & -ja(Z_f - Z_b) \\ ja(Z_f - Z_b) & Z_c + z_{1a} + a^2(Z_f + Z_b) \end{vmatrix}} \quad A \tag{1}
$$

and

$$
I_a = \frac{\begin{vmatrix} z_1 + Z_f + Z_b & V_m \\ ja(Z_f - Z_b) & V_a \end{vmatrix}}{\begin{vmatrix} z_1 + Z_f + Z_b & -ja(Z_f - Z_b) \\ ja(Z_f - Z_{a1}) & Z_c + z_{1a} + a^2(Z_f + Z_b) \end{vmatrix}} \quad A \tag{2}
$$

Substitution of the phasor values of the impedances and applied stator voltages in Eq. 1 yields

$$
I_m = \frac{\begin{vmatrix} 115\underline{/0°} & 28.5\underline{/-43.4°} \\ 115\underline{/0°} & 153.0\underline{/-76.4°} \end{vmatrix}}{\begin{vmatrix} 31.1\underline{/47.55°} & 28.5\underline{/-43.4°} \\ -28.5\underline{/-43.4°} & 153.0\underline{/-76.4°} \end{vmatrix}}
$$

$$
= \frac{115\underline{/0°}(153.0\underline{/-76.4°} - 28.5\underline{/-43.4°})}{(31.1\underline{/47.55°})(153.0\underline{/-76.4°}) + (28.5\underline{/-43.4°})^2}
$$

$$
= \frac{115(35.95 - j148.8 - 20.67 + j19.55)}{4170 - j2290 + 45 - j806}
$$

$$
= \frac{115 \times 130\underline{/-83.3°}}{5240\underline{/-36.3°}}
$$

$$
= 2.86\underline{/-47.0°} = 1.95 - j2.085 \ A
$$

Also,

$$
I_a = \frac{\begin{vmatrix} 31.1\underline{/47.55°} & 115\underline{/0°} \\ -28.5\underline{/-43.4°} & 115\underline{/0°} \end{vmatrix}}{D} \tag{3}
$$

where $D = 5240\underline{/-36.3°}$, the denominator in Eq. 1,

$$
I_a = \frac{115(31.1\underline{/47.55°} + 28.5\underline{/-43.4°})}{5240\underline{/-36.3°}}
$$

$$
= \frac{115(41.63 + j3.36)}{5240\underline{/-36.3°}} = \frac{115 \times 41.9\underline{/4.6°}}{5240\underline{/-36.3°}}
$$

$$
= 0.92\underline{/40.9°} = 0.695 + j0.612 \ A
$$

b. The line current (i.e., the current taken by the motor) is

$$\mathbf{I} = \mathbf{I}_m + \mathbf{I}_a = 1.95 - j2.085 + 0.695 + j0.612$$
$$= 2.65 - j1.47 = 3.03\underline{/-29.1°} \text{ A}$$

c. The power factor is $\cos 29.1° = 0.875$.
d. The voltage across the capacitor is

$$E_c = Z_c I_a = (\sqrt{(2)^2 + (180)^2})0.92 = 166 \text{ V}$$

e. The net power transferred across the air gap, from Eq. 5-89, is

$$P_{gf} - P_{gb} = [(2.86)^2 + (1.18 \times 0.92)^2](17.67 - 1.09)$$
$$+ 2 \times 1.18 \times 2.86 \times 0.92(17.67 + 1.09) \sin 87.9°$$
$$= 155 + 116 = 271 \text{ W}$$

The developed torque, from Eq. 5-90, is

$$T_{em} = \frac{271}{2\pi \times 1800/60} = 1.435 \text{ N-m/rad}$$

f. The developed mechanical power is

$$P_{em} = (1 - s)\omega_{syn}T_{em}$$
$$= (1 - 0.06)271 = 254.5 \text{ W}$$

and the net mechanical power is

$$P_{mech} = P_{em} - P_{rot} = 254.5 - (23 + 15)$$
$$= 216.5 \text{ W}$$

## STUDY QUESTIONS

1. The principle of the induction motor was discovered by Arago in 1824. Describe a simple experiment that might be performed to demonstrate the principle upon which the induction motor operates.

2. Explain why the rotor of an induction motor cannot move as rapidly as the revolving field.

3. Describe the construction of an induction motor by describing the stator, a squirrel-cage rotor, and a wound type of rotor.

4. Describe the advantages of a cast aluminum squirrel-cage rotor.

5. Describe why the rotor bars are skewed in a squirrel-cage rotor.

6. When would a wound-rotor motor be used?

7. List the reasons why a fractional-pitch stator winding would be used in induction motors.

8. Explain how a revolving field is produced in a three-phase induction motor.

9. How can the direction of this revolving field be reversed?

10. How is the speed of the revolving field calculated?

11. What is meant by slip?

12. Assume that a rotor is rotating at synchronous speed. What voltage would be generated in the rotor?

13. What determines the voltage generated in the rotor if the rotor is locked?

14. What frequency of voltage is induced when the rotor is locked? Revolving at normal speed?

15. How is the rotor current in a given machine determined?

16. Why is it possible to represent electrically the mechanical load of an induction motor? Is there any advantage in doing this, and if so, what?

17. What is the difference between the rotor power input and the rotor power developed?

18. Why is the rotor power developed higher than the actual power applied to the mechanical load? (Sometimes referred to as the shaft power.)

19. Describe the following terms: normal torque, starting torque, and maximum torque.

20. What effect does the locked-rotor reactance have on the maximum torque that can be developed by an induction motor?

21. What practices are generally followed in the construction of an induction motor to minimize rotor reactance?

22. Make a list of the losses in an induction motor.

23. Why is it necessary to subtract wattmeter readings when the two-wattmeter method of measuring power is used to measure the no-load input to an induction motor?

24. When performing a no-load test, what losses are measured?

25. Outline a procedure that could be followed to calculate the following important characteristics of an induction motor: horsepower output, torque, percent efficiency, and power factor.

26. Describe how a locked-rotor test is performed.

27. During the performance of a locked-rotor test, what important constants are obtained?

28. At what speed will maximum torque occur for an induction motor?

29. The starting period is the most severe period in the operation of a motor. Why?

30. What methods may be used to reduce the starting current in a squirrel-cage or wound-rotor motor?

31. Would there be any problem with starting a squirrel-cage induction motor at reduced voltage?

32. List methods used to start polyphase squirrel-cage motors.

33. Describe the construction of a double-cage rotor.

34. Tell why the starting current is minimized and the starting torque is high when a double squirrel-cage rotor is used in the construction of an induction motor.

35. Explain why a squirrel-cage rotor motor will have good efficiency under normal operating conditions.

36. Locate a diagram of a full-voltage automatic starter for a squirrel-cage induction motor, and explain its operation.

37. Locate a diagram for a compensator method for starting an induction motor, and explain its operation.

38. Locate a line-resistance starting method for an automatic starter for an induction motor, and explain its operation.

39. If an induction motor is started by a wye-delta method, what voltage is impressed on the stator winding per phase at the starting instant?

40. Locate a diagram for an automatic starter using a wye-delta method and explain its operation.

41. What is meant by part-winding starting of an induction motor?

42. Draw a part-winding automatic starter for an induction motor.

43. Describe the wound-rotor method for starting an induction motor.

44. What value of resistance should be used by the automatic starter when starting a wound-rotor induction motor?

45. List three important factors that affect the operating characteristics of induction motors.

46. List the effects on the starting torque and efficiency of increasing the rotor resistance of an induction motor.

47. Describe the effect upon the starting current and maximum torque if the rotor reactance is increased in an induction motor.

48. What effect will deep narrow rotor slots have on the rotor reactance? Give reasons for your answer.

49. Why is a narrow air gap important to the operation of an induction motor?

50. What factors affect the size of the air gap?

51. List four different classes of squirrel-cage motors and their starting currents and torques when full voltage is impressed.

52. Make a list of the practical application for each of the four classes of squirrel-cage motors.

53. In some cases an induction motor becomes a part of the device itself, and this is sometimes called a shaftless motor. Can you describe the advantages of such motors, and list practical applications.

54. When would one find it advantageous to use a wound-rotor type of induction motor?

55. If a resistance controller is used for induction motors, what happens to the power consumed by this controller?

56. Describe the wound-rotor method of speed control.

57. Explain the principle of the operation of dynamic braking as applied to the stopping of an induction motor.

58. What advantages are possessed by an electric brake system compared with mechanical brakes?

59. Make a sketch of the operation of an automatic controller design for starting and dynamic braking service for an induction motor.

60. Describe how the principle of operation of plugging can be used for the stopping of an induction motor.

61. Speed control of a polyphase induction motor is discussed in this chapter. What are the advantages and disadvantages of variable-frequency speed control?

62. What is meant by a fractional-horsepower motor?

63. Explain why a motor having a rating greater or less than 1 hp is not necessarily classified as a fractional-horsepower motor.

64. Under what conditions must single-phase motors be used?

65. What is a universal motor?

66. Why is it called a universal motor?

67. Explain the principle of operation of a universal motor.

68. Why is high speed often desirable in the operation of a small motor such as the universal motor?

69. What limits the no-load speed of a universal motor?

70. Why does the effect of armature reaction tend to increase the speed of a series motor?

71. For alternating current what factors are responsible for the change in speed when load changes on a universal motor?

72. Why are the rotor slots of universal motors skewed?

73. What is normally meant by small motors?

74. Why is a gearbox frequently used on a universal motor?

75. List as many applications as you can where a gearbox is used with a universal motor.

76. What is a constant-speed governor? Describe the principle of operation when used in conjunction with a universal motor.

77. Give several practical examples of the use of a constant-speed governor in combination with a universal motor.

78. Why does the operation of a universal motor interfere with radio or TV reception? How can this interference be eliminated?

79. Why is a shaded-pole motor an induction motor?

80. Why is the field produced by a shaded-pole motor not a true revolving field in the same way as that created by a polyphase induction motor?

81. Describe the construction of a stator of a shaded-pole motor. Make a sketch of this construction.

82. In what direction will the rotor of a shaded-pole motor rotate?

83. Draw a sketch of a shaded pole and explain how the center of the pole shifts across the face of the pole as the ac sinusoidal excitation of the pole goes through a half-cycle.

84. Explain how a simple shaded-pole motor can be reversed.

85. Make a sketch showing how a shaded-pole motor designed with two sets of shading poles can be reversed.

86. Make a sketch showing how a shaded-pole motor with two sets of stator windings can be reversed.

87. List as many practical applications as you can of shaded-pole motors.

88. How is the speed of a shaded-pole motor controlled?

89. Describe the construction of a stator of a reluctance-start motor.

90. In what direction will the squirrel cage of a reluctance-start motor rotate?

91. Explain the shifting of the field from one side to the other of the pole of the reluctance-start motor.

92. How is the speed of reluctance-start motors controlled?

93. What is meant by a split-phase motor?

94. Describe the construction of the stator of a split-phase motor.

95. Why is a centrifugal switch used in a split-phase motor and what purpose does it serve?

96. Explain why the fluxes created by the main and auxiliary winding are out of phase in both time and space for a split-phase motor.

97. Discuss similarity of the constructional nature of split-phase motors and shaded-pole motors.

98. Compare a split-phase motor with a shaded-pole motor for the following factors: efficiency, starting current, overload capacity, starting torque, cost of manufacture, percent regulation, and percent slip.

99. Explain why it is desirable to disconnect the auxiliary winding in a split-phase motor after the rotor reaches about 75 to 80 percent of rated speed.

100. Explain how the direction of rotation of a split-phase motor may be predetermined.

101. How may a split-phase motor be reversed?

102. Why would it be undesirable to use a centrifugal switch in a hermetically sealed motor?

103. How is it possible to control the speed of a split-phase motor?

104. What is meant by a capacitor-start split-phase motor?

105. What are the advantages of a capacitor-start motor?

106. What type of capacitor is commonly used in capacitive-start motors?

107. How is it possible to convert a standard split-phase motor into a capacitive-start motor? Explain the procedure for doing this properly.

108. What is meant by centrifugal switch flutter and why must it be avoided in capacitor-start motors?

109. Give an example of a two-value capacitor motor, and why would a two-value capacitor motor be used?

110. What types of capacitors are used in two-value capacitor motors?

## PROBLEMS

5-1. For how many poles is a 50-Hz induction motor wound if the nameplate speed is 460 rpm?

5-2. A three-phase 60-Hz 10-pole induction motor has a full-load slip of 0.80. What is the speed of the rotor relative to the revolving field? The revolving field relative to the stator?

5-3. At what speed will a 12-pole 60-Hz induction motor operate if the slip is 0.09?

5-4. The nameplate speed of a 25-Hz induction motor is 720 rpm. If the speed at no load is 745 rpm, calculate slip and the percent regulation.

5-5. A three-phase six-pole 60-Hz 240-V wound-rotor motor has its stator connected in delta, and its rotor in wye. There are 80 percent as many rotor conductors as stator conductors. Calculate the voltage and frequency between slip rings if normal voltage is applied to the stator when: (a) the rotor is at rest; (b) the rotor slip is 0.04; (c) the rotor is driven by another machine at 800 rpm in a direction opposite to that of the revolving field.

**5-6.** Nameplate data for a squirrel-cage induction motor contains the following information: 25 hp, 240 V, three-phase, 60 Hz, 830 rpm, 65 A. If the motor draws 21,000 W when operating at full load, calculate: (a) slip; (b) percent regulation if the no-load speed is 895 rpm; (c) power factor; (d) torque; (e) efficiency.

**5-7.** A 440-V three-phase 40-hp 60-Hz eight-pole wound-rotor induction motor has a double-layer single-circuit winding on the stator and on the rotor. Both windings are connected in wye. There are 96 and 72 slots on the stator and rotor, respectively. The stator has 96 coils of four turns and $\frac{3}{4}$ pitch, while the rotor has 72 coils of four turns and $\frac{8}{9}$ pitch. The dc resistance for a winding temperature of 75°C between terminals is 0.160 Ω for the stator and 0.14 Ω for the rotor. Calculate (a) the transformation ratio $b$ and (b) the resistance of the rotor in ohms per phase referred to the stator.

**5-8.** A 208-V three-phase 10-hp 60-Hz induction motor has its stator winding connected in wye and draws a current of 30 A when delivering its rated load with rated voltage and rated frequency applied to the stator. Calculate the rated values of (a) line-to-line voltage, (b) line current, and (c) horsepower, if the stator winding is changed from a wye connection to a delta connection.

**5-9.** The motor in Prob. 5-8 is to operate from a 50-Hz source with the stator wye-connected. If the magnetic flux density and the current density in the windings are to be the same as for rated conditions in 60-Hz operation, calculate (a) the rated voltage, (b) the rated horsepower, and (c) full-load current for 50-Hz operation.

**5-10.** A 220-V three-phase 15-hp 60-Hz four-pole 1730-rpm wound-rotor induction motor has 147 V between slip rings at standstill with balanced three-phase 60-Hz voltage of 220 V applied between stator terminals. The rotor is coupled directly to a dc motor of which the speed can be varied. (a) Neglect the effects of harmonics in the space wave of the flux and calculate the voltage between rotor slip rings and the frequency of the rotor when the rotor is driven in the same direction as that of the rotating field produced by the stator at (1) 450 rpm, (2) 1730 rpm, (3) 1800 rpm, and (4) 2250 rpm. (b) Repeat part (a) but with the rotor driven opposite the direction of the stator rotating field.

**5-11.** The input to the rotor of a 2200-V three-phase 60-Hz 12-pole induction motor is 242.0 kW. (a) Calculate the developed or electromagnetic torque in newton-meters per radian and in pound-feet per radian. (b) If the current in the rotor is 375 A/phase and the resistance of the rotor 0.0175 Ω/phase, what is the speed and horsepower output of the motor? Neglect rotational losses.

**5-12.** The following constants apply to a 2200-V 50-hp three-phase 60-Hz wye-connected six-pole squirrel-cage induction motor.

$$r_1 = 3.5 \ \Omega/\text{phase} \qquad x_1 = x_2 = 7.2 \ \Omega/\text{phase}$$
$$r_2 = 2.4 \ \Omega/\text{phase} \qquad r_{fe} = 4170 \ \Omega/\text{phase}$$
$$x_M = 328 \ \Omega/\text{phase}$$

Assume $r_{fe}$ to include the rotational losses and calculate, for a slip of 0.019, (1) the torque, (2) horsepower output, (3) efficiency, and (4) power factor on the basis of (a) the equivalent circuit and (b) the approximate equivalent circuit.

**5-13.** Calculate (a) the maximum torque and the slip at which it occurs for the motor of Prob. 5-12 on the basis of the equivalent circuit and (b) the approximate equivalent circuit.

**5-14.** A 2300-V 1000-hp three-phase 60-Hz 16-pole wound-rotor induction motor with stator and rotor wye-connected has constants as follows:

$$r_1 = 0.0725 \ \Omega/\text{phase} \qquad r_{22} = 0.0252 \ \Omega/\text{phase}$$
$$b = 2.025/1 \qquad x_M = 17.7 \ \Omega/\text{phase}$$
$$x_1 = x_2 = 0.625 \ \Omega/\text{phase} \qquad r_{fe} = 200 \ \Omega/\text{phase}$$

Assume $r_{fe}$ to include the rotational losses and calculate, for a slip of 0.019, (1) the torque, (2) horsepower output, (3) efficiency, and (4) power factor on the basis of (a) the equivalent circuit and (b) the approximate equivalent circuit.

**5-15.** A four-pole wound-rotor induction motor has a wye-connected stator and a delta-connected rotor, with the same number of conductors on each. The rotor resistance and reactance at standstill are 0.2 and 0.8 $\Omega$ per phase, respectively. If the voltage impressed on the stator is 110 V and the frequency is 60 Hz, calculate the rotor current at starting and when the speed is 1710 rpm.

**5-16.** The rotor of a wound-rotor induction motor is rewound with twice the number of its original turns and with the cross-sectional area of the conductor material in each turn one-half the original value. Calculate the ratio of the following quantities in the rewound motor to the corresponding original quantities: (a) full-load rotor current, (b) the actual rotor resistance, (c) the rotor resistance referred to the stator, (d) the rated horsepower, (e) full-load efficiency, (f) starting torque, (g) maximum torque, and (h) slip at which the torque is maximum.

**5-17.** Repeat Prob. 5-16 (a), (b), and (c) if the rotor were rewound with the same number as the original number of turns but with one-half the original cross section of the conductor material. Neglect changes in the leakage flux.

**5-18.** Show, on the basis of the approximate equivalent circuit, that the rotor current, torque, and electromagnetic power of a polyphase induction motor vary almost directly as the slip for small values of slip.

**5-19.** Use Fig. 5-7(a) as a basis and (a) calculate the following quantities for the motor in Example 5-4 for $s = 0.022$: (1) $r_{fe}$, (2) the power factor, and (3) the efficiency, if the stray-load loss is 540 W. (Subtract the stray-load loss from the calculated output.) (b) Repeat parts (2) and (3) on the basis of Fig. 5-19(d) and by subtracting the sum of the rotational losses plus the stray-load loss from the calculated output. (c) Calculate the errors in the values of stator current and in the net output for part (b) on the basis that the corresponding values obtained in part (a) are correct.

**5-20.** The following are test results obtained on a 100-hp 440-V 117-A 60-Hz four-pole deep-bar induction motor.

(a)

No-Load Test

| Volts | Amperes | Watts | Frequency, Hz |
|-------|---------|-------|---------------|
| 440[a] | 24.0 | 3200 | 60 |

(b)

**15-Hz Blocked-Rotor Test**

| Volts | Amperes | Watts |
|-------|---------|-------|
| 29.4ᵃ | 120 | 2540 |

(c)

**60-Hz Blocked-Rotor Test**

| Volts | Amperes | Watts |
|-------|---------|-------|
| 440ᵃ | 720 | 148,000 |

ᵃ Line-to-line voltage.

(d) Direct-current resistance between stator terminals = 0.0574 Ω. Calculate the constants for the equivalent circuit of Fig. 5-19(d), (1) for the running range and (2) for starting.

**5-21.** A 5-hp 240-V four-pole 60-Hz three-phase induction motor was tested with the following data. No-load test: $V$ = 240 V, $P$ = 310 W, $I$ = 6.2 A. Load test: $V$ = 240 V, $P$ = 3600 W, $I$ = 11.4 A, rpm = 1720 rpm. The effective ac resistance of the stator per phase is measured to be 0.3 Ω. Calculate: (a) friction, windage, and iron losses; (b) stator copper loss under load; (c) rotor power input; (d) rotor copper loss under load; (e) rotor output in watts; (f) horsepower output; (g) torque; (h) percent efficiency at load; and (i) load power factor.

**5-22.** A locked-rotor test was performed upon the motor of the previous problem, and the following data was obtained: $V$ = 48 V, $I$ = 18 A, $P$ = 610 W. Calculate: (a) the equivalent resistance of the motor per phase; (b) the equivalent resistance of the rotor per phase; and (c) the equivalent locked-rotor reactance per phase.

**5-23.** Use the results of the previous problem and calculate the speed at which the torque will be a maximum.

**5-24.** Calculate the following for the motor in Example 5-6 for $s$ = 0.05: (a) the frequency of the forward and backward rotor currents; (b) the values of the forward and backward rotor currents $I_{2f}$ and $I_{2b}$ referred to the stator; and (c) the developed mechanical power on the basis of

$$(1 - s)I_{2f}^2 \left(\frac{0.5r_2}{s}\right) - I_{2b}^2 \left(\frac{0.5r_2}{2 - s}\right)$$

(d) the ratio of the backward-rotating flux to the forward-rotating flux; (e) the ratio of maximum to minimum value of the combined forward- and backward-rotating fluxes; (f) the space angles between the magnetic axis of the forward rotor mmf, that of the forward-rotating flux, and that between the magnetic axis of the backward rotor mmf and the backward-rotating flux [see Fig. 5-9(a)]; and (g) the ratios of the forward and backward flux to their standstill values.

**5-25.** The no-load test and locked-rotor test on a $\frac{1}{3}$-hp 115-V 60-Hz 1720-rpm split-phase induction motor yielded the following:

| No-load test | | Locked-rotor test | |
| --- | --- | --- | --- |
| $V_0 = 115$ | $I_0 = 3.49$ | $V_L = 115$ | $I_L = 18.2$ |
| $P_0 = 85.0$ | $r_1 = 1.86$ | $P_L = 1600$ | |

Calculate the constants for the equivalent circuit of this motor and the rotational losses.

**5-26.** Show, on the basis of Eq. 5-89, that the net power $P_{gf} - P_{bf}$ transferred across the air gap of a two-phase induction motor with balanced two-phase voltage applied to the stator is $4R_f I_1^2 = 2r_2 I_2^2/s$. What is the value of $P_{gf}$ and that of $P_{bf}$?

**5-27.** The following data was obtained from tests performed upon a 10-hp 220-V six-pole induction motor. No-load test: $V = 220$ V, $P = 410$ W, $I = 8.5$ A. Load test: $V = 220$ V, $P = 8300$ W, $I = 23$ A, rpm $= 1150$ rpm. Effective resistance of stator per phase $= 0.15$ Ω; friction plus windage loss $= 200$ W. Calculate: (a) stator iron loss; (b) rotor power input under load; (c) rotor copper loss under load; (d) horsepower output; (e) load torque; (f) load efficiency; and (g) power factor.

**5-28.** If the starting torque of a 7.5-hp 440-V 1720-rpm motor is 2.5 times its rated full-load torque when rated voltage is impressed, calculate the starting torque when 220 V is applied at the instant the motor is starting.

**5-29.** The starting current of a 15-hp 440-V three-phase induction motor is 125 A when rated voltage is impressed. What voltage should be applied at the starting instant if the current is not to exceed 75 A?

**5-30.** When an induction motor is started directly from the line, the starting torque is 1.5 times full-load torque, and the starting current is 8 times rated value. (a) What will be the motor starting torque and starting current in terms of full-load value if the machine is started by the wye-delta method? (b) What will be the starting current on the line side under these conditions?

**5-31.** Calculate the ratio of the backward-rotating flux to the forward-rotating flux in the capacitor motor in Example 5-7 for $s = 0.06$.

**5-32.** The capacitor motor in Example 5-7 is started with its starting capacitor. Calculate the following on starting: (a) the current in each stator winding, (b) the line current, (c) the power factor, (d) the voltage across the capacitor, and (e) the developed torque.

**5-33.** A two-pole two-phase induction motor is used as a tachometer with its main winding excited at a 60-Hz voltage of 100 V. The constants of the motor at 60 Hz are for both stator windings as follows:

$$r_1 = 500 \ \Omega \qquad x_1 = 70 \ \Omega$$
$$r_2 = 300 \ \Omega \qquad x_2 = 70 \ \Omega$$
$$x_M = 1000 \ \Omega$$

The auxiliary winding supplies an amplifier the input impedance of which may be assumed to be infinite. Calculate the voltage of the auxiliary winding when the rotor is driven at a speed of 3000 rpm.

**5-34.** The motor in Prob. 5-33 is used as a control motor. The main winding is supplied from a 100-V 60-Hz source, and the voltage applied to the auxiliary winding is 20 V, 60 Hz, lagging the voltage across the main winding by 90°. Calculate the developed torque (a) on starting and (b) at $s = 0.8$.

# BIBLIOGRAPHY

Alger, P. L. *Induction Machines,* 2d ed. New York: Gordon and Breach, Inc., 1970.

Brown, David, and E. P. Hamilton III. *Electromechanical Energy Conversion.* New York: The Macmillan Company, 1984.

Crosno, C. Donald. *Fundamentals of Electromechanical Conversion.* New York: Harcourt, Brace and World, Inc., 1968.

Del Toro, V. *Electromechanical Devices for Energy Conversion and Control Systems.* Englewood Cliffs, N.J.: Prentice-Hall, Inc., 1968.

Fitzgerald, A. E., C. Kingsley, Jr., and Alexander Kusko. *Electric Machinery,* 3d ed. New York: McGraw-Hill Book Company, 1971.

Kosow, I. L. *Electric Machinery and Control.* Englewood Cliffs, N.J.: Prentice-Hall, Inc., 1964.

Liwschitz-Garik, W., and R. T. Weil. *Alternating-Current Machines.* New York: John Wiley & Sons, Inc., 1961.

Matsch, Leander W. *Electromagnetic & Electromechanical Machines,* 2d ed. New York: Dun-Donnelley Publishing Co., 1977.

McPherson, George. *An Introduction to Electrical Machines and Transformers.* New York: John Wiley & Sons, Inc., 1981.

Nasar, S. A. *Electromagnetic Energy Conversion Devices and Systems.* Englewood Cliffs, N.J., Prentice-Hall, Inc., 1970.

Puchstein, A. F., et al. *Alternating-Current Machines.* New York: John Wiley & Sons, Inc., 1954.

Siskind, Charles S. *Electrical Machines: Direct & Alternating Current,* 2d ed. New York: McGraw-Hill Book Company, 1959.

Veinott, C. G. *Fractional- and Subfractional-Horsepower Electric Motors.* New York: McGraw-Hill Book Company, 1948.

# Chapter 6

# Direct-Current Machines

Although practically all the electric energy produced commercially is generated and distributed in the form of alternating current, much energy is utilized in the form of direct current in industry. Direct-current motors are well suited for many industrial processes that demand the control of speed and torque. Alternating current lends itself more readily than direct current to the generation and distribution of large blocks of electric power because of the high efficiency and relative simplicity of transformer arrangements for converting voltages from one value to another. For large blocks of power it is therefore more economical to convert alternating current to direct current by means of motor-generator sets and electronic devices than it is to generate and distribute direct current.

Direct-current drives provide: (1) adjustable motor speed over wide ranges, (2) constant mechanical power output or constant torque, (3) rapid acceleration and deceleration, and (4) responsiveness to feedback signals.

## 6-1 STRUCTURAL FEATURES OF COMMUTATOR MACHINES

The usual direct-current machine depends for its operation on the rotation of one winding, called the *armature winding,* in a magnetic field produced by a stationary winding, known as the *field winding.* Figure 6-1 shows a partially wound armature with armature coils displayed in the foreground. The armature winding consists of a number of usually identical coils carried in slots uniformly distributed around the periphery of the rotor iron which is built up with laminations of sheet steel about 0.043 cm thick. The coils are interconnected through

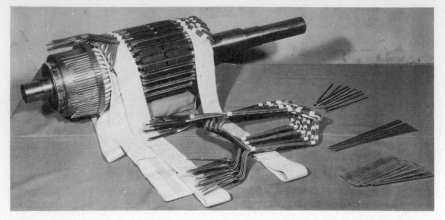

**Figure 6-1**  Partially wound dc armature with coils shown in foreground. (Courtesy Westinghouse Electric Corporation.)

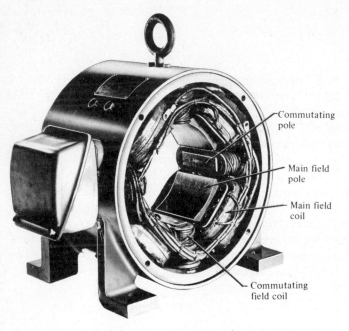

Commutating pole

Main field pole

Main field coil

Commutating field coil

**Figure 6-2**  Stator of a dc machine (Courtesy Westinghouse Electric Corporation.)

the commutator, comprised of a number of bars, sometimes called *commutator segments,* which are insulated from each other. The commutator rotates with the armature and is shown on the left-hand side of the rotor in Fig. 6-1. The commutator serves to rectify the induced voltage and the current in the armature both of which are ac except in the case of acyclic or homopolar machines, which operate without a commutator. A four-pole field structure is shown in Fig. 6-2. The four larger poles carry the field coils. These poles produce the main flux, and the four smaller poles are the commutating poles or interpoles. Commutating poles produce an mmf in opposition to that of the armature in order to achieve commutation with reduced sparking at the brushes. Brushes ride on the commutator to collect the armature current. Commutating poles are generally omitted in small machines, such as fractional-horsepower motors, because the resistance of the armature coils is high enough in relation to other parameters to prevent excessive sparking at the brushes.

The field poles and commutating poles are of rectangular cross section and built up of steel sheets approximately 0.5 mm thick. The laminated pole construction holds to a minimum the eddy currents resulting from the pulsations in the air-gap flux caused by the difference in the reluctance of the armature slots and the intervening armature teeth. The field poles are attached to the yoke, which completes the magnetic circuit.

The number of poles used on a dc machine is governed by the voltage and current rating of the machine. The higher the voltage for a given diameter of armature, the fewer will be the number of poles. This is necessary to provide space for the larger number of commutator bars required for the higher voltage, since there are usually as many brush sets as there are poles, and the space between adjacent brush sets is the same as that between adjacent poles. High current machines require a large number of poles in order to carry the current, which may be as high as 1250 A per brush set.

Figure 6-3 shows a split-frame motor with its frame opened to expose various components of the motor. The split-frame construction is not a common one but is presented here for purposes of illustration.

The schematic diagram in Fig. 6-4 of the magnetic circuit of a four-pole machine shows the approximate paths taken by the flux due to the field excitation.

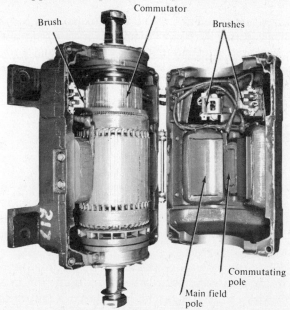

**Figure 6-3** Direct-current split-frame motor. (Courtesy Westinghouse Electric Corporation.)

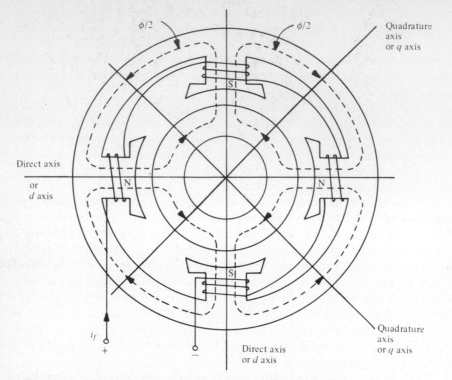

**Figure 6-4**   Schematic representation of a four-pole dc machine.

Magnetic leakage is neglected and the air-gap flux per pole is $\phi$ while that in the yoke is $\phi/2$. This simple structure is shown without commutating poles, an omission that does not affect the main path taken by the flux due to the field mmf. The commutating poles are along an axis in which the mmf due to the field excitation is ideally zero.

## 6-2 ELEMENTARY MACHINE

Consider the elementary two-pole dc machine in Fig. 6-5, which shows an armature coil of one turn the sides of which are designated $a$ and $a'$ and which terminate in segments $s_1$ and $s_2$ of the commutator. The brushes $b_1$ and $b_2$ ride on the commutator and are connected to the armature terminals $T_1$ and $T_2$ in Fig. 6-5(a). The field winding and the portions of the iron that complete the magnetic circuit are not shown.

For clockwise rotation, as indicated, when the armature coil is in the position shown in Fig. 6-5(b) the polarity of the induced emf, in accordance with Lenz's law, is such as to send a current in coil side $a$ toward the observer and in coil side $a'$ away from the observer, as designated by the dot and cross markings.

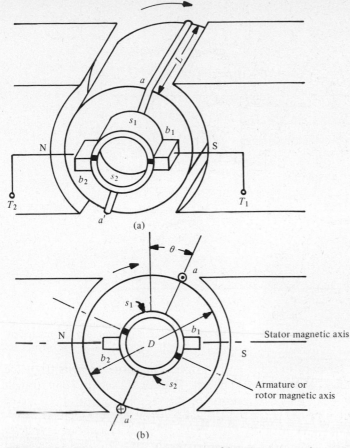

**Figure 6-5** Elementary two-pole dc machine. (a) Partial view. (b) Simplified side view.

At the instant when the angle $\theta$ between the magnetic axis of the stator and rotor is zero or $\pi$, the voltage induced in the armature coil is zero. When the width of the brushes is neglected, brush $b_1$ always rides on that segment which connects to the coil side under the south (S) field pole and brush $b_2$ on that segment which connects to the coil side under the north (N) field pole. As a result, for the assumed polarities of the field poles and the assumed direction of rotation, brush $b_1$ is always positive and brush $b_2$ is always negative.

For motor operation, current enters the armature coil through the positive brush and the action of the commutator is such, during a complete rotation, as to keep the current in the coil side under the south (S) field pole directed away from the observer, while that in the other coil side which is then under the north (N) field pole is directed toward the observer. Consequently, the torque is developed in the clockwise direction throughout a complete revolution of the armature, except at those instants when the current in the armature coil reverses (i.e., when $\theta = 0$ and $\theta = \pi$), which occurs when one commutator segment

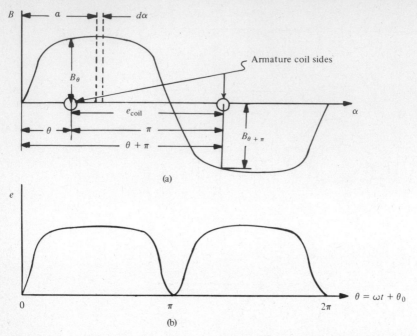

**Figure 6-6** (a) Flux-density space wave. (b) Rectified no-load voltage in a full-pitch coil.

replaces the other segment under a given brush. The torque results from the force† exerted by a field of uniform flux density $B$ along the length $l$ of a straight conductor carrying a current $I$ as expressed by the vector product

$$f = l\mathbf{I} \times \mathbf{B} \tag{6-1}$$

When the conductor is embedded in a slot, the same force relationship is valid if the conductor is regarded as a filament except that the major component of the force is exerted on the iron in the region of the filament.

### 6-2.1 Voltage Induced in a Full-Pitch Armature Coil

A full-pitch armature coil of one or more turns is one that spans a pole pitch (i.e., when one of its two sides lies under the center of a field pole, its other side lies under the center of an adjacent field pole). Thus, in a two-pole machine a full-pitch coil spans the diameter, and in a $P$-pole machine it spans a chord that subtends an angle of $2\pi/P$ on the armature periphery.

For a symmetrical flux-density space wave, constant field current and constant angular velocity, the waveform (time function) of the voltage induced in a full-pitch armature coil is the same as that of the flux-density distribution wave (space function) in the air gap. Figure 6-6(a) shows the approximate flux-density

---

† See, for example, R. P. Winch, *Electricity and Magnetism* (Englewood Cliffs, N.J.: Prentice-Hall, Inc., 1955), p. 442.

distribution curve for a dc machine at no load with the effect of the armature slots neglected. The armature current is zero and the flux is due entirely to the field current. Consider the full-pitch armature coil with its magnetic axis displaced from that of the field by the angle $\theta$. Assume the machine to have two poles similar to that of Fig. 6-5 except that the armature coil has $N_{coil}$ turns instead of one turn embedded in one pair of slots. The flux linkage with the armature coil, due to the flux in the elemental strip of width $d\alpha$ at a point in the air gap displaced $\alpha$ radians from a point midway between field-pole centers, is given by

$$d\lambda = N_{coil} B_\alpha \, dA \tag{6-2}$$

where $dA = (D/2)L \, d\alpha$, the area of the flux path through the elemental strip, and in which $D$ is the diameter and $L$ the axial length of the air gap. Equation 6-2 can then be written as

$$d\lambda = \frac{N_{coil} DL B_\alpha \, d\alpha}{2}$$

and the flux linkage with the armature coil is

$$\lambda = \frac{N_{coil} DL}{2} \int_{\alpha=\theta}^{\alpha=\theta+\pi} B_\alpha \, d\alpha \tag{6-3}$$

Since the flux density space wave is symmetrical because $B_{\theta+\pi} = -B_\theta$ not only at no load but also when there is armature current, as shown in Fig. 6-12, it can be represented by the series

$$B_\alpha = B_1 \sin \alpha + B_3 \sin 3\alpha + \cdots + B_n \sin n\alpha \tag{6-4}$$

where $n$ is odd. When this series is substituted in Eq. 6-3 and integration is carried out between the limits $\alpha = \theta$ and $\alpha = \theta + \pi$, the expression for the flux linkage with the armature coil results in the series

$$\lambda = N_{coil} DL \left( B_1 \cos \theta + \frac{1}{3} B_3 \cos 3\theta + \cdots + \frac{1}{n} B_n \cos n\theta \right) \tag{6-5}$$

For a constant angular velocity of $\omega$ rad/s,

$$\theta = \omega t \tag{6-6}$$

and the voltage induced in the armature coil is

$$e_{coil} = -\frac{d\lambda}{dt} = -\frac{d\lambda}{d\theta} \frac{d\theta}{dt} \tag{6-7}$$

When Eqs. 6-5 and 6-6 are substituted in Eq. 6-7, there results

$$e_{coil} = \omega N_{coil} DL(B_1 \sin \theta + B_3 \sin 3\theta + \cdots + B_n \sin n\theta)$$
$$= \omega N_{coil} DL B_\theta \tag{6-8}$$

which shows that the waveform of the induced voltage $e_{coil}$ is the same as that of the flux-density distribution for a constant angular velocity. However, the commutator rectifies the voltage at the brushes, causing the terminal voltage to be unidirectional as shown in Fig. 6-6(b). Practical dc machines have numerous armature coils connected in series with each other through the commutator, thus producing a smooth voltage [i.e., one in which the dips, as shown at $\omega t = 0, \pi, \ldots, n\pi$ in Fig. 6-6(b), result in a negligible ripple]. Figure 6-7 shows the voltage waveform produced by two dc armature coils displaced 90°, in electrical measure, from each other.

From Fig. 6-7 it is evident that the amplitude of the voltage ripple relative to the average voltage of the two coils is one-half that of a single coil. While the frequency of the ripple increases with an increasing number of distributed armature coils, its amplitude decreases correspondingly and the waveform resulting from the addition of as few as eight coil voltages is quite smooth.

The voltage resulting from several coils in series equals the product of the average coil voltage times the number of coils. The average voltage induced in a coil equals the average of the voltage wave taken over any one-half period $T/2$ s or over $\pi$ rad in Fig. 6-6(b) as follows:

$$E_{(coil)av} = \frac{2}{T} \int_{t_1}^{t_1+T/2} e_{coil} \, dt = \frac{1}{\pi} \int_{\theta_1}^{\theta_1+\pi} e_{coil} \, d\theta \tag{6-9}$$

which is valid for all values of $\theta_1$. As a matter of convenience let $\theta_1 = 0$; then

$$E_{(coil)av} = \frac{1}{\pi} \int_0^{\pi} e_{coil} \, d\theta \tag{6-10}$$

and when Eq. 6-8 is substituted in Eq. 6-10, the result is

$$E_{(coil)av} = \frac{\omega}{\pi} N_{coil} \int_0^{\pi} B_\theta DL \, d\theta$$

in which $DL \, d\theta = 2dA$ or twice the area subtended by the differential angle $d\theta$, since $dA = L(D/2) \, d\theta$, so

$$E_{(coil)av} = \frac{2\omega N_{coil}}{\pi} \int_0^{\pi} B_\theta \, dA \tag{6-11}$$

The integral $\int_0^{\pi} B_\theta \, dA$ yields the flux per pole $\phi_d$, and the average voltage induced in a full-pitch coil is expressed in terms of the flux per pole by

$$E_{(coil)av} = \frac{2\omega N_{coil}\phi_d}{\pi} \tag{6-12}$$

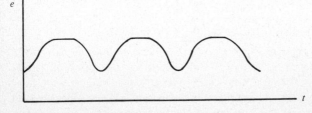

**Figure 6-7**   Dc voltage waveform of two coils displaced 90° in electrical measure.

Equation 6-12 is valid for multipolar machines as well as for two-pole machines if $\omega$ is given in electrical measure. Since the mechanical angular velocity of rotation is

$$\omega_m = \frac{2\pi n}{60}$$

the angular velocity in electrical measure is, according to Eq. 4-1b,

$$\omega = \frac{P}{2}\frac{2\pi n}{60} \qquad \text{rad/s} \tag{6-13}$$

where $n$ is the speed in rpm and therefore, in general,

$$E_{(coil)av} = \frac{2PN_{coil}\phi_d n}{60} \tag{6-14}$$

In the foregoing it was assumed that the brushes were placed in the electrical neutral position, which means that as the armature coil undergoes commutation (in the elementary two-pole machine, the commutator segments under the brushes are interchanged at the instant the magnetic axes of the armature coil and the field poles coincide). Under this condition the sides of the armature coil lie in the neutral zone or halfway between poles, where the flux density is ideally zero or very low in the practical case. If the brushes are shifted from the neutral position, the voltage generated in the coil is not completely rectified. In addition, during commutation, since the brushes are of greater width than the insulation between adjacent commutator segments, the brushes momentarily span this insulation, thus short-circuiting the coil. When the brushes are set on neutral, the short circuit occurs while the induced voltage goes through zero and the resulting short-circuit current in the coil is negligible. On the other hand, if the brushes are far enough away from the neutral position so that appreciable voltage is generated in the coil during commutation, a sizable short-circuit current flows in the coil. This results in sparking at the brushes and consequently in burning of the commutator as well as in reduced efficiency.

## 6-2.2 General EMF Equation for DC Machines

Distributing the armature windings among several slots per pole in dc machines suppresses voltage ripple and torque pulsations and in addition improves the dissipation of heat. The single armature coil in the preceding discussion (i.e., one which occupies one slot per pole) represents a concentrated winding. For a given number of turns and paths in a dc armature, the resultant dc voltage is the same whether the winding is distributed or concentrated since in either case, because of the commutator action, the voltage between brushes is the sum of voltages which have like instantaneous polarities. Although Eq. 6-14 was derived for a full-pitch coil, it applies with negligible error to fractional-pitch coils commonly used in the armature of dc machines. Fractional-pitch coils span less than one pole pitch. However, the pitch of dc armature coils is generally well above two-thirds, and further, the sides of an armature coil when undergoing commutation lie in regions of low flux density so that the maximum flux which links the coil is substantially equal to the flux per pole. Then, if

$a$ = number of parallel current paths in the armature winding
$P$ = number of poles
$N_a$ = number of turns in the armature winding
$n$ = speed of rotation in rpm
$\phi_d$ = flux per pole

the number of armature turns per path (between brushes) is $N_a/a$ and the generated voltage between brushes is

$$E = \frac{2PN_a n\phi_d}{60a}$$

It is customary to express the voltage of dc armatures in terms of the number of armature conductors $Z$, where

$$Z = 2N_a$$

so that

$$E = \frac{PnZ\phi_d}{60a} = \frac{PZ\phi_d\omega_m}{2\pi a} \qquad (6\text{-}15)$$

where $\omega_m$ is the angular velocity of the armature in mechanical measure.

## 6-3 ARMATURE WINDINGS

As mentioned previously, the armature winding in the usual dc machine is placed in slots on the rotor and connected to the commutator, which is comprised of a number of copper segments, or bars insulated from each other and equal to the number of armature coils in conventional machines.

There are two general types of dc armature windings, the *lap* and the *wave* winding. Both of these windings may be arranged as simplex or as multiplex windings. The multiplex windings are, in effect, two or more simplex windings with one commutator and connected in parallel with each other. Only simplex windings are treated in this text.† The simplex lap winding has as many current paths as there are poles (i.e., $a = P$), and the simplex wave winding has only two paths (i.e., $a = 2$), regardless of the number of poles. It is possible to combine a lap and wave winding in one machine. Such a combination is known as a *frog-leg* winding.

The wave winding is far more common than the lap winding and is used in practically all dc machines with ratings of 75 hp or less because it affords greater economy, except in machines with low-voltage and high-current ratings.

---

† For a more complete discussion of armature windings, see A. S. Langsdorf, *Principles of Direct-Current Machines,* 5th ed. (New York: McGraw-Hill Book Company, 1940), pp. 134–186.

From Eq. 6-15 it is evident that, for a given speed, the number of poles, and flux per pole, in all but two-pole machines the wave winding, since $a = 2$, requires fewer turns in the armature winding than does a lap winding. The cross section of the conductors in the wave winding must be correspondingly larger; so the amount of conductor material in both the lap and the wave windings is the same for a given power rating. It is generally cheaper to build a winding of larger cross section and fewer turns. In addition, the total space required by the insulation between turns of a coil is smaller, which also has the advantage of facilitating the transfer of heat away from conductors of the winding. However, in low-voltage high-current armatures a wave winding may require such large cross sections of conductor material as to create mechanical difficulties in constructing the coils and assembling the armature. In such armatures the lap winding, with its larger number of turns and correspondingly smaller conductor cross section, is the more practical.

## 6-3.1 Lap Windings

A four-pole lap winding is shown in developed representations in Fig. 6-8. There are 23 armature coils, of one turn each, and four brushes riding on the commutator of 23 segments or bars. The fact that adjacent coils overlap each other accounts for the term *lap winding*. The four-pole simplex lap winding has four current paths (i.e., $a = 4$) and requires four brushes. It will be found that in tracing through this winding, from a given brush to a brush of opposite polarity, one-fourth of the winding is traversed, showing that there are four paths (i.e., as many paths as there are poles). In the lap winding an armature coil terminates in adjacent commutator segments.

In Fig. 6-8 the positions of the centers of the field poles are indicated by N and S for north and south polarity. It must be kept in mind that the armature winding and the commutator are in motion while the brushes and the field poles are stationary. The brushes are shown wider than one commutator bar, and

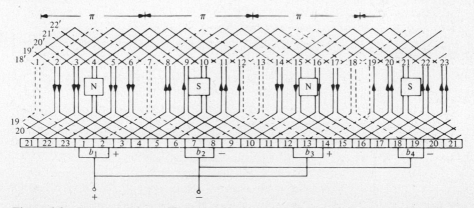

**Figure 6-8** Developed four-pole lap winding. The winding is between the field-pole faces and the observer. Motion of the armature is from left to right for generator action. (The squares indicate the location of the field-pole centers and do not represent the size of the pole faces, which are actually much larger.)

those armature coils shown by the dashed lines are short-circuited by the brushes. In actual machines good commutation requires the brushes to cover several commutator bars. The sides of the short-circuited coils lie approximately halfway between field poles or in regions where the flux density is very low. The rotational emf in these short-circuited coils is therefore small, thus generally producing negligible short-circuit current.

### 6-3.2 Wave Windings

Figure 6-9 shows a four-pole wave winding with the same number of armature coils and commutator bars as the lap winding of Fig. 6-8. The coils have only one turn each. The coils that are short-circuited by the brushes overlapping commutator segments are shown by the dashed lines. The term *wave winding* is due to the wavelike appearance of the single-turn coil. In tracing through the wave winding, from one brush to one of opposite polarity, it is found that one-half the armature winding and one-half the commutator segments are encountered, showing that there are only two current paths in the wave winding regardless of the number of poles. Actually the wave winding requires only one pair of brushes, which is usually sufficient in small machines. Generally, the same number of brushes as poles is used to provide the proper amount of brush area with a shorter commutator. An armature coil in the wave winding terminates in commutator segments approximately two pole pitches apart.

### 6-4 FIELD EXCITATION

Dc machines are excited with direct current in the field windings and are classified as (a) shunt, (b) series, and (c) compound machines, depending on the connection

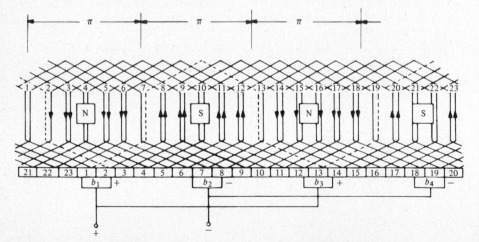

**Figure 6-9**  Developed four-pole wave winding. The winding is between the field-pole faces and the observer. Motion of armature is from left to right for generator action. (The squares indicate the location of the field-pole centers and do not represent the size of the pole faces, which are actually considerably larger.)

of the field circuit relative to the armature circuit. Figure 6-10 shows schematic diagrams of connections for machines in these classifications without including commutating-pole or compensating windings. An adjustable resistance, known as a *field rheostat,* by which the field current may be adjusted, is shown in series with each shunt field. The shunt machine has its field circuit connected in shunt or in parallel with the armature. Shunt generators may be either self-excited or separately excited, as indicated in Fig. 6-10(a) and (b). The field winding of the series machine, as its name implies, is connected in series with the armature as in Fig. 6-10(c), while the compound machine carries a shunt-field winding and a series-field winding both on the same poles. Connections for a compound machine are shown in Fig. 6-10(d). Compound machines may be connected long-shunt as in Fig. 6-10(d) or short-shunt as in Fig. 6-11. Figure 6-12 shows a connection diagram for a compound machine, including commutating-pole and compensating windings.

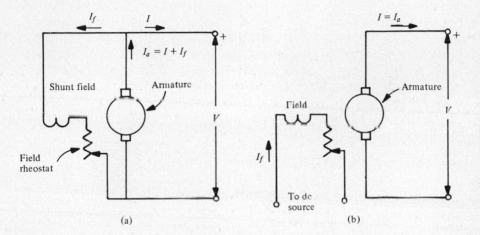

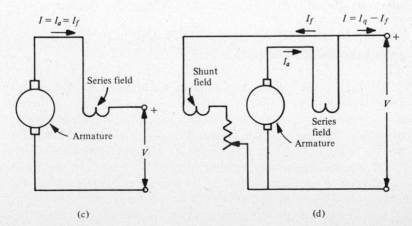

**Figure 6-10** Schematic diagram of connections for dc machines. (a) Self-excited shunt. (b) Separately excited shunt. (c) Series. (d) Compound (Long shunt). Current directions are for generator operation.

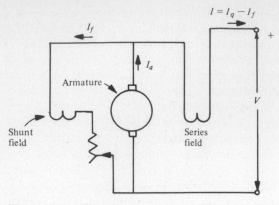

**Figure 6-11** Short-shunt connection for a compound dc machine. Current directions are for generator operation.

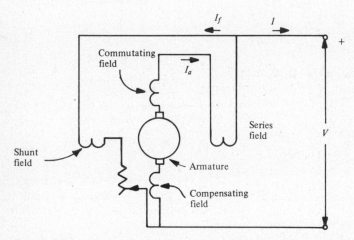

**Figure 6-12** Connections for a dc machine with commutating poles (interpoles) and compensating winding. Current directions are for generator operation.

## 6-5 ARMATURE REACTION—MMF AND FLUX COMPONENTS

Under load conditions there is an mmf due to the armature current the direction of which is determined by the position of the brushes. When the brushes are set on geometric neutral, the armature mmf is directed along an axis midway between field poles. This is called the *quadrature axis.* The magnetic axis of the field winding is called the *direct axis.* Figure 6-13 shows a schematic representation of a four-pole dc generator with the brushes on geometric neutral. The same diagram can be used to represent motor operation by reversing the direction of rotation or by changing the direction of the field current or that of the armature current. The commutator is omitted from the diagram and the armature conductors are shown on the surface of the armature instead of in slots for reasons of simplicity. It should be noted that there are four current paths in the armature.

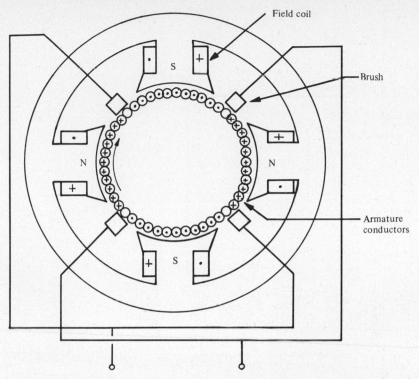

**Figure 6-13** Simplified diagram of a four-pole dc generator with brushes shown on electrical neutral.

Developed views of one pair of field poles and corresponding armature conductors are shown in Fig. 6-14. The flux-density distribution due to the field current alone is shown in Fig. 6-14(a). The space waveforms of the armature mmf and resulting flux density are shown in Fig. 6-14(b). The armature mmf wave approximates a triangular shape which represents the limiting condition as the number of slots is increased indefinitely. The armature mmf thus produces appreciable flux density midway between poles when the brushes are on geometric neutral. Field current and armature current are both present under load, and their combined mmfs produce a flux-density wave somewhat as shown by the dashed line in Fig. 6-15 with the electrical neutral or the region of zero flux density displaced from the geometric neutral in the direction of rotation for a generator. In the motor the electrical neutral is shifted against the direction of rotation.

A component of flux in the quadrature axis causes commutation difficulties in dc machines, and measures are usually taken to minimize them. As mentioned earlier, the sides of the armature coils undergoing commutation should lie in a region of low flux density.

The armature mmf produces armature reaction. While the major component of the armature mmf is in the quadrature axis, the armature reaction in the usual dc machine also has a demagnetizing effect on the field. Two methods

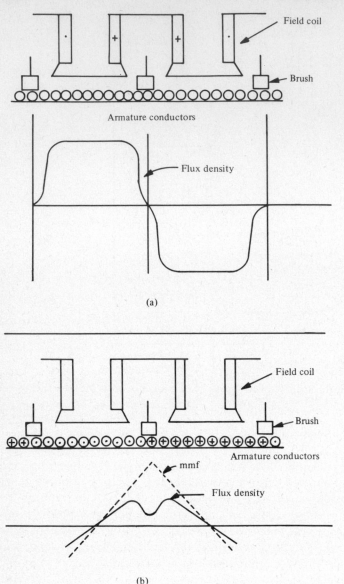

**Figure 6-14** (a) Flux-density space wave due to field current only. (b) Mmf and flux-density space waves due only to armature current with brushes on geometric neutral.

commonly used for overcoming the undesirable effects of armature mmf on commutation will be discussed. The first of these, used in earlier machines and small present-day machines, consists in shifting the brushes toward the electrical neutral zone (i.e., in the direction of rotation for generators and against the direction of rotation for motors). The second method makes use of commutating poles or interpoles placed in the quadrature axis which have windings connected

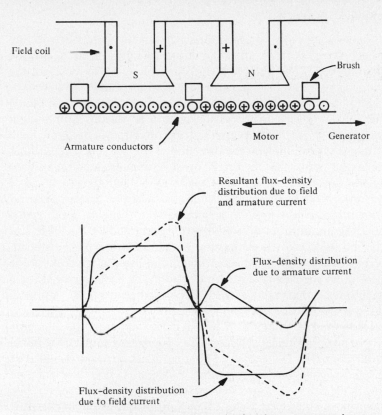

**Figure 6-15**   Flux-density space waves with brushes on geometric neutral.

in series opposition with the armature. The mmf of the commutating poles is thus always proportional to the armature current.

Another effect that influences commutation is the *reactance voltage* produced by the reversal of leakage flux linkage with the short-circuited armature coils as the current reverses during commutation. Unless this voltage is held to a negligible value, sparking occurs across the thin insulation between commutator bars. The reactance voltage is reduced in machines without commutating poles by shifting the brushes past the geometric neutral by an angle slightly greater than that which would set the brushes on electrical neutral for motors as well as for generators. Electric neutral corresponds to a brush position such that the sides of the armature coils undergoing commutation lie in a region of zero flux density. By advancing the brushes beyond this position, an emf is generated in the short-circuited coil which opposes the reactance voltage. The number of turns in commutating poles are adjusted so that their mmf is somewhat greater than that of the armature, and the brushes are set practically on the geometric neutral in machines with commutating poles.

## 6-5.1 Effect of Shifting Brushes from Geometric Neutral

Figure 6-16(a) shows a simplified diagram of a two-pole dc generator with the brushes set on geometric neutral. Then the entire armature mmf is directed along the quadrature axis. The phasors **F** and **A** in Fig. 6-16(b) represent the fundamental components of the field and armature mmf space waves and produce the resultant mmf phasor **R**. Now if the direction of the resultant mmf remained fixed the brushes could be placed on electrical neutral by advancing them through the angle $\delta$ ahead of the geometric neutral in the direction of rotation for a generator. For motor operation, with the direction of the field current and that of rotation unchanged, the armature current must be in a direction opposite to that of Fig. 6-16(a) and the resultant mmf phasor **R** is then ahead instead of behind the field mmf phasor by the angle $\delta$. It follows, therefore, that *the electrical neutral of the motor is displaced from the geometric neutral against the direction of rotation.* Since the armature mmf is advanced or retarded by the angle $\delta$, it would seem that no satisfactory brush position could be achieved. This is not the case, however, because there is an accumulation of flux (so to speak) at the trailing-field pole tip in the case of a generator and at the leading pole tip in a motor, causing the iron in that region of the field pole to become more saturated, so further shifting of the brushes does not significantly increase the angle of the resultant phasor **R** beyond the phasor **F**. A shift in the brushes from the geometric neutral in a dc machine introduces a component of armature mmf in the direct axis as shown in Fig. 6-17. The armature conductors in the belt included in angle $2\alpha$ give rise to *demagnetizing armature reaction* $\mathbf{A}_d$ and the armature conductors lying in the belt $\pi - 2\alpha$ produce *cross-magnetizing armature reaction* $\mathbf{A}_q$. The cross-magnetizing ampere turns cause a distortion in the field and, if there were no saturation, would have no magnetizing or demagnetizing effect.

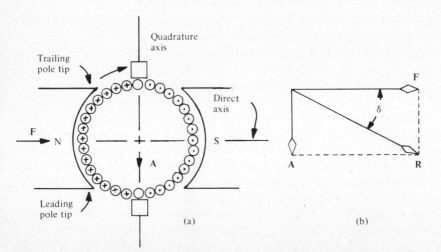

**Figure 6-16** (a) Two-pole dc generator with brushes on geometric neutral. (b) Phasor diagram of fundamental components of mmf space waves.

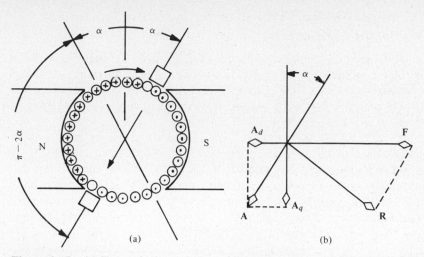

**Figure 6-17** (a) Two-pole dc generator with brushes advanced from geometric neutral. (b) Phasor diagram of fundamental components of mmf space waves.

While the phasor diagrams of Figs. 6-16(b) and 6-17(b) are valid for two-pole machines, they apply to multipolar machines if the mechanical angle is converted to the electrical angle $(P/2)\alpha$, where $P$ is the number of poles.

A certain amount of brush shift in a given direction is ideally correct for only one particular ratio of armature current to field current when variations in magnetic saturation are neglected. Nevertheless, the design of conventional dc machines without commutating poles is such that satisfactory commutation is obtained at rated speed and voltage over the normal range of load. It should be evident that a reversal in the direction of rotation when operating as a motor or as a generator calls for a reversal of brush shift, and that for a given direction of rotation in going from motor to generator operation requires the brushes to be shifted from geometric neutral in the opposite direction. These limitations are not present in machines with commutating poles.

### 6-5.2 Commutating Poles or Interpoles

A simplified diagram of a four-pole machine with two commutating poles is shown in Fig. 6-18. The winding of the commutating poles is in series with the armature. Some machines have one-half as many commutating poles as main field poles while others have the same number. When using as many commutating poles, the number of turns in the commutating-pole winding is such as to produce an mmf from 1.2 to 1.4 times that of the armature. When only half the number of commutating poles is used, this mmf is increased to 1.4 to 1.6 times the armature mmf. In some machines the commutating-pole winding is made to carry only a fraction of the armature current by shunting the remainder of the current through a low-resistance shunt. The winding on the commutating poles then has a relatively large number of turns with correspondingly small cross-

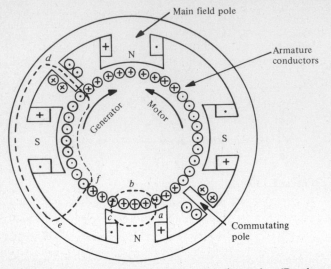

**Figure 6-18**   Dc machine with commutating poles. (Brushes not shown.)

sectional area of conductor material. This shunting arrangement, however, is unsatisfactory in applications where the armature current is subject to considerable fluctuations because the time constant of the shunt is much smaller than that of the interpole winding, and the transient current will not divide properly between shunt and winding. Adjustable nonmagnetic shims are placed between the base of the commutating poles and the frame to obtain the desired value of commutating flux and to prevent the commutating-pole cores from saturating, thus keeping that portion of the magnetic circuit linear.

The commutating poles serve only to provide sufficient field strength in the quadrature axis to assure good commutation. They do not overcome the distortion of the flux resulting from the cross-magnetizing mmf of the armature. This follows from consideration of the line integral of **H** around the path *a-b-c* in Fig. 6-18, which does not link any of the turns on the commutating poles, while the line integral of **H** around the path *d-e-f* links the commutating-pole winding.

**EXAMPLE 6-1**

A six-pole 600-V 600-kW dc generator has a simplex lap winding with 696 armature conductors. There are six commutating poles or interpoles. Calculate the number of turns in the commutating-pole winding if the mmf of the commutating poles is 1.25 times that of the armature.

*Solution*

$$Z = 696 \text{ armature conductors}$$

$$N_a = \frac{Z}{2} = 348 \text{ turns}$$

$$\frac{N_a}{P} = \frac{348}{6} = 58 \text{ armature turns/pole}$$

$$a = P = 6 \text{ parallel current paths in the armature}$$

so $I_a/a = I_a/6$ A/path, which is also the current in the individual turns of the armature winding.

Since the commutating-pole winding is in series with the armature it must carry the armature current $I_a$, and the mmf/pole must therefore be

$$N_{iw}I_a = \frac{1.25 N_a I_a}{Pa} = \frac{1.25 \times 58 I_a}{6}$$

and
$$N_{iw} = 12 \text{ turns/pole}$$

### 6-5.3 Compensating Windings

If the voltage existing between adjacent commutator segments is plotted as a function of angular position around the commutator, the result is a curve which approximates that of the flux-density distribution, the voltage between adjacent segments being highest where the connected coil sides lie in the strongest field. Cross magnetization distorts the voltage distribution around the commutator similar to that of the resultant flux-density distribution shown in Fig. 6-15.

In some applications machines are subject to heavy overloads or rapidly changing loads, as in the case of steel-mill motors. During extreme overloads or sudden load changes, the voltage between adjacent commutator segments may become high enough to cause the commutator to flash over from one brush to the next one of opposite polarity, resulting in a short circuit and sometimes burning the commutator unless steps are taken to overcome the effects of cross magnetization. The cross-magnetizing mmf is neutralized by means of a compensating winding embedded in the faces of the main field poles as indicated in Fig. 6-19(a). The compensating winding is connected in series with the armature,

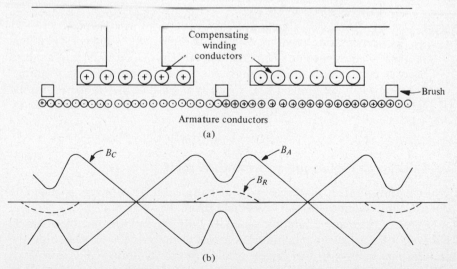

Figure 6-19 (a) Simplified view of a two-pole dc machine with a compensating winding. (b) Flux-density space waves; $B_A$ is due to the armature mmf, and $B_C$ to the mmf of the compensating winding; $B_R$ is the resultant of $B_A$ and $B_C$.

and the number of conductors in it are such as to make its mmf equal to that of the armature conductors under each pole face, thus reducing the armature flux density wave $B_A$ to the resultant $B_R$ as represented approximately in Fig. 6-19(b).

When commutating poles are used in combination with a compensating winding, the sum of the mmfs in the compensating winding and in the commutating-pole winding should approximately equal that of the commutating-pole winding when there is no compensating winding. The addition of a compensating winding therefore requires a corresponding reduction in the turns of the commutating-pole winding.

**EXAMPLE 6-2**

The pole face of the generator in Example 6-1 covers 70 percent of the pole span. (a) Calculate for a compensating winding the number of conductors in each pole face. (b) What should be the number of turns per pole in the commutating-pole winding when the compensating winding of part (a) is in the circuit?

*Solution.*

a. Armature conductors per pole = 696/6 = 116. Armature conductors under each pole face = 116 × 0.7 = 81.2. The compensating winding carries the entire armature current (i.e., 6 times that of the armature conductors) and therefore requires $Z_{cw}$ = 81.2/6 = 13.53 or its nearest whole number of 14 conductors per pole.

b. Good commutation requires 12 turns per pole, of which 14/2 = 7 are provided by the compensating winding. The commutating pole should make up the difference, or

$$N_{iw} = 12 - 7 = 5 \text{ turns/pole}$$

Compensating windings complicate the structure and add greatly to the cost of the machine. They are therefore provided only in machines used in applications which require this feature. Compensating windings are used when the product kW × rpm is about 350,000. In fact, commutating poles are not generally used in dc machines with ratings below 1 kW.

## 6-5.4 Ratio of Field mmf to Armature mmf

The armature mmf in machines without compensating windings is generally held to within about 0.7 to 0.9 of the field mmf. The length of the air gap and the degree of magnetic saturation generally hold the ratio within these limits. However, if the armature mmf at the pole tips were to exceed the field mmf the resulting flux density in that region would reverse its normal direction, as is evident from Fig. 6-15. Such a distortion in the air-gap field usually leads to commutation difficulties, resulting in excessive sparking at the brushes. For that reason dc machines that operate in a range which requires a weak field are usually provided with a compensating winding. Examples are motors designed to operate over wide speed ranges by means of shunt-field control, as well as generators intended to operate over a wide range of voltage.

The average flux density in the air gap of dc machines is roughly 0.8 T. The armature mmf is a function of speed and power rating of the machine and can be determined from empirical relationships.† The length of the air gap is made to correspond with this flux density and the mmf of the field winding that satisfies the above ratio of mmf. When a compensating winding is used, the air gap can be made shorter with a corresponding decrease in the size of the field winding, an advantage that is offset by complications in the field structure and cost associated with the compensating winding.

### 6-5.5 Demagnetization Due to Cross-Magnetizing mmf

If the magnetic circuit were linear, the cross-magnetizing mmf of the armature would not affect the resultant flux per pole since one half the cross-magnetizing ampere turns increase the flux in one half of the face of a given field pole, while the other half of the cross-magnetizing ampere turns would reduce the flux in the other half of the same field pole by an equal amount. Economy of size generally requires operation in the saturated region, and the cross-magnetizing mmf increases saturation in one half of the pole face while it decreases the saturation in the other half. Consequently, there is a greater reduction in the flux in the unsaturated half than the increase of flux in the saturated half of the same pole face.

## 6-6 COMMUTATION

It should be remembered that the current in the armatures of dc commutator machines is alternating while that in the circuit connected externally to the armature through the brushes is unidirectional for normal steady conditions. The direction of the current in each armature coil is reversed as the commutator segments in which the coil terminates pass under a brush. The interval during which this occurs is known as the *commutation period*. Commutation is said to be linear when the current reverses at a uniform rate as shown in Fig. 6-20(a), which is the ideal condition. Four stages of linear commutation are illustrated in Fig. 6-20(b), in which several armature coils are shown in simplified form along with their respective commutator segments and one brush. The current per brush is $I$ amperes, and as the associated leading commutator segment approaches the brush, the current in the coil is $I/2$ amperes. When the trailing commutator segment leaves the brush, the current in the coil is $-I/2$. Com-

---

† For a more complete treatment, see J. H. Kuhlman, *Design of Electrical Apparatus* (New York: John Wiley & Sons, Inc., 1950), Chap. 2. Also A. E. Knowlton (ed.), *Standard Handbook for Electrical Engineers,* 9th ed. (New York: McGraw-Hill Book Company, 1957), Sec. 8.

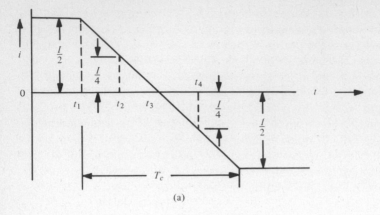

(a)

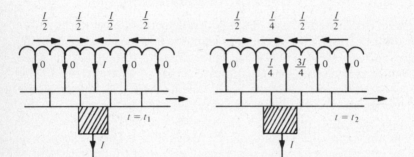

(b)

**Figure 6-20** Linear commutation. (a) Current in a short-circuited armature coil. (b) Simplified diagrams of armature winding undergoing commutation and brush.

mutation is a complicated process,† and the linear time variation of the current in the short-circuited coil is not realized in practice. Unless a voltage is introduced

† Analyses of phenomena underlying commutation are presented by Ragnar Holm, "Contribution to the Theory of Commutation on D-C Machines," *Trans. AIEE* 77, Part III (1958): 1124–1127, and "Theory of the Sparking during Commutation on Dynamos," *Trans. AIEE* 81, Part III (1962): 588–594. See also E. I. Shobert II and J. E. Diehl, "A New Method of Investigating Commutation as Applied to Automotive Generators," *Trans. AIEE* 73, Part III-B (1954): 1592–1603, and E. I. Shobert II, "Commutation," *Trans. AIEE* 81, Part III (1962): 594–600.

in the short-circuited coil either by shifting the brushes in the required direction or by the use of commutating poles, to overcome the emf of self-induction produced by the reversal of the leakage fluxes in the short-circuited armature coils, the current reversal is delayed and *undercommutation* takes place. As a result, the current must change at an excessive rate toward the end of the commutating period. The delayed current reversal also produces excessive current density at the trailing brush tip and may cause the brush to overheat. In addition, if the reversal of the current is not complete when the commutator bar leaves the brush, sparking will result at the trailing brush tip. Excessive sparking burns the brushes and commutator surface. If the brushes are shifted too far in the proper direction or if the commutating field is too strong, an excess voltage is introduced in the coil undergoing commutation, causing the current to reverse prematurely, and *overcommutation* results. The reversed current may be excessive, and in this case the current density may be also excessive at the leading brush tip or at both brush tips. Figure 6-21 shows the characteristic of the short-circuit current for linear commutation, as well as for undercommutation and overcommutation, applying to motors as well as to generators.

Brushes are commonly composed of carbon, graphite, and organic materials. In the case of low-voltage machines such as automobile starting motors, the resistance between brushes and commutator must be quite low, and graphite brushes impregnated with carbon are used.

An interface film of copper oxide with a thin deposition of graphite exists between the brush and commutator surface. This film has an important effect

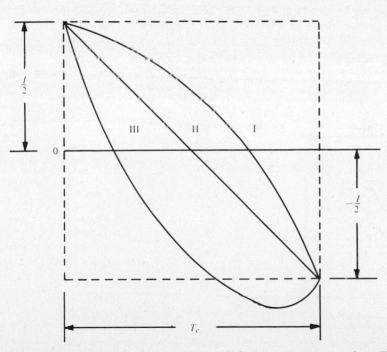

**Figure 6-21**   Current in short-circuited coil for (I) undercommutation; (II) linear commutation; (III) overcommutation.

on the life of the brush, on wear of the commutator, and on commutation itself. The graphite acts as a lubricant, although other ingredients in the brush maintain a polish on the surface of the commutator. The resistance of the film is high enough to limit the short-circuit current, thus making for what is known as *resistance commutation* (i.e., resistance that while affecting the short-circuit current has only a relatively small effect on the resistance $r_a$ of the armature circuit as viewed from the armature terminals).

## 6-7 VOLTAGE BUILDUP IN SELF-EXCITED
##     GENERATORS—CRITICAL FIELD RESISTANCE

A dc generator can furnish its own field excitation at normal speed. However, when building up without load, the resistance of the shunt-field circuit in shunt and in compound generators must be below its critical value if the voltage is to build up to the proper value. The series generator can build up only under load because the load current is also the field current.

When a shunt generator is driven at rated speed with its field circuit open, a small voltage as indicated by *oa* in Fig. 6-22 is generated in the armature if there is residual flux. If the field circuit is now connected across the armature with the correct polarity, field current is initiated, causing the flux to increase.

A resultant increase in the voltage occurs if the resistance of the field circuit is below the critical value. By the time the field current reaches the value *ob*, the terminal voltage, which is impressed on the field circuit, has the value represented by *bd*, while the resistance of the field circuit requires only the voltage *bc*. The point *c* is on the field-resistance line, a plot of the voltage $r_f I_f$ across the field

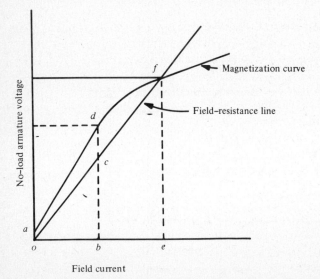

**Figure 6-22** Magnetization curve and field-resistance line of a self-excited shunt generator at constant speed.

circuit versus the field current $I_f$. The slope of this line equals the resistance of the field circuit. The difference in voltage $cd$ between the voltage $bd$ impressed on the field circuit and the resistance drop in the field circuit $bc$ equals the rate of increase in the field flux linkage (i.e., $cd = d\lambda_f/dt$). In fact, during the buildup, the voltage impressed across the field circuit is expressed by

$$v = r_f i_f + \frac{d\lambda_f}{dt} \tag{6-16}$$

where, at the instant under discussion, $v = bd$; $i_f = ob$, the field current; $\lambda_f$ = flux linkage, with the field winding. When the field current reaches the value $oe$, the terminal voltage $ef$ equals the resistance drop in the field circuit and $d\lambda_f/dt$ is zero. The no-load voltage is therefore $ef$, as determined by the intersection of the magnetization curve and the field-resistance line. A decrease in the resistance of the field circuit reduces the slope of the field-resistance line, resulting in a higher voltage if the speed remains constant. An increase in the resistance of the field circuit increases the slope of the field-resistance line, resulting in a lower voltage. The limiting position of the field-resistance line beyond which the voltage does not build up is that of approximate tangency with the lower straight (*unsaturated*) portion $ad$ of the magnetization curve in Fig. 6-22. The slope of this tangent is taken as the *critical resistance* of the field circuit. In Fig. 6-22 the critical field resistance would cause the field-resistance line to intersect the magnetization curve in the vicinity of point $d$.

A reduction in speed causes all points on the magnetization curve to be lowered accordingly, and the point of intersection $f$ would be moved downward until it reached the straight portion of the magnetization curve. The two characteristics are then tangent, and the speed at which this occurs with normal field-circuit resistance is known as the *critical speed*, below which the voltage will not build up.

Voltage fails to build up even when there is residual flux at proper values of field resistance and speed if the polarity of the field winding is incorrect. In that case the field current is in a direction as to reduce the residual flux. It is obvious that the voltage will fail to build up if there is no residual flux. Reasons for failure of voltage buildup then can be summarized as follows:

1. Insufficient residual flux
2. Incorrect polarity of the field winding
3. Speed below critical value if the resistance of the field circuit is normal
4. Field resistance above critical value if the speed is normal

## 6-8 LOAD CHARACTERISTICS OF GENERATORS

The steady-state performance of a dc generator is described by its load characteristic. Voltage drops due to inductive effects are negligible because under steady-state conditions the currents are constant or, at most, slowly varying. While the self-inductance of armature coils undergoing commutation and the mutual inductance between these coils and the rest of the armature winding influence commutation, their effect on the load characteristics of conventional dc machines is negligible.

The voltage at the armature terminals of a generator is given by

$$V = E - r_a I_a \qquad (6\text{-}17)$$

where    $E$ = generated emf

$I_a$ = armature current

$r_a$ = resistance of the armature circuit between terminals of the armature

The generated emf is expressed by Eq. 6-15, which can be abbreviated to

$$\boxed{E = k_E \phi_d \omega_m} \qquad (6\text{-}18)$$

where

$$\omega_m = \frac{2\pi n}{60} \quad \text{and} \quad k_E = \frac{PZ}{2\pi a}$$

At this point it may be well to emphasize that $\phi_d$ is the flux in the direct axis as implied by the subscript $d$. Interpoles and compensating windings are considered as part of the armature circuit, and the resistances of these windings and those of the brushes are included in $r_a$. According to Eq. 6-18, at constant speed, $E$ is proportional to the direct-axis flux $\phi_d$ and is therefore a function of the field current and armature reaction.

### 6-8.1 Separately Excited Generator

Consider a separately excited shunt generator operating at constant speed and constant field current. If there were no armature reaction, the generated emf $E$ would be constant. However, for the general case, assume armature reaction to be present. If the machine has no commutating poles, the brushes are usually shifted from geometric neutral and the armature current produces demagnetizing mmf in addition to cross-magnetizing mmf. The effect of armature reaction then is more pronounced than in a machine of the same design but with interpoles. Unless there is a compensating winding, the cross magnetization of the armature reaction, because of its demagnetizing effect, will cause a falling off of the generated voltage $E$ at constant speed as shown in Fig. 6-23(a), even when commutating poles are present. The generated voltage $E$ does not fall off linearly because the magnetization curve of the machine is nonlinear.

Since the field current of a separately excited machine is furnished by an external source, the armature current equals the load current. Separately excited generators are used on loads that require a wide variation in the output voltage, such as motors that must operate through large speed ranges. The separately excited generator remains stable even at very low field excitations, which is not true for self-excited shunt generators, as these become unstable after the terminal voltage is reduced below a critical value which is usually still a substantial fraction

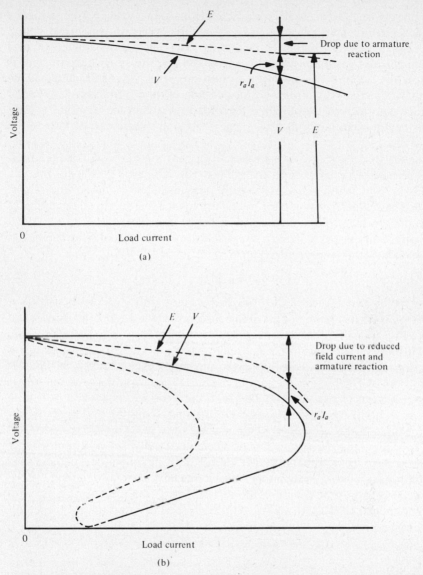

**Figure 6-23** Volt-ampere characteristics of shunt generators at constant speed. (a) Separately excited, constant field current. (b) Self-excited, constant field resistance.

of rated value. The separately excited generator has the disadvantage of requiring an external source for its field excitation.

## 6-8.2 Self-Excited Shunt Generator

Constant field current was assumed for the separately excited generator. In the self-excited shunt generator the terminal voltage is impressed on the field circuit.

Hence, the field current decreases with decreasing terminal voltage if the resistance of the field circuit is constant. As a result, even if there were no armature reaction, the direct-axis flux decreases with increasing load. The load characteristic of a self-excited shunt generator when the speed and field resistance are constant is shown in Fig. 6-23(b). The armature current increases with decreasing load resistance, reaching a maximum and then falling off to a minimum as the load resistance goes to zero. Although at zero load resistance the terminal voltage and the field current are both zero, there is some armature current due to the residual flux.

As the load resistance is increased from zero the terminal voltage builds up but, because of hysteresis, does not follow the decreasing characteristic. Instead, it builds up in a manner somewhat as represented by the dashed line in Fig. 6-23(b). The self-excited generator builds up voltage due to the residual flux as long as the resistance of the field circuit is below its critical value.

In the self-excited generator the armature furnishes the field current in addition to the load current; hence,

$$I_a = I + I_f \tag{6-19}$$

where $I_a$ is the armature current, $I$ the load current, and $I_f$ the field current.

The self-excited generator is suitable for loads which require nearly constant voltage. In a properly designed machine the voltage drop from no load to full load, when driven at constant speed, is relatively small. The self-excited generator is simpler and more economical than the separately excited generator because it requires no external source for its field current. When operating with automatic voltage regulators to adjust the field current, the terminal voltage can be held to very narrow limits for varying loads even if the prime mover undergoes appreciable changes in speed.

To achieve satisfactory values of efficiency, regulation, and temperature rise, the voltage drop in the armature circuit must be low in comparison with rated terminal voltage. For the same reasons, the field current in the self-excited shunt generator must be low in relation to the rated armature current. Accordingly, the resistance of the armature circuit must be low and that of the shunt-field winding high.

### EXAMPLE 6-3

A 10-kW 250-V self-excited generator, when delivering rated load, has an armature-circuit voltage drop that is 5 percent of the terminal voltage and a shunt-field current equal to 5 percent of rated load current. Calculate the resistance of the armature circuit and that of the field circuit.

*Solution*

$$V = 250 \text{ V, rated value}$$
$$I = 10{,}000 \div 250 = 40 \text{ A, rated load current}$$
$$I_f = 0.05 \times 40 = 2 \text{ A, field current}$$

From Eq. 6-19,

$$I_a = I + I_f = 42 \text{ A}$$
$$r_a I_a = 0.05 \times 250 = 12.5 \text{ V}$$
$$r_a = 12.5 \div 42 = 0.298 \ \Omega, \text{ the resistance of the armature circuit}$$
$$r_f I_f = 250$$
$$r_f = 250 \div 2 = 125 \ \Omega, \text{ the resistance of the field circuit}$$

### 6-8.3 Series Generator

The field circuit of a series generator is in series with the armature as shown in Fig. 6-10(c). The no-load voltage is quite low, as it depends on the residual flux. However, as load is added, the flux increases because the load current is also the field current, thus producing the characteristic shown in Fig. 6-24. The terminal voltage of the series generator is lower than the voltage at the armature terminals by an amount equal to the voltage drop in the series-field winding, so

$$V = E - (r_a + r_s)I_a \tag{6-20}$$

where $r_s$ is the resistance of the series-field circuit. The resistance of the series-field winding must be low for good efficiency and for low voltage drop.

Little use is made of the series generator. It was used in early constant-current systems by operating in a range where the terminal voltage fell off very rapidly with increasing current (i.e., in the range from point $b$ to $c$ in Fig. 6-24). After the load current reaches the value $oa$ the voltage falls through the relatively large value $ab$ and the current increases only from $oa$ to $oc$.

### 6-8.4 Compound Generator

The characteristics of compound generators are shown in Fig. 6-25. When the series- and shunt-field windings are aiding, the generator is said to be *cumulatively*

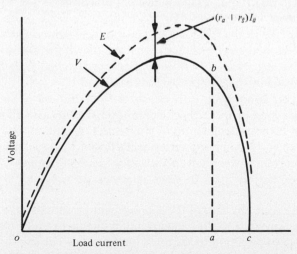

**Figure 6-24** Volt-ampere characteristic of a series generator at constant speed.

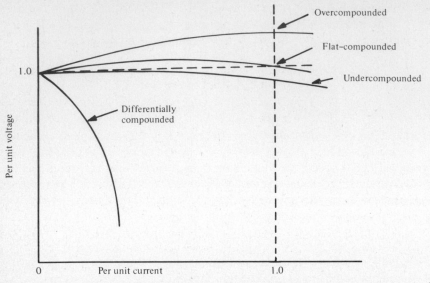

**Figure 6-25**   Volt-ampere characteristic of compound generators at constant speed.

*compounded,* and when the series winding opposes the shunt-field winding, the generator is *differentially compounded.* Cumulatively compounded generators may be overcompounded, flat compounded, and undercompounded. In the overcompounded generator the series field is strong enough to give a rising terminal voltage characteristic in the normal load range at constant speed and may thus counteract the effect of a decrease in the prime-mover speed with increasing load. Overcompounding may also be used to compensate for the line drop when the load is at a considerable distance from the generator. In the flat-compounded generator the full-load voltage and the no-load voltage are practically equal at constant speed. The undercompounded generator has a relatively weak series field and there is a small decrease in the full-load terminal voltage from its no-load value at constant speed, the decrease being somewhat less than if there were no series field (i.e., than if the machine were operating as a self-excited shunt generator). The terminal voltage of the compound generator can be related to the voltage generated in the armature by means of Eqs. 6-18, 6-19, and 6-20, taking into account whether a long or short shunt connection is used. The difference in the characteristics between long-shunt or short-shunt operation of a given machine is small, as in the former the series-field current is somewhat greater and the shunt-field current lower than in the latter mode of operation so that the resultant field excitation in both cases is about the same for a given load.

The shunt-field winding, as in the case of the shunt generator, should have a resistance large in comparison with that of the armature circuit in order to keep the shunt-field current low in relation to the rated load current. The shunt-field winding therefore has a relatively large number of turns with correspondingly small cross section of wire. The series-field winding, as with the series generator, should have low resistance since it carries all or a major portion† of the load current. It therefore has a few turns of relatively large cross section of conductor. *The shunt-field winding and the series-field winding are both placed on the main field poles.*

The differential-compound generator is used in applications where a wide variation in load voltage can be tolerated and where the generator may be exposed to load conditions approaching those of short circuit. Electric power shovels afford a good example because the motor supplied by the generator is frequently subjected to loads which produce stalling.

## 6-9 ANALYSIS OF STEADY-STATE GENERATOR PERFORMANCE

The magnetization curve and field-resistance line play dominant parts in the performance of self-excited dc generators not only in building up voltage but also in the behavior under load.

### 6-9.1 Self-Excited Shunt Generator

The performance of a self-excited shunt generator is determined by (1) speed, (2) magnetization curve, (3) resistance of the field circuit, (4) resistance of the armature circuit, and (5) demagnetizing effect of armature reaction. A self-excited shunt generator with interpoles and compensating winding has negligible armature reaction, and if the resistance of its armature circuit were zero, there would be no decrease in terminal voltage with increasing load as long as the speed and resistance of the field circuit remained constant. Figure 6-26(a) shows the magnetization curve and field-resistance line of a shunt generator that has negligible armature reaction. To simplify the discussion, the magnetization curve is drawn through the origin, thus neglecting the effect of the residual flux, and is shown only for increasing field current, thus neglecting the effect of hysteresis. Under load, the terminal voltage $V$ decreases from its no-load value $ab$ to $o'c$ and the generated voltage $E$ decreases to $d'c$. The difference $d'o'$ between $E$ and $V$ is $r_aI_a$, the armature resistance drop.

If the armature current is known, the terminal voltage is found by constructing the line $od = r_aI_a$ on the voltage axis in Fig. 6-26(a) and then drawing a line from $d$ parallel to the field-resistance line intersecting the magnetization curve at $d'$ and $d''$. The intersections $d'$ and $d''$ determine two values of terminal voltage, and if the graphical construction is carried out for values of armature current ranging from zero through the maximum value, a load characteristic similar to the decreasing portion of that shown in Fig. 6-23(b) is obtained. The maximum armature current is determined by the vertical distance between the field-resistance line and the point at which the parallel line is tangent to the

---

† The proper degree of compounding is sometimes obtained by placing a low resistance known as a *diverter* in parallel with the series-field winding.

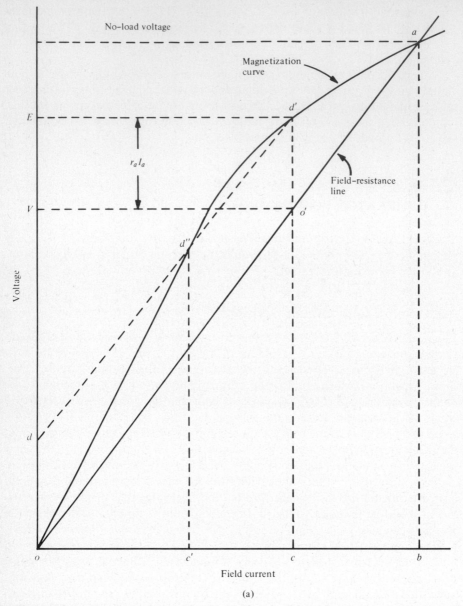

**Figure 6-26**   Graphical relationships for a self-excited shunt generator at constant speed. (a) Without armature reaction. (b) With armature reaction.

magnetization curve. The rated current in conventional shunt and compound generators is usually well below the maximum value and normal operation confined to the upper portion of the load characteristic (i.e., in a range such that the difference between the full-load voltage and no-load voltage is relatively small).

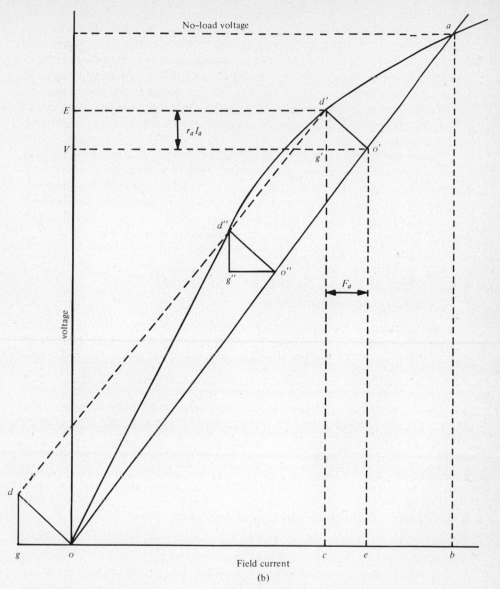

**Figure 6-26** (Continued)

When the demagnetizing effect of armature reaction is appreciable there is a further reduction in the terminal voltage. This is demonstrated in Fig. 6-26(b), where $oc$ represents the field current required to produce the generated voltage $E = d'c$ if there were no demagnetizing armature reaction. To overcome the demagnetizing mmf $F_a$ of armature reaction, the field current must include an additional component $ce = g'o'$. The armature resistance drop $r_a I_a = g'd'$ when subtracted from the generated voltage $E$ again yields the terminal voltage. The armature resistance drop and the demagnetizing mmf are both taken into account by the triangle $o'g'd'$, the sides of which are considered proportional to armature current, since $g'd' = r_aI_a$ and $g'o'$ the demagnetizing mmf of armature reaction $F_a = N_dI_a/N_f$, where $N_d$ is the equivalent demagnetizing armature turns per pole and $N_f$ is the number of turns per pole in the shunt-field winding. The procedure for obtaining the terminal voltage for a given value of armature current is similar to that applied to Fig. 6-26(a) except that the line $od$ is drawn parallel to the hypotenuse of the triangle $o'g'd'$ in Fig. 6-26(b). The test method for evaluating the demagnetizing mmf of armature reaction in terms of equivalent field current is discussed in Sec. 6-10.

While the triangle $ogd$ serves to explain the effects of armature resistance and armature reaction, it does not lend itself readily to accurate computation. For one thing, the voltage across the electric arc which carries the current from the commutator surface to the brush surface is nearly constant over a wide range of currents. There is also the hysteresis effect which gives the magnetization curve a different shape for increasing flux than for decreasing flux.

## 6-9.2 Effect of Speed on Shunt Generator Performance

In order to maintain the armature voltage at a definite value when the speed is increased, it is necessary to increase the resistance of the field circuit. This brings the field-resistance line and the magnetization curve closer to each other in a self-excited shunt generator, and if the speed is made high enough, the field resistance will exceed the critical value, resulting in a loss of excitation. Separate field excitation or special regulating features might therefore be required if the generator is driven well above its rated speed. Also in the case of generators without commutating poles and a compensating winding the armature mmf might be too strong relative to the reduced field mmf, thus causing poor commutation.

## 6-9.3 Series Generator Graphical Analysis

The graphical construction as applied to the performance of series generators is illustrated in Fig. 6-27. Part of the mmf produced by the series field is overcome by the demagnetizing mmf of armature reaction. Since both of these mmfs are proportional to the same current (i.e., the armature current), the resultant can be combined in the triangle $o'g'd'$ in Fig. 6-27 with the direction of its base reversed from that in Fig. 6-26(b).

In Fig. 6-27 the load current, which is also the current in the series field, is represented by $oe$, while the demagnetizing effect of the armature mmf is represented by $F_a = ce$ and the net excitation by $oc$. The vertical distance $d'g'$ is the armature-plus-field-resistance drop $(r_a + r_s)I_a$. The current for any other terminal voltage $V$ is found by drawing line $o''d''$ parallel to $o'd'$ and by completing the triangle $o''d''g''$. The current $I_a = d''g'' \div (r_a + r_s)$.

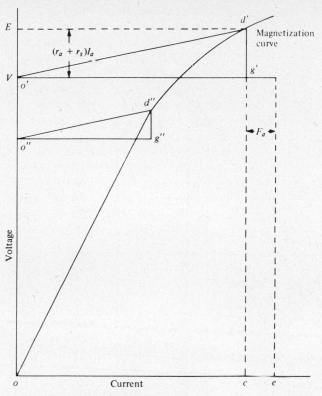

**Figure 6-27** Graphical relationships for a series generator.

## 6-9.4 Compound Generator

The graphical construction for the cumulatively compound generator is obtained by a method which combines the effects of the shunt and series fields. This is done in Fig. 6-28 by including the field-resistance line with the magnetization curve and by making use of the triangle $o'g'd'$ in which the altitude $d'g' = (r_a I_a + r_s I_s)$, where $I_s$ is the current in the series-field winding and $r_s$ is the resistance of the series-field circuit. The base of the triangle $o'g' = hc = F_s - F_a$, where $F_s = N_s I_s / N_f$, in which $N_s$ is the number of series-field turns per pole and $N_f$ is the number of shunt-field turns per pole. The shunt-field current required to overcome the demagnetizing mmf of armature reaction is $ce = F_a$.

The terminal voltage for a given value of armature current is found by drawing the triangle $ogd$ with its vertex $o$ at the origin as shown in Fig. 6-28. The size of the triangle as in previous constructions is determined by the armature current. Then from $d$ a line is drawn parallel to the field-resistance line intersecting the magnetization curve at $d'$. Next, a line is drawn parallel to $od$ from $d'$ on the field-resistance line. The triangles $o'g'd'$ and $ogd$ are equal. The terminal voltage $V = ho'$.

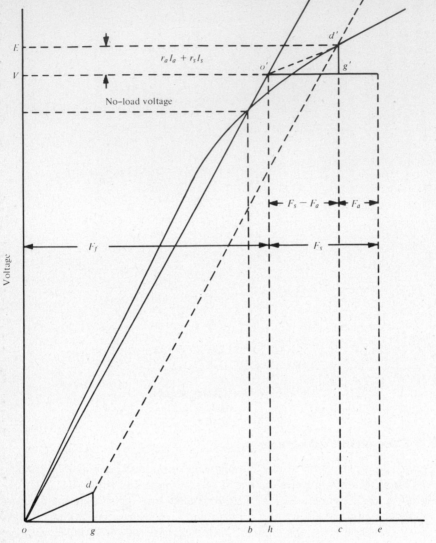

**Figure 6-28**  Graphical relationships for a cumulative-compound generator.

In some cases it is more convenient to plot the field excitation in terms of ampere turns per pole instead of shunt-field current, as shown by the upper horizontal scale in Fig. 6-29.

**EXAMPLE 6-4**

The upper curve in Fig. 6-29 is the magnetization curve of a 500-V 500-kW shunt six-pole dc generator obtained at a speed of 1100 rpm. The resistance of the armature circuit including brushes is 0.024 Ω. This value

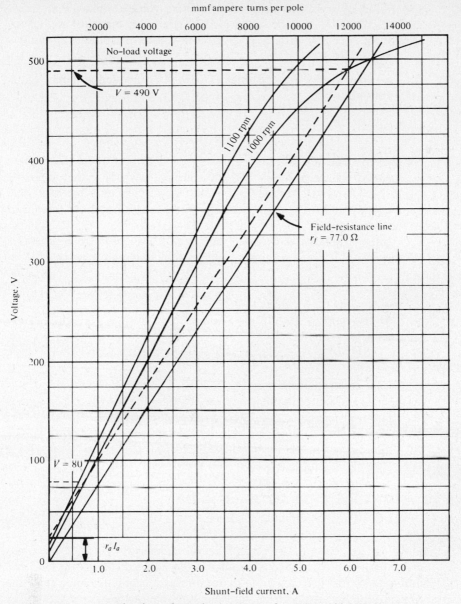

**Figure 6-29**   Determination of terminal voltage of a 500-V 500-kW dc generator, neglecting armature reaction (Example 6.4).

of resistance is about four times that for a practical machine. This exaggerated value is used to lead to larger values of armature resistance drop for purposes of illustration. (a) Calculate the resistance of the shunt-field circuit such that the no-load voltage is 500 V at a speed of 1000 rpm. (b)

Determine, by graphical means, the terminal voltage when the armature current is 1000 A and the speed is 1000 rpm with the resistance of the shunt-field circuit as in part (a). Neglect the effect of armature reaction.

*Solution*

a. Since $E = k_E\phi_d\omega_m$ from Eq. 6-18, the generated voltage $E_{1000}$ at 1000 rpm must be

$$E_{1000} = \frac{1000}{1100} E_{1100}$$

for a given value of field current. Thus, a field current of 6.5 A produces a generated voltage of 550 V at 1100 rpm as shown by extrapolation in Fig. 6-29. The voltage for the same value of field current is 500 V at 1000 rpm. Other points on the magnetization curve for 1000 rpm are calculated in the same manner.

Since a field current of 6.5 A is required to produce a generated voltage of 500 V, the resistance of the shunt-field circuit must be

$$r_f = \frac{500}{6.5} = 77.0 \ \Omega$$

b. The voltage drop in the armature circuit at an armature current of 1000 A is

$$r_a I_a = 1000 \times 0.024 = 24 \ \text{V}$$

This value corresponds to *od* in Fig. 6-26(a). The dashed line drawn parallel to the field-resistance line starting from 24 V on the voltage axis intersects the 1000-rpm magnetization curve at $E = 490$ and at 80 V. Hence, for rated current the terminal voltage is $490 - 24 = 466$ V at 1000 rpm. The 80-V point is below the normal operating range.

## 6-10  ARMATURE CHARACTERISTIC OR FIELD-COMPOUNDING CURVE

The demagnetizing effect of armature reaction is determined for dc machines by test. In that test the machine is usually driven at rated speed with load and the shunt-field current is increased to maintain rated terminal voltage while the armature current is increased from zero to somewhat beyond rated value. The field current is plotted against armature current as shown by the solid curve in Fig. 6-30. If there were no armature reaction, the field current required to maintain constant terminal voltage could be determined on the basis of the generated voltage $E = V + r_a I_a$, from the magnetization curve. This result when plotted against armature current yields the dashed curve in Fig. 6-30. The vertical distance between this dashed curve and the solid upper curve represents the field current required to overcome the demagnetizing effect of armature reaction. This com-

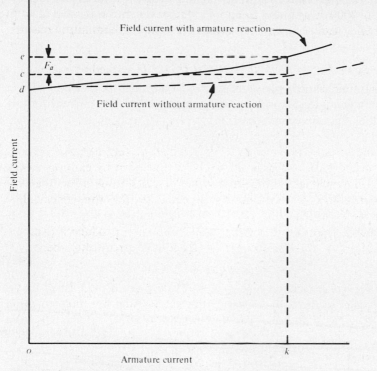

**Figure 6-30** Armature characteristic or compounding curve of a shunt generator with constant terminal voltage at rated value while driven at constant speed.

ponent of field current is also the base of the triangle in Fig. 6-26(b) for the shunt generator. It is shown as $F_a$ in Fig. 6-30 for the shunt generator.

### EXAMPLE 6-5

The armature characteristic of the shunt generator in Example 6-4 obtained at a speed of 1100 rpm is represented by the solid curve in Fig. 6-31. Calculate (a) the field current required to overcome the demagnetizing effect due to an armature current of 1000 A, and (b) the terminal voltage at 1000 rpm for an armature current of 1000 A, taking into account the effect of armature reaction. The resistance of the field circuit is 77.0 Ω.

*Solution*

a. The armature characteristic is obtained at rated voltage. Therefore, the generated voltage for an armature current of 1000 A is

$$E = V + r_a I_a$$
$$= 500 + 0.024 \times 1000 = 524 \text{ V}$$

Figure 6-29 shows that the field current required for this generated voltage at 1100 rpm is 5.60 A. This locates point $x$ on the dashed-line curve in

Fig. 6-31. The total field current for rated terminal voltage at an armature current of 1000 A was found to be 5.93 A from test and is located at point $y$ on the solid (upper) curve. The demagnetizing mmf of armature reaction therefore corresponds to

$$F_a = 5.93 - 5.60 = 0.33 \text{ A}$$

Assuming a linear relationship, which although not strictly correct makes for a good approximation, the component of field current for any other value of armature current $I_a$ is

$$F_a = \frac{0.33}{1000} I_a$$

b. The graphical construction for obtaining the terminal voltage is shown in Fig. 6-32. In accordance with the procedure discussed in Sec. 6-9 and Fig. 6-26(a) the line $od$ is drawn through the origin with a slope corresponding to

$$-\frac{r_a I_a}{F_a} = -\frac{24 \text{ V}}{0.33 \text{ A}} \qquad (6\text{-}21)$$

with the vertical projection $r_a I_a = 24$ V. From point $d$ the dashed line is drawn parallel to the field-resistance line intersecting the magnetization curve at $E = 240$ V and $E = 475$ V. Thus, the terminal voltage falls from

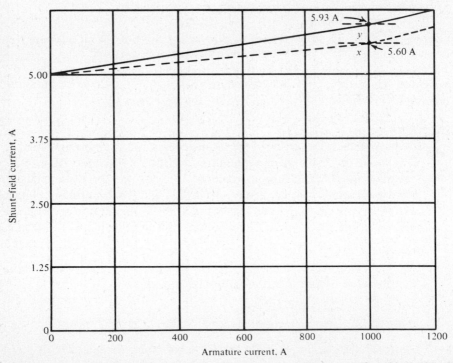

**Figure 6-31**  Armature characteristic or field-compounding curve of a 500-V 500-kW dc generator (Example 6.5).

its no-load value of 500 V to $475 - 24 = 451$ V at full load. The lower value of $240 - 24 = 216$ V is below the normal operating range.

The terminal voltage of 451 V, when armature reaction is taken into account, compares with a value of 466 V in Example 6-4 in which armature reaction is neglected.

Examples 6-4 and 6-5 illustrate the effect of armature-resistance drop and armature reaction on the terminal voltage. While the method of analysis is based

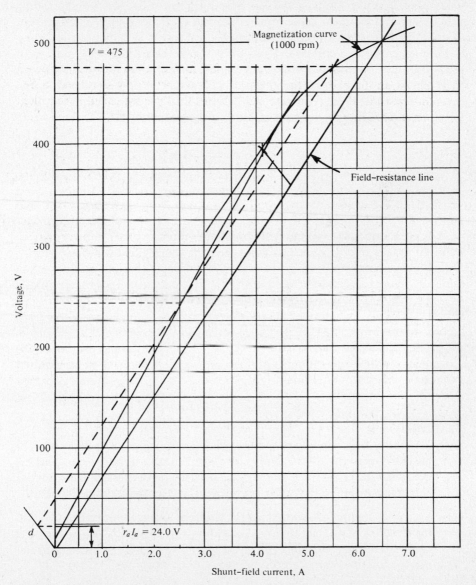

**Figure 6-32** Determination of terminal voltage of a 500-V 500-kW dc generator, taking account of armature reaction (Example 6.5).

on simplifying assumptions, which are not realized exactly in conventional machines, it yields results that are fair approximations.

## 6-11 COMPOUNDING A GENERATOR

It is possible to convert a shunt generator to a series generator of the same full-load rating simply by replacing the shunt-field winding with a series-field winding which develops about the same mmf or ampere turns per pole at rated current as does the shunt-field winding. A more practical situation is that of converting a shunt generator to a compound generator by adding a series-field winding of the appropriate number of turns to the main field poles. Flat compounding can be achieved on the basis of the armature characteristic or field-compounding curve in Fig. 6-31. If the resistance of the series-field winding is neglected, then for flat compounding the number of series turns per pole are

$$N_s = \frac{N_f \Delta I_f}{I_{a(\text{rated})}} \qquad (6\text{-}22)$$

where $\Delta I_f$ is the increase in the shunt-field current from no load to rated load and where $I_{a(\text{rated})}$ is the rated armature current with $N_f$ the number of turns per pole in the shunt-field winding. A slightly larger number of series turns may be required to take into account the resistance of the series-field winding.

For overcompounding, a characteristic similar to that in Fig. 6-31 is obtained by test except that at rated armature current the shunt-field current is adjusted to give a value of terminal voltage corresponding to the desired degree of compounding. The number of series turns are determined from Eq. 6-22 as before. Again, an increase in the series turns may be required because of the resistance drop in the series winding. It is sometimes convenient to provide the series-field winding with a number of turns somewhat in excess of those required and then connect a low-resistance shunt known as a *series-field diverter* to bypass a part of the total current.

**EXAMPLE 6-6**

The shunt-field winding of the 500-V 500-kW generator in Example 6-5 has 2000 turns per pole. Calculate the number of turns per pole in a series-field winding for flat compounding at 1100 rpm on the basis of the armature characteristic in Fig. 6-31. Neglect the resistance of the series-field winding.

*Solution.* Figure 6-31 shows that the shunt-field current must be increased from 5.00 A at no load to 5.93 A at full load ($I_a = 1000$ A). The increase in the shunt-field current is therefore $I_f = 5.93 - 5.00 = 0.93$ A and the number of series turns, according to Eq. 6-22, must be

$$N_s = \frac{2000 \times 0.93}{1000} = 1.86 \text{ turns per pole}$$

The nearest whole number to 1.86 is 2. However, two turns might produce a small amount of overcompounding, which could be corrected by using a series-field diverter to shunt part of the armature current away from the series-field winding. Then $1.86 \times 1000 = 1860$ ampere turns required, and the two-turn series-field winding should carry a current of 1860/2 $= 930$ A and the diverter should bypass $1000 - 930 = 70$ A. The resistance of the diverter should therefore be 930/70, or 13.3 times that of the series-field winding.

Whether the generator is connected long-shunt as in Fig. 6-10(d) or short-shunt as in Fig. 6-11 makes no practical difference in conventional machines as far as the degree of compounding is concerned.

## 6-12 EFFICIENCY AND LOSSES

Efficiency is defined as the ratio of the useful output to the input, and the expression preferred for calculations of efficiency is

$$\boxed{\text{Efficiency} = 1 - \frac{\text{losses}}{\text{input}}} \tag{6-23}$$

Thus, if the losses are known for a given output, the input is the sum of the losses and the output and the efficiency can be calculated to a high degree of accuracy even if there are small errors in the losses.

The losses in rotating electric machines may be classified as electric losses and rotational losses. The electric losses in dc machines (motors as well as generators) include the $I^2R$ losses in the field circuits and armature circuits, while the rotational losses include windage and friction plus brush friction losses, core losses, and stray-load losses. The copper losses[†] are calculated for a temperature of 75°C.

The stray-load losses result from the load on the machines and do not lend themselves to direct measurement or accurate calculations. Effects that contribute to the stray-load losses are distortion of the flux-density distribution wave causing the core losses, which do not vary linearly with flux density, to increase from their no-load value; eddy currents produced in the coils undergoing commutation by the reversal of the flux due to the load current; circulating currents between armature coils by the brushes covering several commutator segments which is normally the case; and so on. According to ANSI Standard C50.4,[‡] a value of 1 percent of the output is assumed as the stray-load loss. Table 6-1 shows the per-unit range of losses for conventional dc machines rated from about 1 to 100 kW. The lower limits of losses apply to the higher ratings (i.e., around 100 kW).

[†] For test procedures for dc machines, see *Rotating Electrical Machinery, ANSI Standards C50.4* (New York: American National Standards Institute); *Test Code for Direct-Current Machines, IEEE No. 113* (New York: Institute of Electrical and Electronic Engineers).

[‡] *Ibid.*

**TABLE 6-1.** DISTRIBUTION OF LOSSES IN DC MACHINES IN RATINGS FROM ABOUT 1 TO 100 kW

| Type of loss | Range per unit |
|---|---|
| No-load rotational | 0.02–0.14 |
| Armature copper including series field, if any | 0.03–0.06 |
| Shunt-field copper | 0.01–0.05 |
| Stray-load | 0.01 |

For efficiency calculations the resistance voltage drop between the brushes and commutator is assumed constant at 2 V, which is sometimes called the *arc drop*. The resistance of the armature circuit then does not include the resistance between brushes and commutator. The brush-contact losses are therefore

$$\text{Brush-contact loss} = 2I_a \tag{6-24}$$

**EXAMPLE 6-7**

The following data applies to a 100-kW 250-V six-pole 900-rpm compound generator (long-shunt connection):

$$\text{No-load rotational losses} = 3840 \text{ W}$$
$$\text{Armature resistance at } 75°C = 0.012 \ \Omega$$
$$\text{Series-field resistance at } 75°C = 0.004 \ \Omega$$
$$\text{Commutating-pole field resistance at } 75°C = 0.004 \ \Omega$$
$$\text{Shunt-field current} = 2.60 \text{ A}$$

Assume a stray-load loss equal to 1 percent of the output and calculate the rated-load efficiency.

*Solution.* The total resistance of the armature circuit, not including brushes, is the sum of the resistance of the armature, series-field, and commutating-pole field windings:

$$r_a = 0.012 + 0.004 + 0.004 = 0.020 \ \Omega$$

The armature current is the sum of the load current and the shunt-field current, or

$$I_a = \frac{100,000}{250} + 2.6 = 402.6$$

The losses may be tabulated and totaled as follows:

| | |
|---|---|
| No-load rotational losses | 3840 |
| Armature-circuit copper losses = $(402.6)^2 \times 0.02$ | 3240 |
| Brush-contact loss $(2I_a) = 2 \times 402.6$ | 805 |
| Shunt-field circuit copper losses = $250 \times 2.6$ | 650 |
| Stray-load loss = $0.01 \times 100,000$ | 1000 |
| Total losses | 9535 |

The input at rated load is therefore

$$100,000 + 9535 = 109,535 \text{ W}$$

and the efficiency is, from Eq. 6-23,

$$\text{Efficiency} = 1 - \frac{\text{losses}}{\text{input}}$$

$$= 1 - \frac{9535}{109,535} = 0.915$$

## 6-13 MOTOR TORQUE

Conventional dc motors fall into the same classification as generators—shunt, series, and compound. In fact, the dc generator may be operated as a dc motor, and vice versa. Whether a dc machine operates as a motor or as a generator, rotation of the armature in a magnetic field generates an emf in the armature just as in the case of the generator. Equations 6-16 and 6-18 apply to motors as well as to generators. In a motor the direction of the armature current is opposite that for generator operation for the same direction of rotation if the voltage polarities are unchanged. If $V$ is the voltage applied to the armature terminals, then

$$V = E + r_a I_a \tag{6-25}$$

in which $E$ is the generated voltage. The power input to the armature of the motor is

$$P_{\text{in}} = V I_a = E I_a + r_a I_a^2$$

where $r_a I_a^2$ is the armature copper loss. The difference between the power input to the armature and the armature copper loss is the electromagnetic power, which is converted into mechanical power. If $T_{em}$ is the developed torque, the developed mechanical power must be

$$P_{em} = T_{em}\omega_m = E I_a \tag{6-26}$$

and the electromagnetic torque, from Eqs. 6-18 and 6-27, is found to be

$$\boxed{T_{em} = k_E \phi_d I_a} \tag{6-27}$$

## 6-14 SPEED-TORQUE CHARACTERISTICS

An understanding of the speed-torque characteristic of a dc motor can be gained from Eqs. 6-18, 6-25, and 6-27. In shunt motors the load current influences the flux only through armature reaction, and its effect is therefore relatively small, while in the series motor the flux is largely determined by the armature current, which is also the field current. In the compound motor the effect of the armature current on the flux depends on the degree of compounding.

### 6-14.1 Shunt Motor

The circuit of the shunt motor is the same as that of the self-excited shunt generator shown in Fig. 6-10(a). However, in the motor the line supplies both the armature and the field so that the directions of the line current and the armature current are reversed from those indicated. The line current is therefore

$$I = I_a + I_f \tag{6-28}$$

where $I_a$ and $I_f$ are the armature and field currents, respectively.

In a shunt motor with commutating poles and compensating winding, the flux per pole is practically unaffected by the armature current and is therefore constant for constant field current. Then, on the basis of Eq. 6-27, the armature current and the torque are proportional to each other. Hence, if the motor is running at a certain speed and the torque demanded by the load increases, the speed decreases until the armature current increases to the value required by the increased torque. The speed-torque characteristic in that case is linear, as shown by curve $A$ in Fig. 6-33. In the absence of interpoles and a compensating winding there is some weakening of the field due to armature reaction and, for a given value of resistance in the armature circuit, the speed is more nearly constant as shown by curve $B$ in Fig. 6-33. In the case of pronounced armature reaction, the speed may actually increase after the torque exceeds a certain value causing the motor to become unstable.

The shunt motor can be made to operate efficiently over a range of speeds by varying the field current, as is evident from the following relationship based

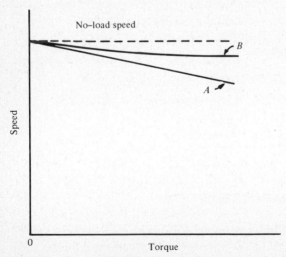

**Figure 6-33**  Speed-torque characteristic of a shunt motor: $A$ without armature reaction; $B$ with armature reaction.

on Eqs. 6-18 and 6-25 when the armature resistance drop $r_aI_a$ is small compared with the voltage $V$ applied to the armature:

$$\omega_m = \frac{V_a - r_aI_a}{\phi_d k_E} \tag{6-29}$$

Since the flux is a function of the field current, as represented by the magnetization curve, a decrease in the field current produces an increase in the motor speed, and vice versa. While speed ranges as high as 6:1 are not uncommon, economic considerations restrict the speed for very large motors to a range of about 2:1.

Shunt motors are also used in applications which require nearly constant speed but do not require high starting torque. Examples are fans, blowers, centrifugal pumps, and machine tools.

## 6-14.2 Series Motor

Since the series motor has its field in series with the armature, the armature current furnishes the field excitation. Consequently, as the armature current increases the flux also increases. In the linear region of magnetization, the flux is almost directly proportional to the armature current and the torque is then approximately proportional to the square of the current. However, when the iron is saturated, there is only a gradual increase in the flux with increasing current, and the torque increases in a lesser proportion than the square of the current but somewhat greater than the current to the first power. When the torque demanded by the load is low, a correspondingly low value of flux is required, and it is evident from Eq. 6-29 that the speed is correspondingly high and may reach destructive values at very light load. At heavy loads the flux is high and the speed is correspondingly low. The speed of the series motor is therefore sensitive to load as shown in Fig. 6-34, and its starting torque is high because the heavy starting current also produces a high value of flux.

The series motor is suitable for electric railways, buses, hoists, cranes, and other applications which require very high starting torques, where varying speed is not objectionable and where the motor, under normal operation, always drives an appreciable load.

## 6-14.3 Compound Motor

The compound motor just as the compound generator has a series- and a shunt-field winding mounted on the main field poles. If the compound motor is to have a starting torque which is high compared with that of an equivalent shunt motor but considerably lower than that of a corresponding series motor, the shunt-field winding predominates and the series-field winding is the smaller of the two. Such a motor also has fairly constant speed and is suited for pulsating loads with flywheel action and as drives for such loads as plunger pumps, punch presses, crushers, conveyors, and hoists. A small series-field winding, known as

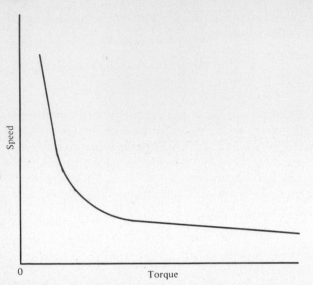

**Figure 6-34**   Speed-torque characteristic of a series motor.

a *stabilizing winding,* is used in some shunt motors to overcome the demagnetizing effect of armature reaction, thus preventing instability. On the other hand, a small shunt-field winding is used in some series motors to prevent excessive speed at or near no load.

Figure 6-35 shows a comparison of the speed characteristics for various types of electric motors in which 1.0 per unit represents rated values.

## 6-15  STEADY-STATE CHARACTERISTICS OF
## THE SHUNT MOTOR

The field current in the shunt motor is unaffected by the load as long as the voltage applied to the motor is constant. It is a fairly simple matter to calculate the steady-state speed, the mechanical power, and the torque of a shunt motor when its magnetization curve and other data are given. The magnetization curve is given for a particular speed usually at or near rated value and can be used to calculate the counter emf of the motor, for a given field excitation, at any other speed. If the magnetization curve is obtained by driving the motor as an unloaded generator at a speed of $n$, and if the voltage on the magnetization curve for a given value of field current is $E$, then at any other speed $n'$, the generated emf or counter emf $E'$ is given by

$$E' = \frac{n'}{n} E \qquad (6\text{-}30)$$

A procedure for calculating the performance of a shunt motor is illustrated in the following example.

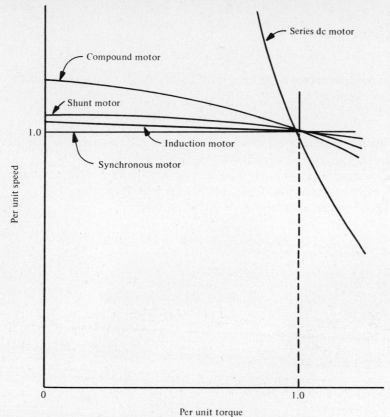

**Figure 6-35** Typical speed-torque characteristic of motors.

### EXAMPLE 6-8

Figure 6-36 shows a portion of the magnetization curve obtained at 1800 rpm on a 25-hp 250-V 84-A shunt motor. The resistance of the shunt-field circuit, including the field rheostat, is 184 $\Omega$ and the resistance of the armature circuit, including the commutating or interpole winding and brushes, is 0.082 $\Omega$. The field winding has 3000 turns per pole.

The demagnetizing mmf $F_a$ of armature reaction, at rated armature current, is 0.09 A in terms of the shunt-field current. The no-load losses, which include windage and friction losses as well as core losses, are 1300 W. The stray-load loss is assumed to be 0.01 of the output. The resistance of the field circuit is assumed constant at 184 $\Omega$. Calculate (a) the speed of the motor when it draws 84 A from the line, (b) the electro-mechanical power, (c) the mechanical power output, (d) the output torque, and (e) the efficiency.

*Solution*

a. The field current is $I_f = 250 \div 184 = 1.36$ A and the armature current is, from Eq. 6-28,

$$I_a = I - I_f = 84 - 1.36 = 82.6 \text{ A}$$

The counter emf is

$$E' = V - r_a I_a = 250 - 0.082 \times 82.6 = 250 - 6.8 = 243.2$$

The net field excitation is the field mmf minus the demagnetizing mmf of armature reaction:

$$I_{f(\text{net})} = 1.36 - 0.09 = 1.27 \text{ A}$$

Since the armature current is near rated value, the value of 0.09 can be used as the demagnetizing mmf with negligible error.

The magnetization curve in Fig. 6-36 shows that a field current of 1.27 A produces a generated voltage of 241.5 V at 1800 rpm.

The motor speed, from Eq. 6-30, is therefore

$$n' = \frac{E'}{E} n = \frac{243.2}{241.5} \times 1800 = 1813 \text{ rpm}$$

b. The electromechanical power is

$$P_{em} = E'I_a = 243.2 \times 82.6 = 20,088 \text{ W}$$

c. The mechanical power is the electromechanical power minus the sum of the rotational losses plus the stray-load losses:

$$P_{\text{mech}} = P_{em} - (P_{\text{rot}} + P_{\text{stray}})$$

$$= 20,088 - 1300 - 0.01 P_{\text{mech}}$$

$$= 18,600 \text{ W} \quad \text{or} \quad \frac{18,600}{746} = 24.9 \text{ hp}$$

d. $\quad \text{Torque} = \dfrac{P_{\text{mech}}}{\omega_m} = \dfrac{18,600}{2\pi \times 1813/60} = 98.02 \text{ N-m/rad}$

or $\qquad 98.02 \times 0.738 = 72.35 \text{ lb-ft/rad}$

e. Losses

| | | |
|---|---|---|
| Rotational | = | 1,300 W |
| Stray load = $0.01 \times 18,560$ = | | 185 W |
| $I_a^2 r_a = (82.6)^2 \times 0.082$ | = | 560 W |
| $I_f^2 r_f = (1.36)^2 (250)$ | = | 462 W |
| Total losses | = | 2,507 W |
| Output | = | 18,600 W |
| Input | | 21,107 W |

$$\text{Efficiency} = 1 - \frac{\text{losses}}{\text{input}} = 1 - \frac{2507}{21,107} = 0.881$$

It is interesting to note that the full-load speed of the motor in Example 6-8 is greater than the no-load (1800 rpm) speed due to $F_a$. This suggests that

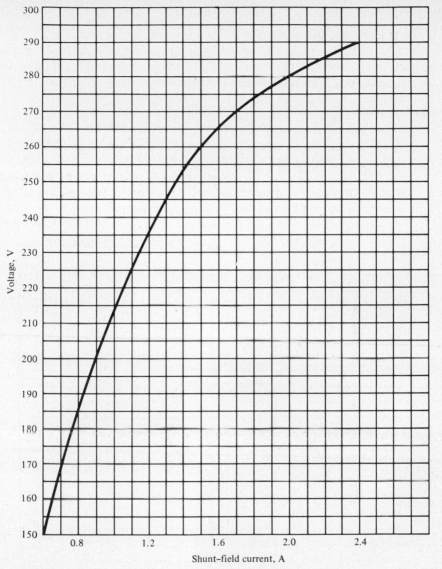

**Figure 6-36**  Portion of magnetization curve for 25-hp 250-V 1800-rpm shunt motor in Example 6.8.

the motor is somewhat unstable. The motor can be given a drooping speed characteristic by adding a series- or stabilizing-field winding of a few turns.

## 6-16  STEADY-STATE PERFORMANCE CHARACTERISTICS OF THE SERIES MOTOR

Since the field of the series motor is in series with the armature, the counter emf must be

$$E = V - (r_a + r_s)I \qquad (6\text{-}31)$$

where $r_a$ and $r_s$ are the resistances of the armature circuit and of the field, respectively. The performance of the series motor is analyzed on the basis of the magnetization curve, the resistances of the armature circuit and series field, and the demagnetizing mmf of armature reaction.

**EXAMPLE 6-9**

Figure 6-37 shows the magnetization curve of a 150-hp 250-V 510-A dc series motor. The resistance of the armature circuit, including the commutating-pole winding and brushes, is 0.0127 Ω and that of the series-field winding is 0.0087 Ω. The field winding has 10 turns per pole. The demagnetizing mmf of armature reaction is 250 ampere turns per pole at rated current and is assumed to vary linearly with current.

Calculate the following for currents of (a) 510 A, (b) 255 A: (1) speed, (2) developed mechanical power, and (3) developed torque.

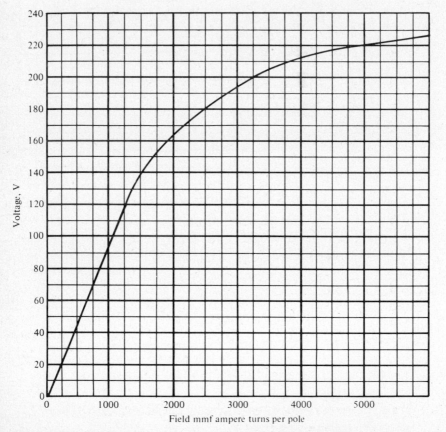

**Figure 6-37**   Magnetization curve for 150-hp dc series motor taken at 900 rpm. Series field has 10 turns per pole.

*Solution*

a. (1)  At 510 A the counter emf is

$$E' = V - (r_a + r_s)I_a = 250 - 510(0.0127 + 0.0087)$$
$$= 250 - 510 \times 0.0214 = 250 - 10.9$$
$$= 239 \text{ V}$$

The field mmf is

$$N_s I = 10 \times 510 = 5100 \text{ ampere turns per pole}$$

The net mmf is obtained by subtracting the demagnetizing mmf of armature reaction from the field mmf, thus:

$$NI = 5100 - 250 = 4850 \text{ ampere turns per pole}$$

According to Fig. 6-37, an mmf of 4850 ampere turns produces an emf of 219.5 V at a speed of 900 rpm. The speed is, therefore, from Eq. 6-30,

$$n' = \frac{E'}{E} n = \frac{239}{219.5} \times 900 = 980 \text{ rpm}$$

(2)  The electromagnetic power is, from Eq. 6-26,

$$P_{em} = E'I_a = 239 \times 510 = 122,000 \text{ W}$$
$$= 122,000 \div 746 = 163.5 \text{ hp}$$

(3)  The electromagnetic torque is, from Eq. 6-26,

$$T_{em} = \frac{P_{em}}{\omega_m} = \frac{122,000}{2\pi \times 980/60} = 1190 \text{ N-m/rad}$$

$$= 1190 \times 0.738 = 878 \text{ lb-ft/rad}$$

b. (1)  At 255 A the counter emf is

$$E' = V - (r_a + r_s)I_a = 250 - 0.0214 \times 255 = 250 - 5.5$$
$$= 244.5 \text{ V}$$

The field mmf is $N_s I = 10 \times 255 = 2550$ ampere turns per pole, the demagnetizing mmf of armature reaction is $(255/510) \times 250 = 125$ ampere turns, and the net mmf is $NI = 2550 - 125 = 2425$ ampere turns per pole, which produces an emf of 178 V at 900 rpm. The speed is therefore

$$n' = \frac{244.5}{178} \times 900 = 1235 \text{ rpm}$$

(2)  The electromagnetic power is

$$P_{em} = E'I_a = 244.5 \times 255 = 62,400 \text{ W}$$
$$= 62,400 \div 746 = 83.6 \text{ hp}$$

(3) The electromagnetic torque is

$$T_{em} = \frac{P_{em}}{\omega_m} = \frac{62{,}400}{2\pi \times 1235/60} = 482 \text{ N-m/rad}$$

$$= 482 \times 0.738 = 356 \text{ lb-ft/rad}$$

## 6-17 COMPOUND-MOTOR STEADY-STATE PERFORMANCE CHARACTERISTICS

The characteristics of the cumulatively compounded motor are a compromise between those of the shunt and the series motor. The analysis of compound motor performance is similar to those of the shunt and series motors and is illustrated below.

**EXAMPLE 6-10**

The field of the series motor in Example 6-9 is replaced by a shunt field of 600 turns and a series field of 4 turns per pole connected cumulatively for compound motor operation at 250 V and connected as shown in Fig. 6-10(d). The armature winding is unchanged. The resistance of the shunt-field winding is 46.5 $\Omega$ and that of the series-field winding is 0.0037 $\Omega$.

Calculate (a) the no-load speed and (b) the (1) speed, (2) developed mechanical power, and (3) developed torque, when the armature current is 510 A.

*Solution*

a. The no-load armature current is negligible. The mmf of the series field is assumed to be zero. The shunt-field current is therefore

$$I_f = \frac{V}{r_f} = \frac{250}{46.5} = 5.38 \text{ A}$$

and the mmf of the shunt field is $N_f I_f = 600 \times 5.38 = 3220$ ampere turns per pole. Then from Fig. 6-37 the generated emf at 900 rpm is 199.5 V. Since the armature current is negligible at no load, the counter emf is 250 V and the no-load speed is

$$n' = \frac{250}{199.5} \times 900 = 1125 \text{ rpm}$$

b. The shunt-field current is independent of the armature current. Hence the mmf of the shunt-field winding is 3220 ampere turns per pole. However, the series-field winding carries a current of 510 A and therefore develops an mmf of $N_s I_s = 4 \times 510 = 2040$ ampere turns. There is also the demagnetizing mmf of 250 ampere turns due to armature reaction; so the net field mmf is

$$F_{\text{net}} = 3220 + 2040 - 250 = 5010 \text{ ampere turns per pole}$$

The counter emf is

$$E' = V - (r_a + r_s)I_a = 250 - (0.0127 + 0.0037)510$$
$$= 250 - 0.0164 \times 510 = 250 - 8.4 = 241.6 \text{ V}$$

(1)  The generated emf at 900 rpm and a net field mmf of 5010 ampere turns per pole is 220.5 V from Fig. 6-37, and the speed is therefore

$$n' = \frac{241.6}{220.5} \times 900 = 986 \text{ rpm}$$

(2)  The electromagnetic power is

$$P_{em} = E'I_a = 241.6 \times 510 = 123,000 \text{ W}$$
$$= 123,000 \div 746 = 165 \text{ hp}$$

(3)  The electromagnetic torque is

$$T_{em} = \frac{P_{em}}{\omega_m} = \frac{123,000}{2\pi \times 986/60} = 1193 \text{ N-m/rad}$$

$$= 1193 \times 0.738 = 880 \text{ lb-ft/rad}$$

## 6-18 MOTOR STARTING

Since the motor is at a standstill on starting and the counter emf is zero, the armature starting current is limited by the resistance of the armature circuit. Except for small motors, external resistance is introduced into the armature circuit to limit the current to about 1.5 or 2 times rated value, although in some cases the starting current may be as high as four times rated value. Good commutation during starting as well as the prevention of overloads on supply feeders to the motor require that the starting current be held to values of the order indicated above. A four-point starting box connected to a shunt motor is shown in Fig. 6-38. The series resistance in that figure is cut out in six steps with values of resistance between steps such as to keep the initial current at each step within the proper limits as the motor accelerates.

The rheostat is held in the running position by force exerted by the holdup magnet against the torque of the spiral restraining spring. If, as the result of a disturbance, the line voltage drops to too low a value, the restraining spring overcomes the reduced force of the holdup magnet and returns the arm of the starting box to the off position. This prevents an excessive inrush of current should the line voltage suddenly be restored to normal after an interruption or severe reduction of the voltage. The following example illustrates the effect of a starting box on the starting and acceleration of a shunt motor.

**EXAMPLE 6-11**

A 7.5-hp 250-V 26-A 1800-rpm shunt motor is started with a four-point starting box as in Fig. 6-38(a). The resistance of the armature circuit including the interpole winding is 0.48 Ω and the resistance of the shunt-field circuit, including the shunt-field rheostat, is 350 Ω. The resistance of the steps in the starting box are as follows, beginning with the point that connects to the terminal marked $F$ in Fig. 6-38(a): 2.24, 1.47, 0.95, 0.62, 0.40, and 0.26 $\Omega$. Assume that at each point, when the armature current has dropped to its rated value, the starting box is switched to the next point, thus eliminating a step at a time in the starting resistance. Neglect change in field current.

*Solution.* When the motor is delivering rated load at rated speed, the starting resistance in series with the armature circuit is cut out and the counter emf is

$$E = V - r_a I_a$$

and

$$I_a = I - I_f$$

where

$$I = 26.0 \text{ A} \quad \text{and} \quad I_f = \frac{250}{350} = 0.71 \text{ A}$$

$$I_a = 26.0 - 0.71 = 25.3 \text{ A}$$

Hence,

$$E = 250 - 0.48 \times 25.3 = 250 - 12.2 = 237.8 \text{ V}$$

and the speed is then 1800 rpm (i.e., rated value).

*First Step.* At this point the entire resistance of the starting box is in series with the armature circuit, and we have

$$R_{T_1} = 2.24 + 1.47 + 0.95 + 0.62 + 0.40 + 0.26 + 0.48 = 6.42 \ \Omega$$

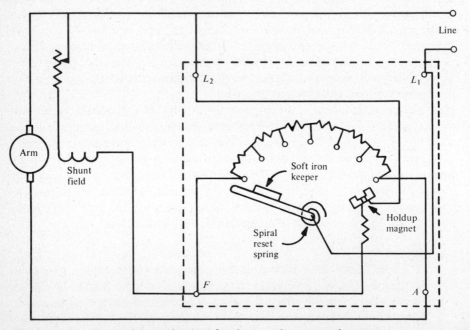

**Figure 6-38a** Four-point starting box for shunt and compound motors.

At starting, the counter emf is zero. The armature starting current is therefore

$$I_{st} = \frac{V}{R_{T_1}} = \frac{250}{6.42} = 38.9 \text{ A}$$

By the time the armature current drops to its rated value of 25.3 A, the counter emf is

$$E = 250 - 6.42 \times 25.3 = 250 - 162.0 = 88.0 \text{ V}$$

and the speed is then

$$n = \frac{88.0}{237.8} \times 1800 = 667 \text{ rpm}$$

*Second Step.* The 2.24-$\Omega$ step is now suddenly cut out, leaving a total resistance of $6.42 - 2.24 = 4.18$ $\Omega$ in the armature circuit. But the motor speed is still at 667 rpm, which means that the counter emf is still 88.0 V if the effect of armature reaction is neglected. Accordingly, the resistance drop in the armature circuit is still 162.0 V:

$$I_a R_{T_2} = 162.0$$

and if the inductance of the armature is neglected, the initial current is

$$I_a = \frac{162.0}{R_{T_2}} = \frac{162.0}{4.18} = 38.7 \text{ A}$$

with the final current at 25.3 A the counter emf is

$$E = 250 - 25.3 \times 4.18 = 250 - 105.7 = 144.3 \text{ V}$$

and the speed is then

$$n = \frac{144.3}{237.8} \times 1800 = 1093 \text{ rpm}$$

When this procedure is followed for the remaining steps, the following quantities are obtained:

| | Current, A | | Speed, rpm | |
| --- | --- | --- | --- | --- |
| Step no. | Initial | Final | Initial | Final |
| 3 | 39.0 | 25.3 | 1093 | 1370 |
| 4 | 39.0 | 25.3 | 1370 | 1557 |
| 5 | 39.0 | 25.3 | 1557 | 1675 |
| 6 | 39.0 | 25.3 | 1675 | 1755 |
| 7 | 39.0 | 25.3 | 1755 | 1800 |

The four-point starter is manual, but automatic starters are used for larger motors. In such installations current relays or relays that respond to counter emf short out the resistance in series with the armature in successive steps.

With automatic motor starters there are a number of additional control actions that can be achieved. They include dynamic braking, reversing, jogging, and plugging. Jogging is a term used for running a machine for a portion of a revolution or a few revolutions without going through the starting sequence. This type of operation is often provided for a positioning application. Plugging is a term that is used to describe the sudden reversal from full speed in one direction to full speed in the opposite direction. This is normally accomplished by a reversal of the armature connections of a motor. Dynamic braking requires the use of a shunt motor, where the shunt field of the motor is left connected to the line after the armature is disconnected from the main supply and is connected across a resistor. This resistor is often called the dynamic braking resistor. It must, of course, be capable of dissipating a large amount of power in a very short period of time. Figure 6-38(b) is a simplified diagram of a dc motor starter that provides for additional features of jogging and plugging.† The following is a description of an operating sequence for jogging forward, based upon Fig. 6-38(b), a diagram of an automatic dc motor starter. The references in parentheses are to line numbers on the diagram. In jogging forward, all relays are initially deenergized. Depression of the jog forward button (13) will energize F-2 (16) by way of the normally closed contact labeled R-2 (15). The armature is connected to the line, with the forward polarity determined appropriately (3, 6). The main contactor is then energized (21). The plugging resistor $R_p$ and the accelerating resistors $R_1$ and $R_2$ will remain in series with the armature during the jog-forward operation.

## 6-19 DYNAMIC AND REGENERATIVE BRAKING OF MOTORS

In some motor applications it is necessary to effect a rapid reversal in the direction of rotation. This can be done by bringing the motor to a quick stop by means of "dynamic braking" and then reversing the voltage applied to the armature. Dynamic braking is effected by connecting a resistor across the armature just after it is disconnected from its supply while the field remains excited. The machine then operates as a generator while slowing down and the kinetic energy stored in the inertia of the armature and connected load is dissipated in the resistor.

The speed of motors driving such loads as electric railway locomotives, elevators, cranes, and hoists, while descending, can be reduced considerably without mechanical braking by using regenerative braking, in which power is returned to the supply. The load in this operation acts as the prime mover by virtue of its potential energy and the motor as a generator. The power thus returned is available for other devices operating from the same source of supply. Another example of dynamic and regenerative braking is found in electric vehicles

---

† George McPherson, Jr., *An Introduction to Electric Machines and Transformers* (New York: John Wiley & Sons, Inc., 1981).

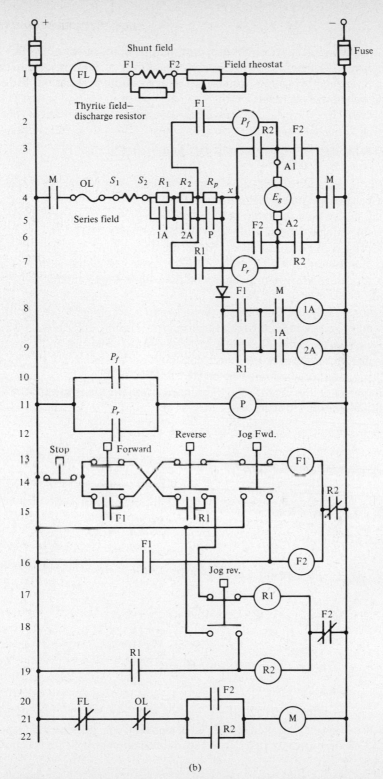

Figure 6-38b  Dc motor starter and control, with two starting resistors, forward and reverse plugging, and forward and reverse jogging.

used in over-the-road and warehouse applications. Solid-state controllers have greatly improved efficiency and response capabilities of these vehicles. Regenerative braking is much more economical than dynamic braking but is limited to speeds well above standstill.

## 6-20 DYNAMIC BEHAVIOR OF DC MACHINES

The dynamic behavior of a machine may be defined as its manner of responding to sudden changes. The effects of inductance and of inertia which are negligible during the steady-state operation are brought into play during a rapid transition from one operating condition to another and must therefore be taken into account when describing the dynamic behavior of the dc machine. Equivalent circuits are therefore presented which include these parameters along with additional symbols regarding the algebraic signs of induced voltages relative to current direction.

Figure 6-39(a) and (b) shows a physical and a simplified representation of an elementary dc machine in which the armature winding and the field winding are indicated by the coils $a$–$a'$ and $f$–$f'$. The voltage polarities and current directions are for *motor* operation, since both coils are treated as loads. The sense of the two windings is assumed to be the same as that in Fig. 6-39(c), which shows the current $i$ flowing through the winding from the lower toward the upper terminal and the flux linkage $\lambda$ as well as the mmf $F$ to be directed axially

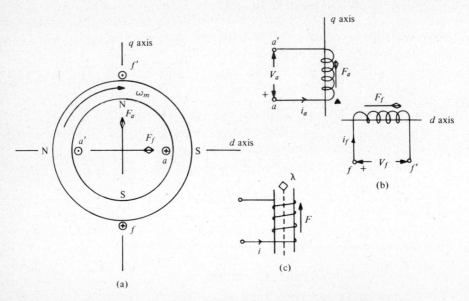

**Figure 6-39** Elementary dc motor. (a) Physical representation. (b) Schematic representation. (c) Coil showing convention relative to assumed directions of current, mmf, and flux linkage.

upward. This convention is applied to the dc machine for the sake of uniformity as it was chosen to systematize the treatment of the phasor diagram of the synchronous machine, in Chap. 5.

The ▲ mark indicates the polarity of the *speed* voltage which is generated in the armature by virtue of its rotation in the presence of the field current $i_f$. The speed voltage has the same polarity as the applied armature voltage $v_a$ in Fig. 6-39(b) for clockwise rotation of the armature. Reversing the direction of the field current or of the armature rotation reverses the polarity of the speed voltage, and in either case the ▲ mark would then be placed near the $a'$ terminal. No speed voltage is induced in the field winding because the armature mmf is stationary relative to the field winding.

## 6-21 BASIC MOTOR EQUATIONS

The speed voltage is expressed by Eq. 6-18 as

$$e_a = k_E \phi_d \omega_m \tag{6-32}$$

where the mechanical angular velocity may be expressed in terms of the mechanical angular displacement $\theta_m$ from an arbitrary reference axis by

$$\omega_m - p\theta_m \tag{6-33}$$

where $p$ represents the differential operator $d/dt$.

A considerable simplification in the analysis of the dynamic behavior is effected by assuming the magnetic circuit to be linear, which in many cases does not involve significant sacrifice in accuracy. The speed voltage may then be expressed in terms of the field current by

$$\boxed{e_a - \mathcal{L}_{af} i_f \omega_m} \tag{6-34}$$

where $\mathcal{L}_{af}$ may be regarded as a fictitious mutual inductance which relates the speed voltage to the field current and the speed and which is determined from the linear or unsaturated portion of the magnetization curve.

The brushes are assumed to be on the geometric neutral with the result that the armature mmf is along the quadrature axis. Although Fig. 6-39(a) indicates a two-pole machine, Fig. 6-39(b) is valid for multipolar machines as well, because the displacement between the $d$ and $q$ axes is $\pi/2$ rad in electrical measure, since a pair of poles subtends $2\pi$ rad in electrical measure. With the brushes on geometric neutral, there is no mutual flux linkage between the field and the armature circuits.

The applied voltages are

$$v_f = r_f i_f + pL_{ff} i_f \tag{6-35}$$

and

$$v_a = r_a i_a + pL_{aa} i_a \tag{6-36}$$

when the armature is at a standstill and where $r_f$ and $r_a$ are the resistances of the field and armature and $L_{ff}$ and $L_{aa}$ are the self-inductances of the field and armature. However, when the armature rotates, the voltage applied to the armature must include the component equal to the speed voltage or counter emf. Then for clockwise rotation which represents operation as a *motor* for the assumed voltage polarities and current directions in Fig. 6-39, the applied armature voltage is

$$v_a = (r_a + pL_{aa})i_a + \mathcal{L}_{af}\omega_m i_f \qquad (6\text{-}37)$$

As in the case of the steady state, the electromagnetic power developed in the armature is

$$p_{em} = e_a i_a = T_{em}\omega_m \qquad (6\text{-}38)$$

and the electromagnetic torque is, on the basis of Eqs. 6-34 and 6-38,

$$T_{em} = \mathcal{L}_{af} i_f i_a \qquad (6\text{-}39)$$

tending to rotate the armature clockwise on the basis of the N and S armature magnetic poles[†] on the quadrature axis and the N and S field poles on the direct axis. If $J_M$ is the polar moment of inertia of the motor rotor, $B_M$ the viscous friction constant used for approximating the rotational losses of the motor, and $T_L$ the torque delivered to the load, then

$$T_{em} = J_M p\omega_m + B_M\omega_m + T_L = \mathcal{L}_{af} i_f i_a \qquad (6\text{-}40)$$

When the inertia and friction constants $J_L$ and $B_L$ of the load are known, these may be combined with the corresponding motor constants so that Eq. 6-40 can be rewritten as

$$(Jp + B)\omega_m + T_l = \mathcal{L}_{af} i_f i_a \qquad (6\text{-}41)$$

where $J = J_M + J_L$, $B = B_M + B_L$, and $T_l$ is the component of load torque not taken into account by $J_L$ and $B_L$. Since the inductances $L_{aa}$ and $L_{ff}$ and $\mathcal{L}_{af}$ are assumed constant, Eqs. 6-35 and 6-37 are conveniently expressed in matrix form:

$$\begin{bmatrix} v_f \\ v_a \end{bmatrix} = \begin{bmatrix} r_f + L_{ff}p & 0 \\ \mathcal{L}_{af}\omega_m & r_a + L_{aa}p \end{bmatrix} \begin{bmatrix} i_f \\ i_a \end{bmatrix} \qquad (6\text{-}42)$$

Equations 6-40 and 6-41 are the equations of motion for the dc machine in Fig. 6-39.

---

[†] The use of magnetic poles in the stator and in the rotor affords an expedient, though artificial, means of accounting for the direction of torque. A critical analysis of various methods in use to account for forces on current-carrying conductors is presented by A. S. Langsdorf, "The Development of Torque in Slotted Armatures," *Trans. IEEE* 82 (1963): 82–87.

**EXAMPLE 6-12**

A 5-hp 240-V dc shunt motor has the following data:

$$r_a = 0.60 \ \Omega \qquad r_f = 240 \ \Omega$$
$$L_{aa} = 0.012 \ H \qquad L_{ff} = 120 \ H$$
$$\mathcal{L}_{af} = 1.8 \ H$$

The torque required by the load is proportional to the speed such that the combined constants of the motor armature and the load are

$$J = J_M + J_L = 1.20 \ \text{kg-m}^2/\text{rad}^2$$

$$B = B_M + B_L = 0.35 \ \text{N-m-s/rad}^2$$

where

$$T_L = B_L \omega_m$$

This motor is started with a 3.40-$\Omega$ resistance in series with the armature to limit the starting current. A constant voltage of 240 V is first applied to the field, and after the field current has practically reached its final value, the same voltage is applied to the armature circuit, including the 3.40-$\Omega$ resistance.

Express, as functions of time, (a) the field current before the armature is energized, and (b) the armature current and the speed after the voltage is applied to the armature circuit. (c) Repeat part (b), neglecting the self-inductance of the armature and compare the maximum value of the armature current with that for part (b).

*Solution*

a. From Eq. 6-35,

$$(L_{ff} p + r_f)i_f = v_f$$

or

$$(120p + 240)i_f = 240$$

The Laplace transform of the first term on the left-hand side of this equation is found from Eq. A-11 in Appendix A to be $120[sI_f(s) - i_f(0^+)]$ and that of the second term is $240I_f(s)$, while that of the right-hand side is $240/s$ according to Eq. A-4. Since the initial field current is zero, $i_f(0^+) = 0$, the Laplace transformed equation is

$$(120s + 240)I_f(s) = \frac{240}{s}$$

from which

$$I_f(s) = \frac{2}{s(s + 2)}$$

for which the inverse transform is found to be, from item 3, Table A-1,

$$i_f = 1.0(1 - \epsilon^{-2t}) \tag{1}$$

b. The final value of the field current is found by letting $t \rightarrow \infty$ in Eq. 1, with the result

$$i_f(\infty) = 1.0 \text{ A}$$

Equation 6-40 includes the inertia constant as well as the other factors that determine the load torque and when the parameters of the motor and load are combined (i.e., $J = 1.20$ and $B = 0.35$). When the other numerical values are substituted, the resulting equation is

$$(1.20p + 0.35)\omega_m = (1.8)(1.0)i_a$$

from which the angular velocity can be expressed by

$$\omega_m = \frac{i_a}{0.667(p + 0.292)} \tag{2}$$

The result of substituting the given numerical values into Eq. 6-37 is, after some algebraic manipulation,

$$i_a = \frac{20,000 - 150\omega_m}{p + 332} \tag{3}$$

Since the armature is energized with the motor at a standstill the initial velocity $\omega_m(0^+)$ is zero because of inertia and the initial armature current $i_a(0^+)$ is zero also, because of the armature inductance. The Laplace transforms of Eqs. 2 and 3 are therefore, respectively,

$$\Omega_m(s) = \frac{I_a(s)}{0.667(s + 0.292)} \tag{4}$$

$$I_a(s) = \frac{(20,000/s) - 150\Omega_m(s)}{s + 332} \tag{5}$$

Substitution of Eq. 4 in Eq. 5 yields, after some algebraic manipulation,

$$I_a(s) = \frac{20,000(s + 0.292)}{s(s + 0.97)(s + 332)} \tag{6}$$

which can be reduced to partial fractions as follows:

$$I_a(s) = \frac{C_1}{s} + \frac{C_2}{s + 0.97} + \frac{C_3}{s + 332}$$

where

$$C_1 = \frac{20,000(s + 0.292)}{(s + 0.97)(s + 332)}\bigg]_{s=0} = 18.1$$

$$C_2 = \frac{20,000(s + 0.292)}{s(s + 332)}\bigg]_{s=-0.97} = 42.3$$

$$C_3 = \frac{20,000(s + 0.292)}{s(s + 0.97)}\bigg]_{s=-332} = -60.4$$

from which the Laplace-transformed armature current is found to be

$$I_a(s) = \frac{18.1}{s} + \frac{42.3}{s + 0.97} - \frac{60.4}{s + 332}$$

and for which the inverse transform is, according to Table A-1,

$$i_a = 18.1 + 42.3\epsilon^{-0.97t} - 60.4\epsilon^{-332t} \tag{7}$$

When Eq. 4 is substituted in Eq. 6, the following Laplace-transformed expression results:

$$\Omega_m(s) = \frac{30,000}{s(s + 0.97)(s + 332)}$$

and by following the same procedure as that for obtaining the armature current, the angular velocity is found to be

$$\omega_m = 93.2 - 93.5\epsilon^{-0.97t} + 0.3\epsilon^{-332t} \tag{8}$$

c. The quantity $\epsilon^{-332t}$ vanishes when the inductance of the armature is neglected and Eqs. 7 and 8 are reduced to

$$\omega_m = 93.2(1 - \epsilon^{-0.97t}) \tag{9}$$

and $$i_a = 18.0 + 42.0\epsilon^{-0.97t} \tag{10}$$

The maximum value of the armature current as expressed by Eq. 10 is $i_a(0) = 60.0$ A. On the other hand, the application of the rules for maxima and minima to Eq. 9 yields a maximum of 59.6 A at $t = 0.0187$ s, by which time the angular velocity is only 1.2 rad/s or 1.3 percent of its final value.

The self-inductance of the armature can often be neglected as illustrated in Example 6-12. An exception is the case of a motor driving a load that has rapid torque pulsations of appreciable magnitude. (See Example 6-14.)

Example 6-12 illustrated the response of a dc shunt motor to the sudden application of a constant voltage when the initial speed and the initial armature current are both zero. The following example deals with the transient response of a dc shunt motor when the initial values are other than zero.

## EXAMPLE 6-13

The shunt motor in Example 6-11 is coupled to a load which consists of inertia only and such that the polar moment of inertia of the load and the motor combined is 2.0 kg-m²/rad². The rotational losses of the motor and the self-inductance of its armature may be neglected.

a. Assume the magnetic circuit of the motor to be linear and calculate the inductance $\mathcal{L}_{af}$ from the full-load data given in Example 6-11.

b. Express the current and speed as functions of time while the motor accelerates on step 3 of the starting box.

c. Calculate the time required for acceleration from the initial speed of 1093 rpm on step 3 to a speed of 1370 rpm in Example 6-11.

*Solution*

a. $\mathcal{L}_{af} = e_a/i_f\omega_m$ from Eq. 6-34, $e_a = 237.8$ V, $n = 1800$ rpm, and $i_f = 0.71$ A from Example 6-11. Hence, $\omega_m = 2\pi \times 1800/60 = 188.5$ rad/s and $\mathcal{L}_{af} = 237.8/(0.71 \times 188.5) = 1.78$ H.

b. If $L_{aa}$ is neglected, Eq. 6-37 expresses the armature current by

$$i_a = \frac{v_a - \mathcal{L}_{af} i_f \omega_m}{r_a} \tag{1}$$

Since the friction constant $B_M$ of the motor is neglected, and since the load is assumed pure inertia (i.e., $T_L = J_L p\omega_m$), the use of Eq. 6-40 results in

$$(J_M + J_L)p\omega_m = \mathcal{L}_{af} i_f i_a \tag{2}$$

Substitution of Eq. 1 in Eq. 2 yields, after some simple algebraic manipulation,

$$\left[ p + \frac{(\mathcal{L}_{af} i_f)^2}{(J_M + J_L)r_a} \right]\omega_m = \frac{\mathcal{L}_{af} i_f v_a}{(J_M + J_L)r_a} \tag{3}$$

Since $i_f$ is a constant, the Laplace transform of Eq. 3,

$$s\Omega_m(s) - \omega_m(0) + \frac{(\mathcal{L}_{af} i_f)^2}{(J_M + J_L)r_a} \Omega_m(s) = \frac{\mathcal{L}_{af} i_f V_a(s)}{(J_M + J_L)r_a} \tag{4}$$

The following numerical values are obtained from Example 6-11 and part (a) of this example.

$$V_a = 250 \qquad \mathcal{L}_{af} i_f = 1.78 \times 0.71 = 1.263$$

$$J_M + J_L = 2.0 \qquad r_a = 6.42 - (2.24 + 1.47) = 2.71$$

And since the initial speed on step 3 is 1093 rpm,

$$\omega_m(0) = \frac{2\pi \times 1093}{60} = 114.3$$

The result of substituting these numerical values in Eq. 4 is

$$\left[ s + \frac{(1.263)^2}{(2.0)(2.71)} \right]\Omega_m(s) = \frac{(1.263)(250)}{(2.0)(2.71)s} + 114.3$$

from which it follows that

$$\Omega_m(s) = \frac{58.2}{s(s + 0.294)} + \frac{114.3}{s + 0.294}$$

which has the inverse transform, according to Table A-1, of

$$\omega_m = 198.0(1 - \epsilon^{-0.294t}) + 114.3\epsilon^{-0.294t}$$
$$= 198.0 - 83.7\epsilon^{-0.294t} \tag{5}$$

When Eq. 5 and the appropriate numerical values are substituted in Eq. 1, the current is found to be

$$i_a = 39.0\epsilon^{-0.294t} \tag{6}$$

c. When the speed accelerates from 1093 to 1370 rpm, the current drops from 39.0 to 25.3 A according to Example 6-11. Hence, if $T$ is the time required for this decrease, then Eq. 6 can be rewritten as

$$25.3 = 39.0\epsilon^{-0.294T}$$

and taking natural logarithms we have

$$3.23 = 3.66 - 0.294T$$

$$T = \frac{0.43}{0.294} = 1.47 \text{ s}$$

Equations 6-40 and 6-42 describe the motion of a dc machine with a field circuit in the $d$ axis and an armature circuit in the $q$ axis. These equations are in general nonlinear due to the product $i_f i_a$ in Eq. 6-40 and the product $\omega_m i_f$ in Eq. 6-42. The analytical solution of these nonlinear equations may be conveniently solved for by means of a personal computer. However, if either $i_f$ or $i_a$ is constant Eqs. 6-40 and 6-42 are linear. A case in point is that of controlling the speed of a separately excited shunt motor by varying the voltage applied to the armature while the field current is constant. This is a feature of the Ward-Leonard system for motor speed control. Then for $i_f = I_f = \text{const.}$, Eqs. 6-40 and 6-42 are reduced to

$$(J_M p + B_M)\omega_m + T_L = \mathcal{L}_{af} I_f i_a \tag{6-43}$$

$$v_a = \mathcal{L}_{af} I_f \omega_m (r_a + L_{aa} p) i_a \tag{6-44}$$

## 6-22 LINEARIZATION FOR SMALL-SIGNAL RESPONSE

Perturbations in the voltage applied to the field or to the armature or in the torque of a dc motor may be regarded as small-signal inputs. The response of the motor to such inputs can be determined to a good degree of approximation by means of linearization techniques by considering the perturbations to be superimposed on quiescent operating points. If the following quantities represent the quiescent operating points for a dc motor,

$$T_{em0} = B_M \Omega_{m0} + T_{L0} = \mathcal{L}_{af} I_{f0} I_{a0} \tag{6-45}$$

$$V_{f0} = r_f I_{f0}$$

$$V_{a0} = r_{a0} I_{a0} + \mathcal{L}_{af} \Omega_{m0} I_{f0} \tag{6-46}$$

a small-signal input may be regarded as producing the quantities designated by the subscript 1 and which are superimposed on the quiescent points, resulting in the following:

$$T_{em} = T_{em0} + T_{em1}$$

$$\omega_m = \Omega_{m0} + \omega_{m1} \tag{6-47}$$

$$i_f = I_{f0} + i_{f1}$$

$$v_a = V_{a0} + v_{a1} \tag{6-48}$$

$$i_a = I_{a0} + i_{a1}$$

The result of substituting Eqs. 6-47 and 6-48 in Eqs. 6-40 and 6-42 and of then subtracting Eqs. 6-45 and 6-46 when the products $i_{f1} i_{a2}$ and $i_{f1} \omega_{m1}$ are neglected, is

$$T_{em1} = (J_M p + B_M)\omega_{m1} + T_{L1} = \mathcal{L}_{af}(I_{f0} i_{a1} + I_{a0} i_{f1}) \tag{6-49}$$

$$v_{f1} = (L_{ff}p + r_f)i_{f1} \tag{6-50}$$
$$v_{a1} = (L_{aa}p + r_a)i_{a1} + \mathcal{L}_{af}(I_{f0}\omega_{m1} + \Omega_{m0}i_{f1})$$

Figure 6-40 shows a block diagram based on Eqs. 6-49 and 6-50.

## 6-23 PHASOR RELATIONSHIPS FOR SMALL OSCILLATIONS

Pulsating loads, as for example compressors, produce oscillations in the torque which are superposed on the average value of torque and are accompanied by oscillations in the speed and armature current. If the torque oscillations are assumed to vary sinusoidally at an angular frequency of $\gamma$ rad/s, the variable torque and the voltage components can be expressed as phasors by substituting the imaginary quantity $j\gamma$ for $p$ in Eqs. 6-49 and 6-50, which results in

$$(B_M + j\gamma J_M)\Omega_{m1} + T_{L1} = \mathcal{L}_{af}(I_{f0}\mathbf{I}_{a1} + I_{a0}\mathbf{I}_{f1}) \tag{6-51}$$
$$\mathbf{V}_{f1} = (r_f + j\gamma L_{ff})\mathbf{I}_{f1}$$
$$\mathbf{V}_{a1} = (r_a + j\gamma L_{aa})\mathbf{I}_{a1} + \mathcal{L}_{af}(I_{f0}\Omega_{m1} + \Omega_{m0}\mathbf{I}_{f1}) \tag{6-52}$$

where $\Omega_{m1}$, $T_{L1}$, $\mathbf{I}_{a1}$, $\mathbf{I}_{f1}$, $\mathbf{V}_{a1}$, and $\mathbf{V}_{f1}$ are phasors.

Since there is no inductive coupling between the armature and the field, $\mathbf{I}_{f1} = 0$ if the voltage applied to the field is constant ($\mathbf{V}_{f1} = 0$) regardless of variations in the armature current.

**EXAMPLE 6-14**

A 200-hp 240-V dc shunt motor has the following data:

$$r_a = 0.02\ \Omega \qquad r_f = 30.0\ \Omega \qquad \mathcal{L}_{af} = 0.48\ \text{H}$$
$$L_{aa} = 0.0012\ \text{H} \qquad L_{ff} = 12.0\ \text{H} \qquad J_M = 18.0\ \text{kg-m}^2/\text{rad}^2$$
$$\text{No-load speed} = 600\ \text{rpm}$$

The rotational losses of the motor are negligible. The polar moment of inertia of the load is

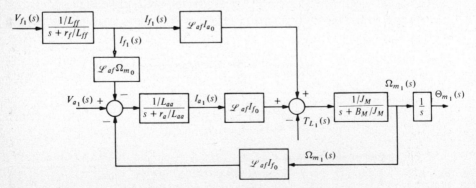

**Figure 6-40** Linearized block diagram for a dc shunt motor.

$$J_L = 20.0 \text{ kg-m}^2/\text{rad}^2$$

and the load torque is

$$T_l = 2000 + 500 \sin 15t$$

If this motor is supplied from a 240-V dc source of negligible impedance, what are (a) the armature current and (b) the speed expressed as functions of time?

*Solution*

(a) Since the torque has a constant component of 2000 N-m/rad and since the voltage to the field and to the armature is constant, the quiescent values of armature current and of the speed may be determined as follows:

$$I_{a0} = \frac{T_{em0}}{\mathcal{L}_{af}I_{f0}} \qquad \text{from Eq. 6-45}$$

where

$$T_{em0} = 2000 \quad \text{and} \quad I_{f0} = \frac{V_{f0}}{r_f} = \frac{240}{30} = 8.0 \text{ A}$$

Hence,

$$I_{a0} = \frac{2000}{(0.48)(8.0)} = 521 \text{ A}$$

$$\Omega_{m0} = \frac{V_{a0} - r_a I_{a0}}{\mathcal{L}_{af}I_{f0}} \qquad \text{from Eq. 6-46}$$

$$= \frac{240 - (0.02)(521)}{(0.48)(8.0)} = \frac{229.6}{(0.48)(8.0)} = 59.8 \text{ rad/s}$$

The variable component $500 \sin 15t$ is treated as the small-signal torque and can be represented in phasor form by

$$T_{l1} = 500\underline{/0°}$$

Since $V_a = V_f = 240$ a constant,

$$v_{a1} = v_{f1} = 0$$

and since $B_M \cong 0$,

$$\mathbf{I}_{a1} = \frac{T_{L1} + j\gamma J\mathbf{\Omega}_{m1}}{\mathcal{L}_{af}I_{f0}} \qquad \text{from Eqs. 6-51 and 6-41} \qquad (1)$$

$$\mathbf{\Omega}_{m1} = -\frac{(r_a + j\gamma L_{aa})\mathbf{I}_{a1}}{\mathcal{L}_{af}I_{f0}} \qquad \text{from Eq. 6-52} \qquad (2)$$

and when Eqs. 1 and 2 are solved simultaneously, the current phasor is found to be

$$\mathbf{I}_{a1} = \frac{\mathcal{L}_{af}I_{f0}T_{L1}}{(\mathcal{L}_{af}I_{f0})^2 - \gamma^2 JL_{aa} + j\gamma Jr_a}$$

which, upon substitution of numerical values, becomes

$$\mathbf{I}_{a1} = \frac{(0.48)(8.0)(500\underline{/0°})}{[(0.48)(8.0)]^2 - (15)^2(38.0)(0.0012) + j(15)(38)(0.02)}$$

$$= \frac{1920\underline{/0°}}{4.5 + j11.4} = \frac{1920\underline{/0°}}{12.27\underline{/68.5°}} = 156.5\underline{/-68.5°} \text{ A}$$

Hence,

$$i_{a1} = 156.5 \sin(15t - 68.5°)$$

and the total armature current is

$$i_a = I_{a0} + i_{a1}$$
$$= 521 + 156.5 \sin(15t - 68.5°) \text{ A}$$

b. The velocity phasor, according to Eq. 2, is

$$\Omega_{m1} = -\frac{(0.02 + j0.018)(156.5\underline{/-68.5°})}{(0.48)(8.0)}$$

$$= -\frac{(0.0269\underline{/42.0°})(156.6\underline{/-68.5°})}{3.84}$$

$$= 1.10\underline{/153.5°} \text{ rad/s}$$

and the speed of the motor is therefore expressed by

$$\omega_m = 59.8 - 1.10 \sin(15t - 26.5°) \text{ rad/s}$$

## 6-24 VARIABLE ARMATURE VOLTAGE, CONSTANT FIELD CURRENT

In applications requiring wide ranges of speed, reversal of rotation, or both, speed control is effected by applying adjustable voltage to the armature. In many such problems the field current is assumed constant and the motor is assumed to drive a pure inertia load ($B_M + B_L = 0$). In that case $T_L = J_L p\omega_m$ and Eq. 6-43 becomes

$$Jp\omega_m = \mathcal{L}_{af}I_f i_a \tag{6-53}$$

where $J = J_M + J_L$.

Then Eqs. 6-44 and 6-53 lead to the following transfer functions:

$$\left.\frac{I_{a(s)}}{V_{a(s)}}\right|_{T_L=J_L p\omega_m} = \frac{s/L_{aa}}{s^2 + (r_a/L_{aa})s + (\mathcal{L}_{af}I_f)^2/JL_{aa}} \tag{6-54}$$

and

$$\left.\frac{\Omega_{m(s)}}{V_{a(s)}}\right|_{T_L=J_L p\omega_m} = \frac{\mathcal{L}_{af}I_f/JL_{aa}}{s^2 + (r_a/L_{aa})s + (\mathcal{L}_{af}I_f)^2/JL_{aa}} \tag{6-55}$$

In many cases where $L_{aa}$ is negligible, Eqs. 6-54 and 6-55 can be simplified to

$$\left.\frac{I_{a(s)}}{V_{a(s)}}\right|_{T_L=J_L p \omega_m} = \frac{s/r_a}{s + 1/\tau_{am}} \tag{6-54a}$$

and
$$\left.\frac{\Omega_{m(s)}}{V_{a(s)}}\right|_{T_L=J_L p \omega_m} = \frac{\mathcal{L}_{af}I_f/Jr_a}{s + 1/\tau_{am}} \tag{6-55a}$$

in which the time constant $\tau_{am} = Jr_a/(\mathcal{L}_{af}I_f)^2$. However, when the friction constant $B = B_M + B_L$ is appreciable, $\tau_{am} = Jr_a/[r_a B_M + (\mathcal{L}_{af}I_f)^2]$.

## 6-25 THE SEPARATELY EXCITED DC MOTOR AS A CAPACITOR

The separately excited dc motor can be represented by a capacitive circuit. The voltage applied to the armature is

$$v_a = (L_{aa}p + r_a)i_a + e_a \tag{6-56}$$

and when Eq. 6-32 is substituted in Eq. 6-56 there results

$$v_a = (L_{aa}p + r_a)i_a + k_E \phi_d \omega_m \tag{6-57}$$

Because of the interpolar space in the direct axis $L_{aa}$ is little affected by saturation and may therefore be considered as being constant whether or not linearization of the magnetic circuit is assumed. The torque is expressed by

$$\boxed{T_{em} = k_E \phi_d i_a} \tag{6-58}$$

When Eq. 6-43 is modified in accordance with Eqs. 6-58 and 6-53, the armature current is shown to be

$$i_a = \frac{T_{em}}{k_E \phi_d} = \left[\frac{J_M p}{(k_E \phi_d)^2} + \frac{B_M}{(k_E \phi_d)^2}\right]e_a + \frac{T_L}{k_F \phi_d} \tag{6-59}$$

The coefficient of $e_a$ in Eq. 6-59 suggests a capacitive circuit in which a capacitance $C_{eq} = J_M/(k_E \phi_d)^2$, or

$$\boxed{C_{eq} = \frac{J_M}{(\mathcal{L}_{af}I_f)^2}}$$

when the circuit is linearized and is in parallel with a resistance

$$\boxed{R_{eq} = \frac{(k_E \phi_d)^2}{B_M}}$$

or
$$\boxed{R_{eq} = \frac{(\mathcal{L}_{af}I_f)^2}{B_M}}$$

The third term on the right-hand side of Eq. 6-59 represents the component of armature current required to deliver the load torque $T_L$. The equivalent capacitive circuit is shown in Fig. 6-41, in which the load-torque component of current is shown flowing through the equivalent impedance of $Z_L$.

A note of caution is in order regarding a limitation of the equivalent circuit as a means of portraying the machine response to sudden changes. For example, a sudden reduction of the field current is analogous to the sudden switching of additional capacitance in parallel with the original capacitor, which corresponds to a sudden decrease in the energy stored in the capacitance based on the conservation of charge [i.e., $q(0) = q(0^+)$], or

$$W(0^+) = \frac{q(0^+)^2}{2(C_{eq} + \Delta C)} \quad \text{and} \quad W(0^-) = \frac{q(0)^2}{2C_{eq}}$$

the final and initial values of stored energy. However, a sudden change in the kinetic energy stored in the rotating parts requires a value of torque approaching infinity with a correspondingly large value of $i_a$. Therefore, the voltage across the capacitor just before and just after switching must satisfy the constant stored-energy relationship,

$$\frac{(C_{eq} + \Delta C)e_a^2(0^+)}{2} = \frac{C_{eq}e_a^2(0^-)}{2}$$

where $\Delta C$ is the increase in the equivalent capacitance due to a decrease in $i_f$.

## 6-26 THE SEPARATELY EXCITED DC GENERATOR

When the field current and the rotation are each in a given direction, the armature current reverses direction when a dc machine goes from motor to generator operation. Figure 6-42 based on Fig. 6-39(b) for motor operation, shows a sche-

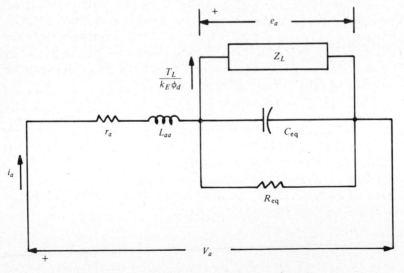

$$C_{eq} = J_M/(k_E\phi_d)^2 \qquad R_{eq} = (k_E\phi_d)^2/B_M$$

**Figure 6-41** Equivalent capacitive circuit of a separately excited dc motor.

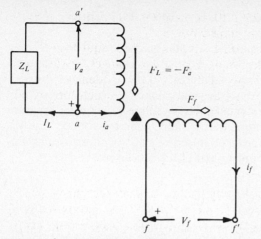

**Figure 6-42**   Schematic circuit for a dc generator.

matic diagram of a dc generator in which the load current $i_L$ flows in the impedance $Z_L$, where

$$i_L = -i_a \tag{6-60}$$

in keeping with the convention adopted for the load current of a transformer in Chap. 3. Then for generator operation it is convenient to replace $i_a$ by $-i_L$ in Eqs. 6-38 and 6-40. There is also the additional equation

$$v_a = Z_L i_L = -Z_L i_a \tag{6-61}$$

which relates the terminal voltage to the armature current and the load impedance.

The dynamic behavior of a generator is brought into play by rapid changes in (1) the load, (2) the voltage applied to the field terminals, and (3) the speed of the prime mover. However, generators are usually driven at nearly constant speed, although sudden changes in the generator load produce at least momentary speed variations which are reflected in the value of the speed voltage $e_a$. The effects of the transients in the speed are generally minor compared with those in the current and are therefore neglected in this chapter.

**EXAMPLE 6-15**

The motor in Example 6-12 is operated as a separately excited shunt generator at a constant speed of 900 rpm while the field current is constant at 1.5 A. The load current is initially equal to zero.

Express the armature current and the armature terminal voltage as functions of time for a suddenly applied load impedance having a resistance of 11.4 Ω and an inductance of 0.088 H.

*Solution.*   The armature current is

$$i_a = -i_L = \frac{V_a - \mathcal{L}_{af}\omega_m i_f}{L_{aa}p + r_a} \qquad \text{from Eq. 6-37} \tag{1}$$

and the terminal voltage is

$$v_a = Z_L i_L = (L_L p + R_L) i_L \qquad \text{from Eq. 6-61} \tag{2}$$

Substitution of Eq. 2 in Eq. 1 yields

$$i_L = \left( \frac{\mathcal{L}_{af} \omega_m i_f}{L_{aa} + L_L} \right) \left[ \frac{1}{p + (r_a + R_L)/(L_{aa} + L_L)} \right] \tag{3}$$

The effect of suddenly connecting the load impedance to the armature is that of the sudden application of the constant generated voltage $\mathcal{L}_{af} i_f \omega_m$ to the armature and load impedance in series. Laplace transforming Eq. 3 with zero initial armature current and $i_f = I_f = $ constant results in

$$I_L(s) = \left( \frac{\mathcal{L}_{af} I_f \omega_m}{L_{aa} + L_L} \right) \left\{ \frac{1}{s[s + (r_a + R_L)/(L_{aa} + L_L)]} \right\}$$

which becomes, after substitution of the numerical values,

$$I_f = 1.5 \text{ A} \qquad \omega_m = \frac{2\pi(900)}{60} = 94.0$$

$$L_{aa} + L_L = 0.012 + 0.088 = 0.10 \qquad r_a + R_L = 0.60 + 11.4 = 12.0$$

$$I_L(s) = \left[ \frac{(1.8)(1.5)(94.0)}{0.10} \right] \left[ \frac{1}{s(s + 12.0/0.10)} \right]$$

$$= \frac{2540}{s(s + 120)} \tag{4}$$

for which the inverse transform is found from item 3 in Table A-1 to be

$$i_L = 21.2(1 - \epsilon^{-120t})$$

The terminal voltage is, from Eq. 6-61,

$$v_a = Z_L i_L = (L_L p + R_L) i_L$$

and the Laplace-transformed terminal voltage is, upon substitution of numerical values,

$$V_a(s) = (0.088s + 11.4) I_L(s) \tag{5}$$

and when Eq. 4 is substituted in Eq. 5, there results

$$V_a(s) = \frac{(0.088s + 11.4)2540}{s(s + 120)} = \frac{223.5(s + 129.5)}{s(s + 120)}$$

$$= \frac{223.5}{s + 120} + \frac{(223.5)(129.5)}{s(s + 120)}$$

The inverse transform for this voltage is, according to items 2 and 3 in Table A-1,

$$v_a = 223.5 \epsilon^{-120t} + \frac{(223.5)(129.5)}{120} (1 - \epsilon^{-120t})$$

$$= 241.2 - 17.7 \epsilon^{-120t}$$

## 6-27 TRANSFER FUNCTIONS FOR THE SEPARATELY EXCITED GENERATOR

The load impedance in Eq. 6-61 may sometimes be treated as a series $RL$ circuit:

$$Z_L = L_L p + R_L \tag{6-62}$$

Then the load current may be expressed, on the basis of Eqs. 6-42, 6-61, and 6-62 by

$$i_L = \frac{\mathcal{L}_{af} v_f \omega_m}{L_{ff}(L_{aa} + L_L)(p + 1/\tau_f)(p + 1/\tau_{ag})} \tag{6-63}$$

where

$$\tau_f = \frac{L_{ff}}{r_f} \quad \text{and} \quad \tau_{ag} = \frac{L_{aa} + L_L}{r_a + R_L}$$

the time constant of the field circuit and that of the armature circuit including the load. The use of the time constants affords a reduction in the number of symbols in this and in subsequent derivations where convenient.

Changes in the generator speed may generally be neglected so that it is convenient to let

$$K_E = \mathcal{L}_{af} \omega_m \quad \text{volts per field ampere} \tag{6-64}$$

It then follows from Eq. 6-63 that the transfer function relating the load current to the applied field voltage is

$$\boxed{\frac{L_{L1}(s)}{V_{f1}(s)} = \frac{K_E}{r_f(r_a + R_L)(1 + s\tau_f)(1 + s\tau_{ag})}} \tag{6-65}$$

The transfer function which expresses the armature terminal voltage in terms of the field voltage obtained from Eqs. 6-62 and 6-65 is

$$\boxed{\frac{V_{a1}(s)}{V_{f1}(s)} = \frac{R_L K_E(1 + s\tau_L)}{r_f(r_a + R_L)(1 + s\tau_f)(1 + s\tau_{ag})}} \tag{6-66}$$

where $\tau_L = L_L/R_L$, the time constant of the load impedance.

The time constants $\tau_L$ and $\tau_{ag}$ are usually much smaller than $\tau_f$ and may be neglected in many cases. Then Eqs. 6-65 and 6-66 can be simplified as follows:

$$\frac{I_{L1}(s)}{V_{f1}(s)} = \frac{K_E}{r_f(r_a + R_L)(1 + s\tau_f)} \tag{6-65a}$$

$$\frac{V_{a1}(s)}{V_{f1}(s)} = \frac{R_L K_E}{r_f(r_a + R_L)(1 + s\tau_f)} \tag{6-66a}$$

It is generally difficult to relate the armature voltage to even small changes in the load impedance. However, an approximate relationship can be obtained in terms of small changes in the load current by assuming $R_L$ to be constant and by taking the ratio of Eq. 6-66 to Eq. 6-65, which results in

$$\frac{V_{a1}(s)}{I_{L1}(s)} = R_L(1 + s\tau_L) \tag{6-67}$$

The block diagram for a separately excited dc generator is shown in Fig. 6-43.

## 6-28 CONTROL OF OUTPUT VOLTAGE

The variations of the terminal voltage in shunt generators due to changes in load or in the speed may be held to small values by manual or automatic adjustment of the field rheostat. Automatic regulators for this purpose generally consist of a solenoid connected across the armature terminals, acting as a relay which causes the resistance of the field circuit to be varied as required. Such systems, however, are inadequate where a high degree of accuracy and rapid rate of response are needed, so that closed-loop systems such as illustrated in Fig. 6-44 are used instead.

The closed-loop arrangement responds to the difference between the output quantity and a reference quantity, and the response is therefore practically independent of drifts in the characteristics of intervening components. The system in Fig. 6-44 is calibrated by making preliminary adjustments in the voltages $av_r$ and $av_a$ which are each proportional by the ratio $a$ to the reference voltage $v_r$ and the armature terminal voltage $v_a$ of the generator. The time constant $\tau_{ag}$ of the armature and load circuit combined is generally negligible. Therefore, the load impedance $Z_L$ is assumed to be equal to the load resistance $R_L$.

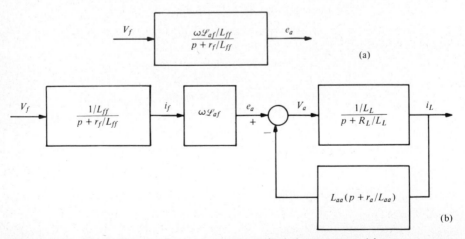

**Figure 6-43** Block diagram for a separately excited dc generator driven at constant speed. (a) Representing generated voltage in terms of field terminal voltage. (b) Complete representation.

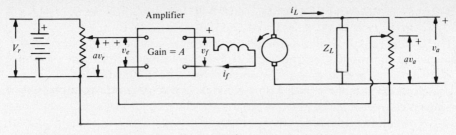

**Figure 6-44**   Closed-loop system for regulating generator voltage.

The error voltage $v_e$ is the difference between the voltages $av_r$ and $av_a$ and is amplified by the amount of the amplifier gain $A$ when loaded and then applied to the field. The amplifier may be an electronic device, magnetic amplifier, or a rotating machine. The voltage applied to the field is therefore

$$v_f = Av_e = Aa(v_r - v_a) \qquad (6\text{-}68)$$

and when this relationship is incorporated in the transfer function of Eq. 6-66a, there results

$$V_a(s) = \frac{K[V_r(s) - V_a(s)]}{1 + s\tau_f}$$

where

$$K = \frac{aAK_ER_L}{r_f(r_a + R_L)}$$

Hence,

$$V_a(s) = \frac{KV_r(s)}{1 + K + s\tau_f}$$

which can be reduced to

$$V_a(s) = \frac{K}{1 + K} \frac{V_r(s)}{1 + s\tau_f'} \qquad (6\text{-}69)$$

where

$$\tau_f' = \frac{\tau_f}{K + 1}$$

This decrease in the time constant of the field by the factor $1/(K + 1)$ results in greater speed of response. Another advantage is that if $K \gg 1$, the effects on the terminal voltage, of variations in rotational speed and in load, are much reduced. For example, if $K = 10$, then the ratio $K/(K + 1) = 10/11$ and a variation in the speed of 5 percent would result in a variation of only about 0.5 percent in the output voltage.

## 6-29 THE WARD-LEONARD SYSTEM

The Ward-Leonard system is a highly flexible arrangement for effecting position and speed control of a separately excited dc motor. Figure 6-45 shows a simple Ward-Leonard system in which a three-phase synchronous motor or a three-phase induction motor drives a separately excited dc generator the armature of which is connected directly (i.e., without intervening circuit breakers or rheostats) to the armature of the separately excited dc motor driving a mechanical load. Some systems make use of rotating amplifiers such as the amplidyne, Rototrol, or Regulex described in Appendix C. A second dc generator, called the *exciter,* furnishes the field excitation for the dc motor and the main dc generator. Precise control of the motor speed and its direction of rotation is achieved by simple and efficient manipulation of the rheostats and switchgear in the field circuits of the motor and the generator.

The systems in Fig. 6-45 are represented by the block diagram in Fig. 6-46 in which

$\omega_{mG}$ = mechanical angular speed of the generator (assumed constant)

$\tau_{fG} = L_{ffG}/r_{fG}$, the time constant of the generator field circuit

$\tau_m = J/B$, the mechanical time constant of the rotor of the motor and of the connected load combined

While Fig. 6-46 assumes constant generator speed, variable generator speed can readily be taken into account by replacing $K_E$ by $\mathcal{L}_{afG}\omega_{mG}$. It is evident that in Fig. 6-46 $e_{aG}$ may be regarded as the voltage applied to the armature of a separately excited motor having an armature circuit resistance and self-inductance of $R$ and $L$. Then the transfer functions are as derived in Sec. 6-21 for the motor.

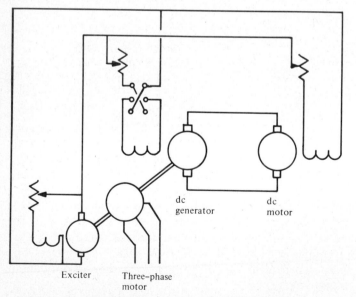

dc
generator

dc
motor

Exciter    Three–phase
motor

**Figure 6-45**   Ward-Leonard system of speed control.

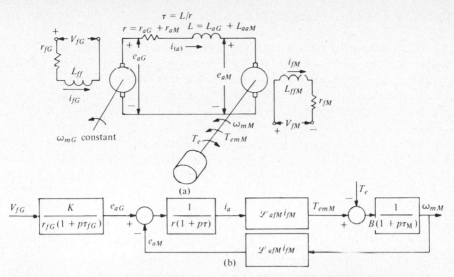

**Figure 6-46** Ward-Leonard system. (a) Schematic diagram. (b) Block diagram.

## 6-30 SOLID-STATE CONTROLS FOR DC MACHINES†

A widely used recent development is the use of silicon-controlled rectifiers (SCR)‡ in the control of dc motors in ratings from $\frac{1}{60}$ to 10,000 hp. In many drives the ac motor–dc generator set is replaced by a circuit which utilizes SCRs to rectify the voltage from a constant-voltage ac source and to control the armature current of the dc motor. Fractional- and low-integral-horsepower motors are usually supplied from single-phase sources, while larger motors are supplied from three-phase sources.

The largest single use for power transistors is in motor drives, a several hundred million dollar industry. Fractional-horsepower dc motors used for home appliances, portable tools, and other low-cost applications constitute the major share of the market in this area in numbers as well as in dollar volume. The increasing use of these drives is largely due to the use of solid-state devices because of their low cost and small sizes.

A semiconductor diode and a thyristor or SCR are represented schematically in Fig. 6-47. The resistance of the diode is very low for current flow in the forward direction, the voltage drop being only about 1 V at rated current, and may therefore be neglected in these applications. On the other hand, the resistance is extremely high for reversed polarity and may then be considered infinite. The thyristor, however, ideally has infinite resistance for both directions unless the

---

† For a more complete discussion of this subject, see Alexander Kusko, *Solid-State D-C Motor Drives* (Cambridge, Mass.: The MIT Press, 1969). See also H. F. Storm, "Solid-State Power Electronics in U.S.A.," *IEEE Spectrum* 6, No. 10 (October 1969): 49–59, and Alexander Kusko, "Solid-State Motor-Speed Controls," *IEEE Spectrum* 9, No. 10 (October 1972): 50–55.

‡ SCRs are discussed by H. E. Stewart in *Engineering Electronics* (Boston: Allyn and Bacon, Inc., 1969), pp. 409–411.

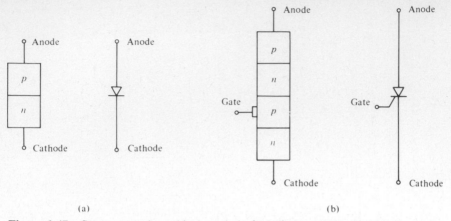

(a)                                                              (b)

**Figure 6-47** Structure and graphical symbol of (a) diode and (b) thyristor.

proper bias is applied to the gate terminal, which causes the thyristor to conduct in the forward direction even after the bias is removed. Conduction ceases in the thyristor upon reversal of the voltage polarity.

Figure 6-48(a) and (b)† illustrates the use of half-wave and full-wave circuits in simple single-phase arrangements. The field circuit in either may be supplied through a half-wave or a full-wave rectifier. The simpler circuit in Fig. 6-48(a) requires only one thyristor, which of course can conduct current only in alternate half-cycles. The portion of the half-cycle (from about 0 to 180°) over which the thyristor conducts is controlled by adjusting the firing angle with the application of the proper value of voltage to the gate terminal $G$ from an auxiliary circuit not shown in Fig. 6-48. Since the single thyristor passes only one pulse of current per cycle which has a duration of 180° or less, there are sizable dips in the armature current resulting in pronounced ripple, which is undesirable. Less ripple results from the full-wave arrangement in Fig. 6-48(b), in which the bridge circuit comprising of thyristors I and II and diodes III and IV accomplishes rectification during both halves of each cycle. The firing angle of the thyristors determines the amount of direct current supplied to the motor armature. The diodes are also in effect connected across the armature, so that in addition to functioning as rectifiers, they provide a path for the armature current, produced during the dips in the voltage ripple, by the energy stored in the inductance of the armature.

Figure 6-49† shows one of a variety of three-phase drives. The firing angle of thyristors I, II, and III controls the current to the motor armature. Diodes IV, V, and VI provide the return path for the three-phase currents. During the dips in the voltage the energy stored in the armature inductance produces a component of current through the path provided by the free-wheeling diode connected across the armature. While the three-phase is more complex than the single-phase, it makes for much less ripple in that it passes six pulses per cycle as compared with one or at most two pulses per cycle. The current ripple is

† Adapted from Kusko, *Solid-State D-C Motor Drives.*

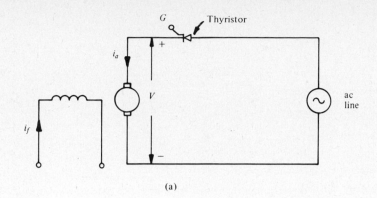

(a)

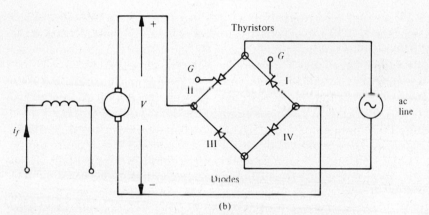

(b)

**Figure 6-48**   Single-phase SCR drives. (a) Half-wave. (b) Full-wave.

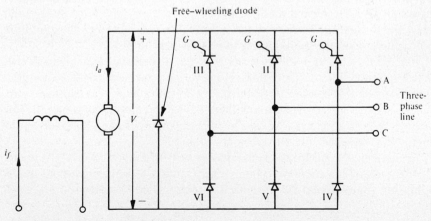

**Figure 6-49**   Three-phase SCR drive.

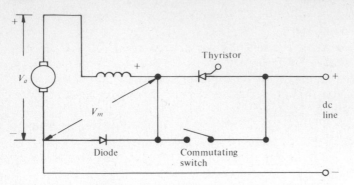

**Figure 6-50** SCR drive from a dc source for a dc series motor.

objectionable because it increases heating of the armature and makes commutation more difficult, particularly at light loads and high speeds, when the pulses are of shorter duration.

SCRs are also used to control dc motors supplied from dc sources. The advantage of the SCR arrangement, where dc sources are available, is due to the much smaller power consumption of the SCRs than that of the resistors that are normally required to control motor speed in the straight dc drives.

A method for controlling a series motor supplied from a dc source is shown in Fig. 6-50.† The dc source arrangement requires a means for turning off the thyristor when the current pulse has attained the desired duration. This is done by means of a commutation circuit which incorporates a capacitor and an auxiliary thyristor. The commutation circuit is represented by the commutating switch in Fig. 6-50. Such an arrangement is not required in the ac circuit, as the thyristor is turned off by the reversal of voltage polarity of the ac voltage.

Problems associated with the commutator and brushes impose limits on the size of conventional dc machines.‡ A serious problem is that of commutator deformation caused by centrifugal forces and by local heating resulting in improper brush contact. At high altitudes there is also rapid brush wear. Another disadvantage is the normal maintenance requirement of commutator and brushes. Solid-state devices have made it possible to replace large dc generators with ac generators using such devices to rectify their output.

For many years and still to a large extent at present, ac generators had their fields excited from dc generators. However, large present-day ac generators require dc power that calls for dc generators with commutators of excessive size, so that ac exciters with solid-state rectifiers have come into increasing use. For instance, a 500-MVA alternator uses a static exciter delivering 2000 kW, consisting of 16 three-phase full-wave bridges in parallel.

In another system an ac exciter (generator) is connected directly to the shaft of the main ac generator. The ac exciter has a stationary dc field and a rotating three-phase armature connected through silicon diodes, rotating with

† Adapted from Kusko, *Solid-State D-C Motor Drives.*

‡ H. F. Storm, "Solid-State Power Electronics in the U.S.A.," *Trans. IEEE* 6, No. 10 (October 1969): 49–59.

the shaft and the field of the main ac generator, thus eliminating the use of brushes.

Conventional diesel-electric locomotives utilized a diesel-driven dc generator supplying the traction motors, usually four to a generator. With increasing speed and horsepower rating of the diesel motors, practical limitations for a commutator were approached. In consequence, diesel-driven ac generators using silicon diodes are being used to supply the dc traction motors.

In adjustable-speed dc motor drive systems in which an ac motor drives a dc generator with adjustable output voltage, the solid-state device replaces the entire motor-generator set. This results in improved overall economy, higher efficiency, higher reliability, and less maintenance.

## 6-31 BASIC SIMILARITIES IN INDUCTION MACHINES, SYNCHRONOUS MACHINES, AND DC MACHINES

A wound-rotor induction motor can be operated as a synchronous machine if direct current is applied to its rotor. Its output, however, would be limited, as far as practical operation is concerned, as a result of the short air gap and corresponding excessive value of synchronous reactance. In a dc machine without interpole and compensating winding, the angular displacement between armature mmf and field mmf can be adjusted by shifting the brushes. The current in an induction motor lags the voltage, whereas the current may be made to lead or lag the voltage in a synchronous motor by overexciting or underexciting the field. Setting the brushes of a dc motor on geometric neutral corresponds to the condition in a synchronous motor when the current is in phase with the generated voltage. When the brushes are shifted away from neutral opposite the direction of rotation, the effect is similar to that of leading current in a synchronous motor or opposite to normal operation of a polyphase induction motor.

While the relationships in the single-phase induction motor are somewhat more complicated than those in the polyphase induction motor, the effects of the forward- and backward-rotating fields are comparable with that of the rotating field in the polyphase motor. In fact, the two-revolving-field theory rests on that premise. Perhaps the most important structural difference between the induction machine and the synchronous machine is that of the air-gap length. The induction machine requires a short air gap to make for high magnetizing reactance, whereas the synchronous machine requires a long air gap so that the synchronous reactance is not excessive.

Another similarity is noteworthy—that the dc machine can be made to operate as a synchronous machine by connecting taps to the armature winding, 120 electrical degrees apart, and connecting them to slip rings on the shaft. Brushes riding on the slip rings would then provide three-phase power to the line. Such an arrangement, however, is inferior to that of the conventional synchronous machine because it is simpler to rotate the smaller field winding than to rotate the much heavier, more complicated high-voltage armature winding in ac machines.

## 6-32 ELECTROMECHANICAL MACHINES AND DEVICE RATINGS

As one might expect, all machines and devices have physical limitations that establish the region of operation. Examples of such limitations are speed, horsepower, flux density, voltage, current, and temperature. These limits are set in the process of design. Device ratings are set below machine limitations that would result in hazardous machine operation or damage to the device itself. They are also set to levels where the maximum capabilities can be achieved for the cost of the device.

Machine ratings are developed to provide a common set of guidelines to assist the user in the purchase and application of electromechanical devices. Such ratings may also have legal implication in that certain explicit and implicit warranties are established by the rating. Use of the device above its rating will invalidate the warranties given with the device. Since most electrical equipment may experience short-time overloads, quite often the limits on overloads can be obtained from the manufacturer as a part of the rating.

There are many codes and standards that speak to the ratings of machines. One of the most important is the National Electric Manufacturers Association (NEMA). The complete specification of machines recommended by NEMA is given in the MG standard. In this standard, machine size is designated by frame size. NEMA standards also establish mounting dimensions, shaft height, and diameter. Further classification is given by motor class that designates general efficiency and starting-torque characteristics of ac motors. Nameplate data normally given are as follows:

**Voltage**  The nominal rating of the machine winding is given in volts. In polyphase machines the voltage is given as line-to-line rms voltage.

**Current**  The steady-state machine current at rated output power and rated speed in amperes. The value is expressed as an rms quantity for ac machines and an average quantity for dc machines. In polyphase machines, the current is line current. In dc machines with commutators a value of field current may be given which is the field current required for maximum torque at rated speed.

**Speed**  Speed is normally expressed in revolutions per minute (rpm). The speed given depends on the type of machine as stated.

Synchronous machine—synchronous speed

Induction machine—speed at rated power = $(1 - s_{\text{rated power}}) \times$ synchronous speed

DC commutator—base speed or maximum speed at which rated torque can be supplied

Universal—no-load or light-load speed

DC control—no-load speed

**Frequency**  This is the supply voltage frequency in Hertz.

**Power** The power is given in horsepower for motors and watts for generators. This term normally refers to the continuous output power that the machine is capable of delivering. This rating is a function of the thermal capacity of the machine which is dependent on the frame size and insulation class.

**Temperature Rise** The maximum safe rise in temperature in degrees Celsius in the "hot spot" of the machine. The "hot spot" is usually in the armature winding. There are basically two types of machine enclosures determined by the environment and the external cooling equipment for the machine. They are:

> Open (dripproof, splashproof, externally ventilated, etc.)

> Totally enclosed (hermetic, nonventilated, dustproof, fan-cooled, etc.)

**Volt-Amperes** For ac motors, this value will have to be calculated from current and voltage ratings. For single-phase rated VA = (rated $V$)(rated $I$). For three-phase rated VA = $\sqrt{3}$ (rated $V$)(rated $I$). For ac generators the VA rating is on the nameplate rather than the current rating. A power-factor rating will also be given for rated field excitation at full load. The current and volt-ampere ratings of machines are established by the winding thermal characteristics.

**Service Factor** This is a number given to indicate how much over nameplate power the machine can be safely and continuously operated without undue overheating at rated voltage and frequency. The number will be in the range of a 10 to 15 percent overload for most machines, and the service factor will be 1.1 to 1.15. Short-time overload ratings can usually be obtained from the manufacturer.

**Efficiency Index** This has been added to nameplates for high-efficiency machines. There are several types of indexes given, and care should be taken to determine if this is a minimum or nominal efficiency. Good references are IEEE 112 Test Method B and NEMA MG1-12.53a.

**Other Ratings** On certain specialty machines, other ratings are given. For example, on control motors, torque may be given in ounce-inches and the inertia might be given on the nameplate.

## 6-33 ENERGY MANAGEMENT AND ECONOMIC CONSIDERATIONS IN MOTOR SELECTION

There is a continuing need in the world to conserve energy resources. Therefore, it is important that motor users understand the selection, application, and maintenance of electric motors in order to manage electric energy consumption. Consider a 100-hp motor operating continuously at 91 percent efficiency at 4 cents/kWh with a purchase price of $3200.

Purchase price          $3200
Usage                   Continuous (8760 h/year)
Efficiency              91 percent

Power cost                    $0.04/kWh
Annual operating cost         $28,725

This example shows that one motor that operates continuously for one year consumes power that costs nine times its original cost.

Energy management is a concept in which all factors of an electric motor drive system are evaluated to reduce energy consumption. One of the factors is the motor itself. The electric motor is basically an energy converter changing electric energy to mechanical energy. For this reason selection of a motor should consider the drive apparatus to which it is to be connected which has specific operating requirements such as starting torque, speed, and load. Proper motor selection for a given application involves a number of factors that may conflict with one another to some degree.

Fractional- and integral-horsepower motors from $\frac{1}{20}$ to 1 hp connected to single-phase power systems are generally found in the home or small businesses. These motors drive household or commercial appliances. Polyphase motors (three-phase), integral-horsepower motors up to several thousand horsepower, range in voltage from 200 to several thousand volts. These motors are typically found in larger businesses or large manufacturing plants.

The system efficiency is the combination of the efficiencies of all the component parts of the system. These may include fans, pumps, pulleys, belts, gears, compressors, and other devices. Other components not a part of the drive system affect the overall efficiency. Such parts are evaporator and condenser coils, piping, ducts, and baffles. Selection of the most efficient motor is based on speed, load versus horsepower ratings, duty cycle, enclosure type, and initial cost. Good energy management is the successful use of a motor and its driven components such that the result is minimum energy consumption.

The design of an electric motor involves a balance between characteristics such as starting and running, thermal performance, and material usage. Operational efficiency requires a careful consideration of these motor characteristics and relating them to the specific requirements of the application and the efficiency of the system of which the motor is a component.

A change in efficiency as a function of load is an inherent property of a motor. Operation at less than the design rating results in a substantial reduction in motor efficiency. Oversizing a motor is a poor choice in any design.

Improvements in motor efficiency can be achieved by matching the voltage and frequency to the supply. Dual rated motors (200 to 230 V or 50 or 60 Hz) should be avoided, as they are of necessity less efficient designs.

In general for a given type of motor, larger motors (larger horsepower) are more efficient. Also motors of higher synchronous speeds are generally more efficient. This does not imply that one should use a high-speed motor. A lower-speed direct-drive motor may be more efficient than using a high-speed motor and lossy speed reduction through gears, pulleys, and belts.

Many motors are used for very short periods of time and for a very low number of hours per year. Examples are can openers, food processors, waste disposers, electric lawn mowers, and home hand tools. In these cases a change in motor efficiency would not have a substantial impact on energy consumption.

In fact the energy used to improve the motor might exceed the lifetime savings.

On the other hand, some motors are used for long periods for a high number of hours per year. Examples are air-handling equipment, circulation pumps, and refrigeration compressors. In these cases an increase in motor efficiency of only a percent or two could substantially reduce the total energy consumption.

In addition to the concern for optimizing the efficiency of motors in many applications, attention should also be given to power factor. There are two driving factors behind this concern. The first is the kvar demand charges that may be levied against the user with a poor power factor. The other is the increased current that a poor power factor causes with the associated increased line losses.

First consider the single-phase motor application problem. The most common single-phase motors are the induction type because of their relatively constant speed, dependability, simplicity, and low cost. Universal motors are also used in specific home applications. A listing of basic characteristics is given in Table 6-2.

In the selection of energy-efficient single-phase motors, several factors must be evaluated. They are:

Duty cycle—short or intermittent duty vs. continuous

Motor speed—gearing versus multipole machine

Loading—rated versus light loaded

Motor type—see Table 6 3

Efficiency

**TABLE 6-2.** SUMMARY OF SINGLE-PHASE MOTOR CHARACTERISTIC AND USES

| Shaded-pole motors: | Split-phase motors: | Permanent split-capacitor |
|---|---|---|
| $\frac{1}{4}$ hp or less (most applications less than $\frac{1}{10}$) | $\frac{1}{12}$ to $\frac{1}{2}$ hp | motors: |
| Low cost | Medium starting torque | $\frac{1}{20}$ to 1 hp |
| Simple construction | High starting current | Low starting torque |
| Rugged and reliable | Medium efficiency | Low starting current |
| Low starting torque | *Uses* | High efficiency |
| Low efficiency | Laundry equipment | *Uses* |
| *Uses* | Furnace blowers | Fans |
| Fans | Attic fans | Business machines |
| Humidifiers | Compressors | Hermetic motor compressors |
| Rotisseries | Pumps | |
| Slide projectors | Grinders | Universal motors: |
| Vending machines | Home-workshop tools | $\frac{1}{10}$ to 1 hp |
| Advertising displays | Capacitor start—capacitor run: | High starting torque |
| Capacitor start—induction run: | $\frac{1}{3}$ hp and larger | Low starting current |
| $\frac{1}{8}$ hp and larger | High starting torque | Medium to low efficiency |
| High starting torque | Low starting current | Varying speed |
| Medium efficiency | High efficiency | *Uses* |
| *Uses* | *Uses* | Vacuum cleaners |
| Where higher starting torques are needed, such as heavy-duty shop tools | Where high starting torques and high efficiency are needed such as machine-shop tools | Hand-held power tools |

**TABLE 6-3.** EXAMPLE APPLICATIONS OF ALTERNATING-CURRENT SINGLE-PHASE FRACTIONAL-HORSEPOWER MOTORS RATED $\frac{1}{20}$ TO 1 HORSEPOWER, 250 V OR LESS

| Application | Motor type | Horsepower | Speed, rpm | Starting torque | Efficiency |
|---|---|---|---|---|---|
| **Fans:** | | | | | |
| Direct drive | Permanent split-capacitor | $\frac{1}{20}-1$ | 1625, 1075, 825 | Low | High |
| | Shaded-pole | $\frac{1}{20}-\frac{1}{4}$ | 1550, 1050, 800 | Low | Low |
| | Split-phase | $\frac{1}{20}-\frac{1}{2}$ | 1725, 1140, 850 | Low | Medium |
| Belted | Split-phase | $\frac{1}{20}-\frac{1}{2}$ | 1725, 1140, 850 | Medium | Medium |
| | Capacitor start–induction run | $\frac{1}{100}-\frac{3}{4}$ | 1725, 1140, 850 | Medium | Medium |
| | Capacitor start–capacitor run | $\frac{1}{100}-\frac{3}{4}$ | 1725, 1140, 850 | Medium | High |
| **Pumps** | | | | | |
| Centrifugal | Split-phase | $\frac{1}{100}-\frac{1}{2}$ | 3450 | Low | Medium |
| | Capacitor start–induction run | $\frac{1}{100}-1$ | 3450 | Medium | Medium |
| | Capacitor start–capacitor run | $\frac{1}{100}-1$ | 3450 | High | High |
| Positive-displacement | Capacitor start–induction run | $\frac{1}{100}-1$ | 3450, 1725 | High | Medium |
| | Capacitor start–capacitor run | $\frac{1}{100}-1$ | 3450, 1725 | High | High |
| **Compressors:** | | | | | |
| Air | Split-phase | $\frac{1}{100}-\frac{1}{2}$ | 3450, 1725 | Low or medium | Medium |
| | Capacitor start–induction run | $\frac{1}{100}-1$ | 3450, 1725 | High | Medium |
| | Capacitor start–capacitor run | $\frac{1}{100}-1$ | 3450, 1725 | High | High |

| Application | Motor type | HP | Speed (rpm) | | |
|---|---|---|---|---|---|
| Refrigeration | Split-phase | $\frac{1}{8}-\frac{1}{2}$ | 3450, 1725 | Low or medium | Medium |
| | Permanent split-capacitor | $\frac{1}{8}-1$ | 3250, 1625 | Low | High |
| | Capacitor start–induction run | $\frac{1}{8}-1$ | 3450, 1725 | High | Medium |
| | Capacitor start–capacitor run | $\frac{1}{8}-1$ | 3450, 1725 | High | High |
| Industrial | Capacitor start–induction run | $\frac{1}{8}-1$ | 3450, 1725, 1140, 850 | High | Medium |
| | Capacitor start–capacitor run | $\frac{1}{8}-1$ | 3450, 1725, 1140, 850 | High | High |
| Farm | Capacitor start–induction run | $\frac{1}{8}-\frac{3}{4}$ | 1725 | High | Medium |
| | Capacitor start–capacitor run | $\frac{1}{8}-\frac{3}{4}$ | 1725 | High | High |
| Major appliances | Split-phase | $\frac{1}{6}-\frac{1}{2}$ | 1725, 1140 | Medium | Medium |
| | Capacitor start–induction run | $\frac{1}{6}-\frac{3}{4}$ | 1725, 1140 | High | Medium |
| | Capacitor start–capacitor run | $\frac{1}{6}-\frac{3}{4}$ | 1725, 1140 | High | High |
| Commercial appliances | Capacitor start–induction run | $\frac{1}{3}-\frac{3}{4}$ | 1725 | High | Medium |
| | Capacitor start–capacitor run | $\frac{1}{3}-\frac{3}{4}$ | 1725 | High | High |
| Business equipment | Permanent split-capacitor | $\frac{1}{20}-\frac{1}{4}$ | 3450, 1725 | Low | High |
| | Capacitor start–induction run | $\frac{1}{8}-1$ | 3450, 1725 | High | Medium |
| | Capacitor start–capacitor run | $\frac{1}{8}-1$ | 3450, 1725 | High | High |

*Source*: Reprinted by permission of the National Electrical Manufacturers Association from NEMA Standards Publication No. MGI-1978, *Motors and Generators*, Copyright © 1985 by NEMA.

To compare two similar motors operating at the same load with different efficiencies, the following equation can be used to calculate the operating savings if motor A is used rather than motor B:

$$S = 0.746 \times \text{hp} \times C \times N \left( \frac{100}{E_B} - \frac{100}{E_A} \right)$$  (6-70)

where   $S$ = savings, dollars per year
　　 hp = horsepower of specified load
　　  $C$ = energy cost, dollars per kWh
　　  $N$ = running time, h per year
　　 $E_A$ = percent efficiency of motor A at specified load
　　 $E_B$ = percent efficienty of motor B at specified load

The equation is valid for motors operating at a specified constant load. If a load cycles, the equation can be used for each portion of the cycle and then summed to give the total savings over a specified time period.

Now that we have Eq. 6-70 for comparing motors with different efficiencies, we need to be sure that we understand the term efficiency. Several terms are currently in use:

*Full-load efficiency.* The efficiency of a motor at full load. Not specific as to sample that meets the stated efficiency.

*Average expected and nominal efficiency.* These mean about the same thing in that the motor user can expect that a large portion of the same motor model will meet the average value. However, individual motors can vary widely from the average.

*Calculated efficiency.* This could be equal to average expected efficiency depending on the relationship between design and test values.

*Apparent efficiency.* This is the product of the motor power factor and efficiency. An apparent efficiency does not tell the motor user exactly what to expect, since the power factor can be high and the efficiency low. Or the reverse can be true, as long as the product of the two meets the stated value.

*Minimum or expected minimum efficiency.* These are more clearly definitive—all motors should have efficiencies equal to or higher than the value specified.

After understanding the language of efficiency, it is important for the user to have confidence that the value of efficiency quoted is determined accurately, since a point or two of error can result in thousands of dollars difference in savings over the life of the motor.

Of course, building high-efficiency motors is not a new art. Most large motors require high efficiency to meet temperature requirements. Various ways used to improve efficiency are: using thinner steel laminations in the stator and rotor core, using steel with better electromagnetic properties, adding more steel, increasing wire volume in the stator, improving the rotor slot design and using smaller, more efficient fans. Each of these involves either more material, increased material costs, higher manufacturing costs, or a combination of these to achieve the desired efficiency.

Let us briefly review the principles of losses and efficiency given elsewhere in this text. Motor efficiency is defined as

$$\text{Efficiency} = \frac{746 \times \text{hp output}}{\text{watts input}} \qquad (6\text{-}71)$$

or

$$\text{Efficiency} = \frac{\text{input} - \text{losses}}{\text{input}} \qquad (6\text{-}72)$$

The only way to improve efficiency is to reduce losses.

Since the motor user must pay for the power consumed by motor losses as well as that used for useful work, motor losses are important. Figure 6-51 shows how motor losses and efficiency vary with horsepower. Although larger motors generally have higher efficiencies, they also have higher total losses. Since these larger horsepower ratings represent such large blocks of power, some people concern themselves only with efficiency of motors that are 400 to 500 hp or larger. This is a mistake, because smaller motors represent large potential savings. Note how ten 10-hp motors waste more power than one 100-hp motor.

Typically, motor losses are categorized as:

No-load losses
    Windage and friction
    Core losses

Motor operating under load
    Stator $I^2R$ losses
    Rotor $I^2R$
    Stray-load losses

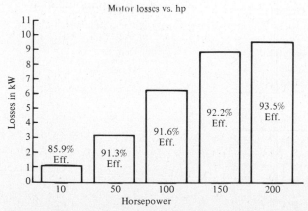

**Figure 6-51** Motor losses versus horsepower. (*Source:* David C. Montgomery, "How to Specify and Evaluate Energy Efficient Motors," General Electric Publication GEA 10951 (50m) R/81)

Treated separately and making correction for temperature, these losses can be precisely determined. Stray-load losses are difficult to measure. Because they are affected by both design and manufacturing processes, they are important and can vary considerably in magnitude depending on

Stator and rotor slot geometry

Number of slots

Air-gap length

Rotor slot insulation

Manufacturing processes

Therefore, it is important that all losses be determined accurately rather than be ignored or assumed to be some value.

NEMA has recognized this problem and after considerable investigation, including having manufacturers test the same motor and comparing the results, has adopted Standard MG1-12.53a, which includes the following:

> Efficiency and losses shall be determined in accordance with the latest revision of IEEE Standard 112. Polyphase squirrel-cage motors rated 1–125 HP shall be tested by dynamometer, Method B. The efficiency will be determined using segregated losses in which stray load loss data is smoothed using a linear regression analysis to reduce the effect of random errors in the test measurements and to adjust $I^2R$ loss for temperature rise.

Even if all NEMA members follow the same procedure, there is still the problem of the British, IEC, and Japanese standards which use an equivalent circuit calculation with assumed values for stray-load loss.

The problem of efficiency determination is illustrated by the comparison of efficiencies for a 7.5- and 20-hp machine using three international standards and the IEEE methods. The results are given as Table 6-4.

NEMA has adopted standard MG1-12.53b, which is an efficiency-labeling standard based on probabilities. The bell-shaped normal or gaussian distributed

**TABLE 6-4.** COMPARISON OF EFFICIENCIES USING INDICATED STANDARDS

|  | Full-load efficiency, % | |
| --- | --- | --- |
| Standard | 7.5 hp | 20 hp |
| International (IEC 34-2) | 82.3 | 89.4 |
| British (BS-269) | 82.3 | 89.4 |
| Japanese (JEC-37) | 85.0 | 90.4 |
| U.S. (IEEE-112, method B) | 80.3 | 86.9 |

*Source:* David C. Montgomery, "How to Specify and Evaluate Energy Efficient Motors," General Electric Publication GEA 10951 (50m) R/81.

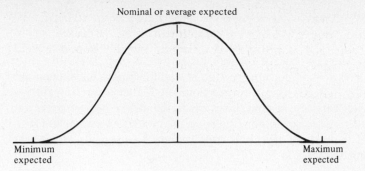

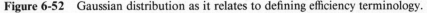

**Figure 6-52** Gaussian distribution as it relates to defining efficiency terminology.

curve in Figure 6-52 shows that once the normal value of efficiency is established for a design, half the motors will be above and half below. The standard, which applies to NEMA designs A and B single-speed, polyphase squirrel-cage integral-horsepower motors in the range of 1 to 125 hp, calls for the full-load nominal efficiency to be identified on the nameplate. This standard recognizes the variations in materials, manufacturing processes, and test results and motor-to-motor efficiency variations for a given design. The full-load efficiency for a large population of motors of a single design is not a unique efficiency but rather a band of efficiencies. The new standard indicates the minimum and nominal efficiency to be expected from a motor design and a population of motors and is to be selected from the values in Table 6-5.

Although the standard establishes nominal efficiency values that are to be used on the motor nameplate, the motor manufacturer selects these values from

**TABLE 6-5.**  NEMA EFFICIENCY-MARKING
STANDARD

| Nominal efficiency | Minimum efficiency | Nominal efficiency | Minimum efficiency |
|---|---|---|---|
| 95.0 | 94.1 | 80.0 | 77.0 |
| 94.5 | 93.6 | 78.5 | 75.5 |
| 94.1 | 93.0 | 77.0 | 74.0 |
| 93.6 | 92.4 | 75.5 | 72.0 |
| 93.0 | 91.7 | 74.0 | 70.0 |
| 92.4 | 91.0 | 72.0 | 68.0 |
| 91.7 | 90.2 | 70.0 | 66.0 |
| 91.0 | 89.5 | 68.0 | 64.0 |
| 90.2 | 88.5 | 66.0 | 62.0 |
| 89.5 | 87.5 | 64.0 | 59.5 |
| 88.5 | 86.5 | 62.0 | 57.5 |
| 87.5 | 85.5 | 59.5 | 55.0 |
| 86.5 | 84.0 | 57.5 | 52.5 |
| 85.5 | 82.5 | 55.0 | 50.5 |
| 84.0 | 81.5 | 52.5 | 48.0 |
| 82.5 | 80.0 | 50.5 | 46.0 |
| 81.5 | 78.5 | | |

the table. The standard does not stipulate a specific efficiency value for each rating.

After motor efficiency is considered, power factor can enter into the motor-selection process. The customer must pay for the watts and the utility must generate volt-amperes. When this ratio gets too low, utilities may charge penalties. A low power factor indicates that a proportionately higher share of total line current is creating distribution system losses which the utility must generate and supply at no cost. As was true of motor efficiency, the motor designer can increase power factor, and the state of the art is well known. The equation for power factor of a three-phase motor is

$$PF = \frac{\text{watts input}}{\sqrt{3}\ V \times A}$$

Most motor design engineers will argue in favor of external or capacitor power-factor correction. They maintain that what is important is motor efficiency, and for the same material cost it is possible to get a slightly higher efficiency if you let the power factor drop below 85 percent (ASHRAE Standard 90-75) on some designs. They point out that it is less costly and better to use power-factor-correction capacitors because the user still gets the benefit of correction even when the motor is lightly loaded and motor power factor drops off sharply. The counterargument is that there are disadvantages in using capacitors. The capacitor is another component that might fail, could introduce a safety problem, and if improperly applied, could damage the motor.

There is one aspect of motor efficiency and power factor deserving mention which is considered very important to many industries where motors are commonly operated at 75 to 80 percent of full load. While variations in load have little effect on efficiency, power factor is affected significantly. Note that the efficiency curve (Fig. 6-53) for high-efficiency motors peaks at approximately 80 percent, and operating at less than full load can actually increase the savings realized. Power factor on all motors is very sensitive to load, and a motor must be operated near full load to realize the benefits of a high-power-factor design.

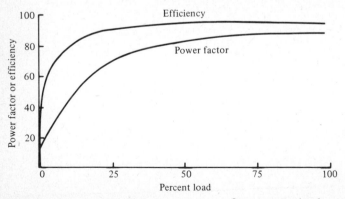

**Figure 6-53**   Typical efficiency and power factor versus load curves.

Once users have fully defined the motor using all the characteristics including efficiency and power factor, they can begin evaluating potential savings based upon their own power costs.

The real cost of electricity is not just the cost per kilowatthour. For industrial customers, power cost may include an energy or fuel adjustment charge, a demand charge, and a power-factor penalty in addition to the kilowatthour charge.

For example, assume that a plant has an average energy cost per kilowatthour of $0.04 and a demand charge of $5 per kW.

Assuming further that penalties are assessed for a power factor below 85 percent, the penalty may be assessed as an additional demand charge using an expression similar to the following:

$$\frac{0.85}{\text{Actual power factor}} - 1$$

During a given month, this plant used 600,000 kWh of electricity, the peak demand was 1500 kW, and the actual power factor was 83.8 percent. The electric bill would include the charges shown in Table 6-6.

Motors with high power factors will contribute to overall improvement of plant performance and help reduce power-factor penalties. However, since power factor is usually corrected on a plantwide basis and the charges cannot be readily attributed to a specific load, penalty charges have been excluded from the remainder of this evaluation. The cost of electricity is calculated as shown in Eq. 6-73. Thus, the real cost of electricity for this example is $0.0537 per kWh instead of the $0.040 per kWh energy charge. Since the real cost is almost 34 percent higher, potential savings from using high-efficiency machinery of any type will be significantly understated if the energy charge alone is used for evaluation.

$$\text{Real energy cost} = \frac{\text{utility charge}}{\text{kWh usage}} = \frac{32,238}{600,000 \text{ kWh}} - \$0.0537 \quad (6\text{-}73)$$

**TABLE 6-6.** TYPICAL MONTHLY PLANT ELECTRIC UTILITY BILL

| | |
|---|---|
| kW energy charges | |
| 600,000 kW at $0.04/kWh | $24,000 |
| Demand charge | |
| 1500 kW at $5/kW | 7,500 |
| Subtotal | $31,500 |
| Taxes 2% | 630 |
| Total without PF penalty | $32,130 |
| Power-factor penalty | |
| $kW = \left(\dfrac{0.85}{0.838} - 1\right) 1500 \text{ kW} = 21.5 \text{ kW at } \$5/kW$ | 108 |
| Total energy costs | $32,238 |

## 6-33.1 Power Factor vs. Efficiency

Power factor and efficiency are often treated as being equally important as they relate to conserving energy or to reducing power costs. This is not the case, since the operating costs associated with a low-efficiency motor cannot be reduced without replacing the motor.

However, low system power factor caused by motors can be easily improved by the addition of power-factor-correction controllers or capacitors, which are relatively inexpensive.

Reactive current increases with a low power factor, and this in turn increases the cost of power distribution. For a value of $100 per kvar (which is probably too high) and $2000 per kW (which is fairly common), it is seen that efficiency has an impact of at least 20 times that caused by power factor. Thus, they can hardly be considered as being equally important. This is the reason the term apparent efficiency, which treats the power factor and efficiency as being equally important, is confusing in making economic comparisons.

## 6-33.2 Calculating Annual Savings

The user should decide whether a simple payback calculation is required or a more comprehensive evaluation is necessary for an energy study. Payback calculations are usually used for small quantities of motors where a basic motor design is being compared with a higher-efficiency design. In this case, the user must consider the individual motors involved, the hours of operation, motor loading, the cost of electricity, and the efficiencies being evaluated.

To determine the savings realized from the difference in operating cost between the high-efficiency motor and a standard-efficiency motor, Eq. 6-70 would be used. The critical part of this relatively simple calculation is that the efficiencies used must be comparable. You must evaluate minimum versus minimum or nominal versus nominal.

As an example, consider a 100-hp 1800-rpm severe-duty motor operating continuously at a constant rated load with a nominal efficiency of 91.6 percent. This motor is to be compared with an efficient motor with a 95 percent efficiency. The cost of power is $0.04 per kWh. The savings using Eq. 6-70 are

$$S = 0.746 \times \text{hp} \times C \times N\left(\frac{100}{E_B} - \frac{100}{E_A}\right)$$

$$= 0.746 \times 100 \times 0.04 \times 8760\left(\frac{100}{91.6} - \frac{100}{95}\right)$$

$$= \$1021.32 \text{ per year}$$

## 6-33.3 Higher-Efficiency Payback

Using Eq. 6-70 savings for a high-efficiency 25-hp dripproof motor using different power costs and hours of operation were calculated. The cost of a high-efficiency motor was divided by the savings to get a rough estimate of payback period in years. The results are plotted in Fig. 6-54.

Hours of operation are plotted on the horizontal axis for 1 to 8000 h per year and power costs on the vertical axis from 1 to 8 cents per kW. Payback curves are shown for 1, 2, and 3 years. Note that good payback is achieved even at low power cost if the hours of equipment operation are high, such as on drives used in the process industries.

Using the curves, consider a specific example. If power costs 4 cents per kW, draw a horizontal line starting at the 4-cent point parallel to the hours-of-operation axis. The line intersects all three curves. The intersecting points indicate that 1300 h of operation are required for a 3-year payback. Similarly, to get a 2-year payback requires 1900 h per year (less than 1 shift per day). One-year payback takes a little over 3800 h, or almost two shifts per day.

The set of curves in Fig. 6-54 recognizes energy-use charge which is proportional to hours of operation. Demand charge is not time-dependent. Therefore, a higher power cost should be used for low-use machines if they are used during peak demand periods.

Another way to look at the economics of buying high-efficiency motors is shown in Fig. 6-55. Two 125-hp 1800-rpm motors are compared. The high-efficiency motor has a full-load efficiency of 95 percent, whereas the industry average is 91.6 percent. Assuming continuous operation using 4 cents per kWh power, there would be an annual savings of $1277. If we assume that the high-efficiency motor costs $838 more than an average motor, the premium price

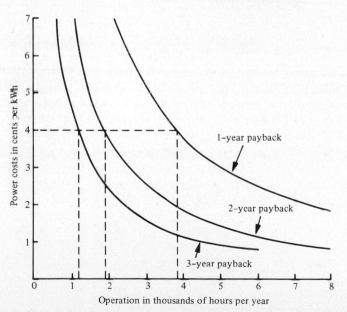

**Figure 6-54** Example payback of premium price versus time of operation. [*Source:* David C. Montgomery, "How to Specify and Evaluate Energy Efficient Motors," General Electric Publication GEA 10951 (50m) 12/81.]

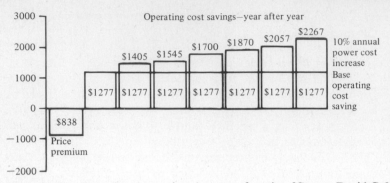

**Figure 6-55** Operating cost savings by year of service. [*Source:* David C. Montgomery, "How to Specify and Evaluate Energy Efficient Motors," General Electric Publication GEA 10951 (50m) 12/81.]

would be recovered in less than 8 months and the $1277 savings would be enjoyed year after year. If we assume an annual 10 percent increase in power cost, the annual savings in the seventh year would be $2267. The $990 savings in this year due to the power-cost increase is greater than the original premium price of the high-efficiency motor. Over the 7-year period, the total net savings for buying the high-efficiency motor would be $11,283, or over 13 times the original premium price.

### 6-33.4 Time Value of Money, Present Worth, and Life Cycle

The simple payback calculation just shown would not be acceptable to those seeking a detailed economic analysis. It does serve to show simply that there is financial leverage to be gained by the use of higher-efficiency motors that sell for premium prices. In order to account for the time value of money, the present worth of that money plus inflation must be taken into account. In addition, the life cycle based on a period of years of operation is often a consideration in motor selection.

In order to incorporate the above factors, a computed integrated present-worth calculation is often used. The first step is to determine an effective interest rate that incorporates inflation and the company's internal rate of return on investments. This is given by Eq. 6-74.

$$i = \frac{1 + R_2}{1 + R_1} - 1 \qquad (6\text{-}74)$$

where    $i$ = effective interest rate
$R_1$ = expected annual rate of inflation (in this case rate of increase in power costs)
$R_2$ = internal (company) expected rate of return on investments

Next a calculation of apparent number of operating years considering present worth is calculated by Eq. 6-75.

$$\text{AOY} = \frac{(1 + i)^n - 1}{i(1 + i)^n} \tag{6-75}$$

where   AOY = apparent operating years considering present worth
$i$ = effective interest rate
$n$ = number of years in life cycle

From Eqs. 6-74 and 6-75 another calculation is made to determine the present-worth factor to be used to estimate the savings of choosing a higher-efficiency motor. This calculation is given by Eq. 6-76.

$$\text{PWF, \$/kW} = C \times N \times \text{AOY} \tag{6-76}$$

where   PWF = present-worth factor
$C$ = average energy cost, \$/kWh
$N$ = running time, h/year
AOY = apparent operating years considering present worth

If one is considering several motors for installation, there are many combinations of vendors. The simplest method would be to compare each vendor's motor with the perfect motor, or $E_A$ would be 100 in Eq. 6-70, and

$$\left(\frac{100}{E_B} - \frac{100}{E_A}\right) = \left(\frac{100}{E_B} - \frac{100}{100}\right) = \left(\frac{100}{E_B} - 1\right)$$

Therefore, the cost of losses over the life cycle of a motor including present worth is given by the following equation:

PWL = present-worth cost of losses

$$= 0.746 \times \text{hp} \times \text{PWF}\left(\frac{100}{E_B} - 1\right) \tag{6-77}$$

This equation accounts for current and forecasted future cost of energy, annual hours of operation, effective years of operation, and rate of return. Adding this value to the first cost (purchase price) of each motor will give the lowest-cost selection.

### EXAMPLE 6-16

Compare the advisability of purchase of 100-hp motor A or motor B under the following circumstances:

Cost of power—\$0.04/kWh

Life cycle—7 years

Continuous duty

Inflation rate—9.5 percent

Rate of return—25 percent

Motor A—95 percent efficient, first cost = \$3660

Motor B—91.6 percent efficient, first cost = \$3150

*Solution.* Calculate the effective interest rate,

$$i = \frac{1 + R_2}{1 + R_1} - 1 = \frac{1 + 0.25}{1 + 0.095} - 1 = 0.142$$

Then determine the apparent operating years,

$$\text{AOY} = \frac{(1 + i)^n - 1}{i(1 + i)^n} = \frac{(1 + 0.142)^7 - 1}{(0.142)(1 + 0.142)^7} = 4.26 \text{ years}$$

Substitute these values into Eq. 6-76 to determine the PWF,

$$\text{PWF} = \$0.04 \times 8760 \times 4.26$$
$$= \$1493 \text{ per kW}$$

Calculation of the cost of motor losses follows:
Motor A:

$$\text{PWL} = (0.746)(100)(1493)\left(\frac{100}{95} - 1\right) = \$5862$$

Motor B:

$$\text{PWL} = (0.746)(100)(1493)\left(\frac{100}{91.6} - 1\right) = \$10,214$$

Total cost A:

$$\text{First cost} + \text{PWL} = \$3660 + \$5862 = \$9522$$

Total cost B:

$$\text{First cost} + \text{PWL} = \$3150 + \$10,214 = \$13,364$$

The best choice is the higher-cost more efficient motor.

### 6-33.5 Other Considerations

Most engineering analysis is done on a before-tax basis. The best alternative is often the same, but other factors may affect and change the before-tax decisions. For example, the value of an investment must be evaluated after taxes, which must include the impact of investment tax credits. Cash-flow effects after taxes are also an important consideration. Cash flow is affected by investment tax credits and credits for depreciation of the new equipment. Not only must a decision be good on the basis of first cost, return on investment, life cycle, and operating costs but it must make good sense from the company viewpoint of cash flow and tax credits.

## STUDY QUESTIONS

1. Name the three general types of dc motors and generators.
2. Describe how the construction of the fields differs in these types of generators.

3. Tell what current the series field carries in the long-shunt compound generator and what current it carries if it is connected as a short-shunt compound generator.

4. For a compound generator the series field coil is wound over what coil?

5. What voltage is developed by a series generator at no load?

6. What voltage is generated by a shunt or compound generator with the shunt field open? Why?

7. Distinguish between separately excited and self-excited shunt generators and motors.

8. What factors affect the no-load voltage of a shunt generator?

9. Explain how the field of a shunt generator is controlled.

10. Is it desirable to operate a generator at a speed lower than the one for which it is designed? Explain your answer.

11. Why is the flux created by the shunt field not directly proportional to the field currents in the normal operation of a shunt generator?

12. Describe what is meant by the saturation curve.

13. Describe the relationship that exists between no-load voltage and field current below the knee of the magnetization curve.

14. Explain how the no-load voltage of a shunt generator varies with changes in speed if the field current is kept constant.

15. For a self-excited shunt generator explain what is meant by a voltage buildup.

16. Before a self-excited shunt generator will build up, what conditions must exist?

17. What factors affect the voltage to which a self-excited generator will eventually build up?

18. Devise a simple test that can be performed to show that a self-excited shunt generator will not build up if the field is reversed.

19. Assume that a self-excited shunt generator does build up its voltage. Will it do so if the field is reversed? If the armature leads are reversed? If the direction of rotation is reversed? If the residual field is reversed? Give answers for each part.

20. Determine what is meant by "flashing the field." Why must one flash the field?

21. What factors cause the voltage of a separately excited shunt generator to drop when it is loaded?

22. What factors cause the voltage of a self-excited shunt generator to drop when it is loaded?

23. Explain why it is important to keep the armature resistance of a generator as low as possible. Explain how this is accomplished in the design of the winding.

24. Describe what is meant by the voltage regulation of a shunt generator.

25. Assume a regulation for a self-excited shunt generator. Would this value be increased or decreased if the same machine were operated as a separately excited shunt generator? Explain your answers.

26. How is the terminal voltage of a shunt generator controlled? What types of voltage regulators are used in dc generators?

27. Why is the regulation of a self-excited shunt generator improved by operating the machine at a lower speed?

28. What is meant by the following terms: cumulative compound, overcompound, undercompound, flat compound, differential compound, as applied to compound generators.

29. Describe how the amount of compounding can be readily adjusted.

30. What is meant by a diverter, and what material might be used for its construction?

31. Tell why the terminal voltage of a series generator will rise from its residual value at no load to some maximum value as load is applied.

32. Why does the terminal voltage of a series generator reach a maximum value as load is increased and then decrease rapidly to zero with further applications of load?

33. Describe or list the practical applications of a series generator.

34. Describe what is meant by armature reaction.

35. What two important effects are produced by armature reaction?

36. Describe what is meant by an interpole.

37. For noninterpole generators tell why the brushes must be shifted when the load increases.

38. How is the interpole winding excited?

39. Describe the zone in which the interpole winding is effective.

40. What is meant by the term "compensating windings"?

41. What current is carried by compensating windings and why?

42. Describe the important functions performed by the commutator and brushes on a dc generator.

43. Define the three basic stages of a commutation process.

44. Describe the reasons for shifting brushes beyond the magnetic neutral on noninterpole generators to achieve sparkless commutation.

45. When interpoles are used, why must they be somewhat stronger than would be necessary to neutralize the effect of armature reaction if sparkless commutation is to be produced?

46. What methods are employed to adjust the voltage of a generator or the speed of a motor?

47. What problems are encountered when motors are started under a load?

48. Is it possible to operate a dc generator as a motor, and vice versa? List the factors that must be considered.

49. What are the three general types of dc motors?

50. Explain how the speeds of the three types of dc motors are affected by an increase in load.

51. What is meant by a constant-speed motor?

52. What type of motor exhibits constant-speed characteristics?

53. What does the term "variable-speed motor" mean?

54. What motors exhibit a variable-speed characteristic?

55. Describe what is meant by an adjustable-speed motor.

56. What conditions would exist for a motor to be called a constant speed–adjustable speed motor? A variable speed–adjustable speed motor?

57. What is meant by the counter emf?

58. Can the counter emf ever be equal to the impressed voltage in a motor? Explain your answer.

59. Explain how the counter emf affects the armature current.

60. What factors affect the counter emf for a given motor?

61. If the load on a shunt motor is increased, what electrical factor affects the speed of the motor?

62. When the load on a compound or series motor increases, what factors affect the speed?

63. Why is the power developed by a dc motor determined by the value of the counter emf?

64. What factor limits the armature current in a dc motor at the instant of starting?

65. How is the armature current kept to a reasonable level when a dc motor is started?

66. Why are small motors normally started directly across the line without external resistors to limit the current?

67. What is the function of a dc motor starter?

68. What are the general types of manual motor starters for dc motors?

69. How are dc motor starters rated?

70. Describe what is meant by a three-point starter or a four-point starter.

71. What is meant by a motor controller?

72. In automatic starter diagrams relays will be described as normally open and normally closed. Explain what these terms mean.

73. How do the counter emf time-limit and current-limit automatic types of starters differ from one another in operation?

74. List the types of dc starters. Describe the advantages of each type of starter.

75. Locate a circuit diagram for a counter emf automatic starter and explain its operation.

76. Find a diagram for a time-limit automatic starter and explain its operation.

77. Locate a current-limit automatic starter diagram and explain its operation.

78. What is meant by the normal speed of a motor?

79. Describe the load-speed characteristics for shunt, compound, and series motors.

80. What is torque?

81. What factors affect the torque of a motor?

82. Why is the torque of a shunt motor proportional to the armature current?

83. Describe how the torque varies with increased load for a compound motor and a series motor.

84. What operating conditions would make it desirable to use a shunt motor, a series motor, a compound motor?

85. Describe what precautions must be taken in the operation of a series motor with a load varying over wide limits.

86. Define what is meant by speed regulation.

87. What variation in speed regulation exists between shunt and compound motors?

88. Why is the term "speed regulation" not applied to a series motor?

89. What is the difference between speed regulation and voltage regulation?

90. Why are shunt motors called constant-speed motors?

**91.** Why are compound and series motors referred to as variable-speed motors?

**92.** List practical applications for shunt motors, compound motors, and series motors.

**93.** What precautions are taken when starting a differential compound motor?

**94.** What happens to the speed of a shunt motor when the shunt field rheostat is adjusted?

**95.** What happens to speed of a shunt motor when the armature rheostat is adjusted?

**96.** What happens to the shunt-motor speed when the armature voltage is adjusted?

## PROBLEMS

**6-1.** An elementary two-pole dc generator has one armature coil of one turn connected to two commutator segments. It is driven at a speed of 1000 rpm. The diameter at the air gap is 0.30 m and the axial length of the armature is 0.20 m. The flux density is distributed sinusoidally with $B_{amp} = 0.60$ T. Neglect the effect of slots and the width of the brushes and (a) plot the waveform of the no-load emf if the brushes are set at the geometric neutral. (b) Calculate the dc emf for part (a).

**6-2.** The purpose of this problem is to show that a small displacement of the brushes from the geometric neutral causes only a slight reduction in the dc voltage. Therefore, prove that, if the brushes are displaced from the geometric neutral in the generator in Prob. 6-1 by an angle of $+\beta$, the voltage is reduced by a factor of cos $\beta$.

**6-3.** A second coil identical to the coil in Prob. 6-1 is added to the armature except that the second coil is displaced 90° from the first. The two coils are connected so that their generated voltages add. For the conditions specified in Prob. 6-1, (a) plot the no-load emf waveform and calculate (b) the dc emf and (c) the minimum and maximum instantaneous values of the emf.

**6-4.** The armature of a six-pole 600-V 600-kW dc generator has a wave winding. The resistance of the armature winding between brushes is 0.10 $\Omega$, not including the resistance of the brushes and contact resistance. The wave winding is replaced by a lap winding of the same pitch, number of turns, length of mean turn, and cross section of armature conductor. For the same value of speed and field mmf, calculate (a) the voltage rating, (b) the current rating, and (c) the resistance of the armature winding between brushes when the machine operates with the lap winding.

**6-5.** Each shunt field coil of a dc generator produces 2700 ampere turns. How many turns are there in the coil if the current is 1.5 A?

**6-6.** The armature of a six-pole 240-V dc generator has 96 slots which carry a simplex lap winding of 384 coils with two turns each. There are 384 commutator segments. (a) Calculate the armature mmf in ampere turns per pole when the armature current is 500 A. (b) If the machine has three commutating poles but no compensating winding, how many turns carrying the full armature current are required if the commutating-pole mmf is to be 1.5 that of the armature? (c) Repeat part (a) if a compensating winding is used in addition to the commutating poles and if the pole faces cover two-thirds of the pole span. (A pole spans $1/P$ of the armature circumference.)

**6-7.** A 230-V shunt motor has an armature resistance of 0.26 $\Omega$. Assuming a 2-V brush drop, calculate the counter emf when the armature current is 36 A.

**6-8.** A 10-kW 250-V 900-rpm four-pole shunt generator with commutating poles has the following points on its magnetization curve for a speed of 900 rpm:

| $I_f$ | 0 | 0.915 | 1.255 | 1.535 | 1.65 | 1.85 | 2.05 | 2.44 | 2.97 | 3.60 |
|-------|-----|-------|-------|-------|------|------|------|------|------|------|
| $E$ | 7.5 | 160 | 210 | 240 | 250 | 265 | 280 | 300 | 320 | 340 |

Other data follows:

| Armature | |
|----------|----------|
| Number of slots | 48 |
| Number of coils | 143 |
| Turns per coil | 3 |
| Type of winding | Simplex wave |
| Resistance | 0.254 Ω |
| Number of commutator segments | 143 |

| Field | |
|-------|----------|
| Number of turns per pole | 2000 |
| Resistance | 125 Ω |

| Commutating winding | |
|---------------------|----------|
| Resistance | 0.016 Ω |

The field is separately excited and the generator delivers 40 A at 250 V while driven at 900 rpm. Calculate (a) the field current neglecting the effect of armature reaction and (b) the ratio of the field mmf to armature mmf.

**6-9.** Repeat Prob. 6-8 for a generator speed of (a) 750 rpm and (b) 1200 rpm.

**6-10.** Calculate (a) the critical field resistance for the generator of Prob. 6-8 when the speed is 900 rpm, (b) the critical speed when the resistance of the field circuit is 125 Ω, and (c) the highest speed at which the generator can deliver a no-load voltage of 250 V when self-excited.

**6-11.** Regulation of a 240-V shunt generator is 5 percent. Calculate the no-load voltage.

**6-12.** A shunt generator is operating at no load. A decrease of 20 percent in the speed results in a decrease of 40 percent in the voltage. Determine the percent change in the flux.

**6-13.** Calculate the resistance of the shunt field rheostat such that the no-load voltage of the generator of Prob. 6-8 is 250 V when the generator is self-excited.

**6-14.** A self-excited dc shunt generator has been operating normally before it is shut down. It is then started up without load but driven in the opposite direction at normal speed and with the resistance of the field circuit unchanged. (a) Will the voltage build up? Explain. (b) If the voltage fails to build up, what change in connections would you make and would the voltage then be of the former polarity? Explain.

**6-15.** A 150-kW 240-V compound generator is connected long-shunt. If the shunt-field resistance is 20 Ω, what is the series-field current at full load?

**6-16.** If the generator of the previous problem is connected short-shunt, what is the full-load series field current?

**6-17.** A compound-wound generator that can be driven in only one direction builds up voltage of the wrong polarity. This generator also has commutating poles. Explain what steps you would take to correct the polarity of the voltage. Would it be necessary to change the polarity of (a) the shunt-field winding, (b) the series-field winding, and (c) the commutating-pole-field winding? Explain.

**6-18.** When the shunt generator in Prob. 6-8 delivers rated load at 250 V and at 900 rpm, the field current is 1.8 A. The shunt-field winding is to be replaced by a series-field winding such that the generator delivers rated current at 250 V and at 900 rpm. Assume that the $I^2R$ losses in the series field at rated load equal those in the field when the machine delivered its rating as a shunt generator. Calculate the number of turns per pole in the series-field winding.

**6-19.** The generator of Example 6-5 is to be flat-compounded (long-shunt) at 1000 rpm. Calculate the number of turns per pole in the series-field winding. The shunt-field winding has 1200 turns per pole. Neglect the resistance of the series-field winding. Is a series-field diverter indicated and, if so, what should be the ratio of its resistance to that of the series-field circuit?

**6-20.** When a motor is operating under load, the armature takes 8600 W and its current is 38 A. If the armature circuit resistance, including brushes, is 0.4 Ω, what horsepower is developed by the motor?

**6-21.** A 10-hp 1750-rpm 600-V shunt motor has an armature resistance of 1.6 Ω. If the armature takes 15 A at full load, calculate the counter emf developed by the motor and the power developed by the motor in watts and in horsepower. Assume a 2-V brush drop.

**6-22.** A dc generator is flat-compounded when driven at its rated speed. Suppose that this generator is driven at a constant reduced speed with the shunt-field current adjusted to a value such that the no-load terminal voltage equals rated value. Is the generator overcompounded or undercompounded when operating at reduced speed? Explain.

**6-23.** The generator in Example 6-7 is to be rebuilt for a rating of 100 kW and 500 V instead of 250 V. The speed, degree of compounding, current density at rated load in the various windings, and magnetic flux densities in the various parts of the magnetic circuit are to be the same as for the generator in Example 6-7. Calculate the resistance of all windings and the efficiency at rated load.

**6-24.** Assume the following for the generator in Example 6-1: (1) kind of field winding = shunt; (2) apparent flux density in air gap = 0.85 T; (3) ratio of field ampere turns to armature ampere turns per pole = 1.10. Estimate (a) the length of the air gap if the reluctance of the iron is equal to two-thirds of the air gap, (b) the number of turns per pole in the field winding if the resistance of the field winding is 250 per unit.

**6-25.** The flux that links an armature coil of one turn, due to the armature current, changes from 0.002 Wb to −0.002 Wb during commutation. The diameter of the commutator is 50 cm and the brush arc is 3.2 cm. The armature rotates at a speed of 900 rpm. Calculate the average value of the reactance voltage (i.e., the voltage produced by the reversal of this flux during commutation).

**6-26.** Calculate the efficiency of the generator in Example 6-7 for three-fourths rated load.

**6-27.** Repeat the calculations in Example 6-8 if a series-field winding of 6.5 turns per pole is added to the 25-hp motor. Neglect the resistance of the series-field winding and the change in rotational losses.

**6-28.** The resistance of the armature circuit of a 240-V shunt motor is 0.11 $\Omega$, including brushes. When the field rheostat is adjusted so that the resistance of the shunt-field circuit is 120 $\Omega$, the current taken by the motor is 82 A and the speed is 900 rpm. Neglect armature reaction and calculate (a) the current and (b) the speed if the load torque is increased 50 percent.

**6-29.** Two shunt generators A and B are to be operated in parallel to supply a common load. Generator A has a no-load voltage of 240 V and a voltage of 220 when it delivers a load of 60 A. Generator B has a no-load voltage of 230 V and a voltage of 220 when it delivers the same current as generator A. For purposes of this problem, assume straight-line characteristics for both machines around the operating point. Calculate: (a) The line voltage and the total load in kilowatts when generator B is floating; (b) the total load delivered by both machines when the line voltage is 225 V.

**6-30.** Refer to Fig. 6-38(b) in the text material. Describe the relay operations when the forward push button is depressed. Description will take the form of that given in the text material.

**6-31.** In a similar manner to the previous problem, describe the contact closing sequences that occur for a plugged reverse operation of the automatic starter and control.

**6-32.** Refer to Fig. 6-38(b). (a) Draw an equivalent circuit diagram for the connections of the motor and resistors before plugging and after plugging. (b) Draw equivalent circuits for forward acceleration and for jogging.

**6-33.** A 550-V long-shunt compound motor has an armature resistance of 0.815 $\Omega$ and a series-field resistance of 0.15 $\Omega$. The full load speed is 1900 rpm when the armature current is 22 A. (a) What is the speed of the motor at no load if the armature current drops to 3 A with a corresponding drop in flux to 88 percent of the full-load value (assume a brush drop of 5 V for full-load operation and 2 V for no-load operation). (b) Calculate the percent speed regulation of this motor.

**6-34.** To reduce the speed of the motor in the previous problem to 1200 rpm it is necessary to insert a resistance of 1.2 $\Omega$ in the armature circuit. Calculate the power loss in the resistor if the current is 20 A. Why does the insertion of a resistance in the armature circuit create a reduction in speed?

**6-35.** The following information is given in connection with a long-shunt compound generator: $E = 220$ V, output $= 2$ kW, stray power loss $= 705$ W. Shunt-field resistance is 110 $\Omega$, the armature resistance is 0.265 $\Omega$, the series-field resistance is 0.035 $\Omega$. The brush drop is approximately 5 V. Calculate the efficiency.

**6-36.** A 25-kW series generator has an efficiency of 85 percent when operating at rated load. If the stray power loss is 20T of the full-load loss, calculate the efficiency of the generator when it is delivering a load of 15 kW, assuming that the stray power loss is substantially constant and the other losses vary as the square of the load.

**6-37.** A 240-V 20-hp shunt motor has a resistance of 0.18 $\Omega$ for the armature circuit, not including brushes (allow for a brush drop of 2 V) and 240 $\Omega$ for the field circuit. The rated speed is 1200 rpm and the full-load efficiency is 0.87. Calculate (a) the torque, (b) the stray-load losses, and (c) the rotational losses.

**6-38.** A 440-V series motor draws a current of 100 A at a speed of 1000 rpm. The resistance of the armature circuit, including brushes, is 0.11 $\Omega$ and that of the field circuit is 0.09 $\Omega$. (a) Assume a linear magnetization curve, neglect armature reaction, and calculate the speed and torque when the current is 50 A. (b) Repeat part (a) on the basis of nonlinearity such that at 100 A the iron requires one-third of the field mmf and that at 50 A the air gap consumes the entire mmf.

**6-39.** A fractional-horsepower series motor draws a current of 4.0 A from a 125-V supply at a speed of 3000 rpm. The resistance of the armature circuit, including brushes, plus that of the field circuit, is 15 $\Omega$ and the rotational loss at 3000 rpm is 60 W. Calculate the no-load speed, assuming the magnetic circuit to be linear and the rotational losses to vary directly as the speed.

**6-40.** The following test data was obtained on a 20-hp 250-V 600-rpm shunt motor:

$r_f = 125\ \Omega$, resistance of the field circuit

$\tau_f = 0.40$ s, time constant of the field circuit

$r_a = 0.14\ \Omega$, resistance of the armature circuit, including brushes

$\tau_a = 0.043$ s, time constant of the armature circuit, including brushes

When the motor is driven at 600 rpm as a generator without load, a field current of 2.0 A produces an armature emf of 250 V. Determine the following constants. (a) $L_{ff}$, the self-inductance of the field circuit; (b) $L_{aa}$, the self-inductance of the armature circuit; (c) $\mathcal{L}_{af}$, the inductance coefficient which relates the speed voltage to the field current; and (d) $B_L$, the friction coefficient of the load at rated load and rated speed.

**6-41.** The field of the motor in Prob. 6-40 is separately excited with a constant field current of 2.0 A while driving a load which has inertia only ($B_L = 0$). The polar moment of inertia of the armature and load combined is 3.0 kg-m²/rad². The motor is initially running at constant speed with an impressed armature voltage of 250 V. Neglect the rotational losses of the motor. (a) Calculate the speed. (b) Neglect the self-inductance of the armature and express the armature current and the speed as functions of time if the applied armature voltage is suddenly reduced from 250 to 240 V. (c) Repeat part (b) but for an increase in the applied armature voltage from 250 to 260 V. (d) Repeat part (c) but include the effect of the armature self-inductance.

**6-42.** Assume the armature of the motor in Example 6-14 to be energized at a constant terminal voltage of 240 V while the voltage applied to the field is expressed by $v_f = 180 + 50 \sin 3.33t$ with $J_L = 20.0$ kg-m²/rad² but with the load torque $T_1 = 0$. (a) Calculate the quiescent values of (1) the field current, (2) the armature current, and (3) the speed. (b) Express (1) the corresponding variable components in part (a) as phasors and as functions of time and (2) the total speed as a function of time. (c) Calculate the armature copper loss ($r_a I_a^2$) and the corresponding dc current that must be supplied to the armature. (d) Verify that the average value of the quantity $-\Omega_0 \mathcal{L}_{af} i_{f1} i_{a1}$ is equal to the armature copper loss. (Note that Eq. 6-49 is an approximation in which the product $i_{f1} i_{a1}$ is omitted.)

**6-43.** Repeat Prob. 6-42 except that the voltage applied to the field is constant at 240-V dc and the voltage applied to the armature is $v_a = 180 + 50 \sin 3.33t$.

**6-44.** Show the equivalent capacitive circuit for the motor in Example 6-12 and its connected load. Indicate the values of the parameters in terms of $\mathcal{L}_{af}$ and $i_f$ and verify on the basis of this circuit the answers to part (b) in Example 6-12.

**6-45.** Show that, if the rotational losses of a shunt motor are negligible, the energy consumed by the resistance in the armature circuit equals the kinetic energy stored in the rotor when the motor is accelerated from standstill with constant field current and constant applied armature voltage. The equivalent capacitive circuit may be used for this purpose.

**6-46.** Show that the time constant of the equivalent capacitive circuit for a shunt motor without load is

$$\tau_{em} = \frac{r_a R_{eq} C_{eq}}{r_a + R_{eq}}$$

when $L_{aa}$ is neglected.

**6.47** The shunt motor in Example 6-12 is operating at rated voltage in the steady state with a field current of 1.0 A and with the resistance in series with the armature reduced to zero. (a) Show the equivalent capacitive circuit, neglecting $L_{aa}$, and calculate the steady armature current. (b) The field current is suddenly reduced to 0.80 A while the applied voltage is constant at 240 V. Calculate the initial armature current after the field current is reduced on the basis that the kinetic energy stored in the rotating parts cannot change instantaneously. (c) Calculate the final armature current for the condition of part (b). (d) Calculate the time constant $\tau_{am}$ of the armature current for the condition of part (b) and express the armature current as a function of time on the basis that

$$i_a = i_a(x) + [i_a(0) - i_a(\infty)]\epsilon^{-1/\tau_{am}}$$

(e) Calculate the speed voltage and the speed on the basis of

$$e_a = V - r_a i_a$$

**6-48.** A 50-hp 250-V dc shunt motor has the following constants:

$$r_f = 62.5\ \Omega \qquad\qquad r_a = 0.06\ \Omega$$
$$L_f = 25.0\ \text{H} \qquad\qquad L_{aa} = 0$$
$$J_M = 1.0\ \text{kg-m}^2/\text{rad}^2 \qquad \mathcal{L}_{af} = 0.63\ \text{H}$$
$$B_M \simeq 0$$

(a) This motor is supplied from a 250-V dc source of negligible impedance and drives a device which has a polar moment of inertia $J_L = 1.25$ kg-m²/rad². The output of the load device under steady conditions is zero; so the friction constant $B_L \cong 0$. Calculate the steady-state armature current and the steady-state speed. (b) After operating in the steady state without load, the driven device is subjected to a load which results in a sudden increase in $B_L$ from zero to 4.2 N-s/rad. Calculate the final steady values of the armature current and speed. (c) Calculate the time constant $\tau'_{am}$ of the armature current for the condition of part (b) and express the armature current as a function of time on the basis that

$$i_a = i_a(\infty) + [i_a(0) - i_a(\infty)]\epsilon^{-1/\tau'_{am}}$$

and the speed from

$$\omega_m = \omega_m(\infty) + [\omega_m(0) - \omega_m(\infty)]\epsilon^{-1/\tau'_{am}}$$

**6-49.** A 200-kW 250-V 600-rpm shunt generator has the following data:

$$r_a = 0.010\ \Omega \qquad\qquad r_f = 12.5\ \Omega$$
$$L_{aa} = 0.0003\ \text{H} \qquad L_{ff} = 10.0\ \text{H}$$
$$\mathcal{L}_{af} = 0.20\ \text{H}$$

This generator is driven at a constant speed of 600 rpm initially without field current while a load impedance of $R_L = 0.30\ \Omega$ in series with $L_L = 0.0007$ H is connected across its armature terminals. (a) Calculate the voltage constant $K_E$. (b) Express the armature current and the armature terminal voltage as functions of time after the field is suddenly connected to a constant dc source of 250 V. (c) Repeat part (b), neglecting the time constant of the armature circuit. (d) Calculate the armature current at $t = 0.3$ s for parts (b) and (c). What is the discrepancy in percent?

## BIBLIOGRAPHY

Adkins, B. *The General Theory of Electrical Machines.* New York: John Wiley & Sons, Inc., 1957.

Brown, David, and E. P. Hamilton III. *Electromechanical Energy Conversion.* New York: The Macmillan Company, 1984.

Chestnut, H., and R. W. Mayer. *Servomechanisms and Regulating System Design.* New York: John Wiley & Sons, Inc., 1951.

Crosno, C. Donald. *Fundamentals of Electromechanical Conversion.* New York: Harcourt, Brace and World, Inc., 1968.

Fitzgerald, A. E., C. Kingsley, and Alexander Kusko. *Electrical Machinery,* 3d ed. New York: McGraw-Hill Book Company, 1971.

Gardner, M. F., and J. L. Barnes. *Transients in Linear Systems.* New York: John Wiley & Sons, Inc., 1942.

Gourishankar, V., and D. H. Kelley. *Electromechanical Energy Conversion,* 2d ed. New York: Intext Educational Publishers, 1973.

Kimball, A. W. "Two-Stage Rototrol for Low-Energy Regulating Systems," *Trans. AIEE,* Part I (1949): 1119–1124.

Kloeffler, R. G., et al. *Direct-Current Machinery.* New York: The Macmillan Company, 1950.

Knowlton, A. E. *Standard Handbook for Electrical Engineers,* 9th ed. New York: McGraw-Hill Book Company, 1957, Sec. 8.

Kosow, I. L. *Electric Machinery and Control.* Englewood Cliffs, N.J.: Prentice-Hall, Inc., 1964.

Langsdorf, A. S. *Principles of Direct-Current Machines,* 5th ed. New York: McGraw-Hill Book Company, 1940.

Liwschitz-Garik, M., et al. *D-C and A-C Machines.* New York: D. Van Nostrand Company, Inc., 1952.

Matsch, Leander W., *Electromagnetic & Electromechanical Machines,* 2d ed. New York: Dun-Donnelley Publishing Co., 1977.

McPherson, George. *An Introduction to Electrical Machines and Transformers.* New York: John Wiley & Sons, Inc., 1981.

Meisel, J. *Principles of Electromechanical Energy Conversion.* New York: McGraw-Hill Book Company, 1966.

Montgomery, T. B. "Regulex-Instability in Harness," *Allis-Chalmers Elec. Rev.* II, Nos. 2 and 3 (1946): 5–9.

Nasar, S. A. *Electromagnetic Energy Conversion Devices and Systems.* Englewood Cliffs, N.J.: Prentice-Hall, Inc., 1970.

Pestarini, J. M. *Metadyne Statics.* New York: The MIT Press/John Wiley & Sons, Inc., 1952.

Riaz, M. "Transient Analysis of the Metadyne Generator," *Trans. AIEE* 72, Part III: 52–62, 1953.

Saunders, R. M. "Measurement of D-C Machine Parameters," *Trans. AIEE* 70 (1951): 700–706.

Siskind, Charles S. *Electrical Machines: Direct & Alternating Current,* 2d ed. New York: McGraw-Hill Book Company, 1959.

Snively, H. D., and P. B. Robinson. "Measurement and Calculation of D-C Machine Armature Circuit Inductance," *Trans. AIEE* 69, Part II (1950): 1228–1237.

Thaler, G. J., and M. L. Wilcox. *Electric Machines: Dynamic and Steady State.* New York: John Wiley & Sons, Inc., 1966.

Tustin, A. *Direct-Current Machines for Control Systems.* New York: The Macmillan Company, 1952.

White, D. C., and H. H. Woodson. *Electromechanical Energy Conversion.* New York: John Wiley & Sons, Inc., 1959.

# Chapter 7

# System Applications of Synchronous Machines

## 7-1 SYNCHRONOUS GENERATOR SUPPLYING AN ISOLATED SYSTEM

An electrical system supplied by one synchronous generator only is considered an isolated system. The concept of the infinite bus does not apply to isolated systems, as there are no other parallel synchronous machines to compensate for changes in field excitation and in prime-mover output to maintain constant terminal voltage and frequency. If the generator is driven at constant speed (constant frequency) and the field current is increased, the terminal voltage increases, which in general is accompanied by an increase in the real- and reactive power output to an isolated system. Similarly, an increase in prime-mover output, with constant field excitation, in general produces an increase in frequency, terminal voltage, and real and reactive power.

## 7-2 PARALLEL OPERATION OF SYNCHRONOUS GENERATORS

Electric power systems are interconnected extensively to promote economy and reliability of operation. Interconnection of ac power systems requires synchronous generators to operate in parallel with each other, and it is common for an electric generating station in which two or more generators are connected in parallel to be connected in parallel itself, by means of transformers and transmission lines, with other generating stations spread over a practically nationwide area. Under normal operating conditions all the generators and synchronous motors in an interconnected system operate in synchronism with each other. The frequencies of all the synchronous machines are exactly equal except during momentary changes in load or excitation. If one or more large synchronous machines pull out of synchronism with the rest of the system, a severe disturbance results, and unless remedial steps are taken immediately the system becomes unstable, a condition that may result in a complete shutdown. The behavior of synchronous generators operating in parallel is therefore of fundamental importance in the study of power systems operation.

## 7-2.1 Requirements for Connecting Synchronous Generators in Parallel

It is common practice to synchronize a large synchronous generator or large synchronous motor with the system before connecting it to the system. Synchronizing requires the following conditions of the incoming machine:

1. Correct phase sequence.
2. Phase voltages must be in phase with those of the system.
3. Frequency must be almost exactly equal to that of the system.
4. Machine voltage must be approximately equal to the system voltage.

The phase sequence of the generator is usually checked carefully at the time of its installation. Conditions (1) and (2) are assured by means of a phase-angle meter known as a *synchroscope,* which compares the voltage from one phase of the incoming machine with that of the corresponding phase of the three-phase system. The synchroscope shows the phase angle between the generator and the system voltages. The frequency and phase positions are controlled by adjustment of the prime-mover input to the incoming generator.

The procedure is the same for synchronizing very large synchronous motors, the motor being brought up to synchronous speed by a smaller auxiliary motor. Smaller synchronous motors are started as induction motors by means of a winding embedded in the pole faces, which is similar to the winding on the rotor of induction motors. As the motor approaches synchronous speed, direct current is applied to the regular field winding, and if the load torque is not excessive, the motor pulls into synchronism with the system.

## 7-2.2 Loading a Synchronous Generator

If all four of the conditions for synchronizing are met exactly, no current will result in the generator armature when it is connected to the system, as the field current is just sufficient to make the generator voltage equal the system or bus voltage with the prime mover furnishing just enough mechanical power to overcome the rotational losses. With the generator now on the system, an increase in its field excitation, without any adjustments of the prime-mover output, will cause the generator to deliver a current which lags the voltage by 90°, as shown in Fig. 7-1(a), if the armature resistance is neglected. However, if the input to the prime mover is now gradually increased, the generator and prime mover will accelerate, causing the generated voltage $\mathbf{E}_{af}$ to lead the terminal voltage or bus voltage $\mathbf{V}$ by the torque angle $\delta$ at a value such that the real-power output matches the mechanical input. This is illustrated by the phasor diagram in Fig. 7-1(b).

Consider two identical three-phase synchronous generators 1 and 2 operating in parallel and supplying a balanced three-phase load as shown schematically in Fig. 7-2(a). One phase of each generator and the equivalent wye of one phase of the load are shown in Fig. 7-2(b). The armature resistance of the generators is neglected. The load may actually be connected in wye or in delta or may be a combination of wye-connected and delta-connected loads. Similarly, the generators may be connected in delta and represented by an equivalent wye. The terminal voltages of the two generators are equal to each other, namely, $\mathbf{V}$, regardless of the division of real- and reactive-power loads between the machines. Further, the phasor sum of the currents from the two machines must equal the load current. Thus, if

$\mathbf{I}_L$ = load current, in amperes per phase or per unit

$\mathbf{I}_1$ = current of generator 1, in amperes per phase or per unit

$\mathbf{I}_2$ = current of generator 2, in amperes per phase or per unit

then

$$\boxed{\mathbf{I}_1 + \mathbf{I}_2 = \mathbf{I}_L} \qquad (7\text{-}1)$$

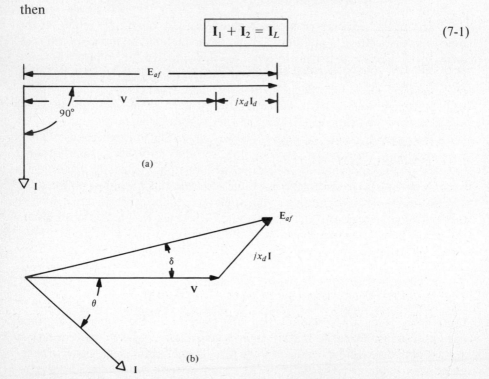

(a)

(b)

**Figure 7-1** Effect of overexcited field in a synchronous generator. (a) Zero prime-mover output. (b) Prime-mover output greater than zero.

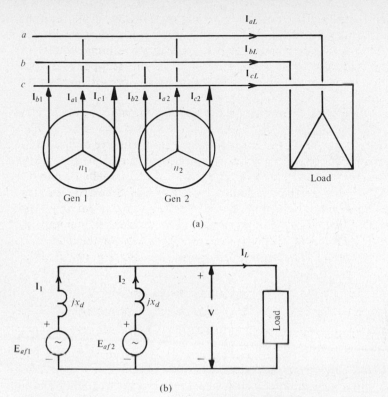

**Figure 7-2** (a) Three-phase representation of two synchronous generators supplying a common load. (b) Schematic representation of one phase of two three-phase generators in parallel.

### 7-2.3 Equal Real-Power Loads and Equal Reactive-Power Loads

When the two generators deliver equal real-power loads and equal reactive loads

$$I_1 = I_2 = \frac{I_L}{2} \tag{7-2}$$

and we have the phasor diagram shown in Fig. 7-3.

### 7-2.4 Loci for Generated Voltage for Constant Terminal Voltage and Constant Frequency

For a constant total load, if the terminal voltage and the frequency are to remain constant and if the real-power output, reactive-, or both real- and reactive-power output of one generator is changed, a corresponding adjustment must be made on the other generator. For example, if the field current of generator 1 is increased, there must be a decrease in the field current of generator 2; otherwise, the terminal

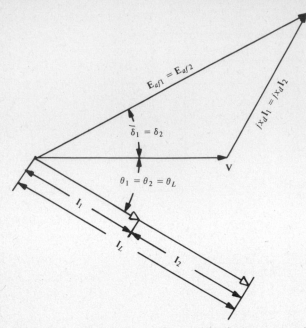

**Figure 7-3** Phasor diagram for two identical generators in parallel carrying equal loads.

voltage **V** increases. Similarly, if the real-power output of generator 1 is increased, the real-power output of generator 2 must be decreased if the frequency and the total real-power load are to remain constant. Let us assume therefore that, when a change in the excitation or in the primer-mover output of generator 1 is made, corrective changes are made in generator 2 to maintain constant bus voltage **V** and constant frequency. A bus that has constant voltage and constant frequency is known as an *infinite bus*.

## 7-2.5 Locus of Generated Voltage for Constant Real Power and Variable Excitation

Suppose that generator 1 is initially carrying the real and reactive load† in VA per phase or in per-unit as given by

$$S_1 = P_1 + jQ_1 = VI_1^*$$

where $I_1^*$ is the conjugate of the current $I_1$. Then in accordance with the phasor diagrams of Figs. 7-3 and 7-4,

$$S_1 = VI_1 \cos \theta_1 + jVI_1 \sin \theta_1$$

Now, if the real-power output of generator 1 is to remain constant,

$$P_1 = VI_1 \cos \theta_1 = \text{constant}$$

---

† Reactive power is considered positive when the current lags the voltage (see Chap. 1).

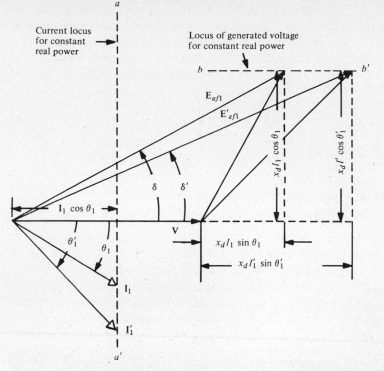

**Figure 7-4** Phasor diagram showing the effect of increasing excitation of a synchronous generator when the real power, frequency, and terminal voltage are constant.

and $\qquad$ $I_1 \cos \theta_1 = \text{constant}$ $\qquad$ since $V$ is constant

The locus of the current for constant real power is therefore represented by line $a$–$a'$ in Fig. 7-4. The synchronous reactance drop is expressed by

$$jx_d\mathbf{I}_1 - jx_d(I_1 \cos \theta_1) - jx_d(jI_1 \sin \theta_1)$$
$$= x_d(I_1 \sin \theta) + jx_d(I_1 \cos \theta_1)$$

and if $x_d$ is constant, the vertical component $x_dI_1(\cos \theta_1)$ of the synchronous reactance drop must remain constant. This vertical component is also related to the generated emf by

$$x_dI_1(\cos \theta_1) = E_{af1} \sin \delta_1$$

and $\qquad$ $$\boxed{P_1 = \frac{VE_{af1} \sin \delta_1}{x_d}}$$ $\qquad$ (7-3)

and if $V$ is constant, then $E_{af1} \sin \delta_1$ must be constant if $P_1$ is constant. The locus of the generated voltage $E_{af1}$ is therefore horizontal line $b$–$b'$ in Fig. 7-4. Now if the field current in generator 1 is increased so as to increase the generated voltage from $E_{af1}$ to $E'_{af1}$ while the field current in generator 2 is decreased so as to keep the terminal voltage $V$ constant, the current $\mathbf{I}_1$ increases to $\mathbf{I}'_1$ and lags by a greater angle, as shown in Fig. 7-4. The loci of the current and of the generated voltage for generator 1 are the same for generator 2 if both generators are identical and carry equal amounts of constant real power. The generated voltage $E'_{af2}$ of generator 2 required to compensate for the change in the generated voltage from $E_{af1}$ to $E'_{af1}$ in generator 1 can be determined from the current $\mathbf{I}'_2$ in generator 2, which according to Eq. 7-1 is

$$\boxed{\mathbf{I}'_2 = \mathbf{I}_L - \mathbf{I}'_1} \tag{7-4}$$

which is shown graphically in Fig. 7-5.

## 7-2.6 Locus of Generated Voltage for Constant Excitation and Variable Real Power

Beginning with initial conditions of equal real power and equal reactive power delivered by the two generators (Fig. 7-3), let the input to the prime mover of generator 1 be increased gradually while the field current of generator 1 is held constant. It is again assumed that adjustments are made in generator 2 to keep the frequency, total load, and terminal voltage constant. As the prime-mover input to generator 1 is increased, it will accelerate momentarily, causing the torque angle or power angle to increase from $\delta_1$ to $\delta''_1$. At $\delta''_1$ the real power output of generator 1 has increased sufficiently to match the increased input to the prime mover. The locus of $\mathbf{E}_{af1}$ being constant in magnitude because of the constant-field current in generator 1 must now be a circle as shown in Fig. 7-6.

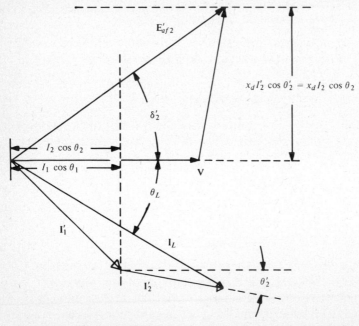

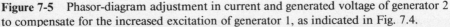

**Figure 7-5**  Phasor-diagram adjustment in current and generated voltage of generator 2 to compensate for the increased excitation of generator 1, as indicated in Fig. 7.4.

Since the locus of the synchronous reactance drop $jx_d\mathbf{I}_1$ is the same as that of $\mathbf{E}_{af1}$, namely, a circle, the locus for the current $\mathbf{I}_1$ must also be a circle, as shown in Fig. 7-6. It is obvious from Fig. 7-6 that as $\delta_1$ increases with constant $E_{af}$, the current increases and lags by a smaller angle $\theta_1''$ than before and that the reactive power decreases as the real power increases. Therefore, not only must the real-power output of generator 2 be decreased, but its reactive-power output must be increased. The procedure for determining the adjustments required of generator 2 is the same as before, i.e., from

$$\mathbf{I}_2'' = \mathbf{I}_L - \mathbf{I}_1''$$

and $$\mathbf{E}_{af2}'' = \mathbf{V} + jx_d\,\mathbf{I}_2''$$

## 7-3 RMS CURRENT ON THREE-PHASE SHORT CIRCUIT

General relationships were derived in terms of inductances for the cylindrical-rotor machine in Sec. 4-8 and are extended in this chapter to include the salient-pole machine. These are useful for describing not only the steady-state perfor-

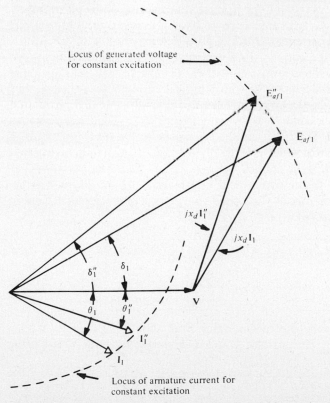

Locus of generated voltage for constant excitation

$\mathbf{E}_{af1}''$

$\mathbf{E}_{af1}$

$jx_d\mathbf{I}_1''$

$jx_d\mathbf{I}_1$

$\delta_1''$   $\delta_1$

$\theta_1$   $\theta_1''$

$\mathbf{V}$

$\mathbf{I}_1''$

$\mathbf{I}_1$

Locus of armature current for constant excitation

**Figure 7-6** Phasor diagram showing the effect of increasing prime-mover input when field current, terminal voltage, and frequency are constant.

mance but also the performance under transient and unbalanced operation. R. H. Park† introduced methods for transforming three-phase quantities to direct-axis and quadrature-axis quantities, a process sometimes known as the *dq transformation.*

This work uses a modification by W. A. Lewis‡ which makes Park's equations consistent with the reciprocity relationship generally used in coupled circuit equations.

Before entering into the more detailed analysis using Park's equations, three-phase short-circuit phenomena will be discussed in the simpler aspects. In Sec. 4-17 the steady-state armature current of the salient-pole machine was divided into the $d$-axis and $q$-axis components. Since these components are constant under steady-state balanced operation, they have no effect on the field current, although the $d$-axis armature current produces an mmf in the direct axis. However, during faults on the power system to which the machine is connected, or during sudden changes in load, the armature current and the armature mmf undergo time variations during which the $d$-axis armature mmf induces a transient current in the field circuit, which in turn reacts on the armature. This mutual effect is taken into account by using the transient reactance $x'_d$ instead of $d$-axis synchronous reactance $x_d$, the former being much lower than the latter. In fact, $x'_d$ is actually smaller than the $q$-axis reactance $x_q$. Since there is no interaction between the $q$-axis and the field winding because of the 90° displacement, the transient $q$-axis reactance is equal to $x_q$. However, the time-varying armature mmfs induce currents in rotor circuits that may link both the $d$ and $q$ axes—as, for example, the iron in cylindrical rotors and dampers in salient-pole structures. These induced currents follow paths of relative high resistance and therefore decay much faster than the transient field current. Nevertheless, they add another component to the armature current, thus giving rise to the armature subtransient reactances $x''_d$ and $x''_4$, both being lower than $x'_d$. During transients, the armature current contains yet another component, the dc component—a phenomenon similar to that which is generally the case when an inductive circuit is subjected to a change in the applied voltage.

The rms value of the ac component of armature current during a three-phase short circuit on an initially unloaded generator is plotted against time in Fig. 7-7 and is expressed by

$$I_{ac} = (I''_d - I'_d)\epsilon^{-t/T''_d} + (I'_d - I_d)\epsilon^{-t/T'_d} + I_d \qquad (7\text{-}5)$$

where

$$I''_d = \frac{E_{af}}{x''_d} \qquad I'_d = \frac{E_{af}}{x'_d} \qquad I_d = \frac{E_{af}}{x_d}$$

$T''_d$ is the subtransient time constant, ranging from about 0.02 to about 0.05 s; $T''_d$ is the transient time constant, with a range of about 0.5 to about 3.5 s. During

---

† R. H. Park, "Definition of an Ideal Synchronous Machine and Formula for the Armature Flux Linkages," *General Electric Rev.* 31 (1928): 332–334; "Two-Reactance Theory of Synchronous Machines Generalized Method of Analysis," *AIEE Trans.* 48, Part I (July 1929): 716–730.

‡ W. A. Lewis, "A Basic Analysis of Synchronous Machines—Part I," *AIEE Trans.* 77, Part III (1958): 436–453.

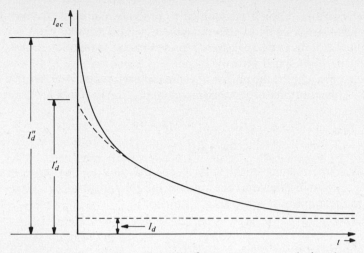

**Figure 7-7**  Ac components (rms) of armature current during three-phase short circuits on a synchronous machine.

a balanced three-phase fault $I_q$ is negligible because $r_a$ is so small that the fault current lags $E_{af}$ by almost 90°. There are other disturbances in which $I_q$ is appreciable, so that $x_q$ must be taken into account.

## 7-4  SALIENT-POLE GENERATOR— GENERAL RELATIONSHIPS

The flux linkages with the armature and field of a cylindrical-rotor synchronous motor are expressed by Eq. 4-68. According to the conventions represented in Fig. 4-13, the direction of the armature currents in a generator is opposite that in the armature of a motor without a change in the direction of the field current. The armature currents then can be represented by $-i_a$, $i_b$, and $-i_c$ in Eq. 4-68. However, the same result can be achieved by changing the signs of the *inductance coefficients* of the *armature currents* instead and representing all currents as positive quantities, as follows:

$$
\begin{bmatrix} \lambda_a \\ \lambda_b \\ \lambda_c \\ \lambda_f \end{bmatrix} = \begin{bmatrix} -L_{aa} & L_{ab} & L_{ca} & -L_{af} \\ L_{ab} & -L_{bb} & L_{bc} & -L_{bf} \\ L_{ca} & L_{bc} & -L_{cc} & -L_{cf} \\ L_{af} & L_{bf} & L_{cf} & L_{ff} \end{bmatrix} \begin{bmatrix} i_a \\ i_b \\ i_c \\ i_f \end{bmatrix} \tag{7-6}
$$

### 7-4.1  Inductances of Salient-Pole Machines

Because of the salient-pole construction, the self-inductances $L_{aa}$, $L_{bb}$, and $L_{cc}$ of phases $a$, $b$, and $c$, as well as the mutual inductances $L_{ab}$, $L_{bc}$, and $L_{ca}$, vary with the position of the rotor. To determine the effect of rotor position on these inductances, consider the elementary three-phase, salient-pole generator in Fig.

7-8(a) and (b), in which each phase is assumed to produce a sinusoidal space mmf. The relationships that apply to this two-pole machine are also valid for multipolar machines as long as all angles are represented in electrical measure. If $i_b = i_c = i_f = 0$, the mmf space wave in the air gap is due only to $i_a$ and its amplitude may be represented by the phasor $\mathcal{F}_a$ coinciding with $a$-phase magnetic axis as in Fig. 7-8. This mmf phasor has a component $\mathcal{F}_{ad}$ in the $d$ axis and one $\mathcal{F}_{aq}$ in the $q$ axis:

$$\mathcal{F}_{ad} = \mathcal{F}_a \cos \sigma$$
$$\mathcal{F}_{aq} = -\mathcal{F}_a \sin \sigma$$

If $\mathcal{P}_d$ and $\mathcal{P}_q$ are permeance coefficients associated with the flux paths in the corresponding axes and defined as the ratio of the fundamental flux component per pole to the amplitude of $\mathcal{F}_d$ and $\mathcal{F}_q$, then

$$\phi_{ad} = \mathcal{F}_{ad}\mathcal{P}_d = \mathcal{F}_a\mathcal{P}_d \cos \sigma$$
$$\phi_{aq} = \mathcal{F}_{aq}\mathcal{P}_q = -\mathcal{F}_a\mathcal{P}_q \sin \sigma \tag{7-7}$$

where $\phi_{ad}$ and $\phi_{aq}$ are the fundamental flux components in the $d$ and $q$ axes. The flux which links $a$ phase is the sum of the projections of phasors $\phi_{ad}$ and $\phi_{aq}$ on the magnetic axis of $a$ phase, as shown in Fig. 7-8(b):

$$\phi_{aa} = \phi_{ad} \cos \sigma - \phi_{aq} \sin \sigma \tag{7-8}$$

and, from Eq. 7-7,

$$\phi_{aa} = \mathcal{F}_a(\mathcal{P}_d \cos^2 \sigma + \mathcal{P}_q \sin^2 \sigma)$$
$$= \mathcal{F}_a\left(\frac{\mathcal{P}_d + \mathcal{P}_q}{2} + \frac{\mathcal{P}_d - \mathcal{P}_q}{2} \cos 2\sigma\right) \tag{7-9}$$

Since inductance is defined as flux linkage per ampere and $\mathcal{F}_a$ is in ampere turns, the self-inductance of $a$ phase must have two components proportional to the right-hand side of Eq. 7-9, which may therefore be expressed by

$$L_{aa} = L_{sa} + L_{sv} \cos 2\sigma$$

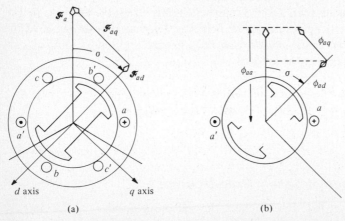

| (a) | (b) |

**Figure 7-8** (a) Elementary three-phase generator with current in $a$ phase only. (b) Simplified representation with phasor diagram of fluxes linking $a$ phase.

Because of their $120°$ displacements from $a$ phase, the angle $\sigma$ must be replaced by $\sigma - 2\pi/3$ for $b$ phase and $\sigma - 4\pi/3$ for $c$ phase, and the self-inductances for the three phases are then defined by

$$
\begin{array}{l}
L_{aa} = L_{sa} + L_{sv} \cos 2\sigma \\[2mm]
L_{bb} = L_{sa} + L_{sv} \cos \left(2\sigma - \dfrac{4\pi}{3}\right) \\[2mm]
L_{cc} = L_{sa} + L_{sv} \cos \left(2\sigma - \dfrac{2\pi}{3}\right)
\end{array}
\tag{7-10}
$$

The mutual inductances may be found by again assuming current in $a$ phase only. The flux which then links $b$ phase is

$$
\begin{aligned}
\phi_{ab} &= \phi_{ad} \cos \left(\sigma - \frac{2\pi}{3}\right) - \phi_{aq} \sin \left(\sigma - \frac{2\pi}{3}\right) \\
&= \mathcal{F}_a \left[ \mathcal{P}_d \cos \sigma \cos \left(\sigma - \frac{2\pi}{3}\right) + \mathcal{P}_q \sin \sigma \sin \left(\sigma - \frac{2\pi}{3}\right) \right] \\
&= \mathcal{F}_a \left[ -\frac{\mathcal{P}_d + \mathcal{P}_q}{4} + \frac{\mathcal{P}_d - \mathcal{P}_q}{2} \cos \left(2\sigma - \frac{2\pi}{3}\right) \right]
\end{aligned}
$$

Then by following a similar procedure for $\phi_{bc}$ and $\phi_{ca}$ and associating the minus signs with the mutual inductances in Eq. 7-6 we may write

$$
\begin{array}{l}
L_{ab} = L_{ma} - L_{sv} \cos \left(2\sigma - \dfrac{2\pi}{3}\right) \\[2mm]
L_{bc} = L_{ma} - L_{sv} \cos 2\sigma \\[2mm]
L_{ca} = L_{ma} - L_{sv} \cos \left(2\sigma - \dfrac{4\pi}{3}\right)
\end{array}
\tag{7-11}
$$

The self-inductance $L_{ff}$ of the field is constant because it is independent of rotor position. The mutual inductance between armature and field is defined by Eq. 4-73.

## 7-4.2 *d*-Axis, *q*-Axis, and Zero-Sequence Quantities, Currents in Damper Circuits Negligible

It should be recalled that the $d$-axis synchronous reactance $x_d$ and the $q$-axis synchronous reactance $x_q$ are associated with $\mathbf{I}_d$ and $\mathbf{I}_q$, respectively. Then if $\mathbf{I} = \mathbf{I}_d$ and Eqs. 7-6, 7-10, and 7-11 are used in a process similar to that in Sec. 4-9, the result is

$$
L_d = \frac{x_d}{\omega} = L_{sa} + L_{ma} + \tfrac{3}{2} L_{sv}
\tag{7-12}
$$

A similar procedure for the case of $\mathbf{I} = \mathbf{I}_q$ leads to

$$L_q = \frac{x_q}{\omega} = L_{sa} + L_{ma} - \tfrac{3}{2}L_{sv} \tag{7-13}$$

From Eqs. 4-73 and 7-6, the field-flux linkage is found to be

$$\lambda_f = L_{ff}i_f + L_{afm}\left[ i_a \cos \sigma + i_b \cos \left( \sigma - \frac{2\pi}{3} \right) + i_c \cos \left( \sigma - \frac{4\pi}{3} \right) \right] \tag{7-14}$$

Park defined a current proportional to the term in the brackets as

$$i_d = K_d\left[ i_a \cos \sigma + i_b \cos \left( \sigma - \frac{2\pi}{3} \right) + i_c \cos \left( \sigma - \frac{4\pi}{3} \right) \right] \tag{7-15}$$

for leading current, where $K_d$ is unspecified for the time being. By analogy, the $d$-axis flux linkage is defined as

$$\lambda_d = K_d\left[ \lambda_a \cos \sigma + \lambda_b \cos \left( \sigma - \frac{2\pi}{3} \right) + \lambda_c \cos \left( \sigma - \frac{4\pi}{3} \right) \right] \tag{7-16}$$

Similarly,

$$i_q = K_d\left[ i_a \sin \sigma + i_b \sin \left( \sigma - \frac{2\pi}{3} \right) + i_c \sin \left( \sigma - \frac{4\pi}{3} \right) \right] \tag{7-17}$$

$$\lambda_q = K_d\left[ \lambda_a \sin \sigma + \lambda_b \sin \left( \sigma - \frac{2\pi}{3} \right) + \lambda_c \sin \left( \sigma - \frac{4\pi}{3} \right) \right] \tag{7-18}$$

Then on the basis of Eqs. 4-73, 7-10, 7-11, 7-12, and 7-13 it is found after some algebraic manipulation that $d$-axis and $q$-axis flux linkages are

$$\begin{aligned} \lambda_d &= (L_{sa} + L_{ma} + \tfrac{3}{2}L_{sv})i_d + \tfrac{3}{2}K_d L_{afm}i_f \\ &= L_d i_d + \tfrac{3}{2}K_d L_{afm}i_f \end{aligned} \tag{7-19}$$

$$\begin{aligned} \lambda_q &= (L_{sa} + L_{ma} - \tfrac{3}{2}L_{sv})i_q \\ &= L_q i_q \end{aligned} \tag{7-20}$$

Substitution of Eq. 7-15 in Eq. 7-14 yields

$$\lambda_f = \frac{L_{afm}}{K_d} i_d + L_{ff}i_f \tag{7-21}$$

Differentiation of Eqs. 7-19 and 7-20 results in the following pair of simultaneous equations:

$$p\lambda_d = L_d p i_d + \tfrac{3}{2}K_d L_{afm} p i_f$$

$$p\lambda_f = \frac{L_{afm}}{K_d} p i_d + L_{ff} p i_f \tag{7-22}$$

which have the same form as the coupled-circuit equations 3-18 and 3-19 when the resistance terms are neglected. The coefficients $\frac{3}{2}K_d L_{afm}$ and $L_{afm}/K_d$ in Eq. 7-22 correspond to $L_{12}$ and $L_{21}$ in Eqs. 3-18 and 3-19. The mutual inductance between linear circuits is reciprocal (i.e., $L_{12} = L_{21}$), and if Eq. 7-22 is to satisfy this condition, $\frac{3}{2}K_d L_{afm} = L_{afm}/K_d$, for which $K_d$ must equal $\sqrt{\frac{2}{3}}$. Park assigned the value $\frac{2}{3}$ to $K_d$ and in addition used per-unit quantities. Lewis introduced the coefficient $\sqrt{\frac{2}{3}}$ and used the physical quantities instead of per-unit. Then for $K_d = \sqrt{\frac{2}{3}}$, the mutual-inductance term in Eq. 7-22 becomes $\sqrt{\frac{3}{2}}\,L_{afm}$ and Eqs. 7-19 and 7-21 can be rewritten as

$$\lambda_d = L_d i_d + \sqrt{\tfrac{3}{2}} L_{afm} i_f$$
$$\lambda_f = \sqrt{\tfrac{3}{2}} L_{afm} i_d + L_{ff} i_f \qquad (7\text{-}23)$$

The voltage induced in the $d$ axis may be expressed in the same form as $i_d$ and $\lambda_d$. Hence, in summary,

$$i_d = \sqrt{\tfrac{2}{3}}\left[ i_a \cos\sigma + i_b \cos\left(\sigma - \tfrac{2\pi}{3}\right) + i_c \cos\left(\sigma - \tfrac{4\pi}{3}\right)\right] \qquad (7\text{-}24)$$

$$\lambda_d = \sqrt{\tfrac{2}{3}}\left[ \lambda_a \cos\sigma + \lambda_b \cos\left(\sigma - \tfrac{2\pi}{3}\right) + \lambda_c \cos\left(\sigma - \tfrac{4\pi}{3}\right)\right] \qquad (7\text{-}25)$$

$$e_d = \sqrt{\tfrac{2}{3}}\left[ e_{an} \cos\sigma + e_{bn} \cos\left(\sigma - \tfrac{2\pi}{3}\right) + e_{cn} \cos\left(\sigma - \tfrac{4\pi}{3}\right)\right] \qquad (7\text{-}26)$$

The corresponding $q$-axis quantities may be defined, by analogy, as

$$i_q = \sqrt{\tfrac{2}{3}}\left[ i_a \sin\sigma + i_b \sin\left(\sigma - \tfrac{2\pi}{3}\right) + i_c \sin\left(\sigma - \tfrac{4\pi}{3}\right)\right] \qquad (7\text{-}27)$$

$$\lambda_q = \sqrt{\tfrac{2}{3}}\left[ \lambda_a \sin\sigma + \lambda_b \sin\left(\sigma - \tfrac{2\pi}{3}\right) + \lambda_c \sin\left(\sigma - \tfrac{4\pi}{3}\right)\right] \qquad (7\text{-}28)$$

$$e_q = \sqrt{\tfrac{2}{3}}\left[ e_{an} \sin\sigma + e_{bn} \sin\left(\sigma - \tfrac{2\pi}{3}\right) + e_{cn} \sin\left(\sigma - \tfrac{4\pi}{3}\right)\right] \qquad (7\text{-}29)$$

The $d$ and $q$ quantities in Eqs. 7-24 and 7-29 are expressed in terms of the three quantities carrying the subscripts $a$, $b$, and $c$. A third relationship is necessary to make the transformation from the $a$, $b$, $c$ quantities complete for the general case. For example, if there is no neutral current, $i_d$ and $i_q$ are sufficient to represent the currents in the three phases. However, the presence of neutral current can be taken into account by means of the zero-sequence component, defined as

$$i_0 = \frac{i_a + i_b + i_c}{3} \qquad (7\text{-}30)$$

a convention that is used in the method of symmetrical components used for the solution of unbalanced three-phase circuits.† By analogy the zero-sequence flux linkage and voltage are

$$\lambda_0 = \frac{\lambda_a + \lambda_b + \lambda_c}{3} \tag{7-31}$$

$$e_0 = \frac{e_{an} + e_{bn} + e_{cn}}{3} \tag{7-32}$$

Equation 7-30 shows the zero-sequence current to equal $\frac{1}{3}$ of the neutral current. If Eqs. 7-11 and 7-12 are substituted in Eq. 7-6 and these in turn are substituted in Eq. 7-31, the zero-sequence flux linkage becomes

$$\lambda_0 = \frac{(L_{sa} - 2L_{ma})(i_a + i_b + i_c)}{3}$$

$$= L_0 i_0$$

where the zero-sequence inductance is

$$L_0 = L_{sa} - 2L_{ma} \tag{7-33}$$

To express $e_d$ and $e_q$ in terms of $\lambda_d$ and $\lambda_q$, differentiate $\lambda_d$ and $\lambda_q$ and rearrange terms. Differentiation of $\lambda_d$ in Eq. 7-25 yields

$$p\lambda_d = \sqrt{\frac{2}{3}} \left\{ (\cos \sigma)p\lambda_a + \left[ \cos \left( \sigma - \frac{2\pi}{3} \right) \right] p\lambda_b \right.$$

$$\left. + \left[ \cos \left( \sigma - \frac{4\pi}{3} \right) \right] p\lambda_c \right\} - \sqrt{\frac{2}{3}}$$

$$\times \left[ \lambda_a \sin \sigma + \lambda_b \sin \left( \sigma - \frac{2\pi}{3} \right) + \lambda_c \sin \left( \sigma - \frac{4\pi}{3} \right) \right] p\sigma \tag{7-34}$$

The second bracketed term contains $\lambda_q$, as shown by Eq. 7-28. For the assumed generator voltage polarities,

$$e_a = -p\lambda_a \qquad e_b = -p\lambda_b \qquad e_c = -p\lambda_c$$

These voltages are induced by the entire flux linking the respective phases and therefore include the component voltages $e_{an}$, $e_{bn}$, and $e_{cn}$, due to the field current alone. The first bracketed term in Eq. 7-34 equals $p\lambda_d$ in accordance with Eq. 7-25. Hence,

$$p\lambda_d = -e_d - \lambda_q p\sigma$$

A similar process applied to Eq. 7-28 yields

$$p\lambda_q = -e_q + \lambda_d p\sigma$$

† See C. F. Wagner and R. D. Evans, *Symmetrical Components* (New York: McGraw-Hill Book Company, 1933).

Differentiation of the zero-sequence flux linkage yields the zero-sequence voltage; thus,

$$p\lambda_0 = -e_0 \tag{7-35}$$

Since synchronous generators normally operate at constant speed and the change in speed during faults is slight, $p\sigma = \omega$; and when the resistance drops are subtracted from the induced voltages, the armature terminal voltages are found to be

$$
\begin{aligned}
v_d &= -r_a i_d - p\lambda_d - \lambda_q \omega \\
v_q &= -r_a i_q - p\lambda_q + \lambda_d \omega \\
v_0 &= -r_a i_0 - p\lambda_0
\end{aligned}
\tag{7-36}
$$

The field circuit constitutes a load and its terminal voltage is therefore

$$v_f = r_f i_f + p\lambda_f \tag{7-37}$$

## 7-5 INSTANTANEOUS THREE-PHASE SHORT-CIRCUIT CURRENT

Equations 7-36 and 7-37 are particularly useful in dealing with transient conditions, an example of which is a three-phase short circuit suddenly applied to the terminals of an initially unloaded generator. In the following, the generator is assumed to have salient poles without damper bars; so no subtransient effects are present. In addition, the resistance of the armature and the field circuit are neglected to begin with. The resistances have negligible effect on the magnitudes of the currents immediately following short circuiting. The predominant effect of the winding resistances is that of determining the rate at which the transient currents in the armature and in the field fall off. Further, if the resistance of the field circuit is zero, the voltage across the field terminals before short circuit must be zero and the field circuit may then be considered as being short-circuited. The flux linkage with a short-circuited winding having zero resistance cannot change, as dictated by the *law of constant flux linkage*. Hence, during the transition from open circuit to three-phase short circuit, $\lambda_f$ is constant and $p\lambda_f = 0$. On that basis, differentiation of Eq. 7-23 leads to

$$
\begin{aligned}
p\lambda_d &= \left( L_d - \frac{3}{2}\frac{L_{afm}^2}{L_{ff}} \right) p i_d \\
&= L_d' p i_d
\end{aligned}
\tag{7-38}
$$

where the transient inductance $L_d'$ is defined as

$$L_d' = L_d - \frac{L_{df}^2}{L_{ff}} = L_d - \frac{3}{2}\frac{L_{afm}^2}{L_{ff}} \tag{7-39}$$

When the resistances are neglected, the first two equations of Eq. 7-36 become

$$v_d = -p\lambda_d - \lambda_q \omega \tag{7-40}$$

$$v_q = -p\lambda_q + \lambda_d \omega \tag{7-41}$$

The third equation of Eq. 7-36 is not needed, because, for a balanced three-phase fault, there is no neutral current and consequently no zero-sequence current, whether the generator is wye-connected or delta-connected.†

**Before Short Circuit** $v_d = 0$, or may be verified from Eqs. 4-75 and 7-26. The field current $i_{fo}$ is constant. Also $i_d = i_q = 0$, since $i_a = i_b = i_c = 0$. Further,

$$\lambda_d = L_{df} i_{fo} = \sqrt{\tfrac{3}{2}} L_{afm} i_{fo}$$

whence

$$v_q(0^-) = \sqrt{\tfrac{3}{2}} \omega L_{afm} i_{fo}$$

**After Short Circuit** $v_d = v_q = 0$, because

$$e_a = e_b = e_c = 0$$

and Eq. 7-40 becomes

$$p\lambda_d + \lambda_q \omega = 0 \tag{7-42}$$

Setting $v_q$ in Eq. 7-41 to zero does not lead to an evaluation of $i_d$ or $i_q$. However, the effect of suddenly reducing $v_q$ from $\sqrt{\tfrac{3}{2}} \omega L_{afm} i_{fo}$ to zero is the same as if the field circuit were closed but initially unexcited and the voltage $-\sqrt{\tfrac{3}{2}} L_{afm} i_{fo}$ suddenly applied to the generator terminals. This is in line with Thévenin's theorem. Then, in accordance with Eq. 7-41 the following is valid after short circuit:

$$-p\lambda_q + \lambda_d \omega = -\sqrt{\tfrac{3}{2}} \omega L_{afm} i_{fo} \tag{7-43}$$

Substitution of Eq. 7-42 in Eq. 7-43 yields

$$p^2 \frac{\lambda_d}{\omega} + \lambda_d \omega = -\sqrt{\tfrac{3}{2}} \omega L_{afm} i_{fo} \tag{7-44}$$

Then from Eqs. 7-38, 7-39, and 7-44 it follows that

$$(p^2 + \omega^2) i_d = -\sqrt{\frac{3}{2}} \frac{\omega^2 L_{afm} i_{fo}}{L'_d} \tag{7-45}$$

When both sides of Eq. 7-45 are Laplace-transformed, the result is

$$(s^2 + \omega^2) I_d(s) = -\sqrt{\frac{3}{2}} \frac{\omega^2 L_{afm} i_{fo}}{s L'_d}$$

† *Ibid.*

for which the inverse transform is found to be, from item 6a in Table A-1,

$$i_d = - \sqrt{\frac{3}{2} \frac{L_{afm}i_{fo}}{L'_d}} (1 - \cos \omega t) \tag{7-46}$$

Since

$$\omega L_{afm}i_{fo} = \sqrt{2}\, E_{af} \quad \text{and} \quad x'_d = \omega L'_d$$

it follows that

$$i_d = - \frac{\sqrt{3}\, E_{af}}{x'_d} (1 - \cos \omega t) \tag{7-47}$$

When Eqs. 7-20 and 7-38 are substituted in the right-hand side of Eq. 7-40 and the result ($v_d$) equated to zero, the $q$-axis current is found after some algebraic manipulation to be

$$i_q = - \frac{L'_d p i_d}{\omega L_q} \tag{7-48}$$

From Eqs. 7-46 and 7-48 we have

$$i_q = \sqrt{\frac{3}{2} \frac{L_{afm}i_{fo}}{L_q}} \sin \omega t \tag{7-49}$$

and since $x_q = \omega L_q$,

$$i_q = \frac{\sqrt{3}\, E_{af} \sin \omega t}{x_q} \tag{7-50}$$

It should be recalled that for a symmetrical three-phase fault

$$i_0 = 0 \tag{7-51}$$

It now remains to express the phase currents as functions of time, and with that end in view, $i_d$, $i_q$, and $i_0$ are expressed in matrix form on the basis of Eqs. 7-24, 7-27, and 7-30 as follows:

$$\begin{bmatrix} i_d \\ i_q \\ i_o \end{bmatrix} = \sqrt{\frac{2}{3}} \begin{bmatrix} \cos \sigma & \cos\left(\sigma - \dfrac{2\pi}{3}\right) & \cos\left(\sigma - \dfrac{4\pi}{3}\right) \\ \sin \sigma & \sin\left(\sigma - \dfrac{2\pi}{3}\right) & \sin\left(\sigma - \dfrac{4\pi}{3}\right) \\ \dfrac{1}{\sqrt{6}} & \dfrac{1}{\sqrt{6}} & \dfrac{1}{\sqrt{6}} \end{bmatrix} \begin{bmatrix} i_a \\ i_b \\ i_c \end{bmatrix} \tag{7-52}$$

from which

$$\begin{bmatrix} i_a \\ \\ i_b \\ \\ i_c \end{bmatrix} = \sqrt{\frac{2}{3}} \begin{bmatrix} \cos \sigma & \sin \sigma & \sqrt{\frac{3}{2}} \\ \\ \cos\left(\sigma - \frac{2\pi}{3}\right) & \sin\left(\sigma - \frac{2\pi}{3}\right) & \sqrt{\frac{3}{2}} \\ \\ \cos\left(\sigma - \frac{4\pi}{3}\right) & \sin\left(\sigma - \frac{4\pi}{3}\right) & \sqrt{\frac{3}{2}} \end{bmatrix} \begin{bmatrix} i_d \\ \\ i_q \\ \\ i_o \end{bmatrix} \qquad (7\text{-}53)$$

When Eqs. 7-47, 7-50, and 7-51 are substituted in Eq. 7-53, $a$-phase fault current is found to be

$$i_a = -\frac{\sqrt{2}\,E_{af}}{x'_d}\cos(\omega t + \sigma_0) + \frac{\sqrt{2}\,E_{af}}{2x'_d x_q}(x'_d + x_q)\cos\sigma_0$$

$$+ \frac{\sqrt{2}\,E_{af}}{2x'_d x_q}(x_q - x'_d)\cos(2\omega t + \sigma_0) \qquad (7\text{-}54)$$

where $\sigma_0$ is the value of $\sigma$ at the instant of short circuit. Phase currents $i_b$ and $i_c$ can be determined by replacing $\sigma_0$ in Eq. 7-54 with $\sigma_0 - 2\pi/3$ and $\sigma_0 - 4\pi/3$.

The field current is also of interest and may be found from Eq. 7-46 and the second equation of Eq. 7-23 on the basis of constant field flux linkage $\lambda_f = L_{ff}i_{fo}$, whence

$$i_f = i_{fo} - \sqrt{\frac{3}{2}}\frac{L_{afm}}{L_{ff}}i_d$$

$$= i_{fo} + \frac{3}{2}\frac{L_{afm}^2 i_{fo}}{L'_d L_{ff}}(1 - \cos\omega t) \qquad (7\text{-}54a)$$

and, on the basis of Eq. 7-39,

$$\boxed{i_f = i_{fo}\left[1 + \frac{x_d - x'_d}{x'_d}(1 - \cos\omega t)\right]} \qquad (7\text{-}55)$$

The $d$-axis and $q$-axis currents and the field current are plotted as functions of time for a three-phase short circuit when all resistances are neglected in Fig. 7-9. When resistances and subtransient effects are neglected, Eq. 7-5 for the rms value of the ac component becomes

$$I_{ac} = I'_d = \frac{E_{af}}{x'_d} \qquad (7\text{-}56)$$

which corresponds to the first term on the right-hand side of Eq. 7-54. It might appear at first glance that Eqs. 7-54 and 7-56 would lead to a significant difference in the calculated rms values of the ac component of fault current. However, the second term in Eq. 7-54 represents the dc component, which is not included in evaluating the ac component; so Eq. 7-56 neglects only the double-frequency ac component in Eq. 7-54, which does not result in a large discrepancy, as may be shown by using the following typical per-unit values:

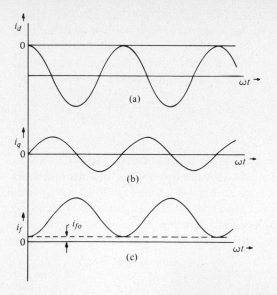

**Figure 7-9**  Currents during three-phase short circuit when resistances are neglected. (a) $d$-axis current. (b) $q$-axis current. (c) Field current.

$$x'_d = 0.30 \qquad x_q = 0.70$$

Then, from Eq. 7-56,

$$I'_d = \frac{E_{af}}{0.30} = 3.33E_{af}$$

and the rms values of the first and third terms in Eq. 7-54 are

$$3.33E_{af} \quad \text{and} \quad \frac{E_{af}}{2 \times 0.30 \times 0.70}(0.70 - 0.30) = 0.95E_{af}$$

which, when combined, result in an rms value of

$$E_{af}\sqrt{(3.33)^2 + (0.95)^2} = 3.45E_{af}$$

as compared with $3.33E_{af}$, the difference being less than 4 percent. Nevertheless, Eq. 7-54 and its derivation provide a better insight into the transient phenomena that can be gained from Eq. 7-5.

## 7-5.1 Subtransient Reactance

Although the expressions and their derivations† for the subtransient reactances $x''_d$ and $x''_q$ of salient-pole machines with dampers are somewhat more involved

---

† *Ibid.* See also Charles Concordia, *Synchronous Machines* (New York: John Wiley & Sons, Inc., 1957) and K. B. Menton, "An Accurate Method of Calculation of Subtransient Reactances of Synchronous Machines," *Trans. AIEE* 78, Part IIIA (1959): 371–378, and also the discussions of this paper, pp. 378–379. A method that uses a computer program to determine various constants from tests on a cylindrical-rotor machine is described by D. Harrington and J. I. Whittlesey in "The Analysis of Sudden-Short-Circuit Oscillograms of Steam-Turbine Generators," *Trans. AIEE* 78, Part IIIA (1959): 551–562.

than those for the transient reactance, they do lend themselves to analytical approaches. However, the paths of the currents induced in the iron of cylindrical rotors are too complicated to make for a straightforward analysis of the subtransient reactance of cylindrical-rotor machines. The subtransient reactances $x_d''$ and $x_q''$ as well as the transient reactance can be determined by test.†

## 7-6 TIME CONSTANTS

Equation 7-5 shows that the armature short-circuit current falls off from its initial value $I_d''$ to its final value of $I_d$ at a rate determined by the time constants $T_d''$ and $T_d'$. The double-frequency term in Eq. 7-54 is due to the ac component in the field current in Eq. 7-55 but is neglected in Eq. 7-5, and the resulting ac component in the armature short-circuit current derives only from the dc component in the field current. The ac component of the field current is therefore disregarded in the following.

### 7-6.1 Direct-Axis Open-Circuit Time Constant, $T_{do}'$

When the armature is open-circuited, a constant voltage $V_f$ suddenly applied to the field produces the field current

$$i_f = \frac{V_f}{r_f}(1 - \epsilon^{-t/T_{do}'}) \qquad (7\text{-}57)$$

where $T_{do}' = L_{ff}/r_f$, the open-circuit time constant.

### 7-6.2 Direct-Axis Short-Circuit Transient Time Constant, $T_d'$

Equation 7-54a shows that a sudden three-phase short circuit causes the dc component of the field current to jump from $i_{fo}$ before short circuit to

$$i_{fo} + \frac{3}{2}\frac{L_{afm}}{L_d' L_{ff}} i_{fo}$$

and to remain at that value if the field resistance were zero. The resistance of the armature is generally small enough, as compared with the transient reactance, that its effect on the magnitude of the armature short-circuit current is negligible and therefore has no appreciable effect on the dc component of the field current. However, the resistance of the field circuit causes this dc component to decrease to $i_{fo}$ with the armature short-circuited. Since the flux linkage with the field winding is $L_{ff}i_{fo}$ before short circuit, it must have the same value just after short circuit, in accordance with the law of constant flux linkage. The transient inductance of the field during the short circuit is therefore

---

† See *Electrical Transmission and Distribution Reference Book,* 4th ed. (East Pittsburgh, Pa., Westinghouse Electric Corporation, 1950), pp. 159–161.

$$L_{fsc} = \frac{L_{ff}i_{fo}}{i_{fo}\left(1 + \frac{3}{2}\frac{L_{afm}^2}{L_d'L_{ff}}\right)} = \frac{L_{ff}L_d'}{L_d' + \frac{3}{2}\frac{L_{afm}^2}{L_{ff}}} \qquad (7\text{-}58)$$

and, from Eq. 7-39,

$$L_{fsc} = \frac{L_d'}{L_d} L_{ff} = \frac{x_d'}{x_d} L_{ff}$$

Hence, the short-circuit transient time constant is

$$\boxed{T_d' = \frac{L_{fsc}}{r_f} = \frac{x_d'}{x_d} T_{do}'} \qquad (7\text{-}59)$$

### 7-6.3 Direct-Axis Short-Circuit Subtransient Time Constant, $T_d''$

$$T_d'' = \frac{x_d''}{x_d'} T_{do}''$$

where $T_{do}''$ is the direct-axis subtransient time constant. It is the time constant of the field circuit if it is assumed that the armature winding and all damper circuits are short-circuited. However, this time constant is small, being about 0.05 s.

### 7-6.4 Armature Short-Circuit Time Constant, $T_a$

Since the time constant $T_d'$ is based on the dc component in the field current, it applies only to the fundamental frequency component of the armature current. This is so because the dc component of field current gives rise only to fundamental frequency armature voltage. For the same reason, $T_d''$ also applies only to the fundamental component of armature current.

The dc and the double-frequency components in Eq. 7-54 die out because of the armature resistance $r_a$ at a rate determined by the armature short-circuit time constant. In the case of a three-phase short circuit, at least two of the phases have a dc component determined by $\sigma_0$. The dc components in the armature current produce an mmf that is stationary in space and of a magnitude independent of $\sigma_0$. The value of $\sigma_0$ merely fixes the location of the axis along which this mmf is directed. Since the dc armature mmf is stationary, it reacts alternately on the $d$ axis and the $q$ axis during the rotation of the field structure. The inductance associated with the dc component of armature current may therefore be regarded as a sort of average of the $L_d'$ and $L_q$ and is expressed as $x_2/\omega$, where $x_2$ is called the *negative-sequence reactance*. The definition of $x_2$ is not straightforward†; however, it may be taken as $(x_q + x_d')/2$ to determine the armature time constant. But because of subtransient effects

† For a more thorough discussion of sequence reactances, see R. H. Park and B. L. Robertson, "Reactances of Synchronous Machines," *Trans. AIEE* 47 (1928): 514–535.

$$T_a = \frac{x''_q + x''_d}{2\omega r_a} \tag{7-60}$$

where $x''_q$ is the quadrature-axis subtransient reactance, and when there are no damper circuits in the quadrature axis, $x''_q = x_q$. The motion of the field structure relative to the stationary component of the armature mmf produces an ac component of fundamental frequency in the field current.

This ac component of field current in turn is responsible for the double-frequency component in the armature short-circuit current. The time constant $T_a$ therefore applies not only to the dc component of armature current but to the double-frequency component as well.

Equation 7-54, when rewritten to include the time constants $T'_d$ and $T_a$, becomes

$$\begin{aligned}
i_a &= \frac{x_d - x'_d}{x_d x'_d} \sqrt{2}\, E_{af} \epsilon^{-t/T'_d} \cos(\omega t + \sigma_0) \\
&\quad + \frac{\sqrt{2}\, E_{af}}{x_d} \cos(\omega t + \sigma_0) \\
&\quad - \frac{x_q - x'_d}{2x'_d x_q} \sqrt{2}\, E_{af} \epsilon^{-t/T_a} \cos(2\omega t + \sigma_0) \\
&\quad - \frac{x_q + x'_d}{2x'_d x_q} \sqrt{2}\, E_{af} \epsilon^{-t/T_a} \cos \sigma_0
\end{aligned} \tag{7-61}$$

On a similar basis, Eq. 7-55 becomes

$$i_f = \frac{x_d - x'_d}{x'_d} (\epsilon^{-t/T'_d} - \epsilon^{-t/T_a} \cos \omega t) + i_{fo} \tag{7-62}$$

Typical waveforms of armature current during a three-phase short circuit are shown in Fig. 7-10.

In the foregoing, the three-phase short circuit was assumed to be at the generator terminals. However, if the short circuit occurs at some point on the system away from the generator so that there is appreciable reactance $x_e$ between the generator and point of fault, the value of $x_e$ should be added to those of the machine reactances $x_d$, $x_q$, $x'_d$, and $x''_d$ to determine the fault current. It should also be stated that other kinds of faults, such as a short circuit on only one phase, occur more frequently on power systems than does the three-phase short circuit. However, the three-phase short circuit is treated here because the underlying phenomena are simpler than those in the other types of short circuit, and yet they serve as a means for gaining insight into the nature of electrical transients in synchronous machines. Treatment of other faults are too involved to fall within the scope of this book.†

---

† For a comprehensive treatment of faults on electric power systems, see C. F. Wagner and R. D. Evans, *op. cit.*

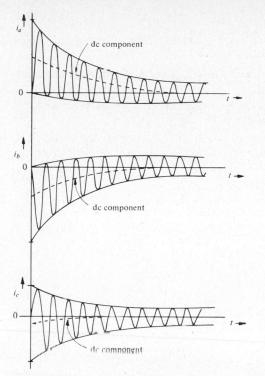

**Figure 7-10** Armature currents during a three-phase short circuit on a synchronous machine.

## 7-7 THREE-PHASE SHORT CIRCUIT FROM LOADED CONDITIONS

Short circuits at the terminals of unloaded generators are extremely rare. They generally occur because of insulation failure or accidental damage on some part of the power system supplied by the generator. It is therefore of importance to deal with the case of the generator carrying load previous to the short circuit. Just as in the case of the unloaded generator, the flux linkage with the field just before the fault determines the initial values of short-circuit current. However, in the unloaded generator, the field-flux linkage is due to the field current only, whereas in the loaded generator it is due to the field current and the armature current, as shown by Eq. 4-84. For the unloaded generator, the voltage that determines the initial short-circuit currents is $E_{af}$, being proportional to the field current. Accordingly, for loaded conditions the corresponding voltage must be lower than $E_{af}$, as shown in the following.

The generator operates under steady load before the fault delivering the armature current of $I$ amperes per phase, for which $I_d = I \sin \theta_i$, as illustrated in Fig. 4-41. Further, for a steady balanced load, it is found from Eq. 7-24 that

$$i_d = -\sqrt{\tfrac{2}{3}}(\tfrac{3}{2})\sqrt{2}\, I_d = -\sqrt{3}\, I_d \qquad (7\text{-}63)$$

In addition, the field current is constant:

$$i_f = i_{fo} \tag{7-64}$$

Substitution of Eqs. 7-63 and 7-64 in Eq. 7-23 yields the constant flux linkages before the fault:

$$\lambda_d = -\sqrt{3}\, L_d I_d + \tfrac{3}{2} L_{afm} i_{fo} \tag{7-65}$$

$$\lambda_f = -\frac{3}{\sqrt{2}} L_{afm} I_d + L_{ff} i_{fo} \tag{7-66}$$

Dividing Eq. 7-66 by $L_{ff}$ results in

$$\frac{\lambda_f}{L_{ff}} = i_{fo} - \frac{3}{\sqrt{2}} \frac{L_{afm}}{L_{ff}} I_d \tag{7-67}$$

which may be regarded as the equivalent field current to produce $\lambda_f$ in the absence of armature current. And the generated voltage due to such a value of field current would be

$$E'_{af} = \frac{\omega L_{afm}}{\sqrt{2}} \frac{\lambda_f}{L_{ff}}$$

$$= \frac{\omega L_{afm}}{\sqrt{2}} i_{fo} - \frac{3}{2} \frac{\omega L_{afm}^2}{L_{ff}} I_d \tag{7-68}$$

However, according to Eq. 7-39,

$$\frac{3}{2} \frac{L_{afm}^2}{L_{ff}} = L_d - L'_d$$

and when this is substituted in Eq. 7-68 there results

$$E'_{af} = \frac{\omega L_{afm}}{\sqrt{2}} i_{fo} - \omega(L_d - L'_d)I_d$$

$$= E_{af} - (x_d - x'_d)I_d \tag{7-69}$$

Figure 7-11 shows a phasor diagram which is a modification of that in Fig. 4-41 to take into account Eq. 7-69 when armature resistance is neglected.

For conventional machines it can be assumed with small error that the magnitude of $E'_{af}$ is equal to that of $E'_i$, where

$$\boxed{E'_i = V + jx'_d I} \tag{7-70}$$

as illustrated in Fig. 7-12. Similarly, when subtransient effects are appreciable,

$$\boxed{E''_i = V + jx''_d I} \tag{7-71}$$

Then the currents in Eq. 7-5 are

$$\boxed{I''_d = \frac{E''_i}{x''_d} \qquad I'_d = \frac{E'_i}{x'_d} \qquad I_d = \frac{E_{af}}{x_d}}$$

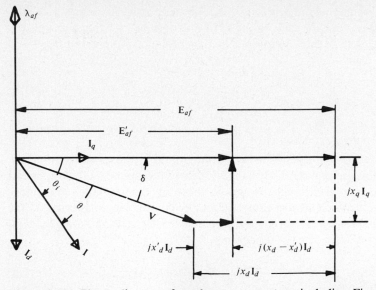

**Figure 7-11**  Phasor diagram of synchronous generator, including $E'_{af}$ and $X'_d$.

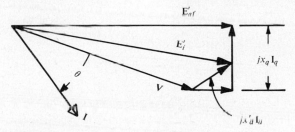

**Figure 7-12**  Phasor diagram of synchronous generator, including $E'_i$.

## 7-8  TRANSIENT STABILITY†

If the load on a synchronous machine is increased gradually until a value of $\delta$ is reached for which the real power becomes a maximum, the steady-state stability limit is said to be reached. In the case of a cylindrical-rotor machine, real power is a maximum when $\delta = \pi/2$ if armature resistance is neglected and the maximum power per phase or per unit is, according to Eq. 4-103,

$$P_{\max} \simeq \frac{E_{af}V}{x_d}$$

However, disturbances such as are due to sudden changes in load or faults on the electric power system cause the torque angle $\delta$ to change. The resulting

† For a comprehensive treatment of stability, see E. W. Kimbark, *Power System Stability* (New York: John Wiley & Sons, Inc.): Vol. I, *Elements of Stability Calculations,* 1948; Vol. II, *Power Circuit Breakers and Protective Relays,* 1950; Vol. III, *Synchronous Machines,* 1956. See also S. B. Crary, *Power System Stability* (New York: John Wiley & Sons, Inc.), Vol. I, *Steady-State Stability,* 1945; Vol. II, *Transient Stability,* 1947.

transients are usually of interest only for time periods so short that $\lambda_f$ is substantially constant and for slow enough time variations of $\delta$ that subtransient effects are negligible. Hence, $x'_d$ instead of $x_d$ is brought into play, so Eq. 4-133 is modified to read

$$P_{em} = \frac{E'_{af}V}{x'_d} \sin \delta + \frac{x'_d - x_q}{2x'_d x_q} V^2 \sin 2\delta \tag{7-72}$$

where for stability studies it is generally convenient to express quantities in per-unit.

While $E'_{af}$ may be regarded as a definite quantity, the terminal voltage $V$ is generally influenced by the disturbance, an extreme case being that of a short circuit at the generator terminals. However, there are cases in which the voltage $V_e$ at some point near a large generator capacity is relatively unaffected by the disturbance, and if the reactance between the generator terminals and the constant-voltage point is $x_e$, Eq. 7-72 becomes

$$P_{em} = \frac{E'_{af}V_e}{X'_d} \sin \delta + \frac{x'_d - x_q}{2X'_d X_q} V_e^2 \sin 2\delta \tag{7-73}$$

where $\delta$ is now the angle between $E'_{af}$ and $V_e$, $X'_d = x'_d + x_e$, and $X_q = x_q + x_e$. The second term in Eq. 7-73 is usually neglected in transient stability studies and further $E'_{af}$ is replaced by $E'_i$, which results in

$$P_{em} = \frac{E'_i V_e}{X'_d} \sin \delta \tag{7-74}$$

where $\delta$ is the angle between $E'_i$ and $V_e$. It should be recalled that $E'_i$ leads $V_e$ in a generator and $V_e$ leads $E'_i$ in a motor; so Eq. 7-74 may be applied to synchronous motors as well as to synchronous generators.

## 7-8.1 Equal-Area Criterion

A sudden change in $E'_i$, $V_e$, or in the load during the operation of a synchronous machine will produce a change in $\delta$, which, however, cannot be sudden because of the rotor inertia. Consider a three-phase synchronous motor supplied from an infinite bus (i.e., a source of constant voltage $V$ and constant frequency regardless of load). Neglect losses and assume the motor to be delivering a steady load $P_{sh1}$. For that condition

$$P_{sh1} = P_{em1} = \frac{E'_i V}{x'_d} \sin \delta_1 \tag{7-75}$$

Now, if the load is suddenly increased from $P_{sh1}$ to $P_{sh2}$ without any change in $E'_i$ or $V$, the initial electrical power input to the motor is still $P_{em1}$ as defined by Eq. 7-75, with an initial deficiency of $P_{sh2} - P_{sh1}$. As a result, the motor starts to decelerate. If

$P_{em}$ = electric power input

$P_{sh}$ = power delivered by the motor or shaft power

$P_a$ = accelerating power (i.e., power that stores kinetic energy of rotation)

then

$$P_{em} = P_{sh} + P_a \qquad (7\text{-}76)$$

On the basis of Eqs. 7-74 and 7-76,

$$\boxed{P_a = \frac{E_i'V}{x_d'} \sin \delta - P_{sh}} \qquad (7\text{-}77)$$

as illustrated in Fig. 7-13. It should be noted that the accelerating power is negative in Fig. 7-13, which means that to meet the increased output power, the motor must decelerate, causing energy to be abstracted from the kinetic energy of rotation stored in the rotor. However, by the time the torque angle reaches the value $\delta_f$ as in Fig. 7-14, at which point $P_{em} = P_{sh2}$, the rotor is running slightly below synchronous speed because of the reduction in its kinetic energy, causing a further increase in the torque angle. But as $\delta$ becomes greater than $\delta_f$, the power input exceeds the power output and the motor accelerates (i.e., $P_a$ is now positive, a process that continues until the rotor attains synchronous speed at the angle $\delta_m$, after which further acceleration causes the torque angle to decrease). In the absence of losses or damping, the motor speed fluctuates, with the torque angle oscillating between $\delta_1$ and $\delta_m$ unless the suddenly applied load is so excessive as to cause $\delta$ to increase indefinitely, in which case the transient stability limit

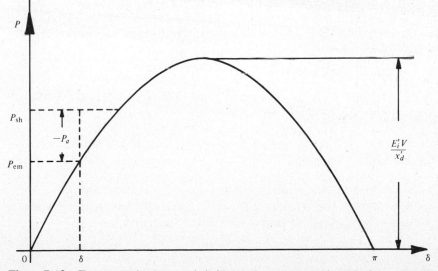

**Figure 7-13** Torque-angle characteristic illustrating power relationship for a synchronous motor while decelerating.

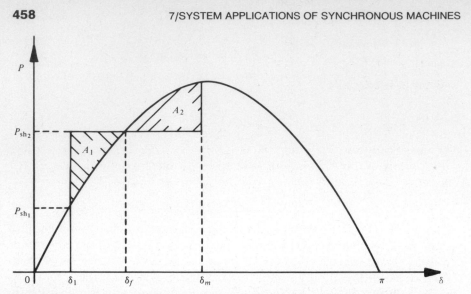

**Figure 7-14**   Torque-angle characteristic illustrating effect of sudden change in load on a synchronous motor.

is exceeded. In practical situations damping causes the oscillations to gradually die out and the torque angle reaches its final value of $\delta_f$, at which point the input matches the output of the motor while running at synchronous speed.

The areas $A_1$ and $A_2$ in Fig. 7-14 are proportional, respectively, to the kinetic energy abstracted from the rotor during deceleration and that absorbed by the rotor during acceleration. These two areas are equal to each other as long as operation is stable. If, on the other hand, $A_2 < A_1$ when $\delta > \pi - \delta_f$, the operation is unstable and $\delta$ increases indefinitely. Then for stable operation the motor runs at synchronous speed at $\delta_1$ and $\delta_m$, and during the oscillations between these two values the electrical angular velocity is

$$\omega = \omega_{\text{syn}} + \frac{d\delta}{dt} \tag{7-78}$$

and the accelerating torque is

$$T_a = \frac{P}{2} \frac{P_a}{\omega}$$

However, as long as the transient stability limit is not exceeded, the velocity component $d\delta/dt$ is small compared with $\omega_{\text{syn}}$ (usually less than 2 percent of $\omega_{\text{syn}}$). Hence, the accelerating torque is expressed approximately by

$$T_a = \frac{P}{2} \frac{P_a}{\omega_{\text{syn}}}$$

and during deceleration,

$$A_1 = \int_{\delta_1}^{\delta_f} P_a \, d\delta = \frac{2}{P} \omega_{\text{syn}} \int_{\delta_1}^{\delta_f} T_a \, d\delta \qquad (7\text{-}79)$$

and during acceleration,

$$A_2 = \frac{2}{P} \omega_{\text{syn}} \int_{\delta_f}^{\delta_m} T_a \, d\delta \qquad (7\text{-}80)$$

The integrals in Eqs. 7-79 and 7-80 represent the decrease and increase in the stored energy of rotation, and since the speed is synchronous at $\delta_1$ and $\delta_m$, the resultant change in the stored energy between $\delta_1$ and $\delta_m$ must be zero:

$$\int_{\delta_1}^{\delta} T_a \, d\delta + \int_{\delta}^{\delta_m} T_a \, d\delta = 0$$

which means that $A_1 = A_2$.

## 7-8.2 Transient Stability Limit

The transient stability of the previously unloaded motor is said to be reached when $A_1 = A_2$ for $\delta_m = \pi - \delta_1$, as indicated in Fig. 7-15. If the sudden application of load increases $\delta$ to a value greater than $\pi - \delta_1$, the motor does not recover its synchronous speed but again decelerates, increasing $\delta$ still further, and the motor pulls out of synchronism.

A sudden increase in the mechanical input to a previously unloaded generator connected to an infinite bus causes the generator to respond in much the same manner as the synchronous motor to a sudden increase in load except that

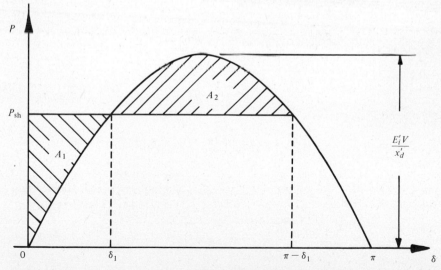

**Figure 7-15**  Torque-angle characteristic illustrating value of load suddenly applied to an unloaded synchronous motor for which the transient stability limit is reached.

the generator accelerates from $\delta_1$ to $\delta$ and decelerates from $\delta$ to $\delta_m$. However, the more general case is that of the generator delivering load when a disturbance on the electrical system to which it is connected produces a sudden change in the load of the generator. Disturbances on power systems are commonly initiated by short circuits (faults) on the system, the sudden loss of one or more generators connected to the system, or the loss of a system interconnection.

## 7-9 SWING CURVES

A *swing curve* represents graphically the angular position of the rotor of a synchronous machine relative to a synchronously rotating reference axis as a function of time during a system disturbance. Swing curves are useful for determining the adequacy of relay protection on power systems with regard to the clearing of faults before one or more machines become unstable and fall out of synchronism, thus compounding the disturbance.

### 7-9.1 The Swing Equation

The behavior of a synchronous machine during transients is described by the *swing equation.* As an introduction to this equation, consider all quantities in their physical dimensions rather than in per-unit. The accelerating torque is then in newton-meters per radian, being expressed by

$$T_a = J \frac{d\omega_m}{dt} = \frac{2}{P} J \frac{d\omega}{dt} \qquad (7\text{-}81)$$

where $J$ is the polar moment of inertia of the rotor of the generator and its prime mover and $P$ is the number of poles:

$$\omega = \omega_{\text{syn}} + \frac{d\delta}{dt}$$

Since $\omega_{\text{syn}}$ is constant,

$$\frac{d\omega}{dt} = \frac{d^2\delta}{dt^2}$$

Further, since in the region of interest, $\omega \cong \omega_{\text{syn}} + d\delta/dt$, Eq. 7-81 can be rewritten, with negligible error, as

$$T_a = \frac{2}{P} J \frac{d^2\delta}{dt^2}$$

The accelerating power is

$$P_a = \frac{2}{P} \omega T_a = \left(\frac{2}{P}\right)^2 J\omega \frac{d^2\delta}{dt^2}$$

$$= M \frac{d^2\delta}{dt^2} \tag{7-82}$$

where $M$ is the inertia constant, expressed in joule-seconds per radian.

Equation 7-77 applies to a motor; the corresponding equation for generator operation is

$$P_{sh} = P_{em} + P_a \tag{7-83}$$

Then, on the basis of Eqs. 7-74, 7-82, and 7-83 the swing equation of a generator is given by

$$M \frac{d^2\delta}{dt^2} = P_{sh} - 3 \frac{E'_i V_e}{X'_d} \sin \delta \tag{7-84}$$

where $P_{sh}$ is in watts, $E'_i$ and $V_e$ are in volts, and $X'_d = x'_d + x_e$ are in ohms.

Equation 7-84 applies to a three-phase machine and uses physical terms. To convert to per-unit quantities, the multiplier 3 in the second term is dropped and use is made of the quantity $H$,† which is the stored kinetic energy at rated speed in megajoules per megavolt-ampere. Then the inertia constant is expressed in per-unit by

$$M = \frac{H}{\pi f}$$

and on that basis the swing equation becomes

$$M \frac{d^2\delta}{dt^2} = P_{sh} - \frac{E'_i V_e}{X'_d} \sin \delta \tag{7-85}$$

where $\delta$ is in radians.

## 7-9.2 Swing Curves

A plot of $\delta$ (usually in electrical degrees) versus time in seconds, based on the swing equation, is known as a *swing curve*. The formal solution of Eq. 7-86 involves elliptic integrals. Prior to the advent of the digital computer, its solution and that for more complex situations were obtained by means of point-by-point calculations. Present practice generally makes use of the digital computer, as in the following rather simple illustration.

† Typical values of $H$ are given in *Electrical Transmission and Distribution Reference Book,* 4th ed. (East Pittsburgh, Pa.: Westinghouse Electric Corporation, 1950), p. 189.

**EXAMPLE 7-1**

Figure 7-16(a) shows a one-line diagram of a three-phase power system in which the generator $G$ supplies power to an infinite bus over two identical ties, each consisting of a transmission line with a transformer $T$ at each end. There are four circuit breakers, $A$, $B$, $C$, and $D$. A three-phase fault occurs near the high-voltage terminals of the transformer, which is supplied on its low-voltage side through circuit breaker $C$. Prior to the fault, the generator $G$ delivers 90 percent of its rated current over this system to the infinite bus, which operates at 0.95 per-unit voltage. The power factor of the delivered load at that bus is 0.96, current lagging.

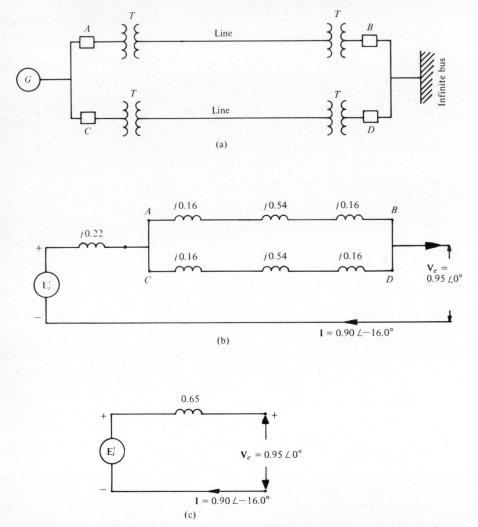

**Figure 7-16** (a) One-line diagram for power system in Example 7.1. (b) Schematic representation before fault with per-unit impedance values. (c) Reduced equivalent circuit.

The data for the generator and the reactances of the system, expressed in per-unit on a 150-MVA base, follow. All resistances are neglected.

Generator $G$:  13.8 kV, 150 MVA, 0.95 power factor
$x_d - 0.65$    $x'_d = 0.22$
$H = 3$
Transformers:  $x = 0.16$
Lines:  $x = 0.54$

Protective relays are provided for clearing the fault by causing circuit breakers $C$ and $D$ to open. Plot the swing curves for the condition (a) that the fault is not cleared, and (b) the fault is cleared in (1) 0.16 s, (2) 0.18 s, (3) 0.20 s.

*Solution.* The generator voltage to be used is $E'_i$, that behind transient reactance, and the initial value of $\delta$ is that existing, before the fault, between $E'_i$ and $V_e$, the voltage at the infinite bus. Figure 7-16(b) shows a schematic representation of one phase with the reactance values of the system components. Figure 7-16(c) shows the corresponding circuit reduced to a single equivalent impedance. Then, on the basis of Fig. 7-16(c),

$$\mathbf{E}'_i = \mathbf{V}_e + jX'_d \mathbf{I}$$

where

$$\mathbf{V}_e = 0.95\underline{/0°} \qquad \mathbf{I} = 0.90\underline{/-16.0°}$$

and   $$X'_d = 0.22 + 0.43 = 0.65$$

Hence,

$$\mathbf{E}'_i = 0.95 + j0 + j0.65 \times 0.90(0.96 - j0.275)$$
$$= 1.11 + j0.56 = 1.24\underline{/26.9°}$$

from which it follows that at $t = 0$, $\delta = 26.9°$.

*During the fault* the system is represented by the circuit diagram in Fig. 7-17(a). A delta-wye transformation† of the circuit *ABDFC* results in the equivalent wye ($j0.08$, $j0.35$, and $j0.065$) in Fig. 7-17(b). And when the wye ($j0.30$, $j0.35$, and $j0.065$) is converted to its equivalent delta, the result is as shown in Fig. 7-17(c). The shunt reactances $j0.42$ and $j0.49$ in Fig. 7-17(c) place no real power load on the system and are therefore ignored in the torque-angle equation. Hence, during the fault

$$P_{em} = \frac{(1.24)(0.95)}{2.26} \sin \delta = 0.521 \sin \delta$$

† Delta–wye transformations are treated in standard textbooks on electric circuit theory. See, for example, Norman Balabanian, *Fundamentals of Circuit Theory* (Boston: Allyn and Bacon, Inc., 1961), p. 294.

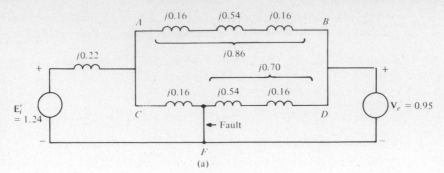

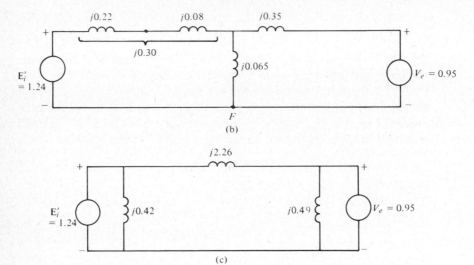

**Figure 7-17** Schematic diagram of power system in Example 7.1 during fault. (a) Complete equivalent circuit. (b) Equivalent wye of (a). (c) Equivalent delta of (b).

and

$$M = \frac{H}{\pi f} = \frac{3}{60\pi} = \frac{1}{20\pi}$$

Further, the input power is assumed constant; so when losses are neglected,

$$P_{sh} = 0.95 \times 0.90 \times 0.96 = 0.82$$

When these values are substituted in Eq. 7-85, the swing equation becomes

$$\frac{1}{20\pi} \frac{d^2\delta}{dt^2} = 0.82 - 0.521 \sin \delta$$

or

$$\frac{d^2\delta}{dt^2} = 16.4\pi - 10.42\pi \sin \delta$$

*After the fault is cleared* (circuit breakers *C* and *D* open), the equivalent circuit of the system is as shown in Fig. 7-18. The swing equation for this condition is

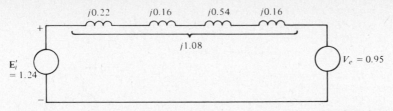

**Figure 7-18**   Schematic diagram of system in Example 7.1 after fault is cleared.

$$\frac{d^2\delta}{dt^2} = 16.4\pi - 21.8\pi \sin \delta$$

The digital-computer program for calculating the swing curves is shown in Fig. 7-19 and the curves plotted by the computer are shown in Fig. 7-20. In Fig. 7-20 the vertical axis represents time and the 100 divisions correspond to a period of 1 s. The horizontal axis represents the torque angle $\delta$ and 100 divisions correspond to 200°. The curves drawn through points $A$, $B$, $C$, and $D$ show the results of (a) the fault not cleared, (b) the fault cleared in 0.16 s, (c) the fault cleared in 0.18 s, and (d) the fault cleared in 0.20 s. All curves except that resulting from a clearing time of 0.16 s (points $B$) indicate instability (i.e., generator $G$ pulls out of synchronism with the infinite bus and continues to accelerate). When the fault is cleared after 0.16 s, the system remains stable, since the generator accelerates for about $\frac{1}{2}$ s ($\delta$ reaching a maximum value of about 111°), after which deceleration takes place, as shown by curve $B$.

Stability is a function of the acceleration $d^2\delta/dt^2$. The greater the acceleration, the shorter must be the fault-clearing time to maintain stability. From Eq. 7-85 it is evident that the acceleration is lower for a higher field excitation (i.e., larger value of $E_i'$), a lower value of $X_d'$, and a larger inertia constant $M$. The design of a power system must take these and numerous other factors into account.

## 7-10 DYNAMIC STABILITY

The term *dynamic stability* applies to synchronous machines with automatically controlled excitation systems commonly known as *quick-response systems*. These incorporate exciters of quick response as well as quick-acting regulators which can adjust the flux within a synchronous machine at a faster rate than that caused by the system falling out of step, thus improving the stability limits.

### 7-10.1 Dual Excitation

The *dual-excitation system*, in its simplest aspect,† is one in which the rotor of a synchronous machine carries a field winding on its quadrature axis in addition to the conventional field winding on the rotor direct axis, as illustrated in Fig.

---

† See, for instance, A. M. El-Serafi and M. A. Badr, "Analysis of Dual-Electric Synchronous Machine," *Trans. IEEE Power Apparatus and Systems* PAS-92 (January-February 1973): 1–13.

```
              PROGRAM SWCURV(INPUT,OUTPUT)
    3         DIMENSION A(5,100), YI(2), Y(2),  B(5,100)
    3         COMMON M
    3         DRC = (2. * 3.14)/ 360.
    7         DO 2  M = 1,4
   10         YI(1) = 26.9* DRC
   13         YI(2) = 0.
   16         DT = 0.010
   20         T = 0.
   21         DO 1  I = 1,100
   22         CALL MXRK4(T,YI,2,T +DT,5,Y)
   30         A(M,I) =(Y(1)/DRC) * 0.5
   41         B(M,I) = Y(1)/DRC
   50         YI(1) = Y(1)
   55         YI(2) = Y(2)
   61         T = T + DT
   63    1    CONTINUE
   65    2    CONTINUE
   67         PRINT 10
   73   10    FORMAT(1H1)
   73         CALL PRINT5(B,4,100)
   76         PRINT 10
  102         CALL PLOT5(A,4,100,100)
  105         STOP
  107         END


              SUBROUTINE GN(T,Y,G)
    6         DIMENSION  G(2), Y(2)
    6         COMMON M
    6         B = .12  + 0.020 * M
   11         IF(M .GT.  1)  GO TO 2
   14         GO TO 3
   15    2    IF(T .GE. B) GO TO 4
   20    3    G(1) =  Y(2)
   25         G(2) =   16.4 * 3.14 -10.42* 3.14* SIN(Y(1))
   41         GO TO 5
   41    4    G(1) =  Y(2)
   46         G(2) =   16.4 * 3.14 -21.8 * 3.14* SIN(Y(1))
   62    5    CONTINUE
   62         RETURN
   63         END
```

**Figure 7-19**  Digital-computer program for swing curves in Example 7.1.

7-21. The individual currents in the two field windings can be regulated automatically to place the resultant magnetic axis of the rotor mmf at any desired angular position relative to the mechanical direct axis of the field pole. Stability is said to be improved by retarding the torque angle $\delta$, which also forces damping of the rotor oscillations during and subsequent to a system disturbance.

Operation of a synchronous generator at a leading power factor lowers its stability limit for a given real power output. Recent developments attending the rapid growth of power systems require the installation of an increasing number

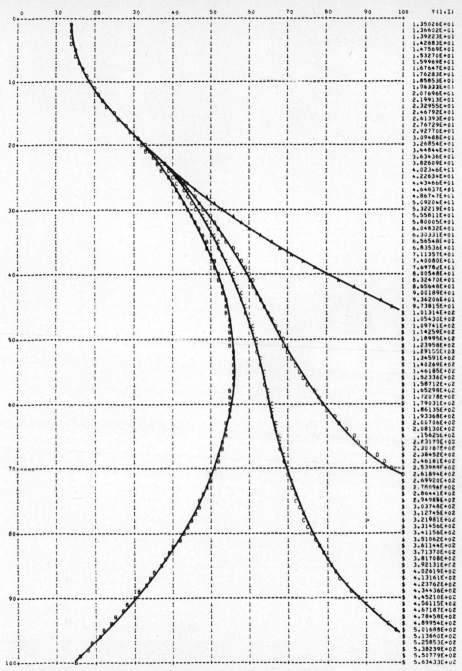

**Figure 7-20** Swing curves for Example 7.1 plotted by digital computer. Points *A*, *B*, *C*, and *D* show the curves. (a) Fault not cleared. (b) Fault cleared in 0.16 s. (c) Fault cleared in 0.18 s. (d) Fault cleared in 0.20 s.

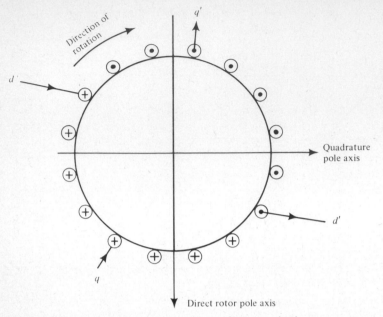

**Figure 7-21**   Schematic representation of dual excitation arrangement; $d$–$d'$ and $q$–$q'$ are the terminals of the direct-axis and quadrature-axis field windings.

of long high-voltage transmission lines and underground cables, resulting in sizable increases of capacitive current. As a result, generators may absorb amounts of reactive power that provide sufficient air-gap flux to operate at a low value of field current with a corresponding reduction in the generated voltage $E_{af}$, resulting in low steady-state and transient-state stability limits, particularly for low real power loads, For example, Eq. 4-103 shows the steady-state stability power limit of a cylindrical-rotor synchronous machine to be $VE_{af}/x_d$, which is reduced to $VE_{af}/(x_d + x_e)$ when the generator is connected through a tie of negligible resistance but with a reactance of $x_e$ to a bus operating at the voltage $V$. The following example illustrates the effect of leading current on the steady-state stability limit.

**EXAMPLE 7-2**

A three-phase 750-MVA 22-kV 60-Hz synchronous generator supplies a 375-kV power system through a 22/375-kV transformer and a long 375-kV transmission line. The synchronous reactance $x_d$ of the generator is 1.90 per unit. On the basis of simplifying assumptions, the reactance of the tie between the generator and its load is 0.80 per unit. The voltage at the load is 1.00 per unit and the load on the generator is $\mathbf{S} = (100 - j140) \times 10^6$. Calculate the steady-state stability limit.

*Solution.*   The per-unit current of the generator is

$$\mathbf{I} = \frac{100 + j140}{750} = 0.133 + j0.187$$

Hence,

$$\mathbf{E}_{af} = \mathbf{V} + j(x_d + x_e)\mathbf{I} = 1.00 + j(1.90 + 0.80)(0.133 + j0.187)$$
$$= 1.00 - 0.505 + j0.360 = 0.611\underline{/36.0°}$$

$$P_{max} = \frac{VE_{af}}{x_d + x_e} = \frac{0.611}{2.7} = 0.226 \text{ or } 170 \text{ MW}$$

Example 7-2 shows that under the given operating conditions, the steady-state stability limit is only about 22 percent of the generator rating. Such a low limit imposes a severe duty on a conventional excitation system in the case of even a moderate system disturbance. While the dual-excitation system has not been adopted to any significant extent, it appears promising and is receiving considerable study.

## STUDY QUESTIONS

1. What would be the advantage of using a single large generator in a power plant, rather than several smaller units operating in parallel?

2. What are the advantages of using several smaller generators operating in parallel, rather than one large unit?

3. List at least four important conditions that must be fulfilled before a generator can be connected in parallel with a bus already supplying a load.

4. Describe how a generator could be paralleled with another generator, assuming that the first generator is already delivering load.

5. What is this process of paralleling generators called?

6. Why will two generators operating in parallel be in stable equilibrium even though the speed of one of them tends to increase?

7. Two generators operating in parallel will be in stable equilibrium even though the voltage of one tends to decrease. Explain why.

8. What is the method by which parallel generators are paralleled?

9. Explain the principles of the method used to parallel generators.

10. What is meant by instantaneous three-phase short-circuit current?

11. What is meant by the term symmetrical short-circuit current?

12. What is meant by the term asymmetrical short-circuit current?

13. Make a list of the various constants used in evaluating the transient performances of synchronous machines.

14. What is the importance of knowing the instantaneous three-phase short-circuit current of a synchronous generator?

15. What is meant by transient stability?

16. What is meant by steady-state stability?

17. Compare the difference between transient stability and steady-state stability.

18. What is meant by a swing curve?

19. How are swing curves used in evaluating the performance of a synchronous machine on a power system?

20. Describe what happens in a power system when a loss of synchronism occurs.

21. Describe what happens in a power system when a loss of steady-state stability occurs.

22. What is meant by the term dynamic stability?

## PROBLEMS

7-1. Two identical three-phase 13.8-kV 100,000-kVA 60-Hz turbine (cylindrical-rotor) generators 1 and 2 operate in parallel, supplying a constant combined load of 150,000-kVA, 0.80 power factor, current lagging, at rated voltage and rated frequency. The synchronous reactance of each machine is 1.10 per unit. The mechanical outputs of the prime movers are adjusted so that the machines share the real power equally, while the field excitations are such that the reactive power output of the two machines is equal. Calculate, for each generator, (a) real power, (b) reactive power, (c) armature current, (d) power factor, (e) generated phase voltage, (f) torque angle, and (g) the angle, in electrical measure, between the armature and field mmfs.

7-2. Repeat Prob. 7-1 but for the condition that the machines deliver equal real power but with the field excitation of generator 1 increased 20 percent above its value in Prob. 4-38 while the field current of generator 2 is adjusted so that the terminal voltage remains constant at rated value. Neglect the losses in the generators and assume the synchronous reactance of each generator to remain constant at 1.10 per unit.

7-3. The excitation of generator 1 remains constant at the value specified in Prob. 7-2 while the input to its prime mover is increased by 20 percent. Adjustments are made in the field current of generator 2 and in its prime mover such that the frequency and terminal voltage remain at their rated values. Neglect generator losses, assume the synchronous reactance of each generator to remain constant at 1.10 per unit, and calculate the quantities called for in Prob. 7-1.

7-4. Repeat Prob. 7-1 for two generators with the same voltage and frequency ratings as in Prob. 7-1 but with generator 1 rated at 80,000 kVA and having a synchronous reactance of 1.00 per unit, while generator 2 is rated at 120,000 kVA with a synchronous reactance of 1.20 per unit.

7-5. Repeat Prob. 7-2 for the generators of Prob. 7-4.

7-6. Repeat Prob. 7-3 for the generators of Prob. 7-4.

7-7. A three-phase 240-V 5-kVA 60-Hz Y-connected synchronous generator has a synchronous reactance of 1.00 per unit. This generator supplies an isolated load comprised of three identical wye-connected noninductive resistors with rated current at rated voltage and at rated frequency. The mutual inductance $L_{afm}$ between the field and one phase of the armature is 0.245 H. (a) Calculate the generated emf $E_{af}$ in volts per phase and the field current in amperes. (b) The prime-mover output is held constant while the field current is varied. Neglect changes in rotational losses and in saturation and plot the frequency as a function of field current. (c) Calculate the minimum value of field current at which the generator can deliver its rating to the isolated load.

7-8. The three-phase generator in Prob. 7-7 supplies an identical machine operating as a synchronous motor. Three inductive reactors, one in each phase, are connected in series between the armatures as shown in the illustration. The value of each reactance is 0.20 per unit in terms of the machine rating and the resistance is negligible. The motor drives a mechanical load in which the torque may be considered independent of the speed and of a value such that the mechanical power is 2.0 kW at rated speed. (a) The field currents of the two machines are adjusted so that rated voltage is applied to the motor terminals at rated frequency with a motor power factor of 0.80, current lagging. Neglect losses and calculate the field current in each machine. (b) What is the minimum value of motor field current (on the basis of cylindrical-rotor theory) for which the machines will remain in synchronism at synchronous speed with the generator field current as in part (a)?

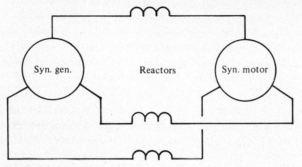

**Prob. 7-8**   Synchronous generator supplying a synchronous motor through reactors.

**7-9.** A 40,000-kVA three-phase 13.2-kV 60-Hz generator has per-unit values of saturated synchronous reactances $x_d = 1.00$ and $x_q = 0.60$. (a) Assume the generator to be connected to an infinite bus and calculate the maximum real power the generator can deliver without losing synchronism when the field circuit is open. (b) Calculate the reactive power, power factor, and current. Does the current lead or lag the terminal voltage?

**7-10.** A 200-hp three-phase 60-Hz synchronous motor has a full-load power factor of 0.85, current leading. The armature resistance is 0.05 per-unit and the reactances are $x_p = 0.12$, $x_{du} = 1.10$, and $x_q = 0.60$ per-unit unsaturated. The rated-load efficiency, not including the exciter, is 0.92. The data for the open-circuit characteristic per unit is as follows:

| Field excitation | 0.0 | 0.7 | 1.0 | 1.2 | 1.3 | 1.4 | 1.5 | 1.6 | 1.8 | 2.0 | 2.3 |
|---|---|---|---|---|---|---|---|---|---|---|---|
| Armature voltage | 0.0 | 0.70 | 0.92 | 1.04 | 1.08 | 1.115 | 1.14 | 1.165 | 1.215 | 1.25 | 1.28 |

Use two-reactance theory (use $x_d$ and $x_q$) and the saturation-factor method to calculate the torque angle $\delta$ and the field current for rated load, 0.85 power factor, current leading. The field current required for rated open-circuit voltage on the air-gap line is 12.2 A.

**7-11.** The purpose of this problem is to illustrate the effect of field excitation on the pull-out torque of a synchronous motor. A synchronous motor has a rated power factor of 0.80, current leading, and $x_d = 0.90$ per unit. Neglect the effect of saliency and losses and calculate the ratio of pull-out torque when the field is excited for 0.80 power factor, current leading, at rated mechanical load to that when the field is excited so that the motor delivers its rated load at unity power factor.

**7-12.** A synchronous machine has the following direct-axis information: $X_d = 1.5$ per unit, $X'_d = 0.20$ per unit, $T'_{do} = 4.0$ s. Assume that the machine does not have damper windings. Find the short-circuit time constant $T'_d$.

**7-13.** A 100-MVA 23-kV three-phase 60-Hz synchronous generator is delivering rated voltage open-circuit when a balanced three-phase fault occurs at its terminals. The machine has the following constants expressed in per-unit on a 100-MVA base: $X_d = 1.0$, $X'_d = 0.3$, $X''_d - 0.2$, $T_a = 0.2$, $T''_d = 0.03$ s, $T'_d = 1.0$ s. Neglect dc and higher-frequency components of the current and find the initial current, current at the end of two cycles, and the current at the end of 10 s.

**7-14.** A 400-MVA 23-kV generator has the following reactances and time constants. The reactances are given per-unit on the generator base. $X_d = 1.85$, $X'_d = 0.33$, $X''_d = 0.25$, $T_a = 0.2$ s, $T'_d = 0.8$ s, $T''_d = 0.04$ s. Assume that the machine is operating at no load when a three-phase fault occurs directly on the terminals. Prior to the fault the line-to-line terminal voltage was 23 kV. Write an expression that describes in per-unit the rms fault current as a function of time. Neglect dc and double-frequency components. What is the current after 0.25 s, 0.5 s, 1 s?

**7-15.** A synchronous generator is connected through parallel transmission lines to a large metropolitan system which may be considered as an infinite bus. Each of the parallel transmission lines has a per-unit reactance of $j0.4$. The unit transformer serving the bus that supplies the parallel transmission lines has a reactance of $j0.1$. Assume that the machine is delivering 1.0 per-unit power and both the terminal voltage and the infinite bus voltage are one per unit. The transient reactance of the generator is 0.2 per unit. Determine the power-angle equation for the system applicable to the operating conditions given.

**7-16.** The system in the previous problem is operating at the conditions specified when a three-phase fault occurs in the middle of one of the parallel transmission lines. Determine the power-angle equation for the system before the fault and during the fault. Assume that $H = 5$ MJ/MVA.

**7-17.** The fault on the system described in the previous problem is cleared by simultaneous opening of breakers at each end of the line on which the fault has occurred. Determine the power-angle equation and the swing equation for the postfault condition.

**7-18.** Assume that the machine of the previous example is operating at a power angle of 28 electrical degrees when it is subjected to a slight temporary electrical disturbance. Determine the frequency and period of oscillation of the machine rotor if the disturbance is immediately removed.

**7-19.** Calculate the critical clearing angle and the critical clearing time for the machine in Fig. 7-16. The fault occurs at location $D$ in the system. The critical clearing angle and the critical clearing time is that maximum time for clearing the fault that will allow the system to remain stable.

# BIBLIOGRAPHY

Calabrese, G. O. *Symmetrical Components.* New York: The Ronald Press Company, 1959.

Concordia, C. *Synchronous Machines.* New York: John Wiley & Sons, Inc., 1957.

Crary, S. B. *Power System Stability.* New York: John Wiley & Sons, Inc. Vol. I *Steady-State Stability,* 1945; Vol. II *Transient Stability,* 1947.

Gross, Charles A. *Power System Analysis.* New York: John Wiley & Sons, Inc., 1979.

Kimbark, E. W. *Power System Stability.* New York: John Wiley & Sons, Inc. Vol. I *Elements of Stability Calculations,* 1948; Vol. II *Power Circuit Breakers and Protective Relays,* 1950; Vol. III *Transient Stability,* 1947.

Matsch, Leander W. *Electromagnetic & Electromechanical Machines,* 2d ed. New York: Dun-Donnelley Publishing Co., 1977.

Schultz, M. A., *Control of Nuclear Reactors and Power Plants,* 2d ed. New York: McGraw-Hill Book Company, 1961.

Schultz, Richard D., and Richard A. Smith. *Introduction to Electric Power Engineering.* New York: Harper & Row Publishers, 1985.

Stevenson, W. D., Jr. *Elements of Power System Analysis.* New York: McGraw-Hill Book Company, 1982.

# Chapter 8

# Special Machines

Although practically all the machines treated in this text operate on the same basic principles, those discussed in this chapter have features that distinguish them from the more conventional types and are termed *special machines* for that reason. An attempt to cover all types of such machines would be unrealistic because of their large variety. This chapter is therefore devoted to some which are used to an appreciable extent, such as reluctance motors, inductor machines, ac commutator motors, acyclic dc machines, and magnetic pumps. In addition, we include generators, such as magnetohydrodynamic generators, which as yet have not found wide application but seem to have great potential as efficient power sources. Special machines that have widespread use but whose use is declining are also included such as ac tachometers and selsyns.

## 8-1 RELUCTANCE MOTORS

Reluctance motors are in effect synchronous motors that operate without dc field excitation, depending for their operation on the difference between the reluctances in the direct and quadrature axes. Fractional-horsepower motors are usually single-phase and are used in applications requiring exact synchronous speed for such drives as electric clocks and other timing devices.

 The polyphase reluctance motor, despite its relatively large size, is finding increased use in integral-horsepower drives for which the maintenance of exact synchronous speed is desirable. The advantage of the polyphase reluctance motor over the synchronous motor studied in Chap. 4 lies in its simplicity, in that it does not require dc field excitation and the concomitant rotor windings, slip rings, and brushes. Polyphase reluctance motors have been built in ratings up to 150 hp.

## 8-1.1 Single-Phase Reluctance Motor Torque

Consider the elementary reluctance motor in Fig. 8-1(a). Assume the design to be such that the magnetic reluctance is a function of the angular position of the rotor described by

$$\mathcal{R} = \tfrac{1}{2}(\mathcal{R}_d + \mathcal{R}_q) - \tfrac{1}{2}(\mathcal{R}_q - \mathcal{R}_d) \cos 2\sigma \tag{8-1}$$

and illustrated graphically in Fig. 8-1(b).

The resistance of the winding in practical devices of this kind is usually low enough so that the induced voltage is practically equal to the applied voltage. Therefore,

$$e = v \cong \sqrt{2}V \cos \omega t = N \frac{d\phi}{dt} \tag{8-2}$$

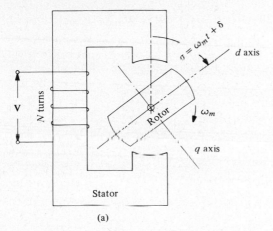

(a)

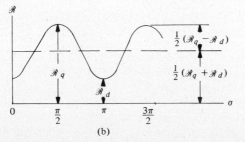

(b)

**Figure 8-1** (a) Elementary reluctance motor. (b) Graphical relationship between reluctance and rotor displacement.

where $\phi$ is the equivalent flux $\lambda/N$, $N$ being the number of turns in the stator winding. Since the winding has some resistance, there is no dc flux component until some time after the motor has been energized, and

$$\phi = \phi_M \sin \omega t \tag{8-3}$$

where, according to Eq. 2-35,

$$\phi_M = \frac{V}{4.44fN} = \frac{\sqrt{2}V}{2\pi fN} = \frac{\sqrt{2}V}{\omega N} \tag{8-4}$$

The torque is expressed in terms of flux and reluctance by Eq. 2-97, as

$$T_{em} = -\tfrac{1}{2}\phi^2 \frac{\partial \mathcal{R}}{\partial \sigma} \tag{8-5}$$

and when Eqs. 8-1 and 8-3 are substituted in Eq. 8-5, the torque is found after some algebraic manipulation to be

$$T_{em} = -\tfrac{1}{4}\phi_M^2(\mathcal{R}_q - \mathcal{R}_d)(\sin 2\sigma - \sin 2\sigma \cos 2\omega t)$$

which can be reduced further to

$$T_{em} = -\tfrac{1}{4}\phi_M^2(\mathcal{R}_q - \mathcal{R}_d)\{\sin 2\sigma - \tfrac{1}{2}[\sin 2(\sigma + \omega t) + \sin 2(\sigma - \omega t)]\} \tag{8-6}$$

If the rotor is made to rotate at $\omega_m$ rad/s,

$$\sigma = \omega_m t + \delta \tag{8-7}$$

and when Eq. 8-7 is substituted in Eq. 8-6, there results

$$T_{em} = -\tfrac{1}{4}\phi_M^2(\mathcal{R}_q - \mathcal{R}_d)[\sin 2(\omega_m t + \delta) - \tfrac{1}{2}\{\sin 2[(\omega_m + \omega)t + \delta] \\ + \sin 2[(\omega_m - \omega)t + \delta]\}] \tag{8-8}$$

Equation 8-8 shows the average torque to be zero for all values of $\omega_m \neq \omega$, and that when $\omega_m = \omega$, the torque is

$$T_{em} = -\tfrac{1}{4}\phi_M^2(\mathcal{R}_q - \mathcal{R}_d)\{\sin (2\omega t + 2\delta) - \tfrac{1}{2}[\sin (4\omega t + 2\delta) + \sin 2\delta]\} \tag{8-9}$$

which has an average value of

$$T_{av} = \tfrac{1}{8}\phi_M^2(\mathcal{R}_q - \mathcal{R}_d) \sin 2\delta \tag{8-10}$$

The average torque is more conveniently expressed in terms of the applied voltage from the following derivation. Substitution of Eq. 8-4 in Eq. 8-10 leads to

$$T_{av} = \frac{1}{4}\frac{V^2}{\omega}\left(\frac{\mathcal{R}_q}{\omega N^2} - \frac{\mathcal{R}_d}{\omega N^2}\right) \sin 2\delta \tag{8-11}$$

and since, according to Eq. 2-64, $L = N^2/\mathcal{R}$, we may write

$$X_{qq} = \frac{\omega N^2}{\mathcal{R}_q} \quad \text{and} \quad X_{dd} = \frac{\omega N^2}{\mathcal{R}_d}$$

which upon substitution in Eq. 8-11 results in

$$T_{av} = \frac{1}{4}\frac{V^2}{\omega}\left(\frac{1}{X_{qq}} - \frac{1}{X_{dd}}\right)\sin 2\delta = \frac{1}{2}\frac{V^2}{\omega}\frac{X_{dd} - X_{qq}}{2X_{dd}X_{qq}}\sin 2\delta \qquad (8\text{-}12)$$

In Eq. 8-12, $X_{dd}$ and $X_{qq}$ are the reactances of the motor for $\sigma = 0$ and for $\sigma = \pi/2$. Further, since the motor in Fig. 8-1 has a two-pole rotor, $\omega = 2\pi n_{syn}/60$. A comparison of Eq. 8-12 with the second term in the right-hand side of Eq. 4-133 shows a pronounced similarity. However, in making this comparison, it is important to note that the nature of the reactances $X_{dd}$ and $X_{qq}$ differs from that of $x_d$ and $x_q$ in the three-phase machine since $x_d$ and $x_q$ include the effect of mutual reactance between phases.

It is possible to operate a motor with a stator similar to that in Fig. 8-1 at subsynchronous speed by shaping the magnetic circuit so that

$$\mathcal{R} = \frac{\mathcal{R}_q + \mathcal{R}_d}{2} - \frac{\mathcal{R}_q - \mathcal{R}_d}{2}\sin 2n\sigma_m$$

in which $\sigma_m$ is in mechanical measure. In that arrangement a nonzero value of torque requires the mechanical angular velocity of the rotor to be $\omega_m = \omega/n$. Substitution of $n\sigma_m$ for $\sigma$ in Eqs. 8-1 and 8-9 yields the expressions of the torque for such an arrangement:

$$T_{em} = -\tfrac{1}{4}\phi_M^2 n(\mathcal{R}_q - \mathcal{R}_d)\{\sin(2\omega t + 2n\delta_m) - \tfrac{1}{2}[\sin(4\omega t + 2n\delta_m) + \sin 2n\delta_m]\}$$
$$(8\text{-}13)$$

and
$$T_{av} = \tfrac{1}{8}\phi_M^2 n(\mathcal{R}_q - \mathcal{R}_d)\sin 2n\delta_m \qquad (8\text{-}14)$$

$$\frac{V^2 n}{2\omega}\frac{X_{dd} - X_{qq}}{2X_{dd}X_{qq}}\sin 2n\delta_m \qquad (8\text{-}15)$$

where $\omega = 2\pi f$ and $\delta_m$ is in mechanical measure and $n\delta_m - \delta$, in which $\delta$ is given in electrical measure. Equation 8-15 may be rewritten in the more familiar form

$$T_{av} = \frac{V^2 n}{2\omega}\frac{X_{dd} - X_{qq}}{2X_{dd}X_{qq}}\sin 2\delta \qquad (8\text{-}16)$$

Since the average torque is zero for $\omega_m \neq \omega/n$, the starting torque is zero and the rotor must be brought up to speed by mechanical means. Electric clocks driven by reluctance motors have been built and are started by accelerating the rotor above synchronous speed by hand, and as the motor approaches synchronous speed while decelerating, it is pulled into synchronism. The need for mechanical starting as well as the noise and vibration due to the double- and quadruple-frequency components of the torque in Eq. 8-13 makes the reluctance motor unsuited for certain drives.

Reluctance motors are usually started as induction motors by making use of a squirrel-cage rotor from which teeth have been removed in locations such as to produce the desired number of salient poles as for example the four-pole and six-pole rotors in Fig. 8-2. In such rotors the teeth are removed leaving all the rotor bars and the entire end rings intact. Any of the single-phase starting

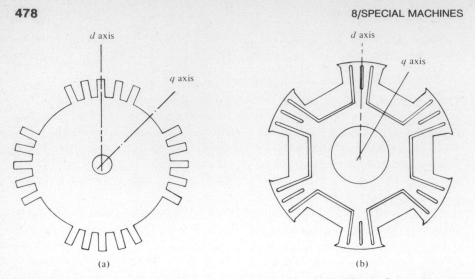

**Figure 8-2** Cross sections of rotors for reluctance motors. (a) Simple four-pole rotor. (b) Six-pole slitted rotor.

arrangements described in Chap. 5 may be used for single-phase reluctance motors. The squirrel-cage winding places an additional load on the single-phase motor at synchronous speed because of the backward torque developed by the backward-rotating field and thus reduces the maximum value of the average torque.

## 8-1.2 Polyphase Reluctance Motors

The stators of polyphase reluctance motors are similar to those of the conventional polyphase machines discussed in Chap. 5. The rotors, however, are of salient-pole configuration with a squirrel-cage winding but without a dc field winding. Figure 8-2(a) shows a rotor of an older design which is essentially a squirrel-cage induction-motor rotor from which teeth have been removed in four locations so as to produce a four-pole salient-pole structure. The rotor bars and end rings, however, are left intact to make for effective starting as an induction motor. The slitted rotor in Fig. 8-2(b) is one of several more recent designs in which the slits direct the flux along more effective paths to produce a reduction in the value of $x_q$ without a corresponding reduction in that of $x_d$, which results in a higher value of pull-out torque. As the motor approaches rated speed, the reluctance torque pulls the rotor into synchronism.

The absence of dc field excitation greatly reduces the maximum output of a synchronous motor, as is indicated in the final paragraph of Sec. 4-21. For that reason, reluctance motors are several times the physical size of synchronous motors, with dc excitation of the same horsepower and speed ratings. In addition, the reluctance motor has a low power factor, since it requires a relatively large amount of reactive power for its excitation. However, these disadvantages are offset in many applications by the simplicity of construction, owing to the lack of slip rings, brushes, and dc field winding, which makes for practically maintenance-free operation.

## 8-2 HYSTERESIS MOTOR

The hysteresis motor starts by virtue of hysteresis losses induced in the rotor which in its simplest form is a ring of permanent-magnet material without teeth or polar projections. This motor operates at synchronous speed as a result of the retentivity of the rotor-core material. The stator windings are usually of the single-value capacitor type and practically achieve two-phase operation. Under ideal conditions (i.e., no eddy-current losses in the rotor and the absence of flux pulsations that might produce reentrant hysteresis loops), the rotating stator field produces hysteresis losses in the permanent-magnet rotor material which are directly proportional to the rotor frequency, according to Eq. 2-33. On that basis the rotor losses are expressed by

$$P_{\text{loss}} = Ksf \tag{8-17}$$

where $K$ is a constant taking into account the constant flux-density distribution and the volume of the magnetic material, $s$ is the slip, and $f$ the stator frequency. Equation 5-11 shows the slip to equal the ratio of the rotor losses to the rotor input, from which it follows that the power input to the rotor is

$$P_{\text{in}} = \frac{P_{\text{loss}}}{s} = Kf \tag{8-18}$$

which is constant for frequency at values of $s \neq 0$. The torque developed by a polyphase induction motor is shown by Eq. 5-49 to equal the ratio of rotor power input to the synchronous angular velocity and is constant in the idealized hysteresis motor when coming up to speed.† The hysteresis torque in conventional induction motors is a negligible part of the total developed torque.

The torque developed by the hysteresis motor at synchronous speed is a function of δ, the angle between the magnetic axes of the stator and the rotor. Figure 8-3(a) shows an elementary two-pole hysteresis motor with the stator magnetic axis passing through stator poles $N_s$ and $S_s$ produced by the stator windings, and with the rotor magnetic axis passing through rotor poles $N_r$ and $S_r$ produced by the retentivity of the rotor-core material. An idealized speed-torque curve of a hysteresis motor is shown in Fig. 8-3(b). Some hysteresis motors depend for their starting torque on shading coils in the stator, as in Fig. 5-22. Because of the relatively small effect of the shading coils, the rotating field undergoes sizable fluctuations, particularly at large values of slip, and the rotor hysteresis losses are too complex to be expressed by the simple relationship in Fig. 8-17. The starting torque is much lower for the shaded-pole motor than for the capacitor motor operating practically as a two-phase machine.

---

† For a more detailed analysis of the hysteresis motor see B. R. Teare, Jr., "Theory of Hysteresis Motor Torque," *Trans. AIEE* 59 (1940): 907; see also H. C. Roters, "The Hysteresis Motor," *Trans. AIEE* 66 (1947): 1419–1430.

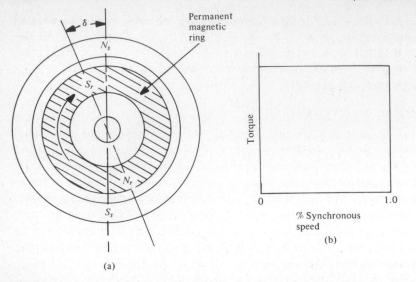

Permanent
magnetic
ring

% Synchronous
speed

(b)

(a)

**Figure 8-3** (a) Simplified representation of a two-pole hysteresis motor. (b) Idealized speed-torque characteristic.

Hysteresis motors have the advantage of quiet operation, a result of the smooth rotor periphery. In the capacitor type there are the added advantages of high-starting, accelerating, and pull-in torques which enable this motor to bring high-inertia loads up to synchronous speed. However, hysteresis motors are limited to small ratings, generally only a few watts, although some commercial ratings are as high as $\frac{1}{7}$ hp. Because of parasitic losses at synchronous speed, such as eddy currents induced by flux pulsations, the torque is lower than the assumed ideal, with a corresponding lowered efficiency. In addition, hysteresis motors require a relatively large magnetizing current to produce the tangential component of flux proportional to sin $\delta$, which is at right angles to the magnetic axis of the rotor. The permanent-magnetic materials used in hysteresis motors have a low value of permeability normal to the direction of magnetization. This material also limits the flux densities to values that are appreciably lower than those in induction motors. As a result of these factors, the hysteresis motor has only from about $\frac{1}{10}$ to $\frac{1}{4}$ the output of an induction motor of the same physical size.

## 8-3 INDUCTOR ALTERNATOR

The *inductor alternator* is a synchronous generator in which the field and armature windings are both stationary and which depends for its operation on a periodic variation in the reluctance of the air gap. This feature makes it possible to operate at high speeds and to generate correspondingly high frequencies, normally from several hundred hertz to 100,000 Hz, for use in such applications as steel and nonferrous melting by induction with high-frequency currents and also for supplying power at radio frequency. The flux through an armature coil of an inductor alternator, in simplest form, is unidirectional and instead of periodically reversing its direction fluctuates between the values $\phi_d$ and $\phi_q$, the design of the air gap being such that the induced voltage is practically sinusoidal. Inductor alternators fall into two general classifications: *homopolar* and *heteropolar*.

### 8-3.1 Homopolar Type

The homopolar-type inductor alternator shown in Fig. 8-4 has two laminated stator cores and two laminated rotor cores. The stator cores are slotted to carry the ac output winding. The field winding consists of a coil that is wound concentric with the axis of the machine and which produces the unidirectional flux $\phi$. The rotor cores are notched with open slots to produce the desired waveform of flux-density distribution. With the rotor in the position as in Fig. 8-4(b), the flux linking the stator coils, at no load, is at its minimum value $\phi_q$. When the rotor is displaced from that position by a distance equal to one-half that between the centers of adjacent teeth, the flux linking the stator coils is at its maximum value $\phi_d$. Then, for an ideal waveform, the flux which links a full-pitch armature coil (i.e., one that spans a rotor tooth), if the coil sides are considered as filaments, may be expressed by

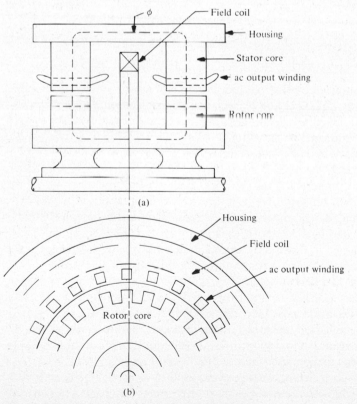

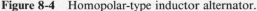

**Figure 8-4**   Homopolar-type inductor alternator.

$$\phi = \frac{\phi_d + \phi_q}{2} + \frac{\phi_d - \phi_q}{2} \cos 2\sigma \qquad (8\text{-}19)$$

where $\sigma = P\sigma_m$, $\sigma_m$ is the mechanical angular displacement between an arbitrary reference stator tooth and rotor tooth, and $P$ is the number of rotor teeth.

One cycle of voltage is generated as the rotor advances through a distance equaling that between the centers of two adjacent rotor teeth and the frequency is therefore

$$f = \frac{Pn_{\text{syn}}}{60} \qquad (8\text{-}20)$$

being twice that of the conventional synchronous machines in Chap. 5, as shown by comparing Eq. 8-20 with Eq. 4-1a, which is an advantage for generating high frequency. Equation 8-19 shows the amplitude of the ac flux component to be $(\phi_d - \phi_q)/2$, which corresponds to the flux per pole $\phi$ in Eq. 4-15 for the synchronous generators used in conventional power-frequency systems. A comparison of these equations shows the voltage induced in one phase of a full-pitch armature winding of an inductor alternator to be

$$E_{ph} = \frac{2.22 f N_{ph}(\phi_d - \phi_q)}{a} \qquad (8\text{-}21)$$

Figure 8-5 shows the principal components of a four-section homopolar inductor alternator, which may be regarded as the equivalent of a combination of two two-section machines such as is shown in Fig. 8-4. The flux paths of such an alternator are indicated in the cross-sectional view of Fig. 8-6. One of the two narrower sections and one-half of the wider middle section of stator core in Fig. 8-5(a) is, in effect, the equivalent of the two stator cores in Fig. 8-4, with a similar equivalence concerning one shorter outer section and one-half of the wider inner section of the rotor in Fig. 8-5(b) to two rotor cores in Fig. 8-4. The armature coil in Fig. 8-5(b) may be looked upon as four series-connected coils such as are associated one each with each stator core in Fig. 8-4(a). The field-coil assembly is shown in Fig. 8-5(d), one such assembly being sufficient for the two-section alternator while two are required for the four-section machine. The field coil is wound on a bobbin generally made of copper, which provides some damping. The holes in the bobbin permit the flow of cooling fluid and are also used to bring out the field leads. The armature winding passes through the large inner opening of the bobbin, near the inside diameter. Figure 8-6(b) shows the flux path in a two-section homopolar inductor alternator of the same basic construction as the four-section alternator in Fig. 8-5.

Typical data for three-phase wye-connected homopolar alternators illustrated in Figs. 8-5 and 8-6 are listed in Table 8-1. Although Table 8-1 does not show voltage ratings, typical values are 240/416 V.

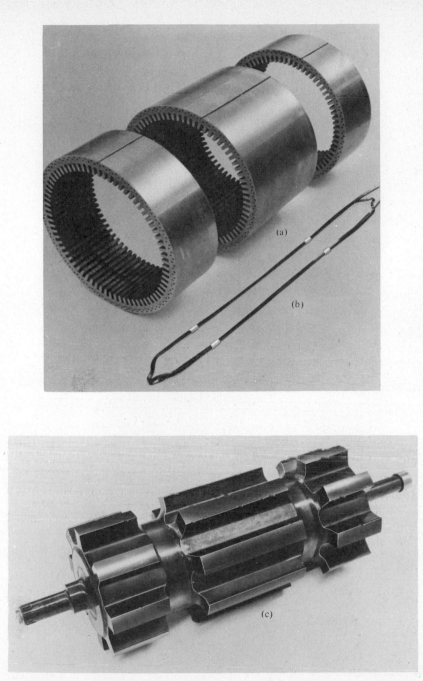

**Figure 8-5** Principal components of a four-section homopolar inductor alternator. (a) Stator cores. (b) Armature coil. (c) Field core (rotor). (d) Field coil. (U.S. Army photographs.) (Continued on page 44.)

(d)

**Figure 8-5**  (Continued)

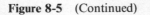

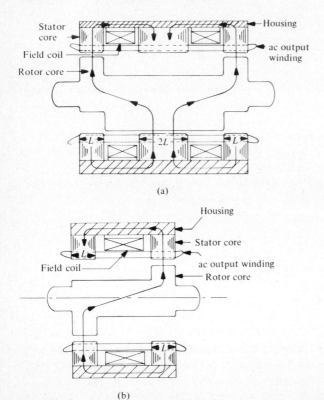

(a)

(b)

**Figure 8-6**  Approximate flux path in homopolar inductor alternators. (a) Four-section alternator. (b) Two-section alternator. (*Source:* Department of the Army.)

**TABLE 8-1**   DATA ON HOMOPOLAR INDUCTOR ALTERNATORS†

| | Alternator 1 | Alternator 2 | Alternator 3 | Alternator 4 |
|---|---|---|---|---|
| kVA | 587 | 520 | 95 | 30 |
| Power factor, % | 95 | 80 | 80 | 50 |
| Frequency, Hz | 3,200 | 3,200 | 3,200 | 3,200 |
| Speed, rpm | 24,000 | 39,000 | 39,000 | 6,000 |
| Number of poles | 16 | 10 | 10 | 64 |
| Subtransient reactance per unit | | 0.50 | | |
| Synchronous reactance per unit | 0.87 | 1.20 | 1.67 | 0.24 |
| Leakage reactance per unit | 0.35 | 0.32 | 0.28 | |
| Efficiency, % | 90 | 90 | 87 | |
| Harmonic content, % | | | | |
|   Line-to-neutral, no load | 10.3 | 3.5 | 5.5 | 18.0 |
|     Rated load | | 5.5 | 8.3 | 21.0 |
|   Line-to-line, no load | 1.6 | 4.6 | 2.0 | 0.5 |
|     Rated load | | 6.8 | 1.1 | 0.5 |
| Length of stator stack, in | 4.0 | 3.06 | 2.4 | 2.5 |
| Number of stator stacks | 4 | 4 | 2 | 2 |
| Stator punching, o.d., in | 11.7 | 10.27 | 7.96 | 18.5 |
|   i.d., in | 9.5 | 7.65 | 6.0 | 15.0 |
| Single air gap, in | 0.125 | 0.12 | 0.08 | 0.06 |
| Number of stator slots | 72 | 60 | 60 | 108 |
| Housing length, in | 36.7 | 29.5 | 14.0 | |
| Diameter, in | 18.3 | 16.5 | 10.0 | |
| Weight, lb | 975 | 630 | 115 | |

† Courtesy of Department of the Army, U.S. Army Mobility Equipment Research and Development Center, Fort Belvoir, Va.

One important application of inductor alternators is that of supplying military power demands, including lighting, electric motors, and equipment that requires precise voltage and frequency regulation and control. They are suitable for direct connection to high-speed prime movers, such as gas turbines, the high speed making for relatively small size and weight.

## 8-3.2 Heteropolar Type

The basic features of the heteropolar-type alternator are illustrated in Fig. 8-7. The stator housing and the cores of the stator and the rotor are similar to those of the squirrel-cage induction motor. The stator punchings are notched to carry the ac output winding. The field winding is wound in several slots on the stator to produce a *multipolar* field as indicated by the circular dashed lines in Fig. 8-7(b), in contrast with the unipolar field produced by the field winding in the homopolar type. Owing to the variations in air-gap length, the field fluctuates in the heteropolar machine exactly as it does in the homopolar type.

The main operating difference between the two types is that the voltage response to changes in the field current is more rapid in the heteropolar machine

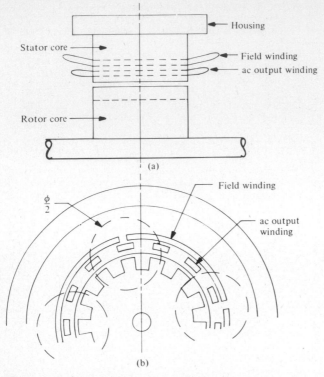

**Figure 8-7**    Heteropolar inductor alternator.

than is the case with the homopolar machine. The damping circuits which are built into some of these machines and their coinciding magnetic axes with the axis of the field windings are omitted from Figs. 8-4 and 8-7 for reasons of simplicity.†

## 8-4 STEP MOTORS‡

The *step motor* responds to a train of impulses to the stator in a manner such that the rotor advances in equal increments. Among their applications are the control of systems such as robots, machine tools, printers, tape drives, capstan drives, memory-access mechanisms of computer equipment, and power control apparatus. Their operation is stable in open-loop systems, with accurate position

---

† For a more complete treatment of inductor alternators, see J. H. Walker, "The Theory of the Inductor Alternator," *Proc. IEE* 89, Part II (June 1942): 227, and M. J. Marchbanks, "Coreless Induction Furnaces," *Proc. IEE* 93, Part II (1946): 520.

‡ See H. D. Chai, "A Mathematical Model for Single-Stack Step Motors," *IEEE Trans. Power Apparatus and Industry PAS-94* (September–October 1975): 1508–1516, and A. E. Snowdon and E. W. Madsen, "Characteristics of a Synchronous Induction Motor," *IEEE Trans. Applications and Industry 81,* Part II (1962): 1. Commercial data are presented in Catalog MD174-1 by Superior Electric Company, Bristol, Conn.

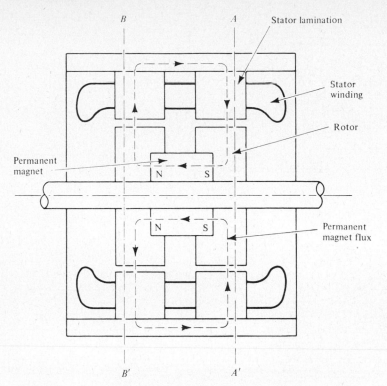

**Figure 8-8**   Axial view of a permanent-magnet step motor.

and speed control. Step motors are also used to operate as synchronous inductor motors.

### 8-4.1 Synchronous Inductor-Motor Operation

The stator winding is two-phase and the rotor carries a permanent magnet positioned axially, although a dc-excited rotor can also be used. There is also the variable-reluctance (VR) motor, in which the rotor has neither a permanent magnet nor dc excitation but receives its flux from the stator mmf. This discussion concerns mainly the permanent-magnet (PM) rotor motor.

The rotor and stator in both the PM and VR motors have numerous teeth which cause operation at low synchronous speed when the stator is excited at constant frequency and two-phase current. Figure 8-8† shows an axial view of one type of PM step motor.

Cross sections $A$–$A'$ and $B$–$B'$ are illustrated in Fig. 8-9. Although the stator has eight polar projections, it is that of a two-phase four-pole motor with each of the two phases wound on alternate projections. Two-phase operation from a

† Reprinted by permission from B. C. Kuo, *Step Motors,* Figs. 1-24, 1-26, 1-27, 1-29, and 1-31. Copyright © 1974 by West Publishing Company. All rights reserved.

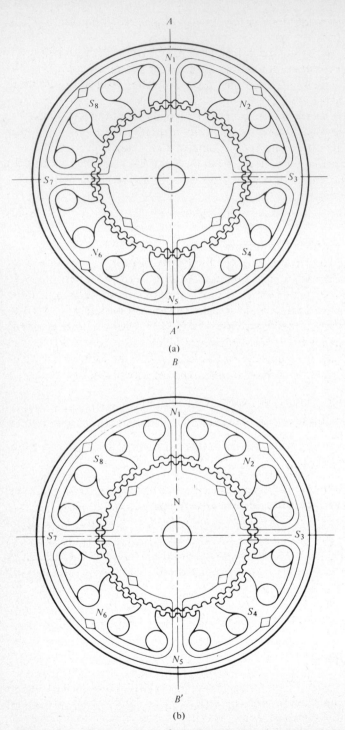

**Figure 8-9** Cross sections of permanent-magnet step motor perpendicular to the shaft. (a) Cross section $A$–$A'$. (b) Cross section $B$–$B'$. (See Fig. 8.8.)

**488**

single-phase source is made possible by the use of a resistor–capacitor phase-shifting network.

The motor illustrated in Figs. 8-8 and 8-9 has 50 rotor teeth and 40 stator teeth, although other tooth combinations for different speeds are possible. The stator tooth pitch is actually 48 since each polar projection is lacking one tooth to provide winding space.

Although the average torque at standstill is zero, the amplitudes of successive rotor oscillations increase on starting, resulting in rotation in about 1.5 cycles. This is due to the slow synchronous speed and relatively low rotor inertia.

The rotor advances one tooth pitch per cycle if it has one more tooth than the stator for each four polar projections, resulting in a speed of

$$n_{\text{syn}} = 60 \, \frac{f}{n_r} \text{ rpm} \tag{8-22}$$

where $n_r$ is the total number of rotor teeth.

In Fig. 8-9 one rotor tooth is under the center of pole $N_1$ and one rotor tooth is one-fourth of a tooth pitch from the center of $N_2$. This corresponds to the no-load rotor position when the ac flux in $N_1$ is a maximum. One cycle later, the ac flux in $N_2$ is a maximum and the rotor advances one-fourth of a tooth pitch with one tooth now under the center of $N_2$. Thus, the synchronous speed is 72 rpm for a frequency of 60 Hz, which also follows from Eq. 8-19. Speeds of 28.8, 72, and 200 rpm are common for 60-Hz motors, depending on the rotor and stator teeth combination. When deenergized, the motor stops in about 1.5 cycles.

Rotor sections $A$ and $B$ in Fig. 8-9 are displaced from each other by one-half tooth pitch, while the stator sections are in alignment with each other.

### 8-4.2 Stepper Operation

Stepping is accomplished by switching dc excitation to the stator in a four-step sequence. In the case of the 50-tooth rotor, each impulse changes the position of the rotor one-fourth of a rotor pitch (i.e., $360°/200 = 1.8°$). Figure 8-10 shows a schematic diagram of connections with the switching cycle for operation in both directions. Both phases are energized at standstill to produce a high value of holding torque. However, the torque decreases with increasing switching rate because the inductance prevents the currents in the windings from building up to their steady-state values. This effect can be reduced by increasing the time constant with the addition of an external resistor and increased voltage.

### 8-4.3 Bifilar Windings

The arrangement in Fig. 8-10 requires a three-wire supply to effect rotation in both directions. The same can be accomplished from the simpler two-wire source by using bifilar windings on the stator, making it unnecessary to reverse the current to the motor. Current flowing in the same direction is switched to an identical winding but wound in the opposite direction. Since each winding oc-

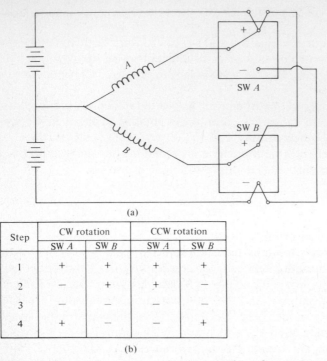

(a)

| Step | CW rotation | | CCW rotation | |
|------|:---:|:---:|:---:|:---:|
|      | SW $A$ | SW $B$ | SW $A$ | SW $B$ |
| 1    | + | + | + | + |
| 2    | − | + | + | − |
| 3    | − | − | − | − |
| 4    | + | − | − | + |

(b)

**Figure 8-10** (a) Connections of stator winding of a synchronous inductor motor as a permanent-magnet step motor. (b) Combination of switching sequences.

cupies one-half the available space, the flux, and consequently the torque and the time constant, are reduced. The switching arrangement for bifilar windings is shown in Fig. 8-11,† with the sequence indicated in Fig. 8-11(b). Both Figs. 8-10(a) and 8-11(a) represent the switching mechanism, which actually consists of a combination of solid-state and other components, by a simple single-pole switch for reasons of clarity.

## 8-5 CERAMIC PERMANENT-MAGNET MOTORS

Motors with permanent-magnet fields are smaller in size for a given power and speed rating than those with wound fields because the permanent magnet requires less space. This is particularly true of motors under 5 cm in diameter but not necessarily of motors over 12 cm in diameter. Alnico perhaps has been the most prominent material, but ceramic magnets, while having lower flux densities and greater temperature sensitivity, are less easily demagnetized. Although the air-gap flux densities in wound field motors are higher than in ceramic PM motors, which in some cases require a larger diameter, the overall diameter need not

† J. R. Ireland, *Ceramic Permanent-Magnet Motors* (New York: McGraw-Hill Book Company, 1968); Figs. 8-8, 8-9, 8-10, and 8-11 adapted from Figs. 13 (p. 26), 14 (p. 32), 16 (p. 35), and 27 (p. 77).

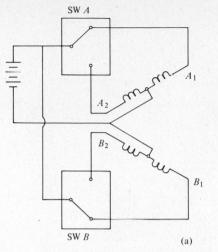

(a)

| Step | CW rotation | | CCW rotation | |
|------|------|------|------|------|
| | SW $A$ | SW $B$ | SW $A$ | SW $B$ |
| 1 | $A_1$ | $B_1$ | $A_1$ | $B_1$ |
| 2 | $A_2$ | $B_1$ | $A_1$ | $B_2$ |
| 3 | $A_2$ | $B_2$ | $A_2$ | $B_2$ |
| 4 | $A_1$ | $B_2$ | $A_2$ | $B_1$ |

(b)

**Figure 8-11** (a) Connections of a synchronous inductor motor with bifilar windings. (b) Combination of switching sequences.

necessarily be increased because of the smaller radial length of the permanent magnet compared with that of the wound field. The length of the air gap in typical ceramic PM motors is about 0.075 cm. Figure 8-12 shows the structural features and photos of a segment and rotor of a ceramic PM motor.

A typical intrinsic demagnetization curve of a ceramic magnet is shown in Fig. 8-13. The large values of $H$ and the relatively low values of $B$ result in a large departure of the intrinsic flux-density characteristic from the normal demagnetization curve so that the former is used for the design of ceramic permanent-magnet motors.

Since $H$ is negative for the demagnetization curve, the intrinsic flux density $B_{int}$ equals the sum $B + \mu_0 H$ instead of the difference. The duty on the permanent magnet is most severe at start and stop, and the $H$ coordinate is customarily chosen to intersect the demagnetization curve at $B_{int} = 0.8B_r$, as shown in Fig. 8-13.

## 8-5.1 Motor Characteristics

The motor is designed so that the demagnetizing effect of cross magnetization is small, particularly when operating from a dc (not rectified) line. The demag-

Rotor armature                                           Stater field poles

**Figure 8-12**   Two-pole motor with ceramic permanent-magnet poles. (a) Rotor armature. (b) Stater field poles.

netization is greatest at standstill (i.e., starting and stall), but even here it is not excessive. The speed and current-versus-torque characteristic is shown in Fig. 8-14.

When operating as a rectified ac motor, the peak current value, not the average value, determines the demagnetizing effect and causes the characteristics shown in Fig. 8-15 to depart from that in Fig. 8-14.

## 8-5.2  Applications

The automobile industry has been one of the largest users of permanent-magnet motors for such drives as heater blowers, windshield wipers, and power windows.

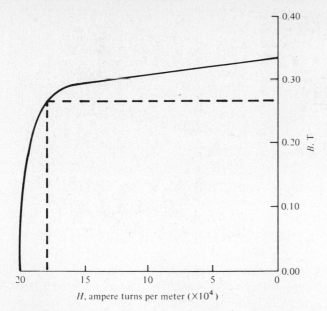

**Figure 8-13** Typical intrinsic magnetization curve of a ceramic permanent-magnet material.

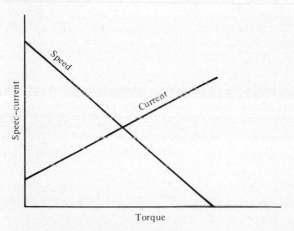

**Figure 8-14** Ceramic permanent-magnet-motor characteristic when supplied from a constant dc voltage source.

Other applications include rectified-ac and battery-operated carving knives, toothbrushes, electric shavers, power tools, and household appliances.

## 8-6 AC COMMUTATOR MOTORS

Ac commutator motors have armatures which are similar to those in dc machines; their field structures, however, are laminated to suppress eddy currents. The most common ac commutator machine is the single-phase series motor, although

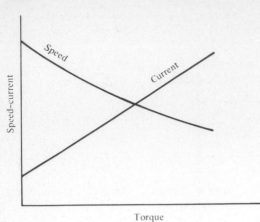

**Figure 8-15** Ceramic permanent-magnet motor when supplied from a rectified ac source.

there are some three-phase series motors and three-phase shunt motors. The ac commutator motor has two advantages over induction motors—a wide speed range and a high starting torque. In addition, series motors may be operated at several times induction-motor synchronous speed.

### 8-6.1 Single-Phase Series Motor

A dc series motor with a laminated field structure can be made to operate on ac; in fact, small series motors up to about $\frac{1}{2}$-hp rating, known as *universal motors,* are designed to operate on either dc or ac. Schematic diagrams for the straight series motor and compensated series motor are shown in Fig. 8-16. The compensating winding in the ac series motor is similar to that in the dc machines discussed in Chapter 6 and serves to overcome the armature mmf. Under steady-state operation the self-inductances of the field and armature circuits have negligible effect when the motor operates on direct current. However, the inductive reactance of the field and armature plays a prominent part when the motor operates on alternating current. Nevertheless, the instantaneous voltage generated in the armature is expressed by Eq. 6-15, where $\phi_d$ is the instantaneous flux in the direct axis produced by the field mmf, whether dc or ac. On that basis the generated voltage $E_a$ is in phase with $\phi_d$, which is in phase with the field current when the effect of the core losses, which are small, is neglected.

Then for a straight series motor as shown in Fig. 8-16(a), which has its brushes on the geometric neutral, the applied voltage is expressed by

$$\boxed{\mathbf{V} = (r_f + jx_f + r_a + jx_a)\mathbf{I} + \mathbf{E}_a} \tag{8-23}$$

which is in accordance with the phasor diagram in Fig. 8-17(a) and where $r_f$ and $r_a$ are the resistances and $x_f$ and $x_a$ the reactances of the field and armature and where $\mathbf{I}$ is the current in the motor. At starting $\mathbf{E}_a$ is zero, and it is evident from the phasor diagram in Fig. 8-17(a) that the angle of lag $\theta$ is quite large, producing a low power factor and correspondingly low starting torque. The compensating

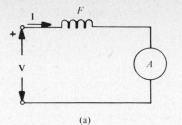

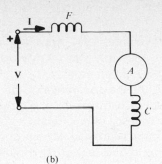

(b)

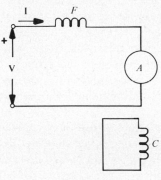

(c)

**Figure 8-16** Ac series motors. (a) Straight series motor. (b) and (c) Conductively and inductively compensated series motors.

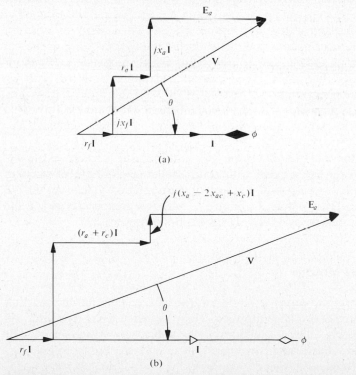

**Figure 8-17** Phasor diagrams. (a) Straight series motor. (b) Compensated series motor.

winding in the compensated series motor effects an improvement in the power factor not only on starting but on running as well, in that it reduces the reactance of the armature circuit from $x_a$ to $x_a - 2x_{ac} + x_c$, where $x_a$ and $x_c$ are the self-reactances of the armature and the compensating winding, and $x_{ac}$ is the mutual reactance between them. Further, $x_c \cong x_a$, and if the coupling were perfect, the reactance of the armature circuit would be nearly zero.

There is the additional advantage of increased starting current due to the lower reactance of the compensated motor. Two methods of compensation are possible. One is the same as that used in dc machines [i.e., conductive compensation in which the compensating winding is connected in series with the armature as shown in Fig. 8-16(b)]. In the other method the compensating winding is short-circuited, depending for its current on transformer action due to the magnetic coupling with the armature. The phasor diagrams for the compensated series motor are shown in Fig. 8-17(b). Conductive compensation has the advantage of operation on both ac and dc and of permitting a small amount of overcompensation to improve commutation.

The advantage of inductive compensation is that the compensating winding is not connected with the supply circuit and can be made for very low voltage, requiring less insulation and thus permitting better use of conductor material with a resulting economy of space. However, inductive compensation cannot be used in motors that are also to be operated on dc, nor can it achieve overcompensation. In the ac motor, currents are induced in the short-circuited armature coils undergoing commutation due to transformer coupling with the field winding. These short-circuit currents do not contribute to the torque but are in fact parasitic and may cause serious heating, particularly on starting. When the armature begins to rotate, these short-circuited turns are opened as they leave the brushes and replaced by other turns, which are momentarily short-circuited and opened. The result of these current interruptions is serious sparking at the brushes and a concentration of heat at the few segments under the brushes. However, as soon as the motor attains appreciable speed, the heating is distributed over all the segments, which, along with the decrease in current, alleviates this condition.

Compensation makes a smaller air gap possible, which results in a lower number of turns in the field winding with a correspondingly lower field reactance, as may be deduced from Eq. 4-46. Because of commutation difficulties the operation of large ac series motors is confined to such low frequencies as $16\frac{2}{3}$ and 25 Hz, and the motors are used mainly for traction purposes.

## 8-6.2 Universal Motors

The speed of a universal motor is somewhat lower for ac than for dc operation because of the reactance voltage drop, especially at heavy loads. However, on ac operation, the rms value of the flux may be appreciably lower than the dc value for the same value of applied voltage, which tends to raise the speed. This is due to the increased saturation of the iron at the peak of the flux wave on ac.

It is possible, however, to design small universal motors that exhibit about the same torque–speed characteristic on ac as on dc. Typical torque–speed characteristics of a universal motor are shown in Fig. 8-18.

Universal motors are generally used for small devices such as vacuum cleaners, food mixers, and portable tools operating at speeds from about 3000 to 11,000 rpm.

### 8-6.3 Repulsion Motor

The repulsion motor is, in effect, a series motor which has its stator and rotor coupled inductively instead of connected in series. The brushes on the commutator of the repulsion motor are short-circuited, as indicated in Fig. 8-19. The currents induced in the rotor by transformer action react on the stator flux, and if the brushes are shifted to the proper position, torque results. When the brushes are in the position shown in Fig. 8-19(b), there is rotor current but no torque, because the magnetic axes of the stator and rotor are in alignment. Also, when the brushes are in the position as in Fig. 8-19(c), there is no torque because the rotor current is zero. Accordingly, the brushes must occupy some position between those in Fig. 8-19(b) and (c) in order for the motor to develop torque. When the brushes are shifted through a position of zero torque, the direction of rotation reverses.

The commutation of the repulsion motor is superior to that of the series motor up to synchronous speed and inferior at higher speeds because of the heavier short-circuit currents in the coils undergoing commutation. These short-

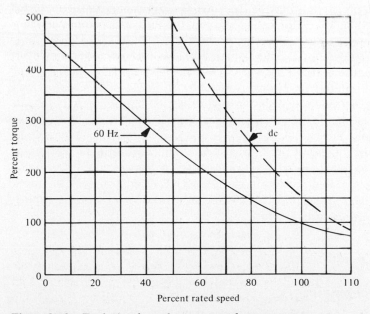

**Figure 8-18**  Typical universal motor speed-torque curves.

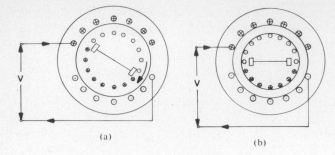

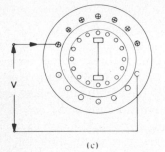

**Figure 8-19** Simplified diagram of repulsion motor. (a) Brushes in position to produce torque. (b) and (c) Brushes in zero-torque positions.

circuit currents act as a load and tend to limit the no-load speed. One of the advantages of the repulsion motor over the series motor is that the armature, not being in the supply circuit, may be designed for any convenient low voltage, thus reducing insulation requirements and making for better commutation.

The repulsion motor has a torque–speed characteristic similar to that of the series motor but with the additional advantage of permitting adjustments in speed by shifting the brushes. Prior to the widespread use of the capacitor-start induction motor, motors known as *repulsion-start, induction-run* motors were used in applications requiring high starting torque and practically constant running speed. These start as repulsion motors, and at about three-fourths rated speed a centrifugal device lifts the brushes and short-circuits the commutator segments, causing the motor to run as a straight single-phase induction motor. There is also the *repulsion-induction* motor, which has a squirrel cage in the bottom of the rotor slots, with the commutated winding occupying the top of the rotor slots. This arrangement makes use of the repulsion motor combined with the induction motor.

## 8-7 CONTROL MOTORS

### 8-7.1 AC Tachometer

A small two-phase machine may be used as a tachometer (i.e., as an instrument for measuring the angular velocity of a shaft). Such a machine is illustrated in

Fig. 8-20 in which the main winding is excited from a constant-voltage, constant-frequency source. The auxiliary winding usually feeds into an amplifier of very high input impedance so that the current in the auxiliary winding is negligible and the magnetic field, known as the *control field,* is produced by the main winding only. The amount of power absorbed from the shaft of which the speed is measured should be small, and for that reason the rotor must have small inertia and small weight. The rotor in many of these tachometers is therefore a thin metallic cup known as a *drag cup.*

The general expression for the voltage induced in the auxiliary winding of a two-phase induction machine is given by Eq. 5-81. However, in the case of the tachometer the current $I_a$ in the auxiliary winding is negligible and the output voltage based on Eqs. 5-80 and 5-81 is expressed by

$$\mathbf{V}_{\text{out}} = \frac{ja\mathbf{V}_m(Z_f - Z_b)}{z_1 + Z_f + Z_b} \tag{8-24}$$

Since $Z_f$ and $Z_b$ are functions of the slip, being equal at standstill (i.e., $s = 1$), and since $\mathbf{V}_m$, the voltage applied to the main winding, is constant, the output voltage is a function of the speed.

### 8-7.2 Two-Phase Control Motors

Two-phase squirrel-cage induction motors are used in some control systems, requiring such motors to deliver an output of from a fraction of a watt to several hundred watts. The two stator windings, phase $m$ and phase $a$, are identical and are displaced from each other by 90° in electrical measure. Figure 8-21 shows a schematic diagram of a two-phase control motor which has phase $m$, the *ref-*

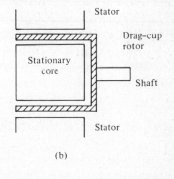

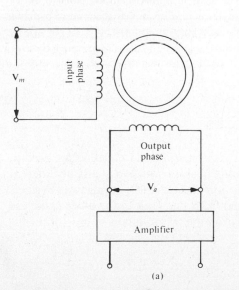

**Figure 8-20**    (a) Ac tachometer with a drag-cup rotor. (b) Cross section of a drag-cup rotor.

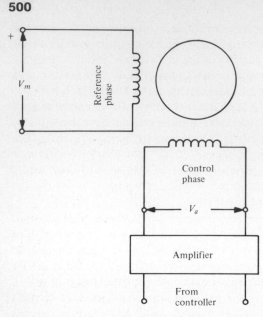

**Figure 8-21**   Schematic diagram of a two-phase control motor.

*erence phase,* connected to a constant-voltage, constant-frequency voltage source of $V_m$ volts. The *control phase,* phase *a,* is supplied with voltage $V_a$ from the controller, usually with an amplifier intervening. The error voltage $V_a$ is proportional to the required amount of correction and is of the same frequency as $V_m$ and displaced from $V_m$ by 90°.

To assure high torque near zero and to prevent the motor from running single-phase after $V_a$ has been reduced to zero, a high-resistance rotor is used. The great advantage of this motor over other types lies in the absence of brushes and sliding contacts as well as the simple and rugged construction of the rotor.

The steady-state performance of such a control motor is governed by Eqs. 5-80 and 5-81 with $Z_c = 0$ and $a = 1$. This is true whether or not $\mathbf{V}_a$ and $\mathbf{V}_m$ are 90° out of phase.

## 8-8  SELF-SYNCHRONOUS DEVICES

Self-synchronous devices known as *selsyns, synchros,* and *autosyns* are used in many systems to synchronize the angular positions of two shafts at different locations where a mechanical interconnection of the shafts is impractical. Systems in which high torque must be transmitted from one shaft to the other make use of three-phase selsyns, while those requiring the transmission of low torque incorporate single-phase selsyns.

### 8-8.1  Three-Phase Selsyns

The three-phase selsyn is comprised of two three-phase wound-rotor induction motors of integral horsepower size having power ratings which in some installations are comparable with that of the driver to be synchronized. One of the motors has its rotor connected to one shaft, and the rotor of the second motor is connected to the other shaft. The stator windings of the two selsyn motors are connected in parallel and energized from a three-phase source, while the rotors are also connected in parallel with each other as shown in Fig. 8-22.

Assume the selsyn motors to be identical and both shafts to rotate against the direction of the stator rotating fields, and the torque transmitted initially by the selsyn from one shaft to the other to be zero. Further, if $\omega_{syn}$ and $\omega$ are the angular velocities of the rotating field and of the rotor, both in electrical measure, then the voltages induced in a given phase of the two rotors are equal:

$$e_{rA} = e_{rB} = \sqrt{2}E_r \sin [(\omega_{syn} + \omega)t + \sigma_0] \qquad (8\text{-}25)$$

If the main drives have drooping speed-load characteristics, a decrease in the load on drive $A$ is accompanied by an increase in speed (i.e., an increase in $\omega$ causing the shaft of $A$ to rotate ahead of shaft $B$ by the electrical angle $\delta$, as represented in Fig. 8-23). As a result, the phase of the voltage induced in selsyn rotor $A$ is advanced by the angle $\delta$ with respect to the voltage induced in selsyn rotor $B$, thus causing a current to circulate between the two rotors with $A$ acting as a generator and $B$ as a motor. This effect is the same as that of advancing the stator voltage phasor of $A$ by an angle $\delta/2$ in electrical measure and retarding that of $B$ by the same angle while the angular displacement between shafts remains at zero. The stator voltage for this condition then would be

$$\mathbf{V}'_A = \mathbf{V}'\left(\cos \frac{\delta}{2} + j \sin \frac{\delta}{2}\right) = \mathbf{V}' \cos \frac{\delta}{2} + j \mathbf{V}' \sin \frac{\delta}{2} \qquad (8\text{-}26)$$

$$\mathbf{V}'_B = \mathbf{V}'\left(\cos \frac{\delta}{2} - j \sin \frac{\delta}{2}\right) = \mathbf{V}' \cos \frac{\delta}{2} - j \mathbf{V}' \sin \frac{\delta}{2} \qquad (8\text{-}27)$$

where

$$\mathbf{V}' = \left(1 - \frac{x_1}{x_M}\right)\mathbf{V}$$

as in Eq. 5-19.

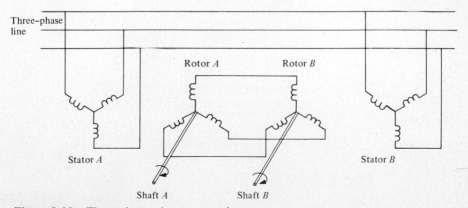

**Figure 8-22**  Three-phase selsyn connections.

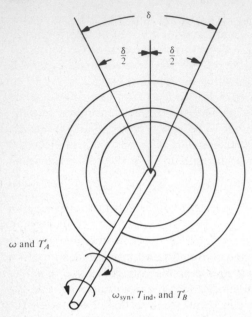

**Figure 8-23**   Two-pole representation of displacement between shafts in a three-phase selsyn.

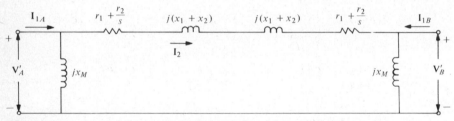

**Figure 8-24**   Approximate equivalent circuit for a three-phase selsyn.

Figure 8-24 shows the voltages $\mathbf{V}'_A$ and $\mathbf{V}'_B$ applied to the approximate equivalent circuit of the selsyn. Let $z = r_1 + r_2/s + j(x_1 + x_2)$ in Fig. 8-24. Then the rotor current referred to the selsyn stators is

$$\mathbf{I}_2 = \frac{\mathbf{V}'_A - \mathbf{V}'_B}{2z} = j\,\frac{\mathbf{V}'\sin(\delta/2)}{z} = j\,\frac{\mathbf{V}'z^*\sin(\delta/2)}{|z|^2} \tag{8-28}$$

where $z^*$ is the conjugate of the complex impedance $z$ and $|z|$ is the magnitude of $z$. The power transferred across the air gap to the rotor of $A$ for all three phases is

$$P_{gA} = 3\,\mathrm{Re}\,\mathbf{E}_{2A}\,\mathbf{I}_2^* \tag{8-29}$$

where

$$\mathbf{E}_{2A} = \mathbf{V}'_A - r_1\mathbf{I}_2 \tag{8-30}$$

and when Eq. 8-28 is substituted in Eq. 8-30 there results

$$\mathbf{E}_{2A} = \mathbf{V}'\left\{\cos\frac{\delta}{2} + j\left[\sin\frac{\delta}{2} - \frac{r_1 z^* \sin(\delta/2)}{|z|^2}\right]\right\} \tag{8-31}$$

$\mathbf{I}_2^*$ is the conjugate of $\mathbf{I}_2$. Then, from Eq. 8-28,

$$\mathbf{I}_2^* = -\frac{j\mathbf{V}^{*\prime}z \sin(\delta/2)}{|z|^2} \tag{8-32}$$

Substitution of Eqs. 8-31 and 8-32 in Eq. 8-29 yields

$$P_{gA} = 3 \text{ Re } \mathbf{V}'\mathbf{V}^{*\prime}\left[\cos\frac{\delta}{2} + j\left(\sin\frac{\delta}{2} - \frac{r_1 z^* \sin(\delta/2)}{|z|^2}\right)\right] \times \left[-j\frac{z \sin(\delta/2)}{|z|^2}\right] \tag{8-33}$$

The product of a complex quantity and its conjugate is a real number equal to the magnitude of the complex quantity squared. Hence, $\mathbf{V}'\mathbf{V}^{*\prime} = V'^2$ and $zz^* = |z|^2$, and when these relationships are applied to Eq. 8-33, the power input to the rotor is found, after some algebraic manipulations, to be

$$P_{gA} = \frac{3V'^2}{|z|^2}\left[\frac{(x_1 + x_2)\sin\delta}{2} + \frac{r_2}{s}\sin^2\frac{\delta}{2}\right] \tag{8-34}$$

The torque is in a direction such as to reduce $\delta$ and is obtained by dividing Eq. 8-34 by the mechanical synchronous angular velocity, which results in

$$\boxed{T_{gA} = \frac{3V'^2[\frac{1}{2}(x_1 + x_2)\sin\delta + (r_2/s)\sin^2(\delta/2)]}{(2\pi n_{syn}/60)[(r_1 + r_2/s)^2 + (x_1 + x_2)^2]}} \tag{8-35}$$

A similar process as in the foregoing when applied to selsyn motor $B$ results in the following expression for its torque in a direction as to reduce $\delta$:

$$\boxed{T_{gB} = \frac{3V'^2[\frac{1}{2}(x_1 + x_2)\sin\delta - (r_2/s)\sin^2(\delta/2)]}{(2\pi n_{syn}/60)[(r_1 + r_2/s)^2 + (x_1 + x_2)^2]}} \tag{8-36}$$

The maximum torque developed for a given value of $s$ by the selsyn motors occurs at a value of $\delta$ for $A$ such that

$$\tan\delta_A = -\frac{s(x_1 + x_2)}{r_2} \tag{8-37}$$

and for $B$ such that

$$\tan\delta_B = +\frac{s(x_1 + x_2)}{r_2} \tag{8-38}$$

Equations 8-37 and 8-38 show $\delta$ to be small for low values of slip with correspondingly small torque capability. For that reason the shafts are driven opposite the direction of the rotating field whenever drives rotate in only one direction.

It is left as an exercise to verify Eqs. 8-37 and 8-38. Figure 8-25 shows the torque-angle characteristics at standstill based on Eqs. 8-35 and 8-36 for a selsyn in which both motors are similar to the 15-hp motor of Example 5-1.

## 8-8.2 Single-Phase Selsyns

Single-phase selsyns are of much smaller size than three-phase selsyns and are capable of transmitting only values of torque in ratings of the order of $7 \times 10^{-4}$ to $14 \times 10^{-3}$ N-m/rad. They are suitable for indicating at some remote distance the positions of some devices, for example, elevators, hoists, generator rheostats, and valves. Another use is in the operation of servomechanisms for controlling the position or motion of larger equipment.

The *selsyn transmitter,* or *generator,* as well as the *selsyn receiver,* or *motor,* has a single-phase winding, usually on the rotor, both connected to the same ac source as shown in Fig. 8-26. The other member of each, usually the stator, has a three-phase **Y**-connected winding, and the two windings are connected in parallel. The rotors are of salient-pole construction, which makes for a simple winding

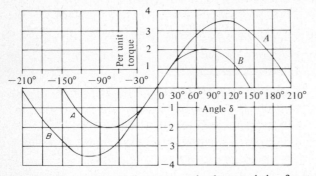

**Figure 8-25**   Zero-speed torque-angle characteristic of system using motors in Example 5.1 for a three-phase selsyn.

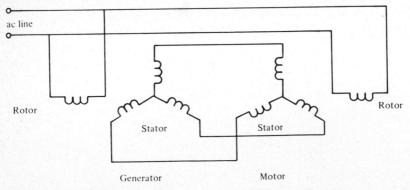

**Figure 8-26**   Single-phase selsyn.

arrangement and introduces a component of torque sometimes called the *reluctance torque*, similar to that in salient-pole synchronous machines. The receiver is generally provided with mechanical damping to reduce oscillations. At standstill and at low angular shaft velocities, the equal and oppositely rotating mmfs produced by the single-phase motor windings produce two components of torque in the generator and in the motor of somewhat the same general nature, as indicated by Eqs. 8-35 and 8-36 for the three-phase selsyn in which the term $\frac{1}{2}(x_1 + x_2) \sin \delta$ is associated with a torque component similar to that in a synchronous machine. From Eqs. 8-35 and 8-36 it is evident that the direction of this component of torque, both in the generator and in the motor, is such as to decrease $\delta$. On that basis, this component resulting from the backward-rotating field in the generator or in the motor is equal to that due to the forward-rotating field and in the same direction. However, the term $r_2/s \sin^2 (\delta/2)$ is associated with a component of torque corresponding to that in a polyphase induction motor, and that produced by the backward-rotating rotor field cancels that due to the forward-rotating field of the rotor at standstill. This is also practically true for low shaft speeds. Therefore, were it not for the effect of rotor saliency, the torque of both the transmitter and the receiver would tend to reduce $\delta$ in proportion to $\sin \delta$.

The addition of a third selsyn, known as a *differential selsyn* (as in Fig. 8-27), produces the rotation of its shaft as a function of the sum or difference of the rotation of two other shafts. The differential selsyn has a cylindrical rotor with a three-phase winding on the rotor and on the stator.

### 8-8.3 Synchro Control Transformers

Small selsyns can be used to supply an error voltage, resulting from angular displacement between two shafts, to a corrective device—for example, a two-phase control motor shown in Fig. 8-21, which has a much greater torque capability than the selsyn system. Such an arrangement is shown in Fig. 8-28.

### 8-9 ACYCLIC MACHINES

The voltage induced in the armatures of machines studied so far in this text are alternating, whether the machine is ac or dc. The distinguishing feature of the

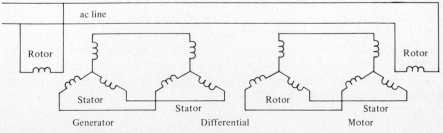

**Figure 8-27**  Selsyn system with differential.

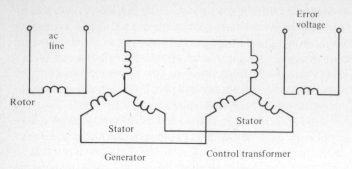

**Figure 8-28**   Selsyn control transformer system.

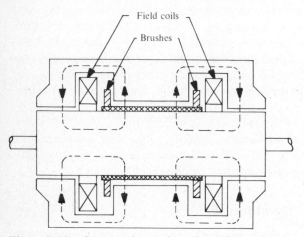

**Figure 8-29**   Cross-sectional view of acyclic generator.

*acyclic machine* is that the voltage induced in its armature is unidirectional, the terminal voltage therefore being dc without the use of a commutator or other rectifying arrangement.

## 8-9.1 Acyclic Generators

A cross-sectional view of an acyclic generator is shown in Fig. 8-29. This generator has two field coils carrying direct current which produce magnetic flux somewhat along the paths indicated by the dashed lines in Fig. 8-29. The armature is shown as a conducting sleeve mounted on the rotor core and rotating with it. The armature may be regarded either as a single conductor or as a large number of conductors of infinitesimal width connected in parallel, and the generated voltage is accordingly expressed on the basis of Eq. 5-24 and the fact that the tangential velocity, the direction of the flux, and the orientation of the armature conductor are mutually perpendicular by

$$E = Blu$$

where $u$ is the tangential velocity, $l$ the effective axial length of the armature, and $B$ the flux density assumed to be uniform throughout the air gap.

Although acyclic generators have been built with multiconductor armatures, they generally have single-conductor armatures and their application is pretty well restricted to loads requiring heavy current at low voltage. To collect these high values of current it is necessary to use a large number of brushes, which in combination with the high peripheral speeds required for the generation of even moderate values of voltage, makes for high brush friction loss, with the attendant reduced efficiency and the problem of heat dissipation. In recent years, however, the problem of brush friction has been ameliorated by making use of a liquid metal conductor, an eutectic alloy of sodium and potassium (NaK) between the rotating and stationary parts of the current collector. A cutaway view of a 10,000-kW 67-V 150,000-A 3600-rpm acyclic generator is shown in Fig. 8-30. The rotor is solid iron, a construction which withstands the centrifugal forces occasioned by the high rotational speed and which makes the rotor a single-conductor armature. A typical acyclic-generator arrangement is one in which a 60,000-kW steam turbine drives six generators directly connected to the turbine shaft to power a 150,000-A 400-V aluminum pot line. Each generator is rated at 67 V and 150,000 A.

### 8-9.2 Linear Acyclic Machines—Conduction Pumps

Linear machines depend for their operation on linear instead of rotary motion. An example of a linear generator is the hydromagnetic flowmeter illustrated in Fig. 8-31(a). It operates on the principle that a voltage is generated in the conducting liquid passing through a magnetic field $B$ which might be due to an electromagnet or a permanent magnet. The velocity $u$ of the liquid is related to the induced voltage by Eq. 5-24:

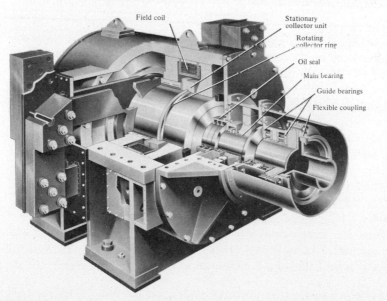

Field coil

Stationary collector unit

Rotating collector ring

Oil seal

Main bearing

Guide bearings

Flexible coupling

**Figure 8-30**  Cutaway view of a 10,000-kW 67-V 150,000-A 3600-rpm acyclic generator. (Courtesy General Electric Company.)

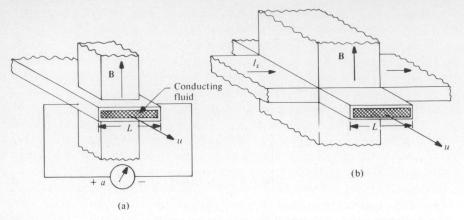

**Figure 8-31** Schematic diagrams. (a) Hydrodynamic flowmeter. (b) Liquid-metal conduction pump.

$$\mathbf{e} = \mathbf{L}(\mathbf{u} \times \mathbf{B}) \tag{8-39}$$

The advantage of this type of flowmeter is mainly its simplicity, as there are no moving parts within the liquid which might be corrosive. The reverse principle (i.e., that of motor action) is utilized in pumping liquid metals such as sodium, sodium potassium, and bismuth used in the operation of nuclear reactors. Figure 8-31(b) shows a schematic diagram of an electromagnetic pump,† known as a *conduction pump,* in which the current $I_s$ is passed through the tube walls and the liquid metal contained therein and at right angles to the magnetic field $B$. Conduction pumps may be dc or ac. In either case the force on a small length of current path **l** through which flows a small current $i$ is expressed by Eq. 6-1

$$\mathbf{f} = i(\mathbf{l} \times \mathbf{B}) \tag{8-40}$$

It is somewhat difficult, even in the case of the dc pump, to relate Eq. 8-40 to the total force on the liquid in terms of the total electrode current $I_s$, part of which flows through the tube walls that enclose the liquid. In addition, part of the current in the liquid flows through the relatively field-free regions near the inlet and the outlet of the pump. Armature reaction introduces a further complication, which can, however, be minimized by a compensating winding operating on somewhat the same principle as that in the rotating machine. The effects of the wall and end currents can be represented in the circuit in Fig. 8-32 in which $I_s$ is the total applied current, $I$ the current in the liquid, $R$ the resistance of the liquid metal within the pole region, $R_t$ the resistance between the electrodes of the tube wall, and $R_0$ the resistance of the liquid metal outside the pole region.

While there is a strong similarity between the ac and the dc pump, the action of the ac pump is complicated further by the effects of eddy currents in the liquid metal, tube walls, and compensating windings. As a result, the ac

---

† For an extensive discussion of electromagnetic pumps, see L. R. Blake, "Conduction and Induction Pumps for Liquid Metals," *Proc. IEE* 104A (1957): 49.

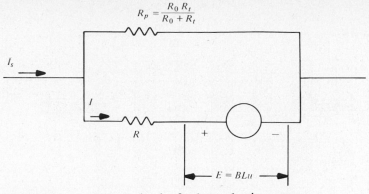

**Figure 8-32**   Equivalent circuit of a dc conduction pump.

conduction pump is restricted to small power applications, since eddy-current losses increase rapidly with size—which in a sense is the equivalent of increasing not only the volume in Eq. 2-41 for eddy-current loss but also the thickness $\tau$. The ac conduction pump also has a low power factor and is usually appreciably larger than a dc pump of similar output, but it has the advantage of getting around the dc supply difficulties when thousands of amperes are required at a 1-V level. A suitable source for the dc pump is the acyclic generator because of its high-current output capability at low voltage. In the case of the ac pump, the proper time-phase relationship between **B** and **I** must be maintained for effective operation.

## 8-9.3 Induction Pumps

The ac and dc conduction pumps require high current and large bus sections— two disadvantages that can be avoided by inducing current in the metal as in the rotor of an induction motor. However, with a linear flow of metal the pump then operates as a linear induction motor, as illustrated in Fig. 8-33. The magnetic field is produced by a polyphase (usually three-phase) winding and is shown traveling from left to right, thus inducing currents in the liquid metal which in their reaction on the traveling field produce forces in the liquid, causing it to be propelled from left to right also. The copper side bars in Fig. 8-33(b) perform the same function as the end rings in the squirrel-cage rotor.

   If the pump were of infinite length, it would operate as an induction motor, and the mmf at a long distance $x$ from the inlet would be

$$\mathcal{F}_{\text{amp}} \cos (\omega t - \psi)$$

where $\psi = 2\pi x/\lambda$, in which $\lambda$ is the wavelength of the mmf. Just as in the rotary machine, one pair of poles correspond to one wavelength of mmf or flux density. The linear velocity of the mmf is $u_{\text{syn}} = \lambda f$, where $f$ is the frequency in hertz.

   If the velocity of the liquid is $u$, the slip is

$$s = \frac{u_{\text{syn}} - u}{u_{\text{syn}}} \tag{8-41}$$

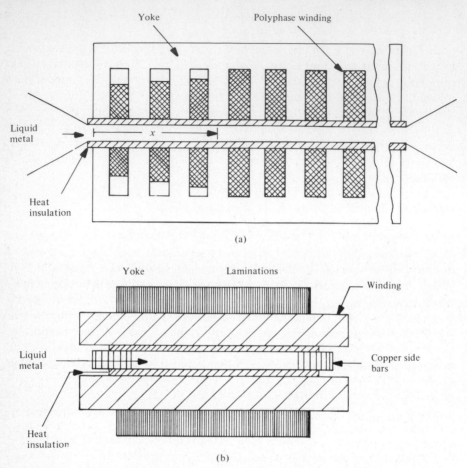

**Figure 8-33**   Linear induction pump. (a) Longitudinal section. (b) Transverse section.

The length of the pump, however, is finite and the mmf at the inlet and output is practically zero. These two discontinuities modify the form of the mmf to

$$\mathcal{F} = \mathcal{F}_{amp}[\cos (\omega t - \psi) - \cos \omega t] \tag{8-42}$$

which satisfies the condition that $\mathcal{F} = 0$ at the inlet where $\psi = 0$ and at the outlet, since $P$ is an even number of poles and here $\psi = P\pi$. The component $\mathcal{F}_{amp} \cos \omega t$ produces flux pulsations, giving rise to parasitic rotor losses which result in a corresponding decrease in efficiency. Methods of grading the flux so as to reduce the flux pulsations are discussed elsewhere.† As a result of these modifications, analytical treatment of the linear induction motor is more complex

† *Ibid.*

than that of the conventional rotary induction motor and is therefore not included in this text.†

Electromagnetic pumps must have large air gaps in the magnetic circuit to accommodate sufficient height of channel not only to avoid high resistance of the liquid but also to make allowance for the thickness of the channel and, in addition, for thermal insulation if the temperature of the liquid is excessive. This means large power requirements for the production of flux. In ac pumps this calls for large amounts of reactive power. Nevertheless, these pumps are practically indispensable for the applications mentioned, since conventional pumps with their maintenance problems and possibilities of leaks, although slight, are not suitable.

## 8-10 MAGNETOHYDRODYNAMIC GENERATORS

*Magnetohydrodynamic* (MHD) *generation* seems a promising means for converting heat to electrical energy at higher temperatures than are feasible with present-day steam systems. Of major interest are land-based applications involving large systems, where costs and high efficiency are of prime importance, and space applications, where weight is of major consideration and efficiency relatively unimportant—which makes it practical to reject heat at high temperature. However, although much effort has been expended toward the development of MHD generators in the United States and abroad since the early 1960s, such machines have not yet found use in the commercial generation of electric power or in space applications. This is largely due to the difficulty of adapting materials for operation at the high temperatures required for this method of energy conversion.

MHD generators depend for their operation on the motion of a fluid in a magnetic field on a principle comparable with that underlying the operation of the hydromagnetic flowmeter in Fig. 8-31(a). However, at present MHD electric power generation appears to depend on the use of hot gases, which are ionized through a combination of high temperature and "seeding" with more readily ionizable metals such as potassium or cesium. Ionization of a gas is a process whereby electrons acquire sufficient energy through thermal agitation to escape from their molecules and become free electrons. The electrons are charged negatively, and the electron-deficient molecules become positive ions. In addition, molecules may become separated into positive ions and negative ions. The degree of ionization largely governs the electrical conductivity of the gas because the current depends on the available number of electrons and ions.

Ionization of gases without seeding requires temperatures that are destructive of all known materials even for short periods of operation. Seeding makes it possible to operate at temperatures of 2500 to 3300 K.

---

† Linear induction motors are treated comprehensively by E. R. Laithwaite, *Induction Machines for Special Purposes* (London: George Newnes Ltd., 1966).

An elementary MHD generator is shown in Fig. 8-34 in which the ionized gas is seen entering the magnetic field at the velocity **u** at right angles to **B**. Gaseous conductors are known as *plasmas,* and MHD generators making use of plasmas are sometimes called *magnetoplasmadynamic* (MPD) *generators.* The forces developed on the charged particles and electrons by virtue of their motion in a magnetic field are in keeping with Eq. 6-1, which when modified to deal with a particle is of the form

$$\mathbf{f} = q\mathbf{u} \times \mathbf{B} \qquad (8\text{-}43)$$

where $q$ is a positive charge. Accordingly, positive ions are driven toward the upper electrode and electrons and negative ions toward the lower electrode, an action which ceases in a very short time if the generator is operating at no load. This is due to the force developed by the electric field in opposition to that developed on the particle by its motion through the magnetic field. The force on a charged particle due to the electric field is

$$\mathbf{f} = q\mathcal{E} \qquad (8\text{-}44)$$

where $\mathcal{E}$ is the electric field intensity in volts per meter.

Connecting a load to the generator results in a sustained current which consists almost entirely of a drift of electrons in a direction opposite to that of the current. Because of their relatively small mass, the response of the electrons to the forces produced by magnetic fields and electric fields is far greater than that of the much heavier positive ions. Therefore, the electrons need only be considered as charge carriers in dealing with MHD generators. The ionized gas, in passing through the generator, gives up some of its thermal energy with an

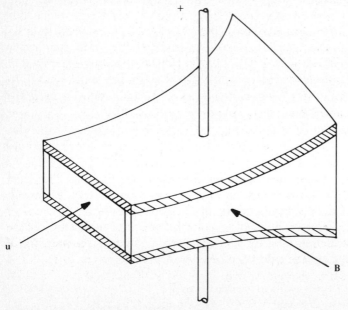

**Figure 8-34**    Elementary MHD generator.

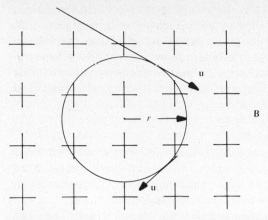

**Figure 8-35**   Trajectory of a constant-velocity electron in a uniform magnetic field directed into the page.

attendant change in pressure counteracted by the force of the current in its reaction on the magnetic field, resulting in a corresponding electrical energy output.

### 8-10.1  Hall Effect

The Hall effect, although negligible in the generation of electric energy in solid or liquid conductors, is appreciable when there is current in a plasma. It is due to the force on the electrons caused by their flow through the magnetic field from the positive electrode of the generator to its negative electrode in accordance with Eq. 8-43 which shows the force on the electron to be normal to the velocity **u**. Hence, in the simple case of an electron that is projected at a constant velocity **u** into a steady uniform magnetic field in free space and normal to direction of the field, its trajectory will be a circle, as shown in Fig. 8-35, as long as there is no electric field. It should be remembered that the charge on an electron is negative. If the velocity has a component parallel to the magnetic field, the trajectory is a helix and the radius† of the circle or the helix is

$$r = \frac{m\,u}{e\,B} \qquad (8\text{-}45)$$

where $m = 9.1 \times 10^{-31}$ kg (the mass of an electron) and $e = 1.6 \times 10^{-19}$ C, the negative charge on an electron and the ratio $m/e = 5.69 \times 10^{-12}$. However, when the motion of the electron is in a conducting medium such as a metal or a plasma, the electron suffers collisions largely with positive ions because of the coulomb force of attraction between opposite charges. Then in the presence of both an electric and magnetic field the trajectory of the electron is somewhat as shown in the simplified diagram in Fig. 8-35 with collisions at $c$, $c'$, and $c''$.

---

† A simple derivation of this relationship is shown in R. E. Lueg, *Basic Electronics* (New York: IEP, A Dun-Donnelley Publisher, 1963), pp. 3–5.

Because of the circular trajectory, the electron progresses in the direction of $u$ by the distance $b$, while the much heavier ions in the plasma travel practically in a straight line through the distance $a$. The electrons then have a component of motion corresponding to $b$–$a$, which results in a component of current known as the *Hall current* $I_H$ in Fig. 8-36 where $I$ is the current delivered to the load.

The shorter the travel of the electron between collisions the smaller is the distance $a$–$b$ and the smaller is the *Hall current.* This explains in a qualitative manner why the *Hall effect* is negligible in a metal conductor since the distance between molecules is much shorter than in a plasma. The electrodes provide a return path for $I_H$ in the same manner as the load does for the current $I$. However, because of their low resistance, the electrodes form a short circuit for $I_H$, giving rise to parasitic losses that may be of such magnitude as to severely reduce the output of an MHD generator using seeded combustion gases.

The Hall effect can be reduced by means of the segmented-electrode construction illustrated in Fig. 8-37 in which the electrode system consists of four pairs of electrodes insulated from each other by means of three insulating barriers. Figure 8-37(a) and (b) shows each pair of electrodes connected to a load and Fig. 8-37(c) the four pairs of electrodes shown connected in series to a common load. When the Hall current $I_H$ is larger than the transverse current, the arrangement shown in Fig. 8-37(d), known as the *Hall generator,* may be used.

## 8-10.2  MHD Steam Power Plants

An open-cycle system in which an MHD generator is placed ahead of a conventional steam-turbine generating system is shown schematically in Fig. 8-38.

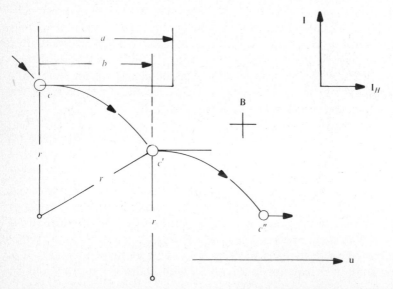

**Figure 8-36**  Electron moving in a conductor through a magnetic field. Collisions with ions at $c$, $c'$, and $c''$. Electric field assumed negligible. [Adapted from K. H. Spring, *Direct Generation of Electricity* (New York: Academic Press, 1965).]

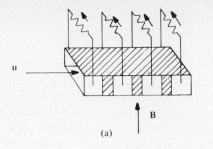

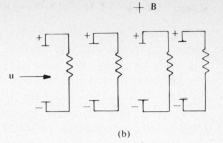

(a)

B

(b)

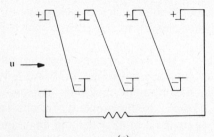

(c)

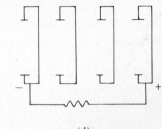

(d)

**Figure 8-37** Segmented-electrode MHD generator. (a) Simplified physical representation. (b) Schematic representation, segments loaded individually. (c) Segments in series with common load. (d) Hall generator.

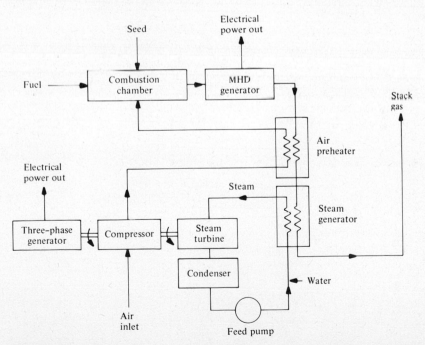

**Figure 8-38** Open-cycle system using MHD generator in topping cycle with steam turbine.

**515**

The inclusion of the MHD generator requires two supplementations—a provision for seeding the combustion products and a compressor, which in addition to the conventional electric generator is shown driven by the steam turbine. The compressor raises the pressure of the air (the air may also be enriched by the introduction of oxygen) to the input to the combustion chamber. After giving up part of their thermal energy for conversion into electrical energy, the seeded combustion products leave the MHD generator and pass through the air preheater and steam generator of the conventional steam-electric system, and are then exhausted through the stack into the atmosphere. For economy of operation it is necessary to recapture most of the seeding material, a provision not indicated in Fig. 8-38.

Figure 8-39 shows a schematic representation of a closed-cycle system in which a nuclear reactor supplies the thermal energy to the plasma which is recirculated, hence the term *closed cycle.*

Large-scale electric power is generated in the form of alternating current. Incorporation of the MHD generator into the schemes of Figs. 8-38 and 8-39 requires the use of an inverter and associated equipment, including switchgear and relays, to convert the dc output into ac. There is also the problem of electrodes for collecting the current, since the electrode material must withstand high temperatures, velocities, and pressures in combination, and in addition must be chemically inert. A good deal of effort is still required for the development of suitable materials to meet these requirements.

To keep the resistance of the plasma within practical limits without introducing excessive mechanical resistance to the axial flow of the plasma, the channel

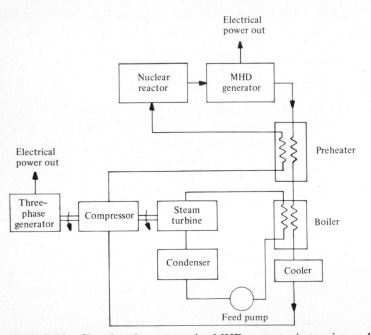

**Figure 8-39**   Closed-cycle system using MHD generator in topping cycle in combination with a nuclear reactor and steam turbine.

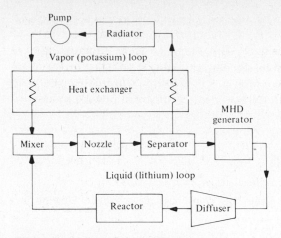

**Figure 8-40** Two-fluid MHD system.

height must be adequate. The field winding must also be insulated thermally from the hot plasma. This makes for a long air gap and a large field mmf with a correspondingly large field winding to which there is the alternative of using cryogenic superconducting field windings.

## 8-10.3 AC MHD

One way of getting around the problem of current collection is to use the induction generator principle. However, the resistivity of the plasma is too great to make for such an arrangement, and the long air gap would require excessive reactive power to produce the necessary flux densities.

A scheme† which has received some study makes use of a two fluid system represented schematically in Fig. 8-40, in which the expanding gas imparts without moving parts a velocity to a conducting liquid, which in turn passes at a speed above synchronous velocity through a magnetic field produced by a three-phase winding somewhat along the principle of the induction pump in Fig. 8-33 except that $u > u_{\text{syn}}$, which is necessary for generator action.

In Fig. 8-40 the potassium, which has a low boiling point, is condensed in the radiator and pumped through the heat exchanger into the mixer, where it combines with the hotter liquid lithium. Lithium has a high boiling point, becomes vaporized, and accelerates the mixture through the nozzle into the separator, where the vapor is removed and the liquid continues through the MHD generator. Although this arrangement appears attractive, there are difficulties to overcome in addition to that of the reactive-power requirement, which is of about the same magnitude as the real-power output of the generator. One of these is that separation of the liquid from the vapor is not complete, and the

† D. G. Elliot, "Two-Fluid Magnetohydrodynamic Cycle for Nuclear-Electric Power Conversion," *Amer. Rocket Soc.* 32 (1962): 924. Also W. D. Jackson et al. "In MHD," *Proceedings of the International Symposium on MHD Electrical Power Generation, Paris* (OECD, Paris, 1964), p. 1311.

liquid droplets that are separated tend to form a foam which is not an adequate working fluid.

Until the difficulties of developing suitable materials and of generating ac directly or finding an economical means for converting from dc to ac output are solved, the MHD generator is not a serious competitor with the conventional turbine-driven generator or hydrogenerator for commercial power production.

## STUDY QUESTIONS

1. What is a repulsion-start motor?
2. What purpose is served by a centrifugal switch in a repulsion-start motor?
3. How do the stators of repulsion-start and split-phase motors differ? How do they resemble each other?
4. Describe the rotor construction for a repulsion-start motor.
5. How would a repulsion-start motor be reversed?
6. Under what operating conditions are repulsion-start motors particularly desirable?
7. List practical applications of repulsion-start motors.
8. Make a distinction between a repulsion motor and a repulsion-start motor with regard to construction and operating performance.
9. What happens if the load is removed from a repulsion motor?
10. What factor determines the direction of rotation of a repulsion motor?
11. How may a repulsion motor be reversed?
12. List the methods that are used to control the speed of a repulsion motor.
13. Distinguish between the rotor constructions of repulsion and repulsion-induction motors.
14. Explain the principle of operation of the reluctance motor.
15. What is meant by a subsynchronous reluctance motor? How does it differ in construction from the standard reluctance motor?
16. How is a hysteresis motor constructed?
17. Describe how hysteresis motors are usually started.
18. Where would one generally apply a hysteresis motor?
19. What is a reluctance motor?
20. Explain how a reluctance motor works.
21. List as many applications of reluctance motors as you can find.
22. What is meant by a fractional-horsepower motor?
23. Explain why a motor having a rating greater or less than 1 hp is not necessarily classified as a fractional-horsepower motor.
24. Under what conditions must single-phase motors be used?
25. What is a universal motor? Why is it called a universal motor?
26. Explain the principle of operation of a universal motor.
27. Why is high speed often desirable in the operation of a small motor such as the universal motor?
28. What limits the no-load speed of a universal motor?

29. Why does the effect of armature reaction tend to increase the speed of the series motor?

30. For alternating current, what factors are responsible for the change in speed when load changes on a universal motor?

31. Why are the rotor slots of universal motors skewed?

32. What is normally meant by small motors?

33. Why is a gearbox frequently used on a universal motor?

34. List as many applications as you can where a gearbox is used with a universal motor.

35. What is a constant-speed governor? Describe the principle of operation when used in conjunction with a universal motor.

36. Give several practical examples of the use of a constant-speed governor in combination with a universal motor.

37. Why does the operation of the universal motor interfere with radio or TV reception?

38. How can this interference be eliminated?

39. Why is a shaded-pole motor an induction motor?

40. Why is the field produced by a shaded-pole motor not a true revolving field in the same way as that created by a polyphase induction motor?

41. Describe the construction of a stator of a shaded-pole motor. Make a sketch of this construction.

42. In what direction will the rotor of a shaded-pole motor rotate?

43. Draw a sketch of a shaded pole and explain how the center of the pole shifts across the face of the pole as the ac sinusoidal excitation of the pole goes through a half-cycle.

44. Explain how a simple shaded-pole motor can be reversed.

45. Make a sketch showing how a shaded-pole motor designed with two sets of shading poles can be reversed.

46. Make a sketch showing how a shaded-pole motor with two sets of stator windings can be reversed.

47. List as many practical applications as you can of shaded-pole motors.

48. How is the speed of a shaded-pole motor controlled?

49. Describe the construction of a stator of a reluctance-start motor.

50. In what direction will the squirrel cage of a reluctance-start motor rotate?

51. Explain the shifting of the field from one side to the other of the pole of a reluctance-start motor.

52. How is the speed of reluctance-start motors controlled?

53. Where are stepper motors used?

54. List as many different permanent-magnet materials used in motors as you can.

55. Where are permanent-magnet motors used?

56. What sizes of permanent-magnet motors are made?

57. Are permanent-magnet machines made to be used as generators? Give examples of such applications.

58. Where are control motors used?

59. List applications of selsyns.

## PROBLEMS

**8-1.** A reluctance motor similar to that in Fig. 8-1(a) has a stator winding of 400 turns. The resistance of the winding is negligible. Tests at 60 Hz yielded the following data:

| Volts | Amperes | $\delta$ |
|-------|---------|----------|
| 120   | 1.25    | 0        |
| 120   | 3.00    | 90       |

Calculate $\mathcal{R}_q$, $\mathcal{R}_d$, and the maximum value of the average torque when the rotor speed is 3600 rpm while the stator is excited at 120 V, 60 Hz. What is the value of $\delta$ at standstill when no torque is applied to the shaft?

**8-2.** The reluctance motor shown here has $X_{dd} = 100$ $\Omega$, $X_{qq} = 60$ $\Omega$, and is energized from a 120-V 60-Hz source. Calculate (a) the synchronous speed and (b) the maximum torque.

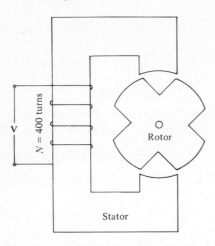

**Prob. 8-2**   Reluctance motor.

**8-3.** The accompanying illustration shows a partial representation showing two teeth of a 12-tooth rotor of a reluctance motor having a stator similar to that in Figs. 8-1 and 8-2. Calculate the synchronous speed of the rotor if the voltage applied to the motor has a frequency of 400 Hz.

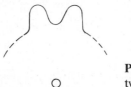

**Prob. 8-3**   Partial representation showing
two teeth of a 12-tooth rotor.

**8-4.** An uncompensated 120-V 60-Hz $\frac{1}{8}$-hp (straight series) universal motor has the following constants:

$$r_f = 10.0\ \Omega \qquad r_a = 7.0\ \Omega$$
$$x_f = 18.0\ \Omega \qquad x_a = 9.0\ \Omega$$

The current and power input to the motor are 1.75 A and 190 W when the motor delivers a mechanical output of 105 W at a speed of 7500 rpm. (a) Calculate (1) the efficiency and (2) the rotational losses of this motor. (b) This motor is redesigned for complete conductive compensation without any changes in the above constants. The resistance of the compensating winding is $r_c = 2.0$ Ω. Assume the rotational losses to remain unchanged and that the motor develops the same value of torque as in part (a) and calculate (1) the current, (2) the real power input, (3) the mechanical power output, (4) the speed, (5) the power factor, and (6) the efficiency.

**8-5.** (a) Calculate the power factor at starting of the two motors in Prob. 8-4(a) and (b) compare the starting torque of the compensated motor with that of the uncompensated motor, neglecting differences in saturation.

**8-6.** The active axial length of the armature in an acyclic generator is 0.30 m and its diameter is 0.50 m. Calculate the no-load voltage for a flux density of 1.65 T and a speed of 3000 rpm.

**8-7.** A three-phase 60-Hz eight-pole linear induction pump has length of 1.0 m along the direction of flow and the channel has a height of 1.7 cm and a width of 0.25 m. The following quantities are known:

> Power input, 8500 W
> Ohmic loss in the tube walls, 400 W
> Winding copper loss, 1350 W
> Hydraulic loss, 750 W
> Stray-load loss, 300 W
> Power factor, 0.24
> Slip $s$, 0.40

Calculate (a) the linear velocity of the liquid metal in kilometers per hour, (b) the rate of flow in liters per minute, (c) the ohmic losses in the liquid, (d) the net power output of the pump, (e) the efficiency, and (f) the reactive-power input to the pump.

# BIBLIOGRAPHY

Angrist, S. W. *Direct Energy Conversion,* 2d ed. Boston: Allyn and Bacon, Inc., 1971.

Chang, S. S. L. *Energy Conversion.* Englewood Cliffs, N.J.: Prentice-Hall, Inc., 1963.

Laithwaite, E. R. *Induction Machines for Special Purposes.* London: George Newnes Ltd. 1966. *Linear Electric Motors.* New York: Crane-Russak Co., 1971. *Propulsion without Wheels.* New York: Hart, 1968.

Levi, E., and M. Panzer. *Electromechanical Power Conversion.* New York: McGraw-Hill Book Company, 1966.

Mather, N. W., and G. W. Sutton. *Engineering Aspects of Magnetohydrodynamics.* New York: Gordon and Breach, Inc., 1964.

Matsch, Leander W. *Electromagnetic & Electromechanical Machines,* 2d ed. New York: Dun-Donnelley Publishing Co., 1977.

McPherson, George. *An Introduction to Electrical Machines and Transformers.* New York, John Wiley & Sons, Inc., 1981.

Spring, K. H. (ed.). *Direct Generation of Electricity.* New York: Academic Press, Inc., 1965.

# Chapter 9

# Direct Conversion to Electrical Energy

One method of direct energy conversion is discussed in Chapter 8—the MHD generator which converts thermal energy directly into electrical energy. Several other methods are considered in this chapter. Some of these have been used in applications requiring only small amounts of power, as, for instance, thermoelectric conversion, in which thermocouples are used for temperature measurements. Others, such as the fuel cell, are rated in kilowatts and have found use in space applications.

## 9-1 FUEL CELLS

A *fuel cell* is, in effect, an electric battery, with the difference that the two electrodes in the battery are the fuel and the oxidant, which are consumed in the battery reaction, while in a fuel cell both fuel and oxidant are supplied by an external source. Although a number of different kinds of fuel cells† are undergoing study and development, only one type, the ion-membrane fuel cell used successfully in the fuel-battery system for the Gemini spacecraft illustrated schematically in Fig. 9-1, is treated in this book. Hydrogen is supplied to a gas chamber on the anode side of the cell and air or oxygen to a gas chamber on the cathode side. The electrodes (anode and cathode) are separated by an ion-exchange membrane about 1 mm thick which allows the passage of positive hydrogen ions $H^+$ but

---

† The following papers on fuel cells are published in *Proc. IEEE* 51 (1963): 784–873: E. W. Justi, "Fuel Cell Research in Europe"; C. G. Peattie, "A Summary of Practical Fuel Cell Technology"; K. V. Kordesch, "Low Temperature Fuel Cells"; E. L. Colichman, "Preliminary Biochemical Fuel Cell Investigations." See also H. A. Liebhofsky and E. J. Cairns, *Fuel Cells and Fuel Batteries* (New York: John Wiley & Sons, Inc., 1968).

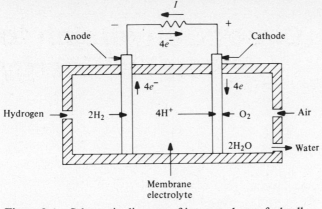

**Figure 9-1** Schematic diagram of ion-membrane fuel cell.

not the neutral oxygen $O_2$ molecules. The two sides of the membrane are coated with a catalyst, which facilitates the chemical reactions in which a number of electrons ($e^-$) are released at the anode from hydrogen atoms, resulting in an equal number of positive hydrogen ions ($H^+$). The free electrons proceed from the anode through the load resistance to the cathode, where they combine with the hydrogen ions that have passed through the membrane and produce water, which is drained from the cell. The chemical reactions are described by the following equations:

*Anode reaction:*

$$2H_2 \rightarrow 4H^+ + 4e^- \tag{9-1}$$

*Cathode reaction:*

$$O_2 + 4H^+ + 4e^- \rightarrow 2H_2O \tag{9-2}$$

This action is the reverse of that by means of which water is decomposed electrolytically into hydrogen and oxygen.

The energy converted by this fuel cell into electrical form is

$$W_{elec} = neE \tag{9-3}$$

where $n$ is the number of electrons and $E$ is the generated voltage between electrodes, which for the cell under discussion cannot exceed 1.23 V.

The fuel cell converts chemical energy isothermally to direct current and does not involve the thermodynamic relation, which limits the efficiency of heat engines, as would be the case if the chemical were first converted into thermal energy and then into electrical energy. This thermodynamic relationship shows the maximum heat engine efficiency to be proportional to the ratio known as the *Carnot efficiency:*

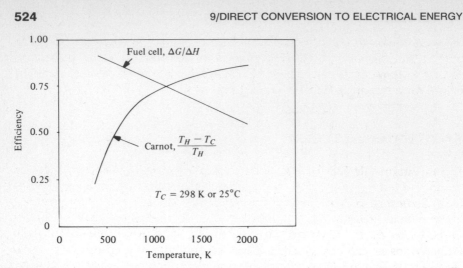

**Figure 9-2** Theoretical efficiency of a hydrogen-oxygen fuel cell and Carnot-cycle efficiency.

$$\text{Efficiency} = \frac{T_H - T_C}{T_H} \tag{9-4}$$

where $T_H$ is the absolute temperature of the incoming working fluid and $T_C$ is that of the working fluid rejected to the cold sink. The electrochemical conversion can be isothermal, whereas the Carnot-cycle limitation requires heat to flow from a higher to a lower temperature and be partially converted to work.

The chemical energy is the free energy, also known as the *Gibbs free energy* of the reaction, expressed by

$$\Delta G = \Delta H - T\Delta S \tag{9-5}$$

where $\Delta H$ is the change in heat content or enthalpy change of reaction and $\Delta S$ is the change in entropy. The energy associated with the entropy is irreversible, and the efficiency based on Eq. 9-5 is

$$\frac{\Delta G}{\Delta H} = 1 - \frac{T\Delta S}{\Delta H} \tag{9-6}$$

According to Liebhofsky and Cairnes,† fuel cells may someday be required to operate over temperature ranges from 240 to 1400 K. It is interesting to compare the efficiency of the fuel cell as expressed by Eq. 9-6 with the Carnot efficiency based on $T_C = 298$ K, as shown graphically in Fig. 9-2, from which it may be observed that the advantage of the fuel cell disappears at temperatures above 1200 K.

The use of fuel cells as electric power sources is at present pretty well restricted to small power requirements, particularly where initial costs are not

† *Ibid.*

of prime importance, as in spacecraft and in military applications. There is practically no likelihood of the fuel cell replacing conventional generating equipment in the power industry, nor are they likely to replace the internal-combustion engine for transport in the foreseeable future.

## 9-2 THERMOELECTRICS

A temperature difference between the two ends of a metal or of a semiconductor produces a voltage between the two ends of the metal. This phenomenon, known as the *Seebeck effect,* results from the higher average kinetic energy of the electrons at the hot end, causing them to diffuse toward the cold end so that the former is at a higher electric potential than the latter, as illustrated in Fig. 9-3(a). The free electrons are mobile while the positive ions are practically locked into position in the crystal lattice of the metal, their response to the heat input manifesting itself by increased vibration.

Although this effect was observed as far back as the early 1820s, its application was limited to the production of very small amounts of power, as, for example, in the field of temperature measurements. However, the development of semiconductors during the 1950s has led to thermoelectric generators with sizable power outputs, ratings in excess of 500 W not being unusual.

The voltage across the two ends is related to their temperature by the Seebeck coefficient as follows:

$$V = \int_{T_C}^{T_H} \alpha \, dT \tag{9-7}$$

where $T_H$ and $T_C$ are the temperatures at the hot and at the cold end. Thermoelectric conversion is achieved by joining two metals having different values

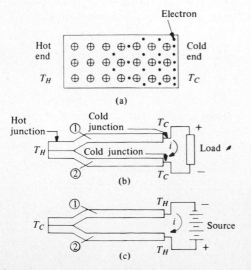

(a)

(b)

(c)

**Figure 9-3** Schematic representations. (a) Electron diffusion in a metal due to temperature gradient. (b) Simple thermoelectric generator and load. (c) Simple thermoelectric refrigerator.

of $\alpha$ or, more effectively, by using a *P-N* junction of semiconductors as shown schematically in Fig. 9-3(b). If the connection is made to the load by means of a third conductor material, two additional junctions are formed at the cold end. However, if these additional junctions are at the same temperature, they do not contribute to the thermoelectric effect. Then, if the upper and lower materials 1 and 2 in Fig. 9-3(b) have Seebeck coefficients of $\alpha_1$ and $\alpha_2$, the resulting coefficient is

$$\alpha = \alpha_1 - \alpha_2 \tag{9-8}$$

which upon substitution in Eq. 9-7 yields the expression for the open-circuit voltage $E_g$. The coefficient $\alpha$ is temperature-dependent for many materials and may be defined for small temperature differences, on the basis of Eq. 9-7, by

$$\lim_{\Delta T \to 0} \alpha = \left. \frac{\Delta V}{\Delta T} \right]_{i=0} \tag{9-9}$$

Curves for *P*-type and *N*-type lead telluride (PbTe) semiconductors, materials which have found considerable use in thermoelectric generators, are shown in Fig. 9-4. It should be mentioned that the electrical resistivity of chemically pure or *intrinsic* semiconductors is too high for use in thermoelectric generators; so impurities are added in a process known as "doping," to give them not only a suitable value of resistivity but also to give them the desired *P*- or *N*-type property. A small amount of lead iodide (PbI$_2$) added to the PbTe semiconductor material increases the number of valence electrons above the value required for

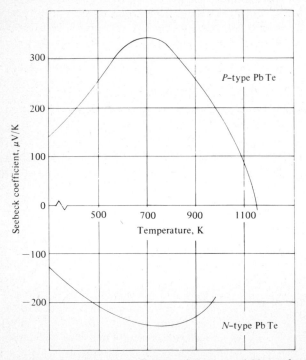

**Figure 9-4** Seebeck coefficient of *P*-type and *N*-type lead telluride.

covalent bonding, resulting in an extrinsic semiconductor of the negative or $N$ type. On the other hand, the addition of a small amount of sodium to PbTe decreases the number of valence electrons, a deficiency resulting in vacancies in the covalent bonds known as holes, to produce the positive or $P$-type extrinsic semiconductor. In the case of the $P$- and $N$-types of Fig. 9-4, the signs of $\alpha$ are opposite and according to Eq. 9-8, their magnitudes add and the resulting Seebeck coefficient of such a junction could be about 600 $\mu$V/$°$K at 675 $°$K.

In the thermoelectric energy-conversion process, heat is absorbed by the hot junction and released at the cold junction when the current flows in response to the Seebeck voltage as indicated in Fig. 9-3(b) (i.e., when the converter acts as a generator of electric energy). This action results from the *Peltier effect.* If the load in Fig. 9-3(b) is replaced by an electric source of voltage as shown in Fig. 9-3(c), the converter operates as a refrigerator since the junction on the left abstracts heat from its surroundings and the junctions on the right release heat. The power converted into reversible heat by the Peltier effect due to the current $i$ is

$$q = \pi i \qquad (9\text{-}10)$$

where $\pi$ is the Peltier coefficient.

A third effect, known as the *Thomson effect,* is evidenced by a voltage between parts of a single homogeneous conductor when at different temperatures. In copper there is a voltage acting from the parts of lower to those of higher temperature, and in iron from parts of a higher to those of a lower temperature. For example, if a copper bar is heated in the middle and a current passed through it, heat is absorbed as the current flows from hotter to colder parts. The reverse effect is true for iron. The Thomson effect is illustrated for copper and iron in Fig. 9-5. The power converted into heat due to the Thomson effect is

$$B = \beta i \Delta T \qquad (9\text{-}11)$$

where $\beta$ is the Thomson coefficient. Further, the relationship between the Seebeck coefficient and the Peltier and Thomson coefficients of a junction at a temperature of $T$ K is

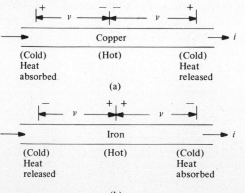

Figure 9-5    Thomson effect in (a) copper and (b) iron.

$$\alpha T = \pi \qquad (9\text{-}12)$$

and
$$\boxed{T \frac{d\alpha}{dT} = \beta_1 - \beta_2} \qquad (9\text{-}13)$$

## 9-2.1 Maximum Output

An exact analysis of the thermojunction as a source of power is unnecessarily complex, and simplifying assumptions are made accordingly. Consider the simple generator in Fig. 9-3(b) and assume the cross-sectional areas $A_1$ and $A_2$ of each of the two materials to be uniform, and assume that the electrical resistivities $\rho_1$ and $\rho_2$, the thermal conductivities $\kappa_1$ and $\kappa_2$, and the Seebeck coefficients $\alpha_1$ and $\alpha_2$, as well as their Peltier coefficients $\pi_1$ and $\pi_2$, are all temperature-independent. On the assumption of temperature-independent properties, the Thomson coefficient must be zero in accordance with Eq. 9-13.

The electrical resistance of the thermocouple is that of the two members in series:

$$r = \frac{l_1\rho_1}{A_1} + \frac{l_2\rho_2}{A_2} \qquad (9\text{-}14)$$

where $l_1$ and $l_2$ are the lengths of the two members.

The thermal conductance is that of the two members in parallel and is therefore

$$K = \frac{A_1\kappa_1}{l_1} + \frac{A_2\kappa_2}{l_2} \qquad (9\text{-}15)$$

and if the current is $i$, the Peltier effect cools the hot junction according to

$$\pi i = \alpha i T_H \qquad (9\text{-}16)$$

The rate of heat flow from the hot to the cold junction by conduction is $K(T_H - T_C)$. There is, in addition, the heat flow $\frac{1}{2}i^2r$ from the hot to the cold junction due to the $i^2r$ heating of the thermocouple, as shown in the following.

Consider an elemental section of thickness $dx$ in the bar of Fig. 9-6 of length $l$, width $w$, and thickness $t$, with current flowing through it in the $x$ direction. If the heat flow normal to the axis of the bar is assumed to be negligible,

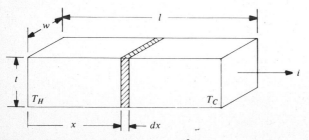

**Figure 9-6**  Heat transfer due to $I^2R$.

the rate of heat flow at $x$ in the $x$ direction is $-\kappa wt(dT/dx)$ and at $x + dx$ it is $-\kappa wt(d/dx)[T + (dT/dx)dx]$, where $T$ and $T + (dT/dx)dx$ are the temperatures of the faces at $x$ and at $x + dx$. The heat generated in the elemental section must be the difference between that which enters and that which leaves the section [i.e., $-\kappa wt(d^2T/dx^2)dx$]. Now the heat generated in this section is due to the current $i$ and must be $(ri^2/l)dx$, where $r$ is the electrical resistance of the bar.

$$-\kappa wt \frac{d^2T}{dx^2} = \frac{r}{l} i^2 \tag{9-17}$$

The result of integrating Eq. 9-17 twice is

$$\kappa wt T = \frac{-ri^2x^2}{2l} + C_1 x + C_2 \tag{9-18}$$

where $C_1$ and $C_2$ are the constants of integration and which are evaluated on the basis that $T = T_H$ at $x = 0$ and $T = T_C$ at $x = l$. Substitution of these values in Eq. 9-18 yields after some simple algebraic manipulation

$$-\kappa wt \frac{dT}{dx} = -\frac{ri^2}{2} + \frac{ri^2x}{l} + \frac{\kappa wt}{l}(T_H - T_C) \tag{9-19}$$

Since $\kappa wt/l = K$, the heat flow into the bar from the hot end where $x = 0$ must be

$$-K\frac{dT}{dx} = -\frac{ri^2}{2} + K(T_H - T_C) \tag{9-20}$$

and the heat flow out of the bar at the cold end, where $x = l$, is

$$K\frac{dT}{dx} = \frac{ri^2}{2} + K(T_H - T_C) \tag{9-21}$$

The heat supplied to the hot junction includes the component required by the Peltier effect in addition to that expressed by Eq. 9-20. Hence,

$$P_{th} = iT_H + K(T_H - T_C) - \frac{ri^2}{2} \tag{9-22}$$

and the generated voltage is

$$E_g = \alpha(T_H - T_C) \tag{9-23}$$

The electric power output to a load resistance $R_L$ is

$$\boxed{P_e = \frac{E_g^2 R_L}{(r + R_L)^2} = \frac{[\alpha(T_H - T_C)]^2 R_L}{(r + R_L)^2}} \tag{9-24}$$

which is a maximum if $R_L$ is made equal to $r$, being

$$P_{e(max)} = \frac{[\alpha(T_H - T_C)]^2}{4r} \tag{9-25}$$

The efficiency for maximum electrical output is therefore

$$\text{Efficiency} = \frac{P_{e(max)}}{P_{th}} = \frac{[\alpha(T_H - T_C)]^2}{4r[\alpha i T_H + K(T_H - T_C) - ri^2/2]} \tag{9-26}$$

The current for maximum output is

$$i = \frac{\alpha(T_H - T_C)}{2r} \tag{9-27}$$

and when this value is substituted into Eq. 9-26, the result is

$$\boxed{\text{Efficiency} = \frac{T_H - T_C}{\frac{3}{2}T_N + T_C/2 + 4rK/\alpha^2}} \tag{9-28}$$

## 9-2.2 Figure of Merit

The quantity $Z = \alpha^2/rK$ is called the *figure of merit* because the greater this value, the higher is the efficiency when the load resistance equals the generator resistance $r$. Equation 9-28 may be rewritten as

$$\text{Efficiency} = \frac{T_H - T_C}{\frac{3}{2}T_H + T_C/2 + 4/Z} \tag{9-29}$$

The greatest output for a given value of $\alpha$ and for a given temperature difference occurs when $rK$, the product of Eqs. 9-14 and 9-15, is a minimum. Now

$$rK = \kappa_1\rho_1 + \frac{A_1 l_2 \kappa_1 \rho_2}{A_2 l_1} + \frac{A_2 l_1 \kappa_2 \rho_1}{A_1 l_2} + \kappa_2 \rho_2 \tag{9-30}$$

The minimum is found by differentiating Eq. 9-30 with respect to $A_1 l_2/A_2 l_1$ and equating the result to zero. It is then found that $rK$ is a minimum when

$$\frac{A_1 l_2}{A_2 l_1} = \left(\frac{\rho_1 \kappa_2}{\rho_2 \kappa_1}\right)^{1/2}$$

for which, according to Eq. 9-30,

$$rK_{min} = [(\rho_1 \kappa_1)^{1/2} + (\rho_2 \kappa_2)^{1/2}]^2 \tag{9-31}$$

The maximum value of the figure of merit for a given value of $\alpha$ is therefore

$$\boxed{Z_{max} = \left[\frac{\alpha}{(\rho_1 \kappa_1)^{1/2} + (\rho_2 \kappa_2)^{1/2}}\right]^2}$$

The nature of the materials that comprise the junction determines the value of $\alpha$.

## 9-2.3 Maximum Efficiency

The output of the generator is expressed by Eq. 9-24 for a given load resistance $R_L$, and the input is given by Eq. 9-22, in which

$$i = \frac{E_g}{r + R_L} = \frac{\alpha(T_H - T_C)}{r + R_L} \tag{9-32}$$

Then on the basis of Eqs. 9-22, 9-25, and 9-32 the efficiency is expressed by

$$\text{Efficiency}_{max} = \frac{\alpha^2(T_H - T_C)R_L}{[\alpha^2(T_H + T_C)/2]r + \alpha^2 T_H R_L + K(r + R_L)^2} \tag{9-33}$$

The value of $R_L$ for which the efficiency is a maximum can be determined by differentiating Eq. 9-33 and equating the result to zero, from which it is found that

$$R_L = r\left[1 + \frac{\alpha^2(T_H + T_C)}{2Kr}\right]^{1/2}$$

$$= r\left[1 + \frac{Z(T_H + T_C)}{2}\right]^{1/2}$$

$$= r(1 + ZT_m)^{1/2} \tag{9-34}$$

where

$$T_m = \frac{T_H + T_C}{2}$$

Substitution of Eq. 9-34 in Eq. 9-33 yields, after some algebraic manipulation,

$$\boxed{\text{Efficiency}_{max} = \frac{(1 + ZT_m)^{1/2} - 1}{(1 + ZT_m)^{1/2} + T_C/T_H} \frac{T_H - T_C}{T_H}} \tag{9-35}$$

where the ratio $(T_H - T_C)/T_H$ is the *Carnot efficiency*.

As an example, consider a thermoelectric generator which has $Z = 3.0 \times 10^{-3}$ and operates between 620 and 300°K. The maximum efficiency is, from Eq. 9-35,

$$\text{Efficiency}_{max} = \frac{\left[1 + 3.0 \times 10^{-3}\left(\frac{620 + 300}{2}\right)\right]^{1/2} - 1}{\left[1 + 3.0 \times 10^{-3}\left(\frac{620 + 300}{2}\right)\right]^{1/2} + \frac{300}{620}} \frac{620 - 300}{620}$$

$$= 0.138$$

If the junction in the foregoing example has a Seebeck coefficient of 200 $\mu$V/°K, its open-circuit voltage would be, from Eq. 9-7,

$$V = \alpha(T_H - T_C) = 200 \times 10^{-6}(620 - 300)$$

$$= 0.064 \text{ V}$$

Such a low value of voltage would seriously limit the use of thermoelectric generators to rather low outputs. However, a higher value of voltage is obtained in practice by cascading several hot and cold junctions as shown schematically in Fig. 9-7. In such an arrangement, although the junctions are in series electrically, the elements are in parallel thermally, a condition that results in reduced effi-

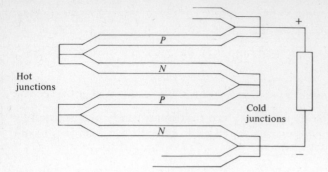

**Figure 9-7**   Thermoelectric generator with junctions in cascade.

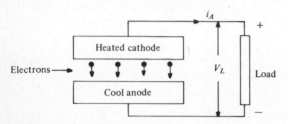

**Figure 9-8**   Elementary thermionic converter.

ciency. As it is, thermoelectric generators are restricted to applications where high efficiency is relatively unimportant but where such features as quiet operation, small size and weight, low maintenance requirements, and long life override the disadvantages of low efficiency.

## 9-3 THERMIONIC CONVERTER

The thermionic converter affords another means for converting heat directly into electrical energy. An elementary thermionic converter is shown in Fig. 9-8, in which energy is supplied in the form of heat to the cathode. If the temperature of the cathode is made high enough, the electrons in the cathode absorb sufficient energy to cause them to leave the metal surface (i.e., electrons are emitted and travel through the intervening space to the anode which then becomes charged negatively). The cathode because of its electron deficiency becomes positive, and if a load is connected between the anode and cathode, an electric current is produced. A two-element arrangement having a cathode and an anode is known as a *diode*.

### 9-3.1 Work Function and Richardson's Equation

The energy required to pull an electron free from a metal surface is equal to the product of the work function of the metal and the charge $e$ of the electron. In order for an electron to escape from the metal surface, it must have a component of velocity $v_n$ normal to the surface that satisfies the energy relationship expressed by

$$\frac{mv_n^2}{2} > \phi e \tag{9-36}$$

where $m$ is the mass of the electron. The current density associated with electron emission is defined by the *Richardson–Dushman equation* as

$$\boxed{J = AT^2 \epsilon^{-\phi/kT}} \tag{9-37}$$

where $A = 1.20$ A/m$^2$, the emission constant, and $T$ is the temperature in kelvins. Equation 9-37 is based on the assumption that the velocity distribution of the gas is such that the average energy of the electrons is $3kT/2$. If the area of the emitting surface is $A_c$ and all the emitted electrons reach the anode, the anode current is

$$i_a = JA_c \tag{9-38}$$

## 9-3.2 Space Charge

The result of thermionic emission in a vacuum is the formation of an electron cloud between the cathode and anode known as a *space charge.* The space charge exerts a force on the emitted electrons, causing some of them to fall back into the cathode material. The space charge thus develops a retarding potential between cathode and anode.

The space charge may be neutralized by the introduction, into the diode, of a gas which has a lower ionization potential than the work function of the cathode. The most commonly used gas is cesium, which has an ionization potential of 3.89 V as compared with a work function of 4.52 V for a tungsten cathode. Other methods† make use of (a) very small cathode–anode spacing and (b) crossed electric and magnetic fields.

## 9-3.3 Efficiency

If the space charge is completely neutralized and an electron progresses from the cathode to the anode, it gives up to the anode an amount of kinetic energy equal to $\phi_a e$, where $\phi_a$ is the work function of the anode material. The remainder of its energy is delivered to the load. Then under the ideal condition that the voltage between the anode and cathode is negligible, the voltage across the load, if $\phi_c$ is the work function of the cathode, is

$$V_L = \phi_c - \phi_a$$

and the power supplied to the load is

---

† For a more complete discussion of these methods, see K. H. Spring, *Direct Generation of Electricity* (New York: Academic Press, Inc., 1965), pp. 230–232; see also S. W. Angrist, *Direct Energy Conversion,* 2d ed. (Boston: Allyn and Bacon, Inc., 1971), Chap. 6.

$$P_L = V_L i_a = (\phi_c - \phi_a) i_a \qquad (9\text{-}39)$$

If $P_l$ is the heat lost by the cathode, the power input to the cathode is

$$P_c = \phi_c i_a + P_l \qquad (9\text{-}40)$$

and the efficiency is

$$\text{Efficiency} = \frac{P_l}{P_c}$$

$$= \frac{\phi_c - \phi_a}{\phi_c + P_l/i_a} \qquad (9\text{-}41)$$

From Eq. 9-41 it is evident that the anode material should have a low work function $\phi_a$ and the cathode heat loss $P_l$ should be small, or that $i_a$ should be large relative to the cathode heat loss. However, as shown by Eqs. 9-37 and 9-38, it is necessary to increase the cathode temperature in order to increase $i_a$. The life of the cathode material places a practical limitation on the operating temperature, and a compromise between efficiency and life of material is necessary. Regardless, however, the efficiency of the thermionic generator cannot exceed the Carnot efficiency $(T_c - T_a)/T_c$, where $T_c$ and $T_a$ are the temperatures of the cathode and anode in kelvins. However, the Carnot efficiency of the thermionic device cannot be increased for a given cathode temperature $T_c$ by decreasing the anode temperature $T_a$ below a value where back emission occurs at the anode (i.e., that which causes a component of current to flow from anode to cathode).

For example, if the work functions are $\phi_c = 1.70$ and $\phi_a = 1.10$ for the cathode and anode of a thermionic converter, the cathode temperature is 1200 K, and the back-emission temperature of the anode is assumed to be

$$T_a = \frac{\phi_a}{\phi_c} T_c = \frac{1.10}{1.70} \times 1200 = 775 \text{ K}$$

then the Carnot efficiency is found to be

$$\frac{T_c - T_a}{T_c} = \frac{1200 - 775}{1200} = 0.35$$

This value of efficiency does not take into account the losses in the converter itself, and an overall efficiency of less than 0.20 would be realized, although future developments may produce efficiencies as high as 0.30.

## 9-3.4 Maximum Output

The thermionic generator is a constant-current source up to a load voltage $V_L = \phi_c - \phi_a$. As $V_L$ is increased above this value, the current falls off exponentially and at a greater rate than that at which $V_L$ increases, and the output is therefore a maximum when the terminal voltage or load voltage $V_L = \phi_c - \phi_a$ if the space charge is neutralized exactly.

### 9-3.5 Applications

Thermionic converters are suitable for use in systems that can utilize solar and nuclear sources of energy, the exhaust of rocket motors, and fossil fuels, where the converter may be used as a topping unit in conjunction with conventional power plants along the lines described for MHD in Sec. 8-10. Because of its simplicity and lightness the thermionic generator is particularly suited as a power supply to spacecraft, and since in addition its operation is quiet, it should lend itself as a portable power source for military field use.

## 9-4 PHOTOVOLTAIC GENERATOR

The photovoltaic generator converts radiant energy directly into electrical energy without the limitation imposed by the Carnot cycle efficiency. The earliest such devices had efficiencies well under 1 percent, but the use of semiconductors has led to photoelectric cells with conversion efficiencies of about 6 percent. The solar cell affects this type of conversion and is used in spacecraft, where it is used to charge electric batteries.

The *P-N* junction of the semiconductor material must be arranged so that it can be illuminated and as a consequence be struck by photons of light. The photons impart energy to the electrons in the semiconductor material, causing them to be released from their valence bonds, which results in a flow of current if an external load is connected across the electrodes attached to the junction materials, as indicated for the simple photovoltaic cell in Fig. 9-9.

### 9-4.1 Photons

Radiation propagated through space possesses both energy and momentum and has the dual properties of both wavelength and mass, a situation that is similar to the *wave-duality* nature of the electron, in that it exhibits properties both of

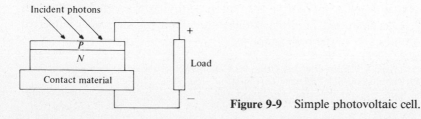

**Figure 9-9** Simple photovoltaic cell.

a particle and of a wave. The photon is a convenient concept which represents a particle of energy expressed by

$$W = \frac{hc}{\lambda} \quad J \tag{9-42}$$

where $h = 6.624 \times 10^{-34}$ J-s, known as *Planck's constant,* and where $c = 3 \times 10^8$ m/s, the velocity of light, and $\lambda$ is the wavelength of the radiant energy.

If a photon strikes a semiconductor with sufficient energy, it can free an electron from its valence bond. The amount of energy required to accomplish this must at least be equal to what is called the *energy gap* of the material. The energy gap varies with different materials, and in the case of semiconductors is reduced by means of "doping" with certain impurity materials. The actual process of photovoltaic conversion is too complex to be taken up in detail within the limited scope of this chapter, and the reader is referred to references cited in the footnote. It would appear that the energy gap should be as small as possible for the best efficiency. This is not true, however, because the open-circuit voltage depends almost directly on the value of the energy gap, and the problem therefore is one of effecting a compromise which must also take into account the spectral distribution of solar radiation.

## 9-4.2 Solar Energy

The power density of solar radiation outside the earth's atmosphere has been determined as 1.350 kW/m²,[†] which is received by cells mounted on orbiting satellites and facing the sun. This value corresponds to $5.8 \times 10^{17}$ photons with an average energy of 1.48 electron-volts. However, at sea level with the sun at zenith these quantities are reduced to 1.06 kW/m², $5.0 \times 10^{17}$ photons, and 1.32 electron-volts per photon. This reduction is due to atmospheric absorption, which results not only in reduced energy but also in difference in the spectral distribution. Further, at sea level with the sun at 60° from zenith, we have 0.88 kW/m², $4.3 \times 10^{17}$ photons, and 1.28 eV per photon. The limit of overall efficiency for cells using silicon has been estimated at 15 percent[‡] and at that value the area of a receptor would need to be about 1 m² to deliver an output 100 W outside the atmosphere.

Other methods of direct energy conversion which at the present time are not as effective or seem not to have as much potential as those discussed in this chapter have been omitted. For some of these other methods, the reader is referred to the footnotes and to the bibliography.

[†] J. J. Loferski, "Theoretical Considerations Governing the Choice of the Optimum Semiconductor for Photovoltaic Solar Energy Conversion," *J. Appl. Phys.* 27 (1956): 777–784.

[‡] T. S. Moss, *Optical Properties of Semiconductors,* 2d ed. (London: Butterworth & Co. Ltd., 1961).

## STUDY QUESTIONS

1. Find a current source that lists the U.S. fuel reserve for gas, petroleum, oil-shale coal, nuclear fission, and nuclear breeders.

2. Find out how many quad of energy are available per year from incident solar energy. What is meant by the term quad?

3. Make a list of the types of solar collectors. Find the collection efficiency of each type of collector.

4. From the results of Question 3 compute the amount of each collector type to supply the energy needs of the average home of 1000 kWh/month usage.

5. How could energy peak demand be met by solar-energy collection?

6. List as many types of solar storage as possible.

7. What is a solar concentrator?

8. What is a solar-position chart?

9. What are the units used to measure the amount of incident solar energy?

10. What is the difference between passive and active solar collectors? Give examples of each.

11. What is a clearness factor in solar terms?

12. Find a source and estimate the global solar heat flux incident upon a collector in Rolla, Missouri (100 miles southwest of St. Louis) between 11 a.m. and solar noon of a typical February day. The collector faces south at a slope of $40°$.

13. Find a source that explains fusion as an energy source. Describe how a fusion reaction can be contained and list the methods that have been used or are proposed.

14. How does a fission reactor operate? How is the fission process controlled? Explain how the reactor control has been designed to be fail-safe.

15. How many fission reactors are currently in operation? How many are under construction currently? What percentage of the U.S. electrical energy supply is currently provided by fission reactors?

16. What is a breeder reactor? How does it breed fuel? How many breeder reactors are in operation worldwide? How many are in operation in the United States?

17. List as many ways as possible that thermal energy can be converted into electricity.

18. Explain the operation of each method listed in Question 17. Give examples of where each method is used.

19. List as many methods as possible for the conversion of chemical energy into electrical energy.

20. Explain how each method listed in Question 19 operates. Give examples of where chemical-to-electrical conversion systems are currently used.

21. How is electromagnetic energy converted to electrical energy?

22. How is mechanical energy converted to electrical energy?

23. What is the current status of MHD system development?

24. What is the status of EGD (electrogas-dynamic) research?

25. What fuels are used in MHD and EGD systems?

26. Make a list of energy-conversion methods that are considered nonconventional such as ocean thermal currents.

27. What is the status of wind-energy development in the United States?

28. What is meant by the hydrogen energy economy? What would be the advantage of hydrogen fuels? How can hydrogen be stored for future use in solid form?

## BIBLIOGRAPHY

Anderson, Edward E. *Fundamentals of Solar Energy Conversion.* Reading, Mass.: Addison-Wesley Publishing Co., 1983.

Angrist, S. W. *Direct Energy Conversion,* 2d ed. Boston: Allyn and Bacon, Inc., 1971.

Chang, S. S. L. *Energy Conversion.* Englewood Cliffs, N.J.: Prentice-Hall, Inc., 1963.

Culp, Archie W., Jr. *Principles of Energy Conversion,* New York: McGraw-Hill Book Company, 1979.

Spring, K. H. (ed.). *Direct Generation of Electricity.* New York: Academic Press, Inc., 1965.

# Laplace Transformation

The operational method of Laplace transformation facilitates the solution of ordinary differential equations and has the advantage over the classical methods of automatically incorporating the initial conditions. The following presents only the most elementary concepts and rules necessary for solving linear differential equations.†

## A-1 THE LAPLACE TRANSFORMATION

The Laplace transform of a known function, $f(t)$, for values of $t > 0$ is defined by the equation

$$\mathcal{L}[f(t)] = \mathbf{F(s)} = \int_0^\infty \epsilon^{-st} f(t)\, dt \qquad (A\text{-}1)$$

The Laplace transform of $f(t)$ exists if

$$\lim_{t \to \infty} \epsilon^{-st} f(t) = 0$$

Further, if $\mathbf{C}$ is a constant, then

$$\mathcal{L}[\mathbf{C}f(t)] = \mathbf{C}\mathbf{F(s)} = \mathbf{C} \int_0^\infty \epsilon^{-st} f(t)\, dt \qquad (A\text{-}2)$$

† For a more complete elementary treatment of the Laplace transformation as applied to problems in engineering and physics, see W. T. Thomson, *Laplace Transformation* (New York: Prentice-Hall, Inc., 1950). See also W. R. LePage, *Complex Variables and the Laplace Transform for Engineers* (New York: McGraw-Hill Book Company, 1961).

and since the Laplace transformation is a linear one, the principle of superposition is valid:

$$\mathcal{L}[C_1 f_1(t) + C_2 f_2(t)] = C_1 F_1(s) + C_2 F_2(s) \tag{A-3}$$

## A-2 TRANSFORMS OF SIMPLE FUNCTIONS

a. The step function $f(t) = A$, a constant. Substitution in Eq. A-1 yields

$$\mathcal{L}[A] = F(s) = A \int_0^\infty \epsilon^{-st}\, dt = \frac{A}{s} \tag{A-4}$$

b. The exponential function $f(t) = A\epsilon^{at}$, where the constant **a** may be real or complex. According to Eqs. A-1 to A-4, we have

$$\mathcal{L}[A\epsilon^{at}] = F(s) = A \int_0^\infty \epsilon^{(a-s)t}\, dt = \frac{A}{s-a} \tag{A-5}$$

c. The trigonometric function $f(t) = A \sin \omega t$. Here we can apply Eqs. A-3 and A-5, since

$$\sin \omega t = \frac{\epsilon^{j\omega t} - \epsilon^{-j\omega t}}{j2}$$

so that

$$\mathcal{L}[A \sin \omega t] = F(s) = \frac{A}{j2}\left[\frac{1}{s - j\omega} - \frac{1}{s + j\omega}\right]$$
$$= \frac{A\omega}{s^2 + \omega^2} \tag{A-6}$$

By a similar process we find that

$$\mathcal{L}[A \cos \omega t] = F(s) = \frac{As}{s^2 + \omega^2} \tag{A-7}$$

d. The transform for the linear function of time, $At$, is given by

$$\mathcal{L}[At] = F(s) = A \int_0^\infty t\epsilon^{-st}\, dt = \frac{A}{s^2} \tag{A-8}$$

e. The indefinite integral $\int f(t)\, dt$ is transformed as follows:

$$\mathcal{L}\left[\int_0^t f(t)\, dt\right] = \int_0^\infty \epsilon^{-st}\left[\int f(t)\, dt\right] dt$$

Let $\int f(t)\, dt = \mu$ and $\epsilon^{-st}\, dt = dv$. Then integration by parts yields

$$
\mathcal{L}\left[\int f(t)\,dt\right] = -\frac{\epsilon^{-st}}{s}\int f(t)\,dt \bigg|_0^\infty + \frac{1}{s}\int_0^\infty \epsilon^{-st} f(t)\,dt
$$

$$
= -\frac{\epsilon^{-st}}{s}\int f(t)\,dt \bigg|_0^\infty + \frac{F(s)}{s}
$$

$$
= \frac{F(s)}{s} + \frac{\displaystyle\int f(0^+)\,dt}{s} \tag{A-9}
$$

where the second term on the right-hand side accounts for the initial condition.

f. The double integration $\int\int f(t)\,dt^2$ is found, by a process similar to that for the single integral, to be

$$
\mathcal{L}\left[\int\int f(t)\,dt^2\right] = \frac{F(s)}{s^2} + \frac{\displaystyle\int f(0^+)\,dt}{s^2} + \frac{\displaystyle\int\int f(0^+)\,dt}{s} \tag{A-10}
$$

g. The time derivative $df(t)/dt$.

$$
\mathcal{L}\left[\frac{df(t)}{dt}\right] = \int_0^\infty \left[\epsilon^{-st}\frac{df(t)}{dt}\right] dt
$$

This can be integrated by parts. Let $u = \epsilon^{-st}$ and $dv = [df(t)/dt]\,dt$, from which we obtain $du = -s\epsilon^{-st}$ and $v = f(t)$. Hence,

$$
\mathcal{L}\left[\frac{df(t)}{dt}\right] = \epsilon^{-st}f(t)\bigg|_0^\infty + s\int_0^\infty \epsilon^{-st}f(t)\,dt
$$

$$
= sF(s) - f(0^+) \tag{A-11}
$$

h. The second time derivative $d^2f(t)/dt^2$. A process similar to that used for the first time derivative in part (g) yields

$$
\mathcal{L}\left[\frac{d^2f(t)}{dt^2}\right] = s^2F(s) - sf(0^+) - f'(0^+) \tag{A-12}
$$

By continuing this process the Laplace transform of the $n$th time derivative is found to be

$$
\mathcal{L}\left[\frac{d^nf(t)}{dt^n}\right] = s^nF(s) - s^{n-1}f(0^+) - s^{n-2}f'(0^+) - \cdots - f^{n-1}(0^+) \tag{A-13}
$$

It is evident from the foregoing that the application of the Laplace transform to an ordinary differential equation results in an algebraic equation in terms of the parameter $s$, which is called the *subsidiary equation.*

Although these few derivations serve as the barest introduction to this powerful operational method, they provide a basis for solving a variety of problems involving transients in linear systems. The following example illustrates the use of some of these derivations.

**EXAMPLE A-1**

The single-pole double-throw switch in Fig. A-1 is originally in position 1 and is assumed to be thrown instantaneously from position 1 to position 2 at $t = 0$, at which instant the current $i = i(0^+) = 0.31$ A and the voltage across the capacitor $e_C = 281$ V. (a) Write the defining differential equation for the current for $t > 0$. (b) Write the subsidiary equation of this differential equation. (c) Find the inverse transform for the equation in part (b) and express the current as a function of time for $t > 0$.

*Solution*

a. The differential equation is

$$L\frac{di}{dt} + Ri + \frac{1}{C}\int i \, dt = V_2 \tag{1}$$

and when the numerical values in Fig. A-1 are substituted for the parameters, we have

$$\frac{di}{dt} + 100i + 10^5 \int i \, dt = 100 \tag{2}$$

b. To obtain the Laplace transform of the first term in the left-hand side of Eq. 2, we make use of Eq. A-11, which results in

$$\mathcal{L}\left[\frac{di}{dt}\right] = s\mathbf{I(s)} - i(0)$$

$$= s\mathbf{I(s)} - 0.31$$

For the second term, since $R$ is a constant, we obtain

$$\mathcal{L}[100i] = 100\mathcal{L}[i] = 100\mathbf{I(s)}$$

The third term on the left-hand side of Eq. 2 is obtained from Eq. A-9 as follows:

$$\mathcal{L}\left[10^5 \int i \, dt\right] = \frac{\mathbf{I(s)}}{s} \times 10^5 + \frac{e_C(0^+)}{s}$$

$$= \frac{\mathbf{I(s)}}{s} \times 10^5 + \frac{281}{s}$$

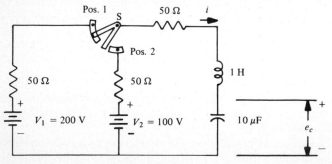

**Figure A-1**　Circuit for Example A.1.

The Laplace transform of the constant voltage on the right-hand side of Eq. 2 is found to be, from Eq. A-4,

$$\mathcal{L}[100] = \frac{100}{s}$$

When these transforms are combined in accordance with Eq. 2, the result is

$$sI(s) - 0.31 + 100I(s) + \frac{I(s)}{s} \times 10^5 + \frac{281}{s} = \frac{100}{s}$$

from which the Laplace transform of the current is found to be

$$I(s) = \frac{0.31s - 181}{s^2 + 100s + 10^5} \tag{3}$$

The right-hand side of Eq. 3 can be expressed in terms of partial fractions as follows:

$$I(s) = \frac{C_1}{s + 50 - j312} + \frac{C_2}{s + 50 + j312} = \frac{0.31s - 181}{s^2 + 100s + 10^5} \tag{4}$$

which can be reduced to

$$(s + 50 + j312)C_1 + (s + 50 - j312)C_2 = 0.31s - 181$$

To solve for the constants $C_1$ and $C_2$ we equate the coefficients of s to yield

$$C_1 + C_2 = 0.31 \tag{5}$$

and by equating the constant terms we get

$$C_1 - C_2 = j0.63 \tag{6}$$

$C_1$ and $C_2$ are conjugates because their sum is real and their difference is imaginary. Accordingly,

$$\text{Re } C_1 = \text{Re } C_2 = \frac{0.31}{2} = 0.155$$

and

$$\text{Im } C_1 = -\text{Im } C_2 = \frac{0.63}{2} = 0.315 \tag{7}$$

so that

$$C_1 = 0.155 + j0.315 \quad \text{and} \quad C_2 = 0.155 - j0.315$$

When these complex values are substituted in Eq. 4, the Laplace transform of the current is found to be

$$I(s) = \frac{0.155 + j0.315}{s + 50 - j312} + \frac{0.155 - j0.315}{s + 50 + j312} \tag{8}$$

c. Each term on the right-hand side of Eq. 8 corresponds to the term on the right-hand side of Eq. A-5, and we find the current expressed as a function of time by taking the inverse transform,

$$i(t) = \mathcal{L}^{-1}[I(s)] = (0.155 + j0.315)\epsilon^{-(50-j312)t} + (0.155 - j0.315)\epsilon^{-(50+j312)t}$$

which after manipulation is reduced to

$$i(t) = (0.31 \cos 312t - 0.63 \sin 312t)\epsilon^{-50t}$$
$$= 0.70\epsilon^{-50t} \cos (312t + 63.8°) \tag{9}$$

Equation 9 shows the initial and final values of the current to be 0.31 A and zero. The case in which the final value of a quantity (i.e., the voltage across the capacitor), as well as its initial value, is other than zero is treated in Example A-2.

**EXAMPLE A-2**

Express the voltage across the capacitor in Example A-1 as a function of time.

*Solution.* The third term on the left-hand side of Eq. 2 in Example A-1 expresses the voltage across the capacitor. Its Laplace transform is

$$\mathcal{L}\left[ 10^5 \int i \, dt \right] = \frac{I(s)}{s} \times 10^5 + \frac{281}{s} = E_C(s) \tag{1}$$

Substitution of Eq. 3, Example A-1, in Eq. 1, Example A-2, yields

$$E_C(s) = \frac{281s^2 + 59,100s + 100 \times 10^5}{s(s^2 + 100s + 10^5)} \tag{2}$$

which can be reduced by means of partial fractions as follows:

$$\frac{C_1}{s} + \frac{2}{s + 50 + j312} + \frac{C_3}{s + 50 - j312} = \frac{281s^2 + 59,100s + 100 \times 10^5}{s(s^2 + 100s + 10^5)}$$

$$C_1(s^2 + 100s + 10^5) + C_2[s^2 + (50 - j312)s]$$
$$+ C_3[s^2 + (50 + j312)s] = 281s^2 + 59,100s + 100 \times 10^5$$

Equating coefficients of like powers of s results in

$$C_1 + C_2 + C_3 = 281 \tag{3}$$

$$100C_1 + 50(C_2 + C_3) - j312(C_2 - C_3) = 59,100 \tag{4}$$

$$C_1 = 100 \tag{5}$$

Substitution of Eq. 5 in Eq. 3 results in

$$C_2 + C_3 = 181 \tag{6}$$

and substitution of Eqs. 5 and 6 in Eq. 4 yields

$$C_2 - C_3 = j128.4 \tag{7}$$

Since Eq. 6 shows the sum of $C_2$ and $C_3$ to be real and Eq. 7 shows their difference to be imaginary, $C_2$ and $C_3$ must be conjugates. Hence,

$$C_2 = 90.5 + j64.2 \tag{8}$$

and $$C_3 = 90.5 - j64.2 \tag{9}$$

According to Eqs. 5, 8, and 9, we have

$$E_C(s) = \frac{100}{s} + \frac{90.5 + j64.2}{s + 50 + j312} + \frac{90.5 - j64.2}{s + 50 - j312}$$

for which the inverse transform is

$$e_C(t) = 100 + (90.5 + j64.2)\epsilon^{-(50+j312)t} + (90.5 - j64.2)\epsilon^{-(50-j312)t}$$
$$= 100 + (181 \cos 312t + 128.4 \sin 312t)\epsilon^{-50t}$$
$$= 100 + 222\epsilon^{-50t} \cos (312t - 35.4°) \tag{10}$$

It may be observed from Eq. 3 in Example A-1 and Eq. 2 in Example A-2 that the Laplace transform has the general form

$$f(s) = \frac{A(s)}{B(s)} \tag{A-14}$$

When the denominator has repeating factors, the transform is said to have poles of higher order:

$$B(s) = (s - a_1)^k(s - a_2)(s - a_3) \cdots (s - a_m)$$

and Eq. A-14 can be reduced by means of partial fractions as follows:

$$\frac{A(s)}{B(s)} = \frac{C_{11}s^{k-1} + C_{12}s^{k-2} + \cdots + C_{1n}}{(s - a_1)^k} + \frac{C_2}{s - a_2}$$
$$+ \frac{C_3}{s - a_3} + \cdots + \frac{C_m}{s - a_m} \tag{A-15}$$

One way of determining the various constants is to multiply both sides of Eq. A-15 by $B(s)$ and equate coefficients for like powers of $s$. Other methods sometimes involve simpler procedures, as illustrated in Example A-3.

**EXAMPLE A-3**

Find the inverse Laplace transform of

$$f(s) = \frac{A(s)}{B(s)} = \frac{s^2 + 2s + 1}{(s + 3)^3(s + 4)(s + 5)}$$

*Solution.* From Eq. A-15,

$$\frac{A(s)}{B(s)} = \frac{s^2 + 2s + 1}{(s + 3)^3(s + 4)(s + 5)}$$
$$= \frac{C_{11}s^2 + C_{12}s + C_{13}}{(s + 3)^3} + \frac{C_2}{(s + 4)} + \frac{C_3}{(s + 5)} \tag{1}$$

Multiplying by $(s + 3)^3(s + 4)(s + 5)$ results in

$$s^2 + 2s + 1 = (s + 4)(s + 5)(C_{11}s^2 + C_{12}s + C_{13})$$
$$+ (s + 5)(s + 3)^3C_2 + (s + 4)(s + 3)^3C_3 \tag{2}$$

Equation 2 is valid for all values of $s$, and for $s = -4$ we have

$$9 = (1)(-1)^3C_2$$

from which we find

$$C_2 = -9$$

Similarly, from $s = -5$ we obtain

$$C_3 = 2$$

and from $s = 0$ it follows that

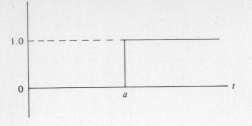

**Figure A-2**   Unit step function $f(t) = U(t - a)$.

$$C_{13} = 50$$

Equation 1 can now be rewritten as

$$s^2 + 2s + 1 = (s^2 + 9s + 20)(C_{11}s^2 + C_{12}s + 50) - (7s + 37)(s + 3)^3 \quad (3)$$

Let us now differentiate both sides of Eq. 3 with respect to s to obtain

$$2s + 2 = (s^2 + 9s + 20)(2C_{11}s + C_{12}) + (2s + 9)(C_{11}s^2 + C_{12}s + 50)$$
$$- 7(s + 3)^3 - 3(7s + 37)(s + 3)^2$$

Again let $s = 0$, and we have

$$C_{12} = 37$$

Next let $s = -3$, from which it follows that

$$C_{11} = 7$$

When the numerical values of these constants are substituted in Eq. 1, there results

$$f(s) = \frac{7s^2 + 37s + 50}{(s + 3)^3} - \frac{9}{s + 4} + \frac{2}{s + 5} \quad (4)$$

The first term on the right-hand side of Eq. 4 can be reduced further to

$$\frac{7s^2 + 37s + 50}{(s + 3)^3} = \frac{(7s^2 + 42s + 63) - (5s - 15) + 2}{(s + 3)^2}$$

$$= \frac{7}{s + 3} - \frac{5}{(s + 3)^2} + \frac{2}{(s + 3)^2} \quad (5)$$

Then from Laplace transforms 2 and 11 in Table A-1 we obtain

$$f(t) = 7\epsilon^{-3t} - 5t\epsilon^{-3t} - t^2\epsilon^{-3t} - 9\epsilon^{-4t} + 2\epsilon^{-5t}$$
$$= (t^2 + 5t + 7)\epsilon^{-3t} - 9\epsilon^{-4t} + 2\epsilon^{-5t}$$

The constants in Example A-3 are all real. In many cases the constants are complex. However, Example A-3 was selected to illustrate some common techniques for obtaining the inverse Laplace transform.

## A-2.1  Initial- and Final-Value Theorems

The behavior of systems in the neighborhood $t = 0$ and $t = \infty$ can be determined by means of the initial- and final-value theorems directly from the subsidiary equation.

**TABLE A-1.  LAPLACE TRANSFORM
               PAIRS**

| Item | $F(s)$ | $f(t)$ |
|------|--------|--------|
| (1) | $\dfrac{1}{s}$ | $1$ |
| (2) | $\dfrac{1}{s+a}$ | $\epsilon^{-at}$ |
| (3) | $\dfrac{1}{s(s+a)}$ | $\dfrac{1}{a}(1-\epsilon^{-at})$ |
| (4) | $\dfrac{1}{(s+a)(s+b)}$ | $\dfrac{1}{b-a}(\epsilon^{-at}-\epsilon^{-bt})$ |
| (5) | $\dfrac{\omega}{s^2+\omega^2}$ | $\sin \omega t$ |
| (6) | $\dfrac{s}{s^2+\omega^2}$ | $\cos \omega t$ |
| (6a) | $\dfrac{\omega^2}{s(s^2+\omega^2)}$ | $1-\cos \omega t$ |
| (7) | $\dfrac{a}{s^2-a^2}$ | $\sinh at$ |
| (8) | $\dfrac{s}{s^2-a^2}$ | $\cosh at$ |
| (9) | $\dfrac{1}{s^n}$ | $\dfrac{t^{n-1}}{(n-1)!}$ |
| (10) | $\dfrac{1}{(s+a)^2}$ | $t\epsilon^{-at}$ |
| (11) | $\dfrac{1}{(s+a)^n}$ | $\dfrac{t^{n-1}\epsilon^{-at}}{(n-1)!}$ |
| (12) | $\dfrac{a}{s(s+a)}$ | $1-\epsilon^{-at}$ |
| (13) | $\dfrac{\beta}{(s+\alpha)^2+\beta^2}$ | $\epsilon^{-\alpha t}\sin \beta t$ |
| (14) | $\dfrac{s}{(s+\alpha)^2+\beta^2}$ | $\epsilon^{-\alpha t}\cos \beta t$ |
| (15)[a] | $\epsilon^{-as}F(s)$ | $f(t-a)U(t-a)$ $U(t-a)=0, t<0$ $= 1, t>a$ |

[a] See Fig. A-2.

### Initial-Value Theorem   The initial-value theorem is expressed mathematically by

$$\lim_{s\to\infty} sF(s) = \lim_{t=0} f(t) \tag{A-16}$$

The validity of this theorem can be demonstrated by writing the equation for the transform of the derivation $f'(t)$ as follows:

$$\int_0^\infty \epsilon^{-st} f'(t) = sF(s) - f(0) \tag{A-17}$$

Since $s$ is a parameter and not a function of time, it may be allowed to approach infinity before integrating, and the left-hand side of Eq. A-17 is then equal to zero. Hence,

$$0 = \lim_{s\to\infty} [sF(s) - f(0)]$$

from which

$$\boxed{\lim_{s\to\infty} sF(s) = \lim_{t\to 0} f(t)]}$$

since $f(0)$ stands for $\lim_{t\to 0} f(t)$. For example, when this theorem is applied to the subsidiary equation for the voltage across the capacitor in Example A-2, the initial value is 281 V.

**Final-Value Theorem**   The final-value theorem is stated by

$$\lim_{s\to 0} sF(s) = \lim_{t\to\infty} f(t) \tag{A-18}$$

For the proof of this theorem, let $s = 0$ in Eq. A-17 before integrating; then the left-hand side of Eq. A-17 is rewritten as

$$\int_0^\infty f'(t)\, dt = \lim_{t\to\infty} \int_0^t f'(t)\, dt = \lim_{t\to\infty} [f(t) - f(0)]$$

and when substituted into Eq. A-17 along with the condition that $s \to 0$, there results

$$\left[\lim_{t\to\infty} f(t) - f(0)\right] = \lim_{s\to 0} [sF(s) - f(0)] \tag{A-19}$$

Since $f(0)$ is independent of $t$ and of $s$, we have

$$\left[\lim_{t\to\infty} f(t)\right] - f(0) = \left[\lim_{s\to 0} sF(s)\right] - f(0)$$

or

$$\boxed{\lim_{t\to\infty} f(t) = \lim_{s\to 0} sF(s)} \tag{A-20}$$

When this theorem is applied to the subsidiary equation for the voltage across the capacitor in Example A-2, the final value is 100 V.

# Appendix B

# Constants and Conversion Factors

| Constants |
|---|

Permeability of free space, $\mu_0 = 4\pi \times 10^{-7}$ Wb/ampere turn m
Permittivity (capacitivity) of free space, $\epsilon_0 = 8.854 \times 10^{-12}$ C$^2$/N-m$^2$
Acceleration of gravity, $g = 9.807$ m/s$^2$
Rest mass of an electron, $m = 9.107 \times 10^{-31}$ kg
Charge on an electron, $e = 1.602 \times 10^{-19}$ C
Velocity of light, $c = 2.998 \times 10^8$ m/s
Planck's constant, $b = 6.624 \times 10^{-34}$ J-s

| Conversion factors |
|---|

| | |
|---|---|
| Length | 1 m = 3.281 ft |
| | = 39.37 in. |
| | 1 km = 0.6214 mile |
| Mass | 1 kg = 0.0685 slug |
| | = 2.205 lb (mass) |
| Force | 1 newton = 0.225 lb |
| | = 7.23 poundals |
| Torque | 1 newton-meter/rad = 0.738 lb-ft/rad |
| Energy | 1 joule (watt-s) = 0.738 ft-lb |
| Power | 1 watt = 1.341 $\times 10^{-3}$ hp |
| Moment of inertia | 1 kg-m$^2$/rad$^2$ = 0.738 slug-ft$^2$/rad$^2$ |
| | = 23.7 lb-ft$^2$/rad$^2$ |
| Magnetic flux | 1 weber = $10^8$ maxwells (lines) |
| Magnetic flux density | 1 tesla = 1 weber/m$^2$ = 10,000 gauss |
| | = 64.5 kilolines/in$^2$ |
| Magnetizing force | 1 ampere turn/m = 0.0254 ampere turn/in. |

# Metadyne, Amplidyne, and Rotary Regulators

## C-1 EQUATIONS FOR THE METADYNE

The *metadyne,* also known as a *cross-field machine,* represents a class of dc electric machines which includes the amplidyne generator.

The dynamic behavior of the amplidyne and other regulating devices is of utmost importance and the basic relationships which determine the response of a simple amplidyne to changes in the input are therefore treated below. Figure C-1 illustrates an elementary metadyne which may be regarded as an elaboration of the dc machine in Fig. 6-39. Although Fig. C-1 shows only one field coil $f$–$f'$, several independent field windings are frequently used. The coil $q$–$q'$ represents the quadrature-axis circuit of the armature associated with the brushes $q$ and $q'$ in Fig. C-9, and the coil $d$–$d'$ represents the direct-axis circuit of the armature associated with brushes $d$ and $d'$.

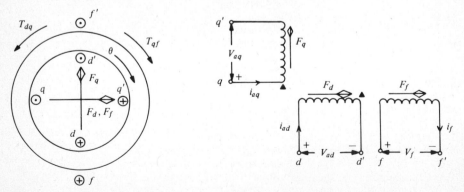

**Figure C-1** Elementary metadyne. (a) Physical representation. (b) Schematic circuit. All circuits represented as loads.

All three circuits in Fig. C-1 are represented as loads with *applied* voltages $v_f$, $v_{aq}$, and $v_{ad}$. The reaction of the quadrature-axis current $i_{aq}$ on the flux produced by the field current results in counterclockwise torque $T_{qf}$ just as in the case of the dc motor in Fig. 6-39. On the other hand, the interaction between the direct-axis armature current $i_{ad}$ and the flux produced by $i_{uq}$ produces counterclockwise torque. It should be kept in mind that the electromagnetic torque is in such a direction as to bring the magnetic axes of two members into alignment.

Brushes $q$–$q'$ are assumed to be on the geometric neutral and displaced from brushes $d$–$d'$ by an angle of 90° in electrical measure. The mutual inductance between the $d$ and $q$ axes is therefore zero if the magnetic circuit is symmetrical. Then for clockwise rotation and for the conventions adopted for the two-brush generator, which led to Eq. 6-42, the equations for the metadyne are

$$
\begin{bmatrix} v_f \\ v_{ad} \\ v_{aq} \end{bmatrix} = \begin{bmatrix} Z_f(p) & L_{df}\,p & 0 \\ L_{df}\,p & Z_d(p) & -\omega_m\mathcal{L}_{dq} \\ \omega_m\mathcal{L}_{qf} & \omega_m\mathcal{L}_{qd} & Z_q(p) \end{bmatrix} \begin{bmatrix} i_f \\ i_{ad} \\ i_{aq} \end{bmatrix}
\tag{C-1}
$$

where

$$ Z_f(p) = r_f + L_{ff}\,p \qquad Z_d(p) = r_d + L_{dd}p \qquad Z_q(p) = r_q + L_{qq}p $$

and where $L_{ff}$, $L_{dd}$, and $L_{qq}$ are the self-inductances of the field circuit, $d$-axis, and $q$-axis armature circuits. $L_{df}$ is the mutual inductance between the $d$ axis of the armature and the field, $\mathcal{L}_{qf}$ and $\mathcal{L}_{dq}$ are the coefficients of the speed voltages generated in the $q$- and $d$-armature axes, and $r_d$ and $r_q$ are the resistances of the $d$- and $q$-armature axes.

## C-1.1  Equations for the Amplidyne

The simple metadyne in Fig. C-1 can be adapted to operate as the amplidyne illustrated in Fig. C-9 by the addition of compensating windings $c$–$c'$ and $s$–$s'$ in the $d$ and $q$ axes of the stator as indicated in Fig. C-2. However, before dealing with the amplidyne as connected in Fig. C-9, consider the more general arrangement of the metadyne in Fig. C-2 in which all five circuits are treated as loads and for which the behavior is defined by the following matrix equation:

$$
\begin{bmatrix} v_f \\ v_c \\ v_{ad} \\ v_s \\ v_{aq} \end{bmatrix} = \begin{bmatrix} Z_f(p) & L_{fc}p & L_{fd}p & 0 & 0 \\ L_{cf}p & Z_c(p) & L_{cd}p & 0 & 0 \\ L_{df}p & L_{dc}p & Z_d(p) & -\omega_m\mathcal{L}_{ds} & -\omega_m\mathcal{L}_{dq} \\ 0 & 0 & 0 & Z_s(p) & L_{sq}p \\ \omega_m\mathcal{L}_{qf} & \omega_m\mathcal{L}_{qc} & \omega_m\mathcal{L}_{qd} & L_{qs}p & Z_q(p) \end{bmatrix} \begin{bmatrix} i_f \\ i_c \\ i_{ad} \\ i_s \\ i_{aq} \end{bmatrix}
\tag{C-2}
$$

The mutual inductances $L_{fc}$ and $L_{cd}$ are between the compensating winding and field winding and between the compensating winding and the $d$ axis of the armature. $L_{sq}$ is the mutual inductance between the series winding and the $q$ axis of the armature. $\mathcal{L}_{ds}$ and $\mathcal{L}_{qc}$ are the speed coefficients associated with the series stator winding and with the compensating winding. On the basis of linearity. $L_{fc} = L_{cf}$; $L_{df} = L_{fd}$; $L_{dc} = L_{cd}$; $L_{qs} = L_{sq}$; and $\mathcal{L}_{qd} = \mathcal{L}_{dq}$.

Figure C-3 shows the metadyne connected to operate as the amplidyne in Fig. C-9. Because of the generator action the actual current directions are as shown in Fig. C-3 and when these are compared with the current directions in Fig. C-2 it is found that

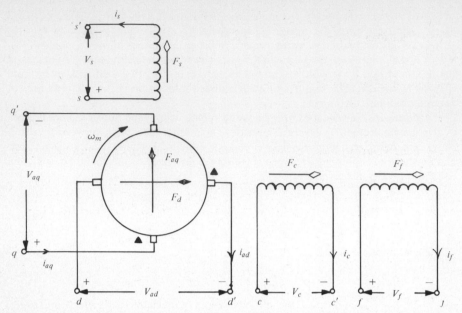

**Figure C-2** Metadyne with compensating winding $c–c'$ and series $s–s'$ winding in the $d$ and $q$ axes of the stator. All circuits represented as loads.

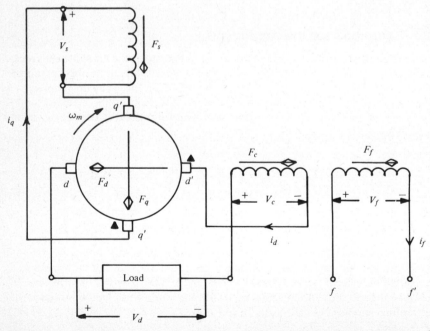

**Figure C-3** Amplidyne connection. Compensating winding opposes armature $d$ axis and series winding aids armature $q$ axis.

$$i_d = i_c = -i_{ad} \tag{C-3}$$

Further, the load voltage is

$$v_d = v_{ad} - v_c \tag{C-4}$$

and because of the short-circuit connection in the $q$ axis,

$$v_{aq} + v_s = 0 \tag{C-5}$$

$$-i_s = i_q = -i_{aq} \tag{C-6}$$

For this arrangement, the behavior of the amplidyne is then defined by

$$\begin{bmatrix} v_f \\ v_d \\ 0 \end{bmatrix} = \begin{bmatrix} Z_f(p) & (L_{fc} - L_{fd})p & 0 \\ (L_{df} - L_{cf})p & -[Z_d(p) + Z_c(p) - 2L_{dc}p] & \omega_m(\mathcal{L}_{dq} + \mathcal{L}_{ds}) \\ \omega_m \mathcal{L}_{af} & \omega_m(\mathcal{L}_{qc} - \mathcal{L}_{qd}) & -[Z_q(p) + Z_s(p) + 2L_{qs}p] \end{bmatrix} \begin{bmatrix} i_f \\ i_d \\ i_q \end{bmatrix} \tag{C-7}$$

If the load impedance is

$$Z_L(p) = R_L + L_L p \tag{C-8}$$

then Eq. C-7 can be reduced to

$$\begin{bmatrix} v_f \\ 0 \\ 0 \end{bmatrix} = \begin{bmatrix} Z_f(p) & (L_{fc} - L_{fd})p & 0 \\ (L_{df} - L_{cf})p & -[Z_D(p) + Z_L(p)] & \omega_m(\mathcal{L}_{dq} + \mathcal{L}_{ds}) \\ \omega_m \mathcal{L}_{af} & \omega_m(\mathcal{L}_{qc} - \mathcal{L}_{qd}) & -Z_Q(p) \end{bmatrix} \begin{bmatrix} i_f \\ i_d \\ i_q \end{bmatrix} \tag{C-9}$$

where

$$Z_D(p) = Z_d(p) + Z_c(p) - 2L_{dc}p$$
$$= r_d + L_{dd}p + r_c + L_{cc}p - 2L_{dc}p \tag{C-10}$$

and

$$Z_Q(p) = Z_q(p) + Z_s(p) + 2L_{qs}p$$
$$= r_q + L_{qq}p + r_s + L_{ss}p + 2L_{qs}p \tag{C-11}$$

The block diagram corresponding to the connections in Fig. C-3 and Eq. C-9 is shown in Fig. C-4.

When compensation is complete the mmfs of the armature $d$ axis and of the compensating winding are equal and opposite so that $F_c = -F_d$, $L_{fc} = L_{fd}$, and $\mathcal{L}_{qc} = \mathcal{L}_{qd}$. These relationships then correspond to opening of the feedback loops in Fig. C-4, which results in the simple block diagram of Fig. C-5.

If, in addition, the leakage flux in the direct axis is assumed negligible, then

$$L_{dd} = L_{cc} = L_{dc} \quad \text{and} \quad Z_D(p) = r_d + r_c \tag{C-12}$$

Although the discussion has been restricted to an amplidyne with only one field winding or control winding, the analysis can be extended to include additional control windings by including an equation for each additional control circuit, taking into account the self-impedance of each such circuit and the mutual inductances between it and the other circuits. The effect of such additional control circuits on the speed voltage in the quadrature axis must also be included in the additional equations.

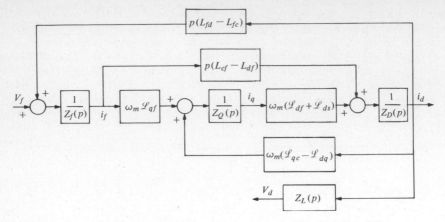

**Figure C-4**   Block diagram for amplidyne in Fig. C.3 based on Eq. C.9.

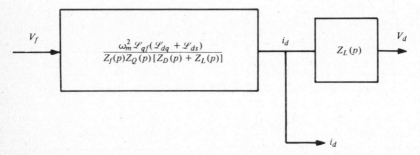

**Figure C-5**   Block diagram for completely compensated amplidyne in Fig. C.4.

Numerous arrangements other than those discussed in the foregoing are in use to control voltage, speed, and torque singly or in combination.

Metadynes are essentially a class of dc machines that use the armature mmf as the magnetizing field. Most are used as control generators to meet a particular need. A wide variety of metadynes have been used in Europe. The only one currently in use in the United States is the amplidyne. As has been shown in this section, an amplidyne is a two-stage rotary amplifier. At the time of this writing amplidynes are being displaced by solid-state power amplifiers. There are a few continuing to be manufactured and thousands still in service, hence their inclusion in this text.

## C-2   THE AMPLIDYNE

The *amplidyne* is a dc generator used in feedback control systems in which the output or regulated quantity is compared with a desired value or reference. The difference is amplified in the amplidyne to regulate the output of the system. It is possible to obtain power amplifications at full voltage of 10,000:1 to 250,000:1 for machines rated from 1 to 50 kW. Accordingly, the power output of such machines can be controlled accurately by less than $\frac{1}{2}$-W input to the field. A few typical examples of applications are (1) in combination with regulators, as an exciter for generators to maintain steady voltage in power systems; (2) as a control to hold constant tension or torque—for greater uniformity—

in rolling, winding, and drawing operations; (3) as a control to synchronize separate machines, or to maintain an exact preset speed for continuous process control; (4) for synchronous motors, to provide automatic power-factor control and to improve system voltage stability; and (5) as a control, to automatically hold the position of moving materials.

The most prominent and perhaps the most basic feature that distinguishes the amplidyne from the more common dc generators is that the main flux is produced by armature reaction. As pointed out in Sec. 6-5, the armature mmf in conventional dc machines is almost as great at rated load as the field mmf. Figure C-6(a) is a simplified

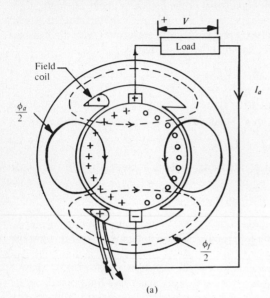

(a)

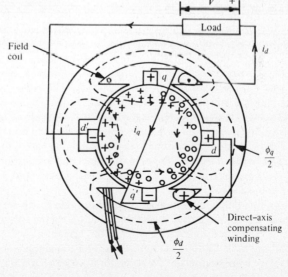

(b)

**Figure C-6** Simplified diagrams. (a) Conventional two-pole dc generator. (b) Two-pole amplidyne generator.

diagram of a conventional two-pole dc generator in which only one field coil (on the left-hand pole) is shown. Current directions in the armature conductors are represented by the dots and crosses within the armature surface. The dashed lines indicate the approximate paths taken by the field flux $\phi_f$ while the solid lines indicate those taken by the armature flux $\phi_a$. If this generator were rated at 25 kW, the field could be designed for an excitation requirement of 250 W or 1 percent of rated output. When considered as a single-stage amplifier, this generator would then have a power amplification of 100. The armature could also be designed for a loss of 250 W at rated current or 1 percent rated output and further, so that at rated current its mmf equals that of the field. If a linear magnetization curve is assumed, there is no demagnetizing mmf of armature reaction, and the field excitation and generated voltage are directly proportional to each other with the brushes set on geometric neutral. The field mmf can then be expressed by

$$F_f = CE$$

where $E = V + r_a I_a$.

On a per-unit basis, at full load $V = 1.00$, $I_a = 1.00$, and $r_a = 0.01$, since the armature losses are 1 percent at rated load. The field mmf is expressed in per-unit by

$$F_f = 1.01C$$

Now if the armature current is at its rated value and the brushes are short-circuited, the generated voltage would be

$$E = r_a I_a = 0.01 \text{ per unit}$$

requiring a power input to the field winding of only 0.0001 rated value. This arrangement thus enables a small amount of field current to produce full excitation but in the quadrature axis.

The armature flux is utilized in the amplidyne as shown in Fig. C-6(b) by adding another pair of brushes $dd'$ with their axis along the direct axis while the brushes $qq'$ in the quadrature axis are short-circuited. The poles are recessed in the vicinity of their axes to reduce the flux in that region in order to promote good communication in the direct axis. The field winding, also known as the *control field,* is shown on the left-hand pole of the amplidyne in Fig. C-6(b) and is much smaller than the field winding in the conventional generator of Fig. C-6(a). The right-hand pole of the amplidyne carries a compensating winding that neutralizes the armature mmf in the direct axis. Two sets of current $i_d$ and $i_q$ are indicated in the armature of the amplidyne. The outer currents represent $i_q$, the current in the quadrature circuit through the short-circuited brushes $qq'$ in the quadrature axis, and the inner currents represent the load current $i_d$ carried by the brushes $dd'$ in the direct axis.† Consequently, the armature conductors between brushes $d'$ and $q$ and between brushes $d$ and $q'$ carry more current than the remaining half of the armature conductors. It should also be noted that the fluxes in the direct and quadrature axes of the amplidyne in Fig. C-6(b) are designated as $\phi_d$ and $\phi_q$ instead of $\phi_f$ and $\phi_a$ for the conventional generator in Fig. C-6(a).

The amplidyne is a two-stage rotating amplifier with its first stage between the control field and the quadrature axis (short-circuited) of the armature and its second stage between the direct and quadrature axes of the armature (i.e., between the two sets of brushes $qq'$ and $dd'$). The power amplification for the amplidyne is 10,000 as compared with 100 for the conventional 25-kW generator from which it was modified in this example.

---

† While the armature has a two-layer winding, the outer currents as represented in the armature of Fig. C-6(b) are not to be taken as flowing in the outer layer only, and the inner currents as flowing in the inner layer only, of the armature winding.

Amplidynes are normally built in ratings from 0.5 to 100 kW. Two-pole structures are used in ratings up to 25 kW and four-pole structures are common for ratings from 10 to 100 kW. A laminated magnetic structure is used throughout to achieve good transient performance, low residual flux, and uniformity of design. A schematic diagram of a two-pole amplidyne is shown in Fig. C-7 with commutating poles in both the direct and quadrature axes and with a compensating winding in the direct axis. The purpose of this compensating winding, which may also be distrubuted among slots in the pole faces as indicated in Fig. 6-19, is to prevent the direct-axis armature mmf from neutralizing the field mmf.

The field winding may have a number of sections each of which may be excited from a different source. In the simpler applications two sections of control field are used, although there are some in which one section suffices. When two sections are used, one

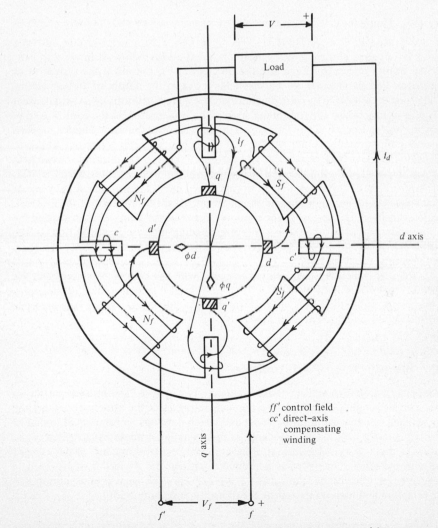

**Figure C-7** Schematic diagram of a two-pole amplidyne generator with a compensating winding in the direct axis.

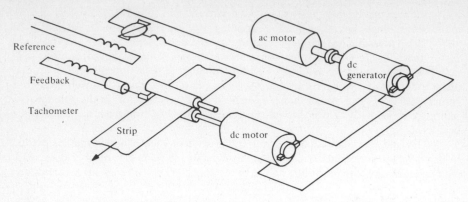

**Figure C-8**  Example of feedback-control system using an amplidyne for controlling speed.

is excited from a reference source and the other from the output of the system to be controlled so that the interaction of these fields regulates the output of the amplidyne, thus adjusting the system to its desired performance. Figure C-8 illustrates an arrangement which uses an amplidyne for controlling speed. The two sections in the control field of the amplidyne oppose each other. The predominating section of the control field is supplied from a constant-current reference and the other from a tachometer whose output is proportional to the speed to be controlled. The difference between the mmfs of the two field sections is amplified by the amplidyne to excite the field of the dc generator, which supplies the armature of the dc driving motor. At normal speed the excitation is such as to deliver normal voltage to the motor armature. When the speed falls below normal, the feedback decreases, causing a net increase in the mmf of the amplidyne field—thus increasing the field excitation of the dc generator with corresponding increase in the voltage to the armature of the driving motor, resulting in increasing speed. The opposite action occurs when the speed exceeds normal.

Systems in which the feedback is compared with the reference externally to the amplidyne require only one section of control field. However, several sections are necessary for comparisons within the amplidyne and where one or more sections may receive signals to provide stabilizing and antihunting effects.†

## C-2.1  Steady-State Performance

An elementary amplidyne is shown schematically in Fig. C-9. The brushes are set on their respective geometric neutrals. The quadrature circuit of the armature includes the stator winding $s$ connected in series aiding and which develops an mmf equal to that of the armature in the $q$ axis. The impedance of the $s$ winding is about equal to that of the armature so that a given value of control-field current results in one-half of the value of $i_q$ but with the same value of $\phi_q$ as is produced without the $s$ winding. As a result, a somewhat larger value of the load current $i_d$ is possible for the same amount of heat loss in the armature and without increasing the difficulty of commutation.

† For a more complete discussion of the functions of the control windings in amplidynes, see "The Amplidyne—Characteristics and Technical Data," *Pamphlet GET1985C* (Erie, Pa.: General Electric Company).

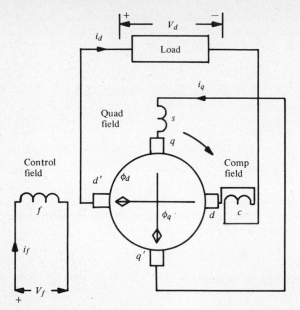

**Figure C-9**   Schematic diagram of an elementary amplidyne connected to a load. (Commutating-pole windings not included.

Under the steady-state conditions (i.e., constant currents and constant speeds) the voltage generated in the quadrature axis is

$$E_q = -K_E \omega_m \phi_d \qquad (\text{C-13})$$

and the voltage generated in the direct axis is

$$E_d = K_F \omega_m \phi_q \qquad (\text{C-14})$$

Flux is considered positive when in the direction of the corresponding *armature* mmf. Thus, in Figs. C-6 and C-9 the flux $\phi_d$ in the direct axis is positive when from right to left and the flux $\phi_q$ in the quadrature axis is positive when in the downward direction. This accounts for the minus sign in Eq. C-13. With all currents zero except $i_d$, then for the clockwise direction of rotation the polarity of the generated quadrature-axis voltage $E_q$ would be such as to make the brush $q$ positive instead of negative as indicated, and thus the minus sign in Eq. C-13. On the other hand, if only the current $i_q$ is present, the polarity of the generated direct-axis voltage $E_d$ would cause brush $d$ to be negative as shown, and no minus sign appears in Eq. C-14.

If linearity is assumed, these fluxes may be represented as the sums of component fluxes produced by the currents in the different windings as follows:

$$\phi_d = \phi_{dd} - \phi_{df} - \phi_{dc} \qquad (\text{C-15})$$

where the fluxes $\phi_{dd}$, $\phi_{df}$, and $\phi_{dc}$ are components of flux produced in the direct-axis armature circuit by the current in the direct axis of the winding itself, the current in the control-field winding, and the current in the compensating winding and direct-axis interpole winding. These currents all equal $i_d$ for the normal series connection of the direct-axis circuits as shown in Fig. C-9. With the brushes on their geometric neutrals, an mmf in one axis produces no flux in the other axis in a symmetrical unsaturated magnetic structure. Then, if compensation is complete, $\phi_{dd} = \phi_{dc}$ and Eq. C-15 can be reduced to

$$\phi_d = -\phi_{df} \tag{C-16}$$

Further, on the basis of linearity, the direct-axis flux can be expressed in terms of the control-field current by

$$\phi_{df} = K_{df}i_f$$

and Eq. C-13 can be simplified to

$$E_q = K_E K_{df}\omega_m i_f \tag{C-17}$$

If $\phi_{qq}$ is the flux due to armature mmf in the quadrature axis and $\phi_{qs}$ that due to mmf in the quadrature-axis winding of the stator including the quadrature-axis interpole winding, the flux in the quadrature axis of the armature is

$$\phi_q = \phi_{qq} + \phi_{qs} \tag{C-18}$$

The quadrature-axis flux can also be expressed by

$$\phi_q = K_q i_q$$

so that Eq. C-14 is reduced to

$$E_d = K_E K_q \omega_m i_q \tag{C-19}$$

When the quadrature-axis series circuit is short-circuited,

$$E_q = r_q i_q \tag{C-20}$$

where $r_q$ is the resistance of the quadrature-axis circuit.

It follows from Eqs. C-17, C-19, and C-20 that

$$E_d = \frac{K_E^2 K_{df} K_q \omega_m^2 i_f}{r_q} \tag{C-21}$$

which can be reduced to

$$E_d = K\omega_m^2 i_f \tag{C-22}$$

by letting

$$K = \frac{K_E^2 K_{df} K_q}{r_q}$$

The output voltage is

$$V = E_d - r_d i_d \tag{C-23}$$

where $r_d$ is the resistance of the armature plus that of the compensating winding, and the direct-axis interpole winding.

Equation C-22 shows that the voltage $E_d$ generated in the direct axis is proportional to the *square of the speed* and a reversal of rotation therefore does not change the sign of $E_d$ for given direction of the control-field current. The square of the speed results from the fact that there are two stages of amplification in the amplidyne.

When $n$ sections of control field are excited independently, Eq. C-21 can be modified to read

$$E_d = \frac{K_E^2 K_q \omega_m^2 (K_{df1} i_{f1} + K_{df2} i_{f2} + \cdots + K_{dfn} i_{fn})}{r\hat{q}} \qquad \text{(C-24)}$$

when all the currents $i_{f1}$, $i_{f2}$, . . . , $i_{fn}$ are flowing in the direction of $i_q$. Generally, some of these currents are flowing in the opposite direction, requiring a change in sign for the corresponding term in Eq. C-24.


## C-3 THE ROTOTROL AND THE REGULEX

There are two other common rotary regulators in addition to the amplidyne. These are the Rototrol and the Regulex, both of which are dc generators of conventional construction driven at a constant speed and their performance characteristics are quite similar to those of corresponding conventional dc generators for like operating conditions. They are used in the same kinds of applications as the amplidyne. However, both operate with their iron circuits unsaturated and the flux path is therefore designed with an unusually large cross-sectional area. A simple form of each of these regulators has three fields: self-energizing field (in series with the armature of the Rototrol, and in shunt with the armature of the Regulex) and two separately excited control fields. One of these control fields is commonly referred to as the *pattern field* (reference field) excited with constant potential from a standard or calibration source. The other control field measures the quantity to be regulated and is commonly referred to as the *pilot field* (the field supplied with feedback). The pilot field is usually connected so that its mmf opposes that of the pattern field, although in some applications the mmfs of these two fields are additive. In the former case the pilot field is sometimes called a *differential field*, while in the latter case it is often known as a *cumulative field*. Figure C-10(a) shows a circuit in which a Rototrol regulates the speed of a motor and Fig. C-10(b) a circuit in which a Regulex regulates the voltage of an ac generator.


### C-3.1 Constant Motor Speed Control

The dc motor in Fig. C-10(a) is designed to operate at constant speed regardless of variations in load or of other factors that normally affect the characteristics of a dc motor. If the series field or self-energizing field of the Rototrol were out of the circuit, the Rototrol would depend for its own excitation on the difference between the mmfs of the pattern field and the pilot field. When the speed of the regulated dc motor drops below normal, the voltage generated by the tachometer falls below normal and the mmf of the pilot field decreases. Since the pilot field opposes the pattern field which is excited by a constant current, the flux in the Rototrol increases and thus increases the field current of the dc generator which supplies the armature of the regulated motor. As a result, the speed of the motor is increased, tending to return it to its normal value. Should the speed of the regulated motor rise above normal, as a result of a decrease in its load or for other reasons, the opposite effect would result and the action would be such as to reduce the voltage to the motor armature. The same action would of course take place if a Regulex were used in the same application.

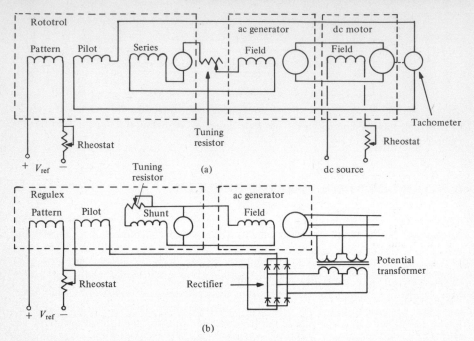

**Figure C-10** (a) Rototrol exciter regulating the speed of a dc motor. (b) Regulex exciter regulating the voltage of a three-phase generator.

If either the Rototrol or Regulex depended for its excitation entirely on the difference between the mmfs of the pattern and pilot fields, the windings of these fields would need to be large and would require relatively large amounts of power from the reference source and from the tachometer. Therefore, self-energizing fields are used—a series field in the Rototrol and a shunt field in the Regulex—to furnish the excitation. In order to obtain a high degree of sensitivity, the resistance of the self-energizing field circuit is adjusted, by means of a tuning resistor, to within plus or minus 5 percent of its critical value. Figure C-11(a) shows the field-resistance line tangent to the lower portion of the magnetization curve, thus representing the condition for critical field resistance. If it were not for the mmfs of the pattern and the pilot fields, the voltage generated in the armature of the Rototrol or of the Regulex might take on any value of the line $OA$ below the knee of the magnetization curve. However, when the pattern field current is adjusted to a certain value, the steady-state response of the system is such that the mmf of the pilot field neutralizes that of the pattern field when the voltage generated in the armature of the Rototrol or Regulex has the value $Oa$. The self-energizing field alone provides the mmf $ab$. The system in Fig. C-10(a) can be adjusted to regulate the speed of the dc motor at a higher or lower value simply by reducing or increasing the current in the pattern field. The terminal voltage of the ac generator in Fig. C-10(b) is controlled by the same process.

If the self-energizing field resistance is appreciably below the critical value as shown by the field resistance line $OB$ in Fig. C-11(b), there is only one value of generated voltage $Oa'$ (in the Rototrol or Regulex) for which the self-energizing field alone provides the excitation in the mmf $a'b'$. A decrease in the load on the regulated motor of Fig. C-10(a) would cause an increase in speed so that the tachometer would increase the current in the pilot field, thus producing the final steady value of generated voltage $Of$ in Fig.

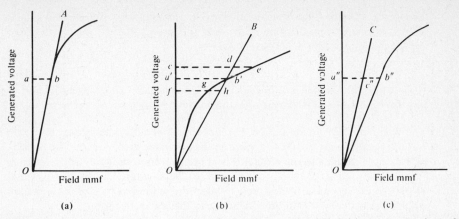

**Figure C-11** Magnetization curves and field-resistance line. (a) Critical value (typical of Rototrol and Regulex). (b) Below critical value. (c) Above critical value.

C-11(b). If the resistance of the Rototrol armature is neglected, the self-energized field furnishes the mmf *fg*, and the pilot field furnishes the mmf *gh* in excess of the pattern field mmf. The regulated motor then runs somewhat above the desired speed by an amount nearly proportional to the mmf *gh*. Similarly, for an increase in the load such that the Rototrol armature generates the voltage *Oc*, the regulated motor now runs below normal speed nearly in proportion to the excess mmf *de* of the pattern field. In that case the self-energizing field provides the mmf *cd*. If, on the other hand, the resistance of the self-energizing field circuit is above the critical value, as indicated by the resistance line *OC*, the relationships in Fig. C-11(c) are valid. In that case the pattern field must be stronger than the pilot field, since the self-energizing field is deficient. For example, when the generated voltage in the Rototrol is *Oa″* in Fig. C-11(c), the self-energizing field produces the mmf *a″c″* and the pattern field must furnish the mmf *c″b″* in excess of the mmf developed by the pilot field. It therefore follows that regulation is most effective when the self-energizing field circuit has its resistance at the critical value. Figure C-11 applies to the Regulex as well as to the Rototrol. Just as in the case of the amplidyne, additional control fields may be used so that the Rototrol or Regulex is responsive to more than one signal.

## PROBLEMS

**C-1.** A 200-kW 250-V 600-rpm shunt generator has the following data:

$$r_a = 0.010 \ \Omega \qquad r_f = 12.5 \ \Omega$$
$$L_{aa} = 0.0003 \ \text{H} \qquad L_{ff} = 10.0 \ \text{H}$$
$$\mathcal{L}_{af} = 0.20 \ \text{H}$$

This generator is driven at a constant speed of 600 rpm and supplies a 200-hp 250-V separately excited shunt motor in a Ward-Leonard system. The motor has constants as follows:

$$r_a = 0.012 \ \Omega \qquad r_f = 12.0 \ \Omega$$
$$L_{aa} \simeq 0 \qquad L_{ff} = 9.0 \ \text{H}$$
$$\mathcal{L}_{af} = 0.18 \ \text{H} \qquad J_M = 10.0 \ \text{kg-m}^2/\text{rad}^2$$

The motor is assumed to drive a load which has inertia, only, of $J_L = 20.0$ kg-m$^2$/ rad$^2$. The voltage applied to the motor field is constant at 250 V and the voltage applied to the generator field is expressed by

$$v_{fG} = 100 \sin 0.5t$$

Neglect the self-inductance $L$ of the armature circuit and express as a function of time (a) the motor speed, (b) the armature current.

**C-2.** The Ward-Leonard system in Prob. C-1 is operating initially in the steady state with a constant dc voltage of 100 V applied to the generator field. Assume the polarity of the voltage applied to the generator field to be reversed suddenly and express as functions of time (a) the motor speed, and (b) the armature current after reversal of the applied generator field voltage. Neglect the self-inductance of the armature circuit. Calculate the maximum magnitude of the armature current.

**C-3.** A 5-kW 250-V amplidyne driven at a constant speed of 1750 rpm has the following constants:

$$r_f = 1060 \ \Omega \qquad r_d = \qquad r_q = r_c = r_s = 0.50 \ \Omega$$
$$L_{ff} = 180 \ \text{H} \qquad L_{dd} = L_{qq} = L_{cc} = L_{ss} = 0.12 \ \text{H}$$
$$\mathcal{L}_{af} = 1.40 \ \text{H} \qquad \mathcal{L}_{dq} = \mathcal{L}_{cq} = \mathcal{L}_{ds} = 0.14 \ \text{H}$$

Assume the machine to be completely compensated and express the open-circuit (no-load) output voltage $v_d$ as a function of time after a constant voltage $v_f = 20.0$ V is applied to the field winding (a) with the quadrature-series field in series with the $q$-axis armature circuit, and (b) with the quadrature-series field disconnected. Which mode of operation—(a) or (b)—will give the higher power output rating of the amplidyne? Why?

**C-4.** The amplidyne in Prob. C-3 is in the closed-loop system of Fig. PC-4 to regulate the output of a 500-kW 250-V dc generator driven at a constant speed of 900 rpm. The amplidyne speed is constant at 1750 rpm. The generator has the following constants:

$$r_f = 12.0 \ \Omega \qquad r_a = \quad 0.0031 \ \Omega$$
$$L_{ff} = 12.6 \ \text{H} \qquad L_{aa} = 0.0025 \ \text{H}$$
$$\mathcal{L}_{af} = 0.20 \ \text{H}$$

The vacuum-tube amplifier which supplies the control field of the amplidyne may be assumed to have infinite input impedance and infinite output impedance when considered as a current source. The amplifier has a gain $g_m$ of 0.001 A/V. (a) Calculate the value of the reference voltage $v_r$ such that the no-load terminal voltage $v_a$ of the generator is 250 V. (b) With the value of $v_r$ as calculated in part (a), what is the steady-state terminal voltage when the generator delivers a current to a resistive load of 0.125 $\Omega$? How does this value compare with that for constant excitation of the generator field? (c) Express the no-load output voltage of the generator as a function of time after the voltage $v_r$ is suddenly increased from zero to the value determined in part (a). (d) The generator is operating initially without load with $v_r$ at the value determined in part (a) when a load resistance of 0.125 $\Omega$ is suddenly connected across the armature of the generator. Calculate the final value of the armature terminal voltage.

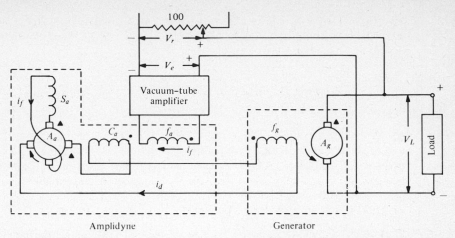

**Prob. PC-4**   Amplidyne in closed-looped voltage regulation of a dc generator.

**C-5**   The motor in Prob. 6-48 is running at 900 rpm when the armature is disconnected from the source and immediately connected across a 0.44-$\Omega$ resistance, for dynamic braking, while the field remains connected to the 250-V source. Use the equivalent capacitance and equivalent resistance expressed in terms of $\mathcal{L}_{af}I_f$ to determine the time required to decrease the motor speed to (a) 90 rpm and (b) 9 rpm.

## BIBLIOGRAPHY

Matsch, Leander W. *Electromagnetic & Electromechanical Machines*, 2d ed. New York: Dun-Donnelley Publishing Co., 1977.

Siskind, Charles S. *Electrical Machines: Direct & Alternating Current*, 2d ed. New York: McGraw-Hill Book Company, 1959.

# Index